Lecture Notes in Computer Science   10987

Commenced Publication in 1973
Founding and Former Series Editors:
Gerhard Goos, Juris Hartmanis, and Jan van Leeuwen

More information about this series at http://www.springer.com/series/7409

Yi Cai · Yoshiharu Ishikawa
Jianliang Xu (Eds.)

Web and Big Data

Second International Joint Conference, APWeb-WAIM 2018
Macau, China, July 23–25, 2018
Proceedings, Part I

 Springer

Editors
Yi Cai
South China University of Technology
Guangzhou
China

Jianliang Xu
Hong Kong Baptist University
Kowloon Tong, Hong Kong
China

Yoshiharu Ishikawa
Nagoya University
Nagoya
Japan

ISSN 0302-9743   ISSN 1611-3349 (electronic)
Lecture Notes in Computer Science
ISBN 978-3-319-96889-6   ISBN 978-3-319-96890-2 (eBook)
https://doi.org/10.1007/978-3-319-96890-2

Library of Congress Control Number: 2018948814

LNCS Sublibrary: SL3 – Information Systems and Applications, incl. Internet/Web, and HCI

This Springer imprint is published by the registered company Springer Nature Switzerland AG
The registered company address is: Gewerbestrasse 11, 6330 Cham, Switzerland

Preface

This volume (LNCS 10987) and its companion volume (LNCS 10988) contain the proceedings of the second Asia-Pacific Web (APWeb) and Web-Age Information Management (WAIM) Joint Conference on Web and Big Data, called APWeb-WAIM. This joint conference aims to attract participants from different scientific communities as well as from industry, and not merely from the Asia Pacific region, but also from other continents. The objective is to enable the sharing and exchange of ideas, experiences, and results in the areas of World Wide Web and big data, thus covering Web technologies, database systems, information management, software engineering, and big data. The second APWeb-WAIM conference was held in Macau during July 23–25, 2018. As an Asia-Pacific flagship conference focusing on research, development, and applications in relation to Web information management, APWeb-WAIM builds on the successes of APWeb and WAIM: APWeb was previously held in Beijing (1998), Hong Kong (1999), Xi'an (2000), Changsha (2001), Xi'an (2003), Hangzhou (2004), Shanghai (2005), Harbin (2006), Huangshan (2007), Shenyang (2008), Suzhou (2009), Busan (2010), Beijing (2011), Kunming (2012), Sydney (2013), Changsha (2014), Guangzhou (2015), and Suzhou (2016); and WAIM was held in Shanghai (2000), Xi'an (2001), Beijing (2002), Chengdu (2003), Dalian (2004), Hangzhou (2005), Hong Kong (2006), Huangshan (2007), Zhangjiajie (2008), Suzhou (2009), Jiuzhaigou (2010), Wuhan (2011), Harbin (2012), Beidaihe (2013), Macau (2014), Qingdao (2015), and Nanchang (2016). The first joint APWeb-WAIM conference was held in Bejing (2017). With the fast development of Web-related technologies, we expect that APWeb-WAIM will become an increasingly popular forum that brings together outstanding researchers and developers in the field of the Web and big data from around the world. The high-quality program documented in these proceedings would not have been possible without the authors who chose APWeb-WAIM for disseminating their findings. Out of 168 submissions, the conference accepted 39 regular (23.21%), 31 short research papers, and six demonstrations. The contributed papers address a wide range of topics, such as text analysis, graph data processing, social networks, recommender systems, information retrieval, data streams, knowledge graph, data mining and application, query processing, machine learning, database and Web applications, big data, and blockchain. The technical program also included keynotes by Prof. Xuemin Lin (The University of New South Wales, Australia), Prof. Lei Chen (The Hong Kong University of Science and Technology, Hong Kong, SAR China), and Prof. Ninghui Li (Purdue University, USA) as well as industrial invited talks by Dr. Zhao Cao (Huawei Blockchain) and Jun Yan (YiDu Cloud). We are grateful to these distinguished scientists for their invaluable contributions to the conference program. As a joint conference, teamwork was particularly important for the success of APWeb-WAIM. We are deeply thankful to the Program Committee members and the external reviewers for lending their time and expertise to the conference. Special thanks go to the local Organizing Committee led by Prof. Zhiguo Gong.

Thanks also go to the workshop co-chairs (Leong Hou U and Haoran Xie), demo co-chairs (Zhixu Li, Zhifeng Bao, and Lisi Chen), industry co-chair (Wenyin Liu), tutorial co-chair (Jian Yang), panel chair (Kamal Karlapalem), local arrangements chair (Derek Fai Wong), and publicity co-chairs (An Liu, Feifei Li, Wen-Chih Peng, and Ladjel Bellatreche). Their efforts were essential to the success of the conference. Last but not least, we wish to express our gratitude to the treasurer (Andrew Shibo Jiang), the Webmaster (William Sio) for all the hard work, and to our sponsors who generously supported the smooth running of the conference. We hope you enjoy the exciting program of APWeb-WAIM 2018 as documented in these proceedings.

June 2018   Yi Cai
                                                            Jianliang Xu
                                                      Yoshiharu Ishikawa

Organization

Organizing Committee

Honorary Chair

Lionel Ni   University of Macau, SAR China

General Co-chairs

Zhiguo Gong   University of Macau, SAR China
Qing Li   City University of Hong Kong, SAR China
Kam-fai Wong   Chinese University of Hong Kong, SAR China

Program Co-chairs

Yi Cai   South China University of Technology, China
Yoshiharu Ishikawa   Nagoya University, Japan
Jianliang Xu   Hong Kong Baptist University, SAR China

Workshop Chairs

Leong Hou U   University of Macau, SAR China
Haoran Xie   Education University of Hong Kong, SAR China

Demo Co-chairs

Zhixu Li   Soochow University, China
Zhifeng Bao   RMIT, Australia
Lisi Chen   Wollongong University, Australia

Tutorial Chair

Jian Yang   Macquarie University, Australia

Industry Chair

Wenyin Liu   Guangdong University of Technology, China

Panel Chair

Kamal Karlapalem   IIIT, Hyderabad, India

Publicity Co-chairs

An Liu   Soochow University, China
Feifei Li   University of Utah, USA

| Wen-Chih Peng | National Taiwan University, China |
| Ladjel Bellatreche | ISAE-ENSMA, Poitiers, France |

Treasurers

| Leong Hou U | University of Macau, SAR China |
| Andrew Shibo Jiang | Macau Convention and Exhibition Association, SAR China |

Local Arrangements Chair

| Derek Fai Wong | University of Macau, SAR China |

Webmaster

| William Sio | University of Macau, SAR China |

Senior Program Committee

Bin Cui	Peking University, China
Byron Choi	Hong Kong Baptist University, SAR China
Christian Jensen	Aalborg University, Denmark
Demetrios Zeinalipour-Yazti	University of Cyprus, Cyprus
Feifei Li	University of Utah, USA
Guoliang Li	Tsinghua University, China
K. Selçuk Candan	Arizona State University, USA
Kyuseok Shim	Seoul National University, South Korea
Makoto Onizuka	Osaka University, Japan
Reynold Cheng	The University of Hong Kong, SAR China
Toshiyuki Amagasa	University of Tsukuba, Japan
Walid Aref	Purdue University, USA
Wang-Chien Lee	Pennsylvania State University, USA
Wen-Chih Peng	National Chiao Tung University, Taiwan
Wook-Shin Han Pohang	University of Science and Technology, South Korea
Xiaokui Xiao	National University of Singapore, Singapore
Ying Zhang	University of Technology Sydney, Australia

Program Committee

Alex Thomo	University of Victoria, Canada
An Liu	Soochow University, China
Baoning Niu	Taiyuan University of Technology, China
Bin Yang	Aalborg University, Denmark
Bo Tang	Southern University of Science and Technology, China
Zouhaier Brahmia	University of Sfax, Tunisia
Carson Leung	University of Manitoba, Canada
Cheng Long	Queen's University Belfast, UK

Lisi Chen	Wollongong University, Australia
Lu Chen	Aalborg University, Denmark
Maria Damiani	University of Milan, Italy
Markus Endres	University of Augsburg, Germany
Mihai Lupu	Vienna University of Technology, Austria
Mirco Nanni	ISTI-CNR Pisa, Italy
Mizuho Iwaihara	Waseda University, Japan
Peiquan Jin	University of Science and Technology of China, China
Peng Wang	Fudan University, China
Qin Lu	University of Technology Sydney, Australia
Ralf Hartmut Güting	Fernuniversität in Hagen, Germany
Raymond Chi-Wing Wong	Hong Kong University of Science and Technology, SAR China
Ronghua Li	Shenzhen University, China
Rui Zhang	University of Melbourne, Australia
Sanghyun Park	Yonsei University, South Korea
Sanjay Madria	Missouri University of Science and Technology, USA
Shaoxu Song	Tsinghua University, China
Shengli Wu	Jiangsu University, China
Shimin Chen	Chinese Academy of Sciences, China
Shuai Ma	Beihang University, China
Shuo Shang	King Abdullah University of Science and Technology, Saudi Arabia
Takahiro Hara	Osaka University, Japan
Tieyun Qian	Wuhan University, China
Tingjian Ge	University of Massachusetts, Lowell, USA
Tom Z. J. Fu	Advanced Digital Sciences Center, Singapore
Tru Cao	Ho Chi Minh City University of Technology, Vietnam
Vincent Oria	New Jersey Institute of Technology, USA
Wee Ng	Institute for Infocomm Research, Singapore
Wei Wang	University of New South wales, Australia
Weining Qian	East China Normal University, China
Weiwei Sun	Fudan University, China
Wen Zhang	Wuhan University, China
Wolf-Tilo Balke	Technische Universität Braunschweig, Germany
Wookey Lee	Inha University, South Korea
Xiang Zhao	National University of Defence Technology, China
Xiang Lian	Kent State University, USA
Xiangliang Zhang	King Abdullah University of Science and Technology, Saudi Arabia
Xiangmin Zhou	RMIT University, Australia
Xiaochun Yang	Northeast University, China
Xiaofeng He	East China Normal University, China
Xiaohui (Daniel) Tao	The University of Southern Queensland, Australia
Xiaoyong Du	Renmin University of China, China
Xike Xie	University of Science and Technology of China, China

Keynotes

Graph Processing: Applications, Challenges, and Advances

Xuemin Lin

School of Computer Science and Engineering,
University of New South Wales, Sydney
lxue@cse.unsw.edu.au

Abstract. Graph data are key parts of Big Data and widely used for modelling complex structured data with a broad spectrum of applications. Over the last decade, tremendous research efforts have been devoted to many fundamental problems in managing and analyzing graph data. In this talk, I will cover various applications, challenges, and recent advances. We will also look to the future of the area.

Differential Privacy in the Local Setting

Ninghui Li

Department of Computer Sciences, Purdue University
ninghui@cs.purdue.edu

Abstract. Differential privacy has been increasingly accepted as the de facto standard for data privacy in the research community. Recently, techniques for satisfying differential privacy (DP) in the local setting, which we call LDP, have been deployed. Such techniques enable the gathering of statistics while preserving privacy of every user, without relying on trust in a single data curator. Companies such as Google, Apple, and Microsoft have deployed techniques for collecting user data while satisfying LDP. In this talk, we will discuss the state of the art of LDP. We survey recent developments for LDP, and discuss protocols for estimating frequencies of different values under LDP, and for computing marginal when each user has multiple attributes. Finally, we discuss limitations and open problems of LDP.

Big Data, AI, and HI, What is the Next?

Lei Chen

Department of Computer Science and Engineering, Hong Kong University
of Science and Technology
leichen@cse.ust.hk

Abstract. Recently, AI has become quite popular and attractive, not only to the academia but also to the industry. The successful stories of AI on Alpha-go and Texas hold 'em games raise significant public interests on AI. Meanwhile, human intelligence is turning out to be more sophisticated, and Big Data technology is everywhere to improve our life quality. The question we all want to ask is "what is the next?". In this talk, I will discuss about DHA, a new computing paradigm, which combines big Data, Human intelligence, and AI. First I will briefly explain the motivation of DHA. Then I will present some challenges and possible solutions to build this new paradigm.

Contents – Part I

Recommender Systems

Information Retrieval

Demo Papers

Contents – Part II

Data Mining and Application

Text Analysis

Abstractive Summarization with the Aid of Extractive Summarization

Yangbin Chen[✉], Yun Ma, Xudong Mao, and Qing Li

City University of Hong Kong, Hong Kong SAR, China
{robinchen2-c,yunma3-c,xdmao2-c}@my.cityu.edu.hk,
qing.li@cityu.edu.hk

Abstract. Currently the abstractive method and extractive method are two main approaches for automatic document summarization. To fully integrate the relatedness and advantages of both approaches, we propose in this paper a general framework for abstractive summarization which incorporates extractive summarization as an auxiliary task. In particular, our framework is composed of a shared hierarchical document encoder, an attention-based decoder for abstractive summarization, and an extractor for sentence-level extractive summarization. Learning these two tasks jointly with the shared encoder allows us to better capture the semantics in the document. Moreover, we constrain the attention learned in the abstractive task by the salience estimated in the extractive task to strengthen their consistency. Experiments on the CNN/DailyMail dataset demonstrate that both the auxiliary task and the attention constraint contribute to improve the performance significantly, and our model is comparable to the state-of-the-art abstractive models.

Keywords: Abstractive document summarization
Squence-to-sequence · Joint learning

1 Introduction

Automatic document summarization has been studied for decades. The target of document summarization is to generate a shorter passage from the document in a grammatically and logically coherent way, meanwhile preserving the important information. There are two main approaches for document summarization: extractive summarization and abstractive summarization. The extractive method first extracts salient sentences or phrases from the source document and then groups them to produce a summary without changing the source text. Graph-based ranking model [1,2] and feature-based classification model [3,4] are typical models for extractive summarization. However, the extractive method unavoidably includes secondary or redundant information and is far from the way humans write summaries [5].

© Springer International Publishing AG, part of Springer Nature 2018
Y. Cai et al. (Eds.): APWeb-WAIM 2018, LNCS 10987, pp. 3–15, 2018.
https://doi.org/10.1007/978-3-319-96890-2_1

The abstractive method, in contrast, produces generalized summaries, conveying information in a concise way, and eliminating the limitations to the original words and sentences of the document. This task is more challenging since it needs advanced language generation and compression techniques. Discourse structures [6,7] and semantics [8,9] are most commonly used by researchers for generating abstractive summaries.

Recently, Recurrent Neural Network (RNN)-based sequence-to-sequence model with attention mechanism has been applied to abstractive summarization, due to its great success in machine translation [22,27,30]. However, there are still some challenges. First, the RNN-based models have difficulties in capturing long-term dependencies, making summarization for long document much tougher. Second, different from machine translation which has strong correspondence between the source and target words, an abstractive summary corresponds to only a small part of the source document, making its attention difficult to be learned.

We adopt hierarchical approaches for the long-term dependency problem, which have been used in many tasks such as machine translation and document classification [10,11]. But few of them have been applied to the abstractive summarization tasks. In particular, we encode the input document in a hierarchical way from word-level to sentence-level. There are two advantages. First, it captures both the local and global semantic representations, resulting in better feature learning. Second, it improves the training efficiency because the time complexity of the RNN-based model can be reduced by splitting the long document into short sentences.

The attention mechanism is widely used in sequence-to-sequence tasks [13, 27]. However, for abstractive summarization, it is difficult to learn the attention since only a small part of the source document is important to the summary. In this paper, we propose two methods to learn a better attention distribution. First, we use a hierarchical attention mechanism, which means that the attention is applied in both word and sentence levels. Similar to the hierarchical approach in encoding, the advantage of using hierarchical attention is to capture both the local and global semantic representations. Second, we use the salience scores of the auxiliary task (i.e., the extractive summarization) to constrain the sentence-level attention.

In this paper, we present a novel technique for abstractive summarization which incorporates extractive summarization as an auxiliary task. Our framework consists of three parts: a shared document encoder, a hierarchical attention-based decoder and an extractor. As Fig. 1 shows, we encode the document in a hierarchical way (Fig. 1 (1) and (2)) in order to address the long-term dependency problem. Then the learned document representations are shared by the extractor (Fig. 1 (3)) and the hierarchical attention-based decoder (Fig. 1 (5)). The extractor and the decoder are jointly trained which can capture better semantics of the document. Furthermore, as both the sentence salience scores in the extractor and the sentence-level attention in the decoder indicate the

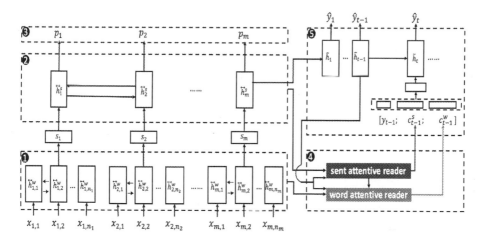

Fig. 1. General framework of our proposed model with 5 components: (1) word-level encoder encodes the sentences word-by-word independently, (2) sentence- level encoder encodes the document sentence-by-sentence, (3) sentence extractor makes binary classification for each sentence, (4) hierarchical attention calculates the word-level and sentence-level context vectors for decoding steps, (5) decoder decodes the output sequential word sequence with a beam-search algorithm.

importance of source sentences, we constrain the learned attention (Fig. 1 (4)) with the extracted sentence salience in order to strengthen their consistency.

We have conducted experiments on a news corpus - the CNN/DailyMail dataset [16]. The results demonstrate that adding the auxiliary extractive task and constraining the attention are both useful to improve the performance of the abstractive task, and our proposed joint model is comparable to the state-of-the-art abstractive models.

2   Neural Summarization Model

In this section we describe the framework of our proposed model which consists of five components. As illustrated in Fig. 1, the hierarchical document encoder which includes both the word-level and the sentence-level encoders reads the input word sequences and generates shared document representations. On one hand, the shared representations are fed into the sentence extractor which is a sequence labeling model to calculate salience scores. On the other hand, the representations are used to generate abstractive summaries by a GRU-based language model, with the hierarchical attention including the sentence-level attention and word-level attention. Finally, the two tasks are jointly trained.

2.1 Shared Hierarchical Document Encoder

We encode the document in a hierarchical way. In particular, the word sequences are first encoded by a bidirectional GRU network parallelly, and a sequence of sentence-level vector representations called sentence embeddings are generated. Then the sentence embeddings are fed into another bidirectional GRU network and get the document representations. Such an architecture has two advantages. First, it can reduce the negative effects during the training process caused by the long-term dependency problem, so that the document can be represented from both local and global aspects. Second, it helps improve the training efficiency as the time complexity of RNN-based model increases with the sequence length.

Formally, let $\mathbf{V}$ denote the vocabulary which contains D tokens, and each token is embedded as a d-dimension vector. Given an input document $\mathbf{X}$ containing m sentences $\{\mathbf{X}_i, i \in 1, ..., m\}$, let n_i denote the number of words in $\mathbf{X}_i$.

Word-level Encoder reads a sentence word-by-word until the end, using a bidirectional GRU network as the following equations:

$$\overleftrightarrow{h}_{i,j}^{w} = [\overrightarrow{h}_{i,j}^{w}; \overleftarrow{h}_{i,j}^{w}] \tag{1}$$

$$\overrightarrow{h}_{i,j}^{w} = GRU(x_{i,j}, \overrightarrow{h}_{i,j-1}^{w}) \tag{2}$$

$$\overleftarrow{h}_{i,j}^{w} = GRU(x_{i,j}, \overleftarrow{h}_{i,j+1}^{w}) \tag{3}$$

where $x_{i,j}$ represents the embedding vector of the jth word in th ith sentence. $\overleftrightarrow{h}_{i,j}^{w}$ is a concatenated vector of the forward hidden state $\overrightarrow{h}_{i,j}^{w}$ and the backward hidden state $\overleftarrow{h}_{i,j}^{w}$. H is the size of the hidden state.

Furthermore, the ith sentence is represented by a non-linear transformation of the word-level hidden states as follows:

$$s_i = tanh(\mathbf{W} \cdot \frac{1}{n_i} \sum_{j=1}^{n_i} \overleftrightarrow{h}_{i,j}^{w} + b) \tag{4}$$

where s_i is the sentence embedding and $\mathbf{W}$, b are learnable parameters.

Sentence-level Encoder reads a document sentence-by-sentence until the end, using another bi-directional GRU network as depicted by the following equations:

$$\overleftrightarrow{h}_{i}^{s} = [\overrightarrow{h}_{i}^{s}; \overleftarrow{h}_{i}^{s}] \tag{5}$$

$$\overrightarrow{h}_{i}^{s} = GRU(s_i, \overrightarrow{h}_{i-1}^{s}) \tag{6}$$

$$\overleftarrow{h}_{i}^{s} = GRU(s_i, \overleftarrow{h}_{i+1}^{s}) \cdot \tag{7}$$

where $\overleftrightarrow{h}_{i}^{s}$ is a concatenated vector of the forward hidden state $\overrightarrow{h}_{i}^{s}$ and the backward hidden state $\overleftarrow{h}_{i}^{s}$.

The concatenated vectors $\overleftrightarrow{h}_{i}^{s}$ are document representations shared by the two tasks which will be introduced next.

2.2   Sentence Extractor

The sentence extractor can be viewed as a sequential binary classifier. We use a logistic function to calculate a score between 0 and 1, which is an indicator of whether or not to keep the sentence in the final summary. The score can also be considered as the salience of a sentence in the document. Let p_i denote the score and $q_i \in \{0,1\}$ denote the result of whether or not to keep the sentence. In particular, p_i is calculated as follows:

$$
\begin{aligned}
p_i &= P(q_i = 1 | \overleftrightarrow{\boldsymbol{h}}_i^s) \\
&= \sigma(\mathbf{W}^{extr} \cdot \overrightarrow{\boldsymbol{h}}_i^s + b^{extr})
\end{aligned}
\tag{8}
$$

where $\mathbf{W}^{extr}$ is the weight and b^{extr} is the bias which can be learned.

The sentence extractor generates a sequence of probabilities indicating the importance of the sentences. As a result, the extractive summary is created by selecting sentences with a probability larger than a given threshold τ. We set $\tau = 0.5$ in our experiment. We choose the cross entropy as the extractive loss function, i.e.,

$$
E_{se} = -\frac{1}{m} \sum_{i=1}^{m} q_i log p_i + (1 - q_i) log(1 - p_i)
\tag{9}
$$

2.3   Decoder

Our decoder is a unidirectional GRU network with hierarchical attention. We use the attention to calculate the context vectors which are weighted sums of the hidden states of the hierarchical encoders. The equations are given as below:

$$
\boldsymbol{c}_t^s = \sum_{i=1}^{m} \boldsymbol{\alpha}_{t,i} \cdot \overleftrightarrow{\boldsymbol{h}}_i^s
\tag{10}
$$

$$
\boldsymbol{c}_t^w = \sum_{i=1}^{m} \sum_{j=1}^{n_i} \boldsymbol{\beta}_{t,i,j} \cdot \overrightarrow{\boldsymbol{h}}_{i,j}^w
\tag{11}
$$

where $\boldsymbol{c}_t^s$ is the sentence-level context vector and $\boldsymbol{c}_t^w$ is the word-level context vector at decoding time step t. Specifically, $\boldsymbol{\alpha}_{t,i}$ denotes the attention value on the ith sentence and $\boldsymbol{\beta}_{t,i,j}$ denotes the attention value on the jth word of the ith sentence.

The input of the GRU-based language model at decoding time step t contains three vectors: the word embedding of previous generated word $\hat{\boldsymbol{y}}_{t-1}$, the sentence-level context vector of previous time step $\boldsymbol{c}_{t-1}^s$ and the word-level context vector of previous time step $\boldsymbol{c}_{t-1}^w$. They are transformed by a linear function and fed into the language model as follows:

$$
\tilde{\boldsymbol{h}}_t = GRU(\tilde{\boldsymbol{h}}_{t-1}, f_{in}(\hat{\boldsymbol{y}}_{t-1}, \boldsymbol{c}_{t-1}^s, \boldsymbol{c}_{t-1}^w))
\tag{12}
$$

where $\tilde{\boldsymbol{h}}_t$ is the hidden state of decoding time step t. f_{in} is the linear transformation function with $\mathbf{W}^{dec}$ as the weight and $\boldsymbol{b}^{dec}$ as the bias.

The hidden states of the language model are used to generate the output word sequence. The conditional probability distribution over the vocabulary in the tth time step is:

$$P(\hat{\boldsymbol{y}}_t | \hat{\boldsymbol{y}}_1, ..., \hat{\boldsymbol{y}}_{t-1}, \boldsymbol{x}) = g(f_{out}(\tilde{\boldsymbol{h}}_t, \boldsymbol{c}_t^s, \boldsymbol{c}_t^w)) \tag{13}$$

where g is the softmax function and f_{out} is a linear function with $\mathbf{W}^{soft}$ and $\boldsymbol{b}^{soft}$ as learnable parameters.

The negative log likelihood loss is applied as the loss of the decoder, i.e.,

$$E_y = \frac{1}{T} \sum_{t=1}^{T} -log(y_t) \tag{14}$$

where T is the length of the target summary.

2.4   Hierarchical Attention

The hierarchical attention mechanism consists of a word-level attention reader and a sentence-level attention reader, so as to take full advantage of the multi-level knowledge captured by the hierarchical document encoder. The sentence-level attention indicates the salience distribution over the source sentences. It is calculated as follows:

$$\boldsymbol{\alpha}_{t,i} = \frac{\boldsymbol{e}_{t,i}^s}{\sum_{k=1}^m \boldsymbol{e}_{t,k}^s} \tag{15}$$

$$\boldsymbol{e}_{t,i}^s = exp\{\mathbf{V}^{s T} \cdot tanh(\mathbf{W}_1^{dec} \cdot \tilde{\boldsymbol{h}}_t + \mathbf{W}_1^s \cdot \overleftrightarrow{\boldsymbol{h}}_i^s + \boldsymbol{b}_1^s)\} \tag{16}$$

where $\mathbf{V}^s$, $\mathbf{W}_1^{dec}$, $\mathbf{W}_1^s$ and $\boldsymbol{b}_1^s$ are learnable parameters.

The word-level attention indicates the salience distribution over the source words. As the hierarchical encoder reads the input sentences independently, our model has two distinctions. First, the word-level attention is calculated within a sentence. Second, we multiply the word-level attention by the sentence-level attention of the sentence which the word belongs to. The word-level attention calculation is shown below:

$$\boldsymbol{\beta}_{t,i,j} = \boldsymbol{\alpha}_{t,i} \frac{\boldsymbol{e}_{t,i,j}^w}{\sum_{l=1}^{n_i} \boldsymbol{e}_{t,i,l}^w} \tag{17}$$

$$\boldsymbol{e}_{t,i,j}^w = exp\{\mathbf{V}^{w T} \cdot tanh(\mathbf{W}_2^{dec} \cdot \tilde{\boldsymbol{h}}_t + \mathbf{W}_2^w \cdot \overleftrightarrow{\boldsymbol{h}}_{i,j}^w + \boldsymbol{b}_2^w)\} \tag{18}$$

where $\mathbf{V}^w$, $\mathbf{W}_2^{dec}$, $\mathbf{W}_2^w$ and $\boldsymbol{b}_2^w$ are learnable parameters for the word-level attention calculation.

The abstractive summary of a long document can be viewed as a new expression of the most salient sentences of the document, so that a well-learned sentence extractor and a well-learned attention distribution should both be able to detect the important sentences of the source document. Motivated by this, we design a constraint to the sentence-level attention which is an L2 loss as follows:

$$E_a = \frac{1}{m} \sum_{i=1}^{m} (p_i - \frac{1}{T} \sum_{t=1}^{T} \alpha_{t,i})^2 \tag{19}$$

As p_i is calculated by the logistic function which is trained simultaneously with the decoder, it is not suitable to constrain the attention with the inaccurate p_i. In our experiment, we use the labels of sentence extractor to constrain the attention.

2.5 Joint Learning

We combine three types of loss functions mentioned above to train our proposed model – the negative log likelihood loss E_y for the decoder, the cross entropy loss E_{se} for the extractor, and the L2 loss E_a as the attention constraint which performs as a regularizer. Hence,

$$E = E_y + \lambda \cdot E_{se} + \gamma \cdot E_a \tag{20}$$

The parameters are trained to minimize the joint loss function. In the inference stage, we use the beam search algorithm to select the word which approximately maximizes the conditional probability [17,18,28].

3 Experimental Setup

3.1 Dataset

We adopt the news dataset which is collected from the websites of CNN and DailyMail. It is originally prepared for the task of machine reading by Hermann et al. [19]. Cheng and Lapata [16] added labels to the sentences for the task of extractive summarization. The corpus contains pairs of news content and human-generated highlights for training, validation and test. Table 1 lists the details of the dataset.

Table 1. The statistics of the CNN/DailyMail dataset. S.S.N. indicates the average number of sentences in the source document. S.S.L. indicates the average length of the sentences in the source document. T.S.L. indicates the average length of the sentences in the target summary.

Dataset	Train	Valid	Test	S.S.N	S.S.L	T.S.L
CNN/DailyMail	277,554	13,367	11,443	26.9	27.3	53.8

3.2 Implementation Details

In our implementation, we set the vocabulary size D to be 50 K and word embedding size d as 300. The word embeddings have not been pretrained as the training corpus is large enough to train them from scratch. We cut off the documents as a maximum of 35 sentences and truncate the sentences with a maximum of 50 words. We also truncate the targeted summaries with a maximum of 100 words. The word-level encoder and the sentence-level encoder each corresponds a layer of bidirectional GRU, and the decoder also is a layer of unidirectional GRU. All the three networks have the hidden size H as 200. For the loss function, λ is set as 100 and γ is set as 0.5. During the training process, we use Adagrad optimizer [31] with the learning rate of 0.15 and initial accumulator value of 0.1. The mini-batch size is 16. We implement the model in Tensorflow and train it using a GTX-1080Ti GPU. The beam search size for decoding is 5. We use ROUGE scores [20] to evaluate the summarization models.

4 Experimental Results

4.1 Comparison with Baselines

We compare the full-length Rouge-F1 score on the entire CNN/DailyMail test set. We use the fundamental sequence-to-sequence attentional model and the words-lvt2k-hieratt [13] as baselines. The results are shown in Table 2.

Table 2. Performance comparison of various abstractive models on the **entire CNN/DailyMail test set** using **full- length F1** variants of Rouge.

Method	Rouge-1	Rouge-2	Rouge-L
seq2seq+attn	33.6	12.3	31.0
words-lvt2k-hieratt	35.4	13.3	32.6
Our method	**35.8**	**13.6**	**33.4**

From Table 2, we can see that our model performs the best in Rouge-1, Rouge-2 and Rouge-L. Compared to the vanilla sequence-to-sequence attentional model, our proposed model performs quite better. And compared to the hierarchical model, our model performs better in Rouge-L, which is due to the incorporation of the auxiliary task.

4.2 Evaluation of Proposed Components

To verify the effectiveness of our proposed model, we conduct ablation study by removing the corresponding parts, i.e. the auxiliary extractive task, the attention constraint and combination of them in order to make a comparison among their

Table 3. Performance comparison of removing the components of our proposed model on the **entire CNN/DailyMail test set** using **full-length F1** variants of Rouge.

Method	Rouge-1	Rouge-2	Rouge-L
Our method	**35.8**	**13.6**	**33.4**
w/o extr	34.3	12.6	31.6
w/o attn	34.7	12.8	32.2
w/o extr+attn	34.2	12.5	31.6

effects. We choose the full-length Rouge-F1 score on the test sets for evaluation. The results are shown in Table 3.

The results demonstrate that adding the auxiliary task and the attention constraint improves the performance of the abstractive method. The performance declines most when both the extractive task and the attention constraint are removed. Furthermore, as shown in the table, the performance declines more when the extractive task is removed, which means that the auxiliary task plays a more important role in our framework.

4.3   Case Study

We list some examples of the generated summaries of a source document(news) in Fig. 2. The source document contains 24 sentences with totally 660 words. Figure 2 presents three summaries: a golden summary which is the news highlight written by the reporter, the summary generated by our proposed model, and the summary generated by the sequence-to-sequence attentional model.

From the figure we can see that all system-generated summaries are copied words from the source document, because the highlights written by reporters used for training are usually partly copied from the source. However, different models have different characteristics. As illustrated in Fig. 2, all the four summaries are able to catch several key sentences from the document. The fundamental seq2seq+attn model misses some words like pronouns, which leads to grammatical mistakes in the generated summary. Our model without the auxiliary extractive task is able to detect more salient content, but the concatenated sentences have some grammatical mistakes and redundant words. Our model without the attention constraint generates fluent sentences which are very similar to the source sentences, but it focuses on just a small part of the source document. The summary generated by our proposed full model is most similar to the golden summary: it covers as much information and keeps correct grammar. It does not just copy sentences but use segmentation. Moreover, it changes the order of the source sentences while keeping the logical coherence.

4.4   Discussions

Our model has the advantages from three aspects. First, summaries generated by our model contain as much important information and perform well

Source Document
a flaw in robots designed to perform surgery has been found that lets them be easily hacked, according to researchers. the experts were able to take control of a so-called tele-robot during surgery by exploiting a simple programming trick. this enabled them to change the speed of the arms of the robot and change their orientation, making it impossible for the machines to carry out a procedure as directed. researchers at the @entity19 studied the @entity20 (shown) they found that robots designed for surgery could be 'easily' hacked in to. this is because they are operated over public networks which allowed the researchers to access them and stop them working the research was carried out by scientists ......
Golden Summary
researchers at the @entity19 studied so-called tele-robots. they found robots designed for surgery could be hacked and manipulated. this is because robots being tested were operated over public networks. it allowed the researchers to access them and stop them working.
Our Method
researchers at the @entity19 studied the @entity20. they found that robots designed for surgery could be 'easily' hacked in to. it enabled them to change the speed of the arms of the robot and change their orientation, making it impossible for the machines to carry out a procedure as directed. this is because they are operated over public networks which allowed the researchers to access them and stop them working.
Seq2seq+attn
researchers at the @entity19 studied the @entity20 (shown) robots designed for surgery could be 'easily' hacked in to. this is because they are operated over public networks which allowed the researchers to access them and stop them working.

Fig. 2. An example of summaries towards a piece of news. From top to down, the first is the source document which is the raw news content. The second is the golden summary which is used as the ground truth. The third is the summary generated by our proposed model. The last is the summary generated by the vanilla sequence-to-sequence attentional model.

grammatically. In practice, it depends on users' preference between the information coverage and condensibility to make a suitable balance. Compared to the low recall abstractive methods, our model is able to cover more information. And compared to the extractive methods, the generated summaries are more coherent logically. Second, the time complexity of our approach is much less than the baselines due to hierarchical structures, and our model is trained more quickly compared to those baselines. Third, as our key contribution is to improve the performance of the main task by incorporating an auxiliary task, in this experiment we just use normal GRU-based encoder and decoder for simplicity. More novel design for the decoder such as the hierarchical decoder can also be applied and incorporated into our model.

5   Related Work

The neural attentional abstractive summarization model was first applied in sentence compression [12], where the input sequence is encoded by a convolutional network and the output sequence is decoded by a standard feedforward Neural Network Language Model (NNLM). Chopra et al. [13] and Lopyrev [23] switched to RNN-type model as the encoder, and did experiments on various values of hyper parameters. To address the out-of-vocabulary problem, Gu et al. [24], Cao et al. [26] and See et al. [25] presented the copy mechanism which adds a selection operation between the hidden state and the output layer at each decoding time step so as to decide whether to generate a new word from the vocabulary or copy the word directly from the source sentence.

The sequence-to-sequence model with attention mechanism [13] achieves competitive performance for sentence compression, but is still a challenge for document summarization. Some researchers use hierarchical encoder to address the long-term dependency problem, yet most of the works are for extractive summarization tasks. Nallapati et al. [29] fed the input word embedding extended with new features to the word-level bidirectional GRU network and generated sequential labels from the sentence-level representations. Cheng and Lapata [16] presented a sentence extraction and word extraction model, encoding the sentences independently using Convolutional Neural Networks and decoding a binary sequence for sentence extraction as well as a word sequence for word extraction. Nallapati et al. [21] proposed a hierarchical attention with a hierarchical encoder, in which the word-level attention represents a probability distribution over the entire document.

Most previous works consider the extractive summarization and abstractive summarization as two independent tasks. The extractive task has the advantage of preserving the original information, and the abstractive task has the advantage of generating coherent sentences. It is thus reasonable and feasible to combine these two tasks. Tan et al. [14] as the first attempt to combine the two, tried to use the extracted sentence scores to calculate the attention for the abstractive decoder. But their proposed model using unsupervised graph-based model to rank the sentences is of high computation cost, and incurs long time to train.

6   Conclusion

In this work we have presented a sequence-to-sequence model with hierarchical document encoder and hierarchical attention for abstractive summarization, and incorporated extractive summarization as an auxiliary task. We jointly train the two tasks by sharing the same document encoder. The auxiliary task and the attention constraint contribute to improve the performance of the main task. Experiments on the CNN/DailyMail dataset show that our proposed framework is comparable to the state-of-the-art abstractive models. In the future, we will try to reduce the labels of the auxiliary task and incorporate semi-supervised and unsupervised methods.

Acknowledgements. This research has been supported by an innovative technology fund (project no. GHP/036/17SZ) from the Innovation and Technology Commission of Hong Kong, and a donated research project (project no. 9220089) at City University of Hong Kong.

References

1. Mihalcea, R., Tarau, P.: Textrank: bringing order into texts. In: EMNLP, Barcelona, pp. 404–411 (2004)
2. Erkan, G., Radev, D.R.: Lexrank: graph-based lexical centrality as salience in text summarization. JAIR **22**, 457–479 (2004)
3. Zhang, J., Yao, J., Wan, X.: Towards constructing sports news from live text commentary. In: ACL, Berlin, pp. 1361–1371 (2016)
4. Cao, Z., Wei, F., Li, S., Li, W., Zhou, M., Wang, H.: Learning summary prior representation for extractive summarization. In: IJCNLP, Beijing, pp. 829–833 (2015)
5. Yao, J., Wan, X., Xiao, J.: Recent advances in document summarization. Knowl. Inf. Syst. **53**, 297–336 (2017). https://doi.org/10.1007/s10115-017-1042-4
6. Cheung, J.C.K., Penn, G.: Unsupervised sentence enhancement for automatic summarization. In: EMNLP, Doha, pp. 775–786 (2014)
7. Gerani, S., Mehdad, Y., Carenini, G., Ng, R.T., Nejat, B.: Abstractive summarization of product reviews using discourse structure. In: EMNLP, Doha, pp. 1602–1613 (2014)
8. Fang, Y., Zhu, H., Muszynska, E., Kuhnle, A., Teufel, S.H.: A proposition-based abstractive summarizer. In: COLING, Osaka, pp. 567–578 (2016)
9. Liu, F., Flanigan, J., Thomson, S., Sadeh, N., Smith, N.A.: Toward abstractive summarization using semantic representations. In: NAACL-HLT, Denver, pp. 1077–1086 (2015)
10. Li, J., Luong, M.T., Jurafsky, D.: A hierarchical neural autoencoder for paragraphs and documents. arXiv preprint arXiv:1506.01057 (2015)
11. Yang, Z., Yang, D., Dyer, C., He, X., Smola, A.J., Hovy, E.H.: Hierarchical attention networks for document classification. In: NAACL-HLT, San Diego, pp. 1480–1489 (2016)
12. Rush, A.M., Chopra, S., Weston, J.: A neural attention model for abstractive sentence summarization. arXiv preprint arXiv:1509.00685 (2015)
13. Chopra, S., Auli, M., Rush, A.M.: Abstractive sentence summarization with attentive recurrent neural networks. In: NAACL-HLT, San Diego, pp. 93–98 (2016)
14. Tan, J., Wan, X., Xiao, J.: Abstractive document summarization with a graph-based attentional neural model. In: ACL, Vancouver, pp. 1171–1181 (2017)
15. Cho, K., Van Merriënboer, B., Bahdanau, D., Bengio, Y.: On the properties of neural machine translation: encoder-decoder approaches. arXiv preprint arxIV:1409.1259 (2014)
16. Cheng, J., Lapata, M.: Neural summarization by extracting sentences and words. arXiv preprint arXiv:1603.07252 (2016)
17. Graves, A.: Sequence transduction with recurrent neural networks. arXiv preprint arXiv:1211.3711 (2012)
18. Boulanger-Lewandowski, N., Bengio, Y., Vicent, P.: Audio chord recognition with recurrent neural networks. In: ISMIR, Curitiba, pp. 335–340 (2013)
19. Hermann, K.M., Kocisky, T., Grefenstette, E., Espeholt, L., Kay, W., Suleyman, M., Blunsom, P.: Teaching machines to read and comprehend. In: NIPS, Montreal, pp. 1693–1701 (2015)
20. Lin, C.Y.: Rouge: a package for automatic evaluation of summaries. In: ACL workshop (2014)
21. Nallapati, R., Zhou, B., Gulcehre, C., Xiang, B.: Abstractive text summarization using sequence-to-sequence RNNS and beyond. arXiv preprint arXiv:1602.06023 (2016)

22. Bahdanau, D., Cho, K., Bengio, Y.: Neural machine translation by jointly learning to align and translate. arXiv preprint arXiv:1409.0473 (2014)
23. Lopyrev, K.: Generating news headlines with recurrent neural networks. arXiv preprint arXiv:1512.01712 (2015)
24. Gu, J., Lu, Z., Li, H., Li, V.O.: Incorporating copying mechanism in sequence-to-sequence learning. arXiv preprint arXiv:1603.06393 (2016)
25. See, A., Liu, P.J., Manning, C.D.: Get to the point: summarization with pointer-generator networks. arXiv preprint arXiv:1704.04368 (2017)
26. Cao, Z., Luo C., Li, W., Li, S.: Joint copying and restricted generation for paraphrase. In: AAAI, San Francisco, pp. 3152–3158 (2017)
27. Luong, M.T., Pham, H., Manning, C.D.: Effective approaches to attention-based neural machine translation. arXiv preprint (2015) arXiv:1508.04025
28. Sutskever, I., Vinyals, O., Le, Q.V.: Sequence to sequence learning with neural networks. In: NIPS, Montreal, pp. 3104–3112 (2014)
29. Nallapati, R., Zhai, F., Zhou, B.: Summarunner: a recurrent neural network based sequence model for extractive summarization of documents. In: AAAI, San Francisco, pp. 3075–3081 (2017)
30. Wu, Y., Schuster, M., Chen, Z., Le, Q.V., Norouzi, M., Macherey, W., Krikun, M., Cao, Y., Gao, Q., Macherey, K., et al.: Google's neural machine translation system: bridging the gap between human and machine translation. arXiv preprint arXiv:1609.08144 (2016)
31. Duchi, J., Hazan, E., Singer, Y.: Adaptive subgradient methods for online learning and stochastic optimization. JMLR **12**, 2121–2159 (2011)

Rank-Integrated Topic Modeling: A General Framework

Zhen Zhang, Ruixuan Li, Yuhua Li$^{(\boxtimes)}$, and Xiwu Gu

School of Computer Science and Technology,
Huazhong University of Science and Technology, Wuhan 430074, China
{zenzang,rxli,idcliyuhua,guxiwu}@hust.edu.cn

Abstract. Rank-integrated topic models which incorporate link structures into topic modeling through topical ranking have shown promising performance comparing to other link combined topic models. However, existing work on rank-integrated topic modeling treats ranking as document distribution for topic, and therefore can't integrate topical ranking with LDA model, which is one of the most popular topic models. In this paper, we introduce a new method to integrate topical ranking with topic modeling and propose a general framework for topic modeling of documents with link structures. By interpreting the normalized topical ranking score vectors as topic distributions for documents, we fuse ranking into topic modeling in a general framework. Under this general framework, we construct two rank-integrated PLSA models and two rank-integrated LDA models, and present the corresponding learning algorithms. We apply our models on four real datasets and compare them with baseline topic models and the state-of-the-art link combined topic models in generalization performance, document classification, document clustering and topic interpretability. Experiments show that all rank-integrated topic models perform better than baseline models, and rank-integrated LDA models outperform all the compared models.

Keywords: Normalized topical ranking · Topic distribution
Rank-integrated topic modeling framework

1 Introduction

With the rapid development of online information systems, document networks, i.e. information networks associated with text information, are becoming pervasive in our digital library. For example, research papers are linked together via citations, web pages are connected by hyperlinks and Tweets are connected via social relationships. To better mine values from documents with link structures, we study the problem of building topic models of document networks.

The most popular topic models include PLSA (Probabilistic Latent Semantic Analysis) [1] and LDA (Latent Dirichlet Allocation) [2]. Traditional topic models assume documents are independent with each other and links among them will

© Springer International Publishing AG, part of Springer Nature 2018
Y. Cai et al. (Eds.): APWeb-WAIM 2018, LNCS 10987, pp. 16–31, 2018.
https://doi.org/10.1007/978-3-319-96890-2_2

not be considered in the modeling process. Intuitively, linked documents should have similar semantic information, which can be utilized in topic modeling.

To take advantage of link structures in document networks, several topic models have been proposed. One line of this work is to build unified generative models for both texts and links, such as iTopic [3] and RTM [4], and the other line is to add regularization into topic modeling, such as graph-based regularizer [5] and rank-based regularizer [6]. As a state-of-the-art link combined topic model, LIMTopic [7] incorporates link structures into topic modeling through topical ranking. However, LIMTopic treats topical ranking as document distribution for topic which causes that topical ranking can only be combined with symmetric PLSA model. Therefore LIMTopic can not be combined with the popular LDA model. To solve this problem, we normalize topical ranking vectors along the topic dimension and treat them as topic distributions for documents. Link structures are then fused with text information by iteratively performing topical ranking and topic modeling in a mutually enhanced framework.

In this paper, we propose a general framework for rank-integrated topic modeling, which can be integrated with both PLSA and LDA models. The main contributions of this paper are summarized as follows.

- A novel approach to integrate topical ranking with topic modeling is proposed, upon which we build a general rank-integrated topic modeling framework for document networks.
- Under this general framework, we construct two rank-integrated PLSA models, namely RankPLSA and HITSPLSA, and two rank-integrated LDA models, i.e. RankLDA and HITSLDA.
- Extensive experiments on three publication datasets and one Twitter dataset demonstrate that rank-integrated topic models perform better than baseline models. Moreover, rank-integrated LDA models consistently perform better than all the compared models.

The rest of this paper is organized as follows. Section 2 reviews the related work, and Sect. 3 introduces the notations used in topic modeling. In Sect. 4, we propose the rank-integrated topic modeling framework and detail the learning algorithm for rank-integrated PLSA and LDA models. Experimental studies are presented in Sect. 5, and we conclude this paper in Sect. 6.

2   Related Work

Topic modeling algorithms are unsupervised machine learning methods that analyze words of documents to discover themes that run through the corpus and distributions on these themes for each document. PLSA [1] and LDA [2] are two most well known topic models. However, both PLSA and LDA treat documents in a given corpus as independent to each other. Since their presence, various kinds of models have been proposed by incorporating contextual information into topic modeling, such as time [8] and links [3,4,7,9]. Several recent works introduce embeddings into topic modeling to improve topic interpretability [10] or reduce computation complexity [11]. To better cope with word sparsity, many short text-based topic models have been proposed [12]. Topic models have also been explored in other research domains, such as recommender system [13]. The most similar work to ours is the LIMTopic framework [7]. The distinguished feature of our work is that we treat topical ranking vectors as topic distributions of documents while LIMTopic treats them as document distributions of topics. Our method is arguably more flexible and can construct both rank-integrated PLSA and LDA model under a unified framework while LIMTopic can only work with symmetric PLSA model.

Our work is also closely related to ranking technology. PageRank and HITS (Hyperlink-Induced Topic Search) are two most popular link based ranking algorithms. Topical link analysis [14] extends basic PageRank and HITS by computing a score vector for each page to distinguish the contribution from different topics. Yao et al. [15] extend pair-wise ranking models with probabilistic topic models and propose a collaborative topic ranking model to alleviate data sparsity problem in recommender system. Ding et al. [16] take a topic modeling approach for preferences ranking by assuming that the preferences of each user are generated from a probabilistic mixture of a few latent global rankings that are shared across the user population. Both of Yao's and Ding's models focus on employing topic modeling to solve ranking problem, while our work incorporates link structures into topic modeling through ranking.

3   Notations

For ease of reference, we list the relevant notations used in topic modeling in Table 1. Based on topic modeling, topical ranking methods [14] compute a score

Table 1. Notations used in topic modeling

Symbol	Description
$\mathcal{D}$	All the documents in the corpus
D, V, K	The number of documents, unique words, topics in the corpus
d	Document index in the corpus
N_d	The number of words in document d
μ	The probability of generating specific documents
θ_d	The topic distribution of document d, expressed by a multinomial distribution of topics
γ_d	The topical ranking vector of document d
w_{dn}	The nth word in document d, $w_{dn} \in \{1, 2, \ldots V\}$
z_{dn}	The topic assignment of word w_{dn}, $z_{dn} \in \{1, 2, \ldots K\}$
β_k	The multinomial distribution over words specific to topic k
α, η	Dirichlet priors to multinomial distribution θ_d, β_k

vector γ_d for each page to distinguish the contribution from different topics. Topical PageRank and Topical HITS are two topical ranking algorithms extended from PageRank and HITS respectively. Since topical HITS computes an authority vector a_d and a hubness vector h_d for each document, we use the average of these two vectors as the overall ranking vector.

4   Rank-Integrated Topic Modeling Framework

In this section, we first introduce the general framework for incorporating topical ranking into topical modeling. Then we instantiate the framework by detailing rank-integrated PLSA model and LDA model. Finally we describe the learning algorithm of the rank-integrated topic modeling framework.

4.1   The Framework for Rank-Integrated Topic Modeling

The intuition behind the idea of incorporating topical ranking into topic modeling is that the more important a document is on a topic, the more likely the document is about that topic. By normalizing the topical ranking vector γ_d such that $\forall d, \sum_{k=1}^{K} \gamma_{dk} = 1$, the ranking vector γ_d can be regarded as parameters for a multinomial distribution, which can further be viewed as topic distribution of document d.

To fully exploit both text information and link structure in topic modeling, we introduce a weight factor λ ranging from 0 to 1 to represent our belief of topical ranking in topic modeling. We propose a general framework for incorporating topical ranking into topic modeling by defining the topic distribution for document d as

$$p(z_{dn} = k|d) = \lambda \gamma_{dk} + (1 - \lambda)\theta_{dk} \tag{1}$$

where λ balances the importance of link structure and text information. When $\lambda = 0$, topic distributions are the same as PLSA or LDA model, and when $\lambda = 1$, topical distributions are completely dependent on topical ranking.

Our approach is more advanced than previous work in LIMTopic. LIMTopic normalizes topical ranking vectors along the document dimension and treats them as document distribution for topics, which leads that LIMTopic can only integrate with symmetric PLSA model. We normalize topical ranking vectors along the topic dimension and fuse ranking into topic distribution, which is more natural and extensible in that our framework can combine with both PLSA and LDA model. Under this framework, we instantiate four models, RankPLSA, HITSPLSA, RankLDA and HITSLDA, of which the first two models combine PLSA with topical PageRank and topical HITS while the last two combine LDA with topical PageRank and topical HITS. Next we detail rank-integrated PLSA and LDA models respectively.

4.2   Rank-Integrated PLSA Model

By combining PLSA model with topical PageRank and topical HITS, we obtain RankPLSA and HITSPLSA model, whose graphical representation is presented in Fig. 1. Expectation maximization (EM) [7] is employed to derive the parameter update rules for the model.

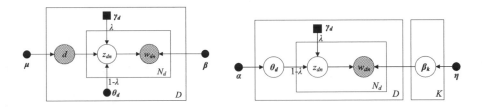

Fig. 1. Rank-integrated PLSA model   **Fig. 2.** Rank-integrated LDA model

The log likelihood of the models is

$$
L = \log p(\boldsymbol{\mathcal{D}}; \boldsymbol{\mu}, \boldsymbol{\gamma}, \boldsymbol{\theta}, \boldsymbol{\beta}) = \sum_{d=1}^{D} \sum_{v=1}^{V} n_{dv} \left(\log \mu_d + \log \sum_{k=1}^{K} \beta_{kv} \big[\lambda \gamma_{dk} + (1-\lambda)\theta_{dk} \big] \right).
\tag{2}
$$

In the E-step, the posterior distribution of latent variable z_{dn} conditioned on document d and word w_{dn} is computed by Eq. (3)

$$
\psi_{dvk} = p(z_{dn} = k | d, w_{dn} = v) = \frac{\beta_{kv} \big[\lambda \gamma_{dk} + (1-\lambda)\theta_{dk} \big]}{\sum_{k=1}^{K} \beta_{kv} \big[\lambda \gamma_{dk} + (1-\lambda)\theta_{dk} \big]}.
\tag{3}
$$

The lower bound of the log likelihood can be obtained by using Jensen's inequality twice as Eq. (4)

$$
\begin{aligned}
L &= \sum_{d=1}^{D} \sum_{v=1}^{V} n_{dv} \left(\log \sum_{k=1}^{K} \psi_{dvk} \frac{\beta_{kv} \big[\lambda \gamma_{dk} + (1-\lambda)\theta_{dk} \big]}{\psi_{dvk}} \right) + \sum_{d=1}^{D} \sum_{v=1}^{V} n_{dv} \log \mu_d \\
&\geq \sum_{d=1}^{D} \sum_{v=1}^{V} n_{dv} \sum_{k=1}^{K} \psi_{dvk} \log \Big(\beta_{kv} \big[\lambda \gamma_{dk} + (1-\lambda)\theta_{dk} \big] \Big) \\
&\quad - \sum_{d=1}^{D} \sum_{v=1}^{V} n_{dv} \sum_{k=1}^{K} \psi_{dvk} \log \psi_{dvk} + \sum_{d=1}^{D} \sum_{v=1}^{V} n_{dv} \log \mu_d \\
&\geq \sum_{d=1}^{D} \sum_{v=1}^{V} n_{dv} \sum_{k=1}^{K} \psi_{dvk} \log \Big(\lambda \log \beta_{kv} \gamma_{dk} + (1-\lambda) \log \beta_{kv} \theta_{dk} \Big) \\
&\quad - \sum_{d=1}^{D} \sum_{v=1}^{V} n_{dv} \sum_{k=1}^{K} \psi_{dvk} \log \psi_{dvk} + \sum_{d=1}^{D} \sum_{v=1}^{V} n_{dv} \log \mu_d
\end{aligned}
$$

$$= \sum_{d=1}^{D} \sum_{v=1}^{V} n_{dv} \sum_{k=1}^{K} \psi_{dvk} \log \beta_{kv} + \sum_{d=1}^{D} \sum_{v=1}^{V} n_{dv} \sum_{k=1}^{K} (1-\lambda)\psi_{dvk} \log \theta_{dk} + const$$

(4)

where we use *const* to represent all the terms that do not depend on parameters θ and β since these are parameters we care about in topic modeling.

In the M-step, we maximize the lower bound of L under the constraint that $\sum_{v=1}^{V} \beta_{kv} = 1$ and $\sum_{k=1}^{K} \theta_{dk} = 1$, which can be achieved using a Lagrange multiplier. By solving the Lagrange multiplier problem, we obtain the closed-form update rules for θ and β as Eqs. (5) and (6)

$$\beta_{kv} = \frac{\sum_{d=1}^{D} n_{dv}\psi_{dvk}}{\sum_{v=1}^{V} \sum_{d=1}^{D} n_{dv}\psi_{dvk}}$$

(5)

$$\theta_{dk} = \frac{\sum_{v=1}^{V} n_{dv}\psi_{dvk}}{\sum_{k=1}^{K} \sum_{v=1}^{V} n_{dv}\psi_{dvk}}.$$

(6)

4.3   Rank-Integrated LDA Model

Following the above mentioned framework, we combine LDA with topical PageRank and topical HITS to obtain RankLDA and HITSLDA model. Figure 2 is the graphical representation of rank-integrated LDA model.

Following the graphical model, the joint distribution over hidden and observed variables for the model is as Eq. (7)

$$p(\theta, z, w, \beta; \alpha, \eta, \gamma, \lambda)$$

$$= \prod_{k=1}^{K} p(\beta_k; \eta) \prod_{d=1}^{D} \left(p(\theta_d; \alpha) \prod_{n=1}^{N_d} p(z_{dn}|\theta_d, \lambda, \gamma_d)p(w_{dn}|z_{dn}, \beta) \right).$$

(7)

Exact inference of rank-integrated LDA model is difficult to compute, we therefore turn to collapsed Gibbs sampling. The sampling procedure of topic assignment z_{dn} is divided into two subroutines according to whether z_{dn} comes from topic modeling θ_d or topical ranking γ_d. It is straightforward to sample z_{dn} from γ_d since it is directly a multinomial distribution. Next we derive the sampling equation for z_{dn} coming from θ_d.

To keep the notation uncluttered, we use i to denote the document-word index (d,n) and $\neg i$ to denote all the indices excluding i. The complete conditional distribution $p(z_{dn} = k|z_{\neg(d,n)}, w)$ can therefore be written as Eq. (8)

$$p(z_i = k|z_{\neg i}, w) \propto p(z_i = k, w_i = v|z_{\neg i}, w_{\neg i})$$

$$= \int \int p(z_i = k, w_i = v, \theta_d, \beta_k|z_{\neg i}, w_{\neg i}; \alpha, \eta)d\theta_d d\beta_k$$

$$= \int p(z_i = k|\theta_d)p(\theta_d|z_{\neg i}; \alpha)d\theta_d \int p(w_i = v|\beta_k)p(\beta_k|z_{\neg i}, w_{\neg i}; \eta)d\beta_k$$

(8)

where we have used the conditional independence implied by the graphical model. Due to the conjugacy relationship between Dirichlet and multinomial distribution, the posteriors $p(\boldsymbol{\theta}_d|\boldsymbol{z}_{\neg i};\boldsymbol{\alpha})$ and $p(\boldsymbol{\beta}_k|\boldsymbol{z}_{\neg i},\boldsymbol{w}_{\neg i};\boldsymbol{\eta})$ are also Dirichlet distribution, which can be written as

$$p(\boldsymbol{\theta}_d|\boldsymbol{z}_{\neg i};\boldsymbol{\alpha}) = Dir(\boldsymbol{\theta}_d;\boldsymbol{n}_{d,\neg i}+\boldsymbol{\alpha}) \tag{9}$$

$$p(\boldsymbol{\beta}_k|\boldsymbol{z}_{\neg i},\boldsymbol{w}_{\neg i};\boldsymbol{\eta}) = Dir(\boldsymbol{\beta}_k;\boldsymbol{m}_{k,\neg i}+\boldsymbol{\eta}). \tag{10}$$

The count $\boldsymbol{n}_{d,\neg i}$ is per-document count of topics and $\boldsymbol{m}_{k,\neg i}$ is per topic count of terms, which are both defined excluding z_i and w_i. The complete conditional $p(z_i = k|\boldsymbol{z}_{\neg i},\boldsymbol{w})$ can be obtained by a product of two posterior Dirichlet expectations, yielding

$$p(z_i = k|\boldsymbol{z}_{\neg i},\boldsymbol{w}) \propto \frac{n_{d,\neg i}^k + \alpha}{\sum_{k=1}^K n_{d,\neg i}^k + K\alpha} \cdot \frac{m_{k,\neg i}^v + \eta}{\sum_{v=1}^V m_{k,\neg i}^v + V\eta} \tag{11}$$

where $\boldsymbol{n}_d$ only counts the sample from $\boldsymbol{\theta}_d$ while $\boldsymbol{m}_k$ counts samples from both $\boldsymbol{\theta}_d$ and $\boldsymbol{\gamma}_d$. By jointly collecting samples from topic modeling $\boldsymbol{\theta}_d$ and topical ranking $\boldsymbol{\gamma}_d$, the topical ranking is thus incorporated into LDA model.

After a sufficient number of sampling iterations, the count $\boldsymbol{n}_d$ and $\boldsymbol{m}_k$ can be used to estimate the parameters $\boldsymbol{\theta}$ and $\boldsymbol{\beta}$, which are given by Eqs. (12) and (13)

$$\theta_{dk} = \frac{n_d^k + \alpha}{\sum_k n_d^k + K\alpha} \tag{12}$$

$$\beta_{kv} = \frac{m_k^v + \eta}{\sum_v m_k^v + V\eta}. \tag{13}$$

Note that the parameter update rules are different from collapsed Gibbs sampling for LDA in that $\boldsymbol{m}_k$ collects samples from both topic modeling $\boldsymbol{\theta}_d$ and topical ranking $\boldsymbol{\gamma}_d$.

4.4   The Learning Algorithm of Rank-Integrated Topic Modeling Framework

By incorporating topical ranking into topic modeling, we build a mutual enhancement framework where we perform topic modeling and topical ranking in an alternative process. The learning algorithm of the rank-integrated topic modeling framework is summarized in Algorithm 1. Next we present the three termination conditions in the algorithm.

Condition 1: This condition is to test whether topical PageRank or topical HITS has converged. We compute the difference of the ranking matrix between the current iteration and the previous iteration, then summarize all the differences in every element. If the summation is less than a previously defined small value (1e−4 in our experiments), this condition is satisfied.

Algorithm 1. The learning algorithm of rank-integrated topic modeling framework

---

   Input: A collection of D documents with V unique words and document
            network G, the expected number of topics K and the weight factor λ
   Output: document-topic distribution θ, topic-word distribution β
1 **Initialization:** Perform PLSA or LDA to obtain θ and β
2 **repeat**
3 | **repeat**
4 | | Perform topical ranking to obtain ranking matrix γ
5 | **until** *Satisfying condition 1*
6 | **repeat**
7 | | Perform rank-integrated PLSA or LDA parameter update rules via Eqs.
        | | (5), (6), (12) and (13)
8 | **until** *Satisfying condition 2*
9 **until** *Satisfying condition 3*

Condition 2: This condition is to test the convergence of the rank-integrated topic modeling process. In each iteration, we compute the log likelihood of the rank-integrated PLSA or LDA model, and then compute the relative change between two continuous iterations as the fraction of the difference between two log likelihoods to the average value of them. If the relative change is lower than a predefined value ($1e-4$ in our experiments), this condition is satisfied. *Condition 3:* This condition is to test whether the whole process has converged. We assess convergence by computing log likelihood of the rank-integrated document-word matrix, which is obtained by performing topical PageRank on the row normalized document-word matrix. If the incremental quality of the log likelihood is smaller than a predefined threshold ($1e-3$ in our experiments), condition 3 is satisfied.

Let D be the number of documents, V be the number of unique words in the corpus and E be the number of links in the document network. After a brief analysis, the time complexity of the framework's learning algorithm is $O(E + D \times V)$ which is liner to the total number of links and words in the documents.

5   Experiments

In this section, we evaluate the performance of our proposed rank-integrated topic models - RankPLSA, RankLDA, HITSPLSA and HITSLDA. The comparing models are two baseline topic models - PLSA and LDA, two state-of-the-art rank based models - RankTopic and HITSTopic which are instances of the LIM-Topic [7] model. We use two genres of datasets, three publication datasets and one Twitter dataset, which come from the LIMTopic paper [7]. The basic statistics of the datasets are shown in Table 2.

Note that in the Twitter dataset, users associated with their published tweets are regarded as documents and the '@' relationship among users as links. Documents in the Cora dataset have been classified into seven categories in advance,

Table 2. Basic statistics of the four experimental datasets

Dataset	Arnetminer	Cora	Citeseer	Twitter
# of documents	6562	2708	3312	814
# of unique words	8815	1433	3703	5316
# of links	8331	5429	4715	4206

and documents in the Citeseer dataset have been labeled one of six classes. The experiments are conducted in several aspects, namely generalization performance, document classification, document clustering and topic interpretability.

5.1 Generalization Performance

In this section, we evaluate the generalization performance of our proposed models by computing the perplexity of a held-out dataset. A lower perplexity indicates better generalization performance. All perplexities are computed by the average of 10-fold cross validation.

First we study how the value of the weight factor λ influences the generalization performance of the models. As shown in Fig. 3, rank-integrated topic models with appropriate values of λ have lower perplexity than the special case of $\lambda = 0$, which degenerates to PLSA or LDA model. Further, we observe that different models have different sensitivities to the value of λ in different datasets. RankPLSA and RankLDA performs best with $\lambda = 0.1$ on Arnetminer, while HITSPLSA and HITSLDA performs best on Twitter with $\lambda = 0.6$ and $\lambda = 0.1$ respectively.

Figure 4 presents perplexity comparison of different models, including two baseline models PLSA, LDA and two link combined models RankTopic, HITSTopic. Results show that all the rank-integrated models have lower perplexities than baseline models. Further we observe that HITSLDA and RankLDA perform best in both Arnetminer and Twitter datasets. The underlying reason is analyzed as follows. By introducing Dirichlet priors, LDA overcomes the overfitting problem, thus performs better than PLSA model. Rank-integrated LDA model further boosts the performance of LDA significantly by incorporating link into LDA model. RankPLSA and RankTopic have similar performance, since they are both rank-integrated PLSA models. Further notice that topical PageRank-integrated models have lower perplexities than topical HITS-integrated models, and HITSLDA has relatively higher perplexity than RankLDA as the number of topic increases. The reason might be that topical PageRank score serves better as document-topic distribution than the average of topical hubness and topical authority in topical HITS.

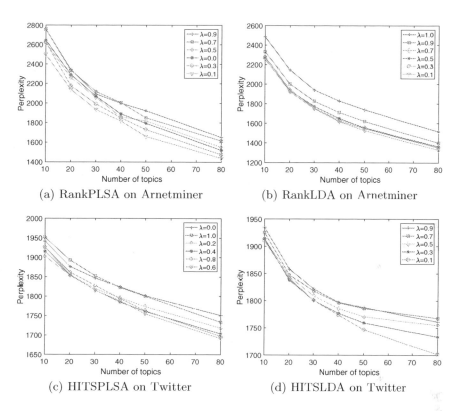

Fig. 3. Parameter study for rank-integrated topic models on Arnetminer and Twitter. The X-axis is number of topics, and the Y-axis is the perplexity. Lower perplexity indicates better generalization performance.

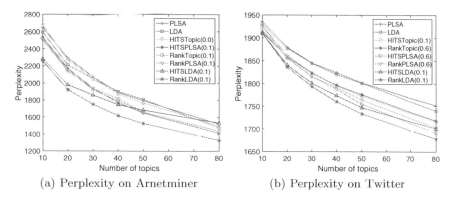

Fig. 4. Perplexity comparison of different topic models on Arnetminer and Twitter

5.2   Document Classification

In addition to generalization performance, topic models can also be evaluated by the performance of their applications, e.g. document classification and clustering. By performing topic modeling, documents are represented as topic distributions, which can be treated as feature vectors used further for classification or clustering. In this section, we study the performance of document classification for different models. Recall that Cora and Citeseer datasets have predefined labels, therefore it's natural to use these two datasets to perform classification.

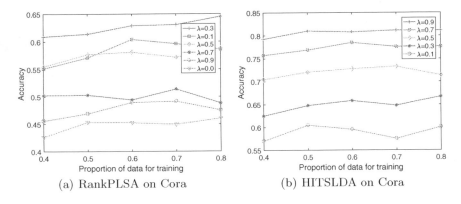

(a) RankPLSA on Cora   (b) HITSLDA on Cora

Fig. 5. Parameter study for RankPLSA and HITSLDA on Cora. The X-axis is proportion of training data, and the Y-axis is the classification accuracy.

Due to the superior performance of support vector machine (SVM), we apply SVM classifier with RBF kernel as the classification algorithm. Parameters study for RankPLSA and HITSLDA on Cora dataset with different proportions of training data are shown in Fig. 5. For RankPLSA, $\lambda = 0.3$ has the highest classification accuracy, and the accuracy decreases when λ is close to 0 or 1. HITSLDA with $\lambda = 0.9$ performs best in terms of classification accuracy, and the accuracy decreases as λ decreases. We observe similar results with Citeseer dataset. Figure 6 presents accuracy comparison on different models, where TopicalRank represents topical PageRank and TopicalHITS represents topical HITS algorithm. From the results, we see that all the rank-integrated models perform better than the baseline models and the corresponding topical ranking models. We also observe that HITSLDA performs best in all the models. Overall, topical HITS-integrated topic models perform better than topical PageRank-integrated models. The underlying reason is that the summation of topical hubness and authority in topical HITS is more discriminated than topical PageRank only.

In addition to classification accuracy, we also compute other classification measures on different topic models, namely precision, recall and F1 score, where higher values indicate better performance. Table 3 reports experimental results on Cora and Citeseer dataset with 50% of the the data as training data. From

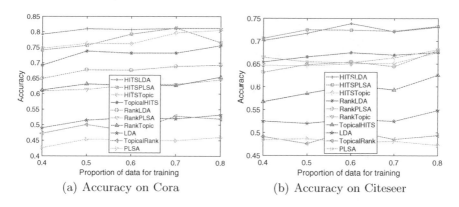

(a) Accuracy on Cora   (b) Accuracy on Citeseer

Fig. 6. Classification accuracy on Cora and Citeseer

Table 3. Classification precision, recall and F1 score of different topic models on Cora and Citeseer

Models	Cora			Citeseer		
	Precision	Recall	F1	Precision	Recall	F1
PLSA	0.3993	0.3739	0.3739	0.4282	0.4392	0.4232
TopicalRank	0.4446	0.4081	0.4040	0.4643	0.4368	0.4249
LDA	0.5002	0.4086	0.4272	0.4757	0.4720	0.4609
TopicalHITS	0.7616	0.7021	0.7257	0.5583	0.5311	0.5273
RankTopic	0.5792	0.5499	0.5569	0.5862	0.5939	0.5798
RankPLSA	0.5809	0.5650	0.5711	0.5804	0.5820	0.5630
RankLDA	0.6922	0.6048	0.6339	0.5994	0.5996	0.5862
HITSTopic	0.7823	0.7183	0.7443	0.6067	0.5871	0.5826
HITSPLSA	0.8039	0.7155	0.7483	0.6590	0.6564	0.6410
HITSLDA	**0.8130**	**0.7784**	**0.7928**	**0.6873**	**0.6619**	**0.6620**

the results, we can clearly see that rank-integrated topic models perform significantly better than baseline models and HITSLDA performs best among all the compared models.

5.3   Document Clustering

This section studies the performance of different topic models on document clustering. For documents with link structures, normalized cut (Ncut) and modularity (Modu) are two well known clustering measures. Lower normalized cut and higher modularity indicate better clustering performance. When the background label is known for documents, normalized mutual information (NMI) can also be used to evaluate clustering performance. The higher the NMI, the better the clustering quality. In these experiments, we apply k-means as the clustering algorithm. Experiments show that Arnetminer has no significant community structure. We therefore conduct clustering experiment on three datasets, whose numbers of cluster are set to 7 for Cora, 6 for Citeseer and 10 for Twitter.

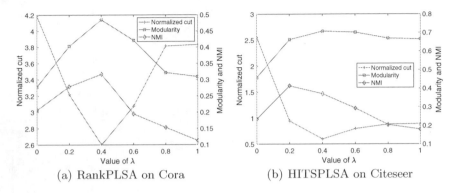

(a) RankPLSA on Cora   (b) HITSPLSA on Citeseer

Fig. 7. Clustering performance for RankPLSA and HITSPLSA with some typical values of λ. Lower normalized cut, higher modularity and NMI indicate better performance. Notice that left Y-axis is for normalized cut, while the right one is for both modularity and NMI.

Similar to the previous experiments, we first study how different values of λ influence the clustering quality. Figure 7 presents parameter study for RankPLSA on Cora and HITSPLSA on Citeseer. From the results, we obtain that RankPLSA achieves best performance on normalized cut, modularity and NMI with $\lambda = 0.4$. HITSPLSA performs best in terms of normalized cut and modularity with $\lambda = 0.4$, while it achieves highest NMI with $\lambda = 0.2$. Overall HITSPLSA achieves best clustering performance with $\lambda = 0.4$. We observe similar result for other models.

We then study the clustering performance of different models, which are reported in Table 4. From the table, we can see that all the rank-integrated models perform better than baseline topic models and topical ranking models, which indicates rank-integrated models have better clustering quality than their ingredients. Further, similar to classification results, topical HITS-integrated models perform better than topical PageRank-integrated models. This indicates the summation of topical hubness and topical authority in topical HITS are more discriminated than topical PageRank in machine learning tasks like classification and clustering. Finally, we observe that rank-integrated LDA models consistently perform better than rank-integrated PLSA models, which justifies the importance of integrating topical ranking with LDA model.

5.4 Topic Interpretability

Our final experiment evaluates the interpretability of different topic models. Topic modeling algorithms discover semantically coherent themes or topics,

Table 4. Clustering performance of different models on Cora, Citeseer and Twitter

Models	Cora			Citeseer			Twitter	
	Ncut	Modu	NMI	Ncut	Modu	NMI	Ncut	Modu
PLSA	4.2335	0.2582	0.1968	2.6271	0.3933	0.2270	5.2398	0.2966
TopicalRank	3.8331	0.3095	0.1129	2.3566	0.4363	0.1164	3.5072	0.5127
LDA	4.2023	0.2595	0.2158	2.0814	0.4724	0.2605	4.4584	0.3156
TopicalHITS	1.7126	0.5734	0.2385	0.8905	0.6646	0.1793	2.0023	0.4935
RankTopic	2.5528	0.4958	0.3167	1.2938	0.6125	0.3025	2.8727	0.5366
RankPLSA	2.6004	0.4850	0.3170	1.2761	0.61089	0.3333	2.7096	0.5581
RankLDA	2.3611	0.5226	0.3235	1.0803	0.6340	0.3554	2.2743	0.6162
HITSTopic	1.4333	0.6312	0.3714	0.7171	0.6858	0.2371	1.5937	0.5690
HITSPLSA	1.1143	0.6566	0.4492	0.5972	0.7078	0.3707	1.5532	0.5458
HITSLDA	**0.7631**	**0.6729**	**0.4809**	**0.3663**	**0.7372**	**0.3955**	**0.9154**	**0.6403**

which are multinomial distribution over words in the vocabulary. By examining the words with top probability in each distribution, the detected topic can be interpreted as meaningful concept. The easier the topics can be interpreted, the better the detected topics are. There exists some quantitative measure to evaluate topic interpretability. Pointwise mutual information (PMI) is such a measure to evaluate topic coherence, where higher PMI indicates better interpretability. In our experiment, we compute the PMI among top ten words in each topic and average them as the PMI of that topic.

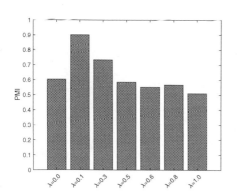

Fig. 8. PMI of RankPLSA on Arnetminer

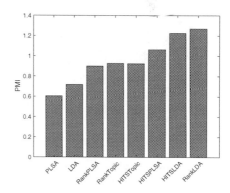

Fig. 9. PMI of all models on Arnetminer

Figure 8 presents PMI values of RankPLSA model on Arnetminer with different weight factor λ, where the topic number is set to 10. We can see that RankPLSA achieves high PMI when λ is set to relatively small value, e.g. 0.1 and 0.3, which have similar result with perplexity of the model. The PMI values of different models on Arnetminer are shown in Fig. 9, from which we observe that rank-integrated topic models perform better than baseline models. Further, RankLDA and HITSLDA have the highest PMI values, which are consistent with the generalization performance, suggesting that there are correlations between generalization performance and topic interpretability.

6 Conclusion and Future Work

In this paper, we bring topical ranking to topic modeling, and propose a general framework for rank-integrated topic modeling. As instances of this framework, two rank-integrated PLSA models, i.e. RankPLSA and HITSPLSA, and two rank-integrated LDA models, i.e. RankLDA and HITSLDA, are presented. To validate the effectiveness of proposed models, extensive experiments are conducted on generalization performance, document classification, document clustering and topic interpretability. Results show that rank-integrated topic models consistently perform better than baseline topic models, and rank-integrated LDA models have the best performance among all the compared models. As future work, we will study how rank-integrated topic modeling influences the ranking results compared to topical ranking algorithms. Moreover, we will implement our algorithm in a distributed computing environment to scale up to big data.

Acknowledgments. This work is supported by the National Key Research and Development Program of China under grants 2016QY01W0202 and 2016YFB08-00402, National Natural Science Foundation of China under grants 61572221, U1401258, 61433006 and 61502185, Major Projects of the National Social Science Foundation under grant 16ZDA092, Science and Technology Support Program of Hubei Province under grant 2015AAA013, Science and Technology Program of Guangdong Province under grant 2014B010111007, and Guangxi High Level Innovation Team in Higher Education Institutions - Innovation Team of ASEAN Digital Cloud Big Data Security and Mining Technology.

References

1. Hofmann, T.: Probabilistic latent semantic analysis. In: Proceedings of the 15th Conference on Uncertainty in Artificial Intelligence, pp. 289–296. Morgan Kaufmann Publishers Inc. (1999)
2. Blei, D.M., Ng, A.Y., Jordan, M.I.: Latent Dirichlet allocation. J. Mach. Learn. Res. **3**(Jan), 993–1022 (2003)
3. Sun, Y., Han, J., Gao, J., Yu, Y.: iTopicModel: information network-integrated topic modeling. In: Proceedings of the 9th IEEE International Conference on Data Mining, pp. 493–502. IEEE (2009)
4. Chang, J., Blei, D.M.: Hierarchical relational models for document networks. Ann. Appl. Stat. **4**(1), 124–150 (2010)
5. Mei, Q., Cai, D., Zhang, D., Zhai, C.: Topic modeling with network regularization. In: Proceedings of the 17th International Conference on World Wide Web, pp. 101–110. ACM (2008)

6. Duan, D., Li, Y., Li, R., Zhang, R., Wen, A.: RankTopic: ranking based topic modeling. In: Proceedings of the 12th IEEE International Conference on Data Mining, pp. 211–220. IEEE (2012)
7. Duan, D., Li, Y., Li, R., Zhang, R., Gu, X., Wen, K.: LIMTopic: a framework of incorporating link based importance into topic modeling. IEEE Trans. Knowl. Data Eng. **26**(10), 2493–2506 (2014)
8. Blei, D.M., Lafferty, J.D.: Dynamic topic models. In: Proceedings of the 23rd International Conference on Machine Learning, pp. 113–120. ACM (2006)
9. Guo, W., Wu, S., Wang, L., Tan, T.: Social-relational topic model for social networks. In: Proceedings of the 24th ACM International Conference on Information and Knowledge Management, pp. 1731–1734. ACM (2015)
10. Yao, L., Zhang, Y., Wei, B., Jin, Z., Zhang, R., Zhang, Y., Chen, Q.: Incorporating knowledge graph embeddings into topic modeling. In: AAAI, pp. 3119–3126 (2017)
11. He, J., Hu, Z., Berg-Kirkpatrick, T., Huang, Y., Xing, E.P.: Efficient correlated topic modeling with topic embedding. In: Proceedings of the 23rd ACM SIGKDD International Conference on Knowledge Discovery and Data Mining, pp. 225–233. ACM (2017)
12. Li, C., Duan, Y., Wang, H., Zhang, Z., Sun, A., Ma, Z.: Enhancing topic modeling for short texts with auxiliary word embeddings. ACM Trans. Inf. Syst. **36**(2), 11:1–11:30 (2017)
13. Liu, C., Jin, T., Hoi, S.C., Zhao, P., Sun, J.: Collaborative topic regression for online recommender systems: an online and Bayesian approach. Mach. Learn. **106**(5), 651–670 (2017)
14. Nie, L., Davison, B.D., Qi, X.: Topical link analysis for web search. In: Proceedings of the 29th International ACM SIGIR Conference on Research and Development in Information Retrieval, pp. 91–98. ACM (2006)
15. Yao, W., He, J., Wang, H., Zhang, Y., Cao, J.: Collaborative topic ranking: leveraging item meta-data for sparsity reduction. In: AAAI, pp. 374–380 (2015)
16. Ding, W., Ishwar, P., Saligrama, V.: A topic modeling approach to ranking. In: Artificial Intelligence and Statistics, pp. 214–222 (2015)

Multi-label Classification
via Label-Topic Pairs

Gang Chen[1,2], Yue Peng[1,2], and Chongjun Wang[1,2(✉)]

[1] National Key Laboratory for Novel Software Technology,
Nanjing University, Nanjing 210023, China
chengang@smail.nju.edu.cn, py198976@gmail.com, chjwang@nju.edu.cn
[2] Department of Computer Science and Technology,
Nanjing University, Nanjing 210023, China

Abstract. The task of learning from multi-label example is rather challenging because of the tremendous number of possible label sets. It has been well recognized that exploiting label relationships in a proper way can facilitate the learning process and boost the learning performance. In this paper, we propose a novel framework called Label-Topic Pairs Multi-Label (LTPML) for multi-label classification. LTPML regards the label set associated with each instance as a document and each class label in the label set as a word and then obtains the topics from the label space by topic models. With the information about label correlations contained by topics, multi-label classification problem is decomposed into a series of single-label classification problems. Based on label-topic pairs which are constructed from relationships among the current label and topics, several multi-class classifiers are built for each class label. Two algorithms named LTPML-α and LTPML-β are derived according to different way of selecting the topics. Experiments on benchmark data sets clearly validate the effectiveness of the proposed approaches.

Keywords: Multi-label classification · Label correlations
Latent Dirichlet Allocation

1 Introduction

As an extension of multi-class classification, multi-label classification allows each example (object) to be assigned with multiple class labels (concepts) simultaneously. Multi-label classification has attracted significant attentions from researchers and has been used in a variety of domains, such as text classification [12,13], image annotation [3,14], music emotions categorization [20,21], social network [18] and bioinformatics [1,19].

The goal of multi-label classification is to induce a function that can assign multiple proper labels for an unseen instance from a given label set. Formally, let $\mathcal{X} \in \mathbb{R}^d$ be the d-dimensional input space and $\mathcal{Y} = \{y_1, y_2, \ldots, y_q\}$ be the output space consisting of q class labels. The task of multi-label classification is

© Springer International Publishing AG, part of Springer Nature 2018
Y. Cai et al. (Eds.): APWeb-WAIM 2018, LNCS 10987, pp. 32–44, 2018.
https://doi.org/10.1007/978-3-319-96890-2_3

to learn a function $h : \mathcal{X} \rightarrow 2^{\mathcal{Y}}$ which can assign each instance $x \in \mathcal{X}$ with a set of relevant labels $h(x) \subseteq \mathcal{Y}$.

The key challenge of learning from multi-label data lies in the potentially overwhelming size of output space, i.e., the number of label sets grows exponentially as the number of class labels increases. Contrary to traditional supervised classification, the labels in multi-label classification are not mutually exclusive but can be correlated often.

In this paper, we propose a novel multi-label classification framework based on label-topic pairs called Label-Topic Pairs Multi-Label (LTPML) which utilizes LDA to exploit global correlations among labels to tackle the problem of exponential-sized output space in multi-label classification. Similar to words which describe a document, labels are semantic units in the description of a instance. Concretely, we regard the label set associated with each instance as a document and each label in the label set as a word, and utilize LDA to discover the topics from the label space. Moreover, the topics associated with each instance are the higher-level of description of the labels and have the capability of introducing the information about label correlations for label prediction. For each class label, LTPML builds several multi class learners coupling with topics for the final prediction and determines the bipartition threshold correspondingly. Labels of unseen instance are predicted separately according to multi-class classifiers built for different labels. Two new methods, LTPML-α and LTPML-β, are proposed based on different way of choosing the topics.

The rest of the paper is organized as follows. Section 2 briefly introduces related work. Section 3 presents the technical details of LTPML. Section 4 reports the experimental results of comparative studies. Finally, Sect. 5 concludes.

2   Related Work

In the past decades, a large number of well-established methods have been proposed to solve multi-label classification problems including simple methods based on binary classifiers and advanced methods that exploit label correlations. These learning algorithms can be categorized into two major types [15]: Problem Transformation Methods and Algorithm Adaptation Methods. The methods constructed by the former strategy, such as Binary Relevance (BR) [3] and Classifier Chains (CC) [11], fit data to algorithm by transforming the problem into one or more single-label classification problems. The methods constructed by the latter strategy, such as Multi-Label k-Nearest Neighbor (ML-kNN) [24], fit algorithm to data by modifying single-label classification algorithms for multi-label classification directly.

To cope with the challenge of exponential-sized output space, current studies on multi-label classification try to incorporate label correlations of different orders [25]. Based on the order of correlations considered, the methods can be roughly grouped into three categories [23]: First-order approaches, Second-order approaches and High-order approaches. First-order approaches ignore dependencies among labels. Second-order approaches, such as Calibrated Label Ranking (CLR) [4], utilize pairwise label correlations to improve the performance of the classifiers. Usually, only considering pairwise relationships makes the methods more expensive to compute, especially with massive label space. High-order approaches, such as Ensembles of Classifier Chains (ECC) [11] and Random k-Labelsets (RAkEL) [16], mainly exploit global label correlations through some certain label learning structures which are based on all the labels or subsets of the labels.

Topic modeling approaches, such as Latent Dirichlet allocation (LDA) [2], have been investigated in recent years. Several methods based on topic models aim to solve multi-label classification problem were proposed recently. Labeled-LDA (L-LDA) [10] adapts LDA to multi-label classification by putting latent topics in one-to-one correspondence with category labels. Dirichlet process with mixed random measures (DP-MRM) [6] extends L-LDA using non-parametric methods. Both of L-LDA and DP-MRM assume that labels are independent. Labelset topic model (LsTM) [8] extends L-LDA by using two labelset layers to capture label dependencies and allowing words to be assigned to combinations of labels rather than a single label. Flat-LDA, Prior-LDA and Dependency-LDA were proposed in [12]. Prior-LDA extends Flat-LDA by accounting for the differences in frequencies of the observed labels. Dependency-LDA extends Prior-LDA by considering the observations of label dependencies. Two novel supervised topic models called FLDA and DFLDA [7] extend Flat-LDA via frequencies of the labels and dependencies among different labels respectively.

However, existing models connecting topic model with multi-label classification mostly aim to improve correlative topic models or propose new topic modeling algorithms, which is not the purpose of LTPML. LTPML merely utilizes topic model as a tool to exploit the information about label correlations contained by topics and introduces the information into the final prediction of labels.

3   Algorithms

As mentioned in introduction section, given a multi-label training set $\mathcal{D} = \{(x_i, Y_i) \mid x_i \in \mathcal{X}, Y_i \subseteq \mathcal{Y}\}_{i=1}^{N}$ with N examples, the task of multi-label classification is to induce a multi-label classifier $h : \mathcal{X} \to 2^{\mathcal{Y}}$ from training set $\mathcal{D}$. This is equivalent to learn q real-value functions $f_j : \mathcal{X} \to \mathbb{R} \, (1 \leq j \leq q)$, each accompanied by a thresholding function $t_j : \mathcal{X} \to \mathbb{R}$. For any instance $x \in \mathcal{X}$, $f_j(x)$ returns the confidence of associating x with class label y_j, and the predicted label set is determined according to:

$$h(x) = \{y_j \mid f_j(x) > t_j(x), 1 \leq j \leq q\} \tag{1}$$

Our approach to the problem of multi-label classification is to exploit label correlations by introducing topic modeling, specifically in the form of Latent Dirichlet Allocation (LDA). Unlike traditional methods that learn label correlations in multi-label classification by label pairs or label subsets, LTPML achieves this goal by discovering abstract topics from the label space. Then $f_j(x)$ and $t_j(x)$ can be obtained from the topics and the training set.

Notice that the selection of topics is crucial to the performance of LTPML. We propose two ways to choose topics in LTPML obtaining two algorithms named LTPML-α and LTPML-β. Without loss of generality, we denote the current topic number as K_t and K_t $(1 \leq t \leq p)$ is chosen from an increasing sequence S, i.e., $S = \{K_1, K_2, \ldots, K_p \mid K_1 = 2, K_i \in \mathbb{Z}^+, 1 \leq i \leq p\}$, where p is a iteration number. Overall, LTPML is composed of the following six parts.

Exploit the Topics from the Label Space. We introduce LDA into training set $\mathcal{D}$. Each instance x_i denotes a document and each label y_{ij} denotes the j-th label of the i-th instance. The generative process of the topics is as follows:

1. Choose the label topic number K_t and the Dirichlet distribution parameters α and β;
2. Choose $\theta_i \sim Dir(\alpha)$, where $i \in \{1, 2, \ldots, N\}$;
3. Choose $\phi_k \sim Dir(\beta)$, where $k \in \{1, 2, \ldots, K_t\}$;
4. For each of label y_{ij}, where $i \in \{1, 2, \ldots, N\}$ and $j \in \{1, 2, \ldots, q\}$
   - Choose a topic $z_{ij} \sim Dir(\theta_i)$;
   - Choose a label $y_{ij} \sim Dir(\phi_{z_{ij}})$;

Then, we can compute the instance-topic probability distribution matrix Θ, where Θ_{ij} denotes the probability of the i-th instance in the j-th topic.

Discretize the Distribution of Topics. After calculating Θ, we obtain the probability (confidence) that each instance belongs to each topic. To determine which topic each instance exactly belongs to, we require a discrete value (e.g. 0 or +1) instead of the probability. After discretizing the distribution of topics, we can get a new training set $\mathcal{D}_T^t = \{(x_i, Y_i, T_i^t) \mid x_i \in \mathcal{X}, Y_i \subseteq \mathcal{Y}, T_i^t \subseteq \mathcal{T}^t\}_{i=1}^N$, where $\mathcal{T}^t = \{topic_1, topic_2, \ldots, topic_{K_t}\}$ denotes the topic space of K_t topics. We define the method for discretization in Algorithm 1.

Though LTPML can support different topic selection strategies for the following steps, we mainly consider two topic selection methods in this paper. For simplicity, K defines the total number of topics that the current algorithm exploits, $\mathcal{T} = \{topic_1, topic_2, \ldots, topic_K\}$ defines the whole topic space in the current algorithm and $\mathcal{D}_T = \{(x_i, Y_i, T_i) \mid x_i \in \mathcal{X}, Y_i \subseteq \mathcal{Y}, T_i \subseteq \mathcal{T}\}_{i=1}^N$ defines the new training set obtained by the current algorithm.

For LTPML-α, $K = K_p$ and $\mathcal{T} = \mathcal{T}^p = \{topic_1, topic_2, \ldots, topic_{K_p}\}$. LTPML-$\alpha$ only needs to execute Algorithm 1 once in the whole procedure and gains the new data set $\mathcal{D}_T = \mathcal{D}_T^p = \{(x_i, Y_i, T_i) \mid T_i = T_i^p \subseteq \mathcal{T}\}_{i=1}^N$.

For LTPML-β, $K = \sum_{i=1}^p K_i$ and $\mathcal{T} = \mathcal{T}^1 \cup \mathcal{T}^2 \cup \cdots \cup \mathcal{T}^p$. Different from LTPML-$\alpha$, LTPML-$\beta$ selects the value from K_1 to K_p in turn as the topic number, then executes Algorithm 1 iteratively and obtains the new data set $\mathcal{D}_T = \{(x_i, Y_i, Ti) \mid T_i = T_i^1 \cup T_i^2 \cup \cdots \cup T_i^p \subseteq \mathcal{T}\}_{i=1}^N$.

LTPML-α only considers the information about label correlations contained by the current topics while LTPML-β considers the information about label correlations contained by all the former topics and the current topics.

Algorithm 1. Discretization of the instance-topic probability distribution matrix

Input:
    K_t: the current topic number
    $\mathcal{D}$: the multi-label training set, $\mathcal{D} = \{(x_i, Y_i)\}_{i=1}^{N}$
    $\Theta[N][K_t]$: the instance-topic probability distribution matrix

Output:
    $\mathcal{D}_T^t$: the new train data set, $\mathcal{D}_T^t = \left\{\left(x_i, Y_i, T_i^t\right)\right\}_{i=1}^{N}$

1: **for** $i = 1$ to N **do**
2:    $T_i^t = \varnothing$
3:    $max = MAX\left(\Theta[i][1], \ldots, \Theta[i][K_t]\right)$
4:    **for** $j = 1$ to K_t **do**
5:      **if** $(max - \Theta[i][j]) < 1/K_t$ **then**
6:        $T_i^t = T_i^t \cup \{topic_j\}$
7:      **end if**
8:    **end for**
9: **end for**
10: **return** $\mathcal{D}_T^t = \left\{\left(x_i, Y_i, T_i^t\right)\right\}_{i=1}^{N}$

Couple the Labels with Topics. To incorporate label correlations into the learning process, we couple each label y_j with K topics as follows. Given a label-topic pair $(y_j, topic_k)$, a multi-class training set $\mathcal{D}_{jk}$ can be derived from $\mathcal{D}_T$:

$$\mathcal{D}_{jk} = \{(x_i, \psi(Y_i, T_i, y_j, topic_k))\}_{i=1}^{N} \tag{2}$$

$$\text{where } \psi(Y_i, T_i, y_j, topic_k) = \begin{cases} 0, & \text{if } y_j \notin Y_i \text{ and } topic_k \notin T_i \\ +1, & \text{if } y_j \notin Y_i \text{ and } topic_k \in T_i \\ +2, & \text{if } y_j \in Y_i \end{cases}$$

In total, there are four possible classes, which are the four possible combinations of y_j and $topic_k$. However, we only focus on the positive assignment for y_j, so the combinations where $y_j \in Y_i$ can be merged together. Thus, the multi-class data $\mathcal{D}_{jk}$ can be transformed into three classes.

Build the Multi-class Learners. Applying a multi-class classifier $\mathcal{M}$ to multi-class data $\mathcal{D}_{jk}$ transformed above, a multi-class learner g_{jk} can be built, i.e., $g_{jk} \leftarrow \mathcal{M}(\mathcal{D}_{jk})$. For each class label y_j, there are K multi-class learners. Accordingly, label-topic pairs based multi-class learners are constructed.

Calculate the Final Predictive Confidence. For each class label y_j, we can calculate the final predictive confidence from each multi-class learner g_{jk}. In LTPML, we merge the confidence of K multi-class learners as follows:

$$f_j(x) = 1/K \sum_{k=1}^{K} g_{jk}(+2 \mid x) \tag{3}$$

Algorithm 2. LTPML

Input:
   $\mathcal{D}$: the multi-label training set, $\mathcal{D} = \{(x_i, Y_i)\}_{i=1}^{N}$
   K: the topic number
   $\mathcal{M}$: the multi-class classifier
   $\hat{x}$: the test example

Output:
   $\hat{Y}$: the set of predicted labels for $\hat{x}$

1: Obtain the new training set $\mathcal{D}_T$ according to Algorithm 1 and the current topic
   selection strategy
2: **for** $j = 1$ to q **do**
3:    **for** $k = 1$ to K **do**
4:       Construct the tri-class training set $\mathcal{D}_{jk}$ with the label-topic pair $(y_j, topic_k)$
         according to Eq.(2)
5:       Build the multi-class learner $g_{jk} \leftarrow \mathcal{M}(\mathcal{D}_{jk})$
6:    **end for**
7:    Build the real-valued function $f_j(\cdot)$ according to Eq.(3)
8:    Construct the binary training set $\mathcal{D}_j$ according to Eq.(4)
9:    Calculate the threshold $t_j(\cdot) = a_j$ which can maximize the learning performance
      according to Eq.(5)
10: **end for**
11: **return** $\hat{Y} = h(\hat{x})$ according to Eq.(1)

The g_{jk} is a multi-class learner which is constructed with the label-topic pair $(y_j, topic_k)$. Correspondingly, let $g_{jk}(+2 \mid x)$ denote the predictive confidence that x should have positive assignment w.r.t. y_j, regardless of x having positive or negative assignment w.r.t. $topic_k$. Here, we calculate the average confidence of K multi-class classifications to obtain the final predictive confidence of label y_j.

Choose the Bipartition Threshold for Each Label. In traditional methods, the threshold $t_j(x)$ is always set as a fixed value, and x is predicted to be positive for y_j if the confidence value $f_j(x) > t_j(x)$, and negative otherwise.

In LTPML, the thresholding function is set as a constant function $t_j(\cdot) = a_j$, which changes with different labels. Additionally, $f_j(x) > a_j$ means that x is predicted to be positive for y_j. The problem lies in how to determine the value of the threshold a_j. Based on certain metric, the goodness of a_j can be evaluated straightforwardly. As the harmonic mean of precision and recall and a popular metric for binary classifiers, F-measure can be employed to measure how well $f_j(\cdot)$ classifies examples in D_j by using a_j as the bipartition threshold, where D_j is derived from $\mathcal{D}$ for the j-th class label y_j:

$$\mathcal{D}_j = \{(x_i, \phi(Y_i, y_j))\}_{i=1}^{N} \tag{4}$$

$$\text{where } \phi(Y_i, y_j) = \begin{cases} +1, & \text{if } y_j \in Y_i \\ -1, & \text{otherwise} \end{cases}$$

Therefore, the goal is to find a real value of the threshold that can maximize the value of the F-measure for the corresponding label. The threshold a_j is determined as follows:

$$a_j = \arg\max_{a \in \mathbb{R}} F(f_j, a, \mathcal{D}_j) \tag{5}$$

where $F(f_j, a, \mathcal{D}_j)$ denotes the value of F-measure achieved by applying $\{f_j, a\}$ to the binary training set $\mathcal{D}_j$.

Algorithm 2 summarizes the complete procedure of the proposed LTPML framework.

4 Experiments

4.1 Data Sets

Six benchmark multi-label data sets provided by the MULAN [17] have been collected for experimental studies and the detailed characteristics of these data sets are summarized in Table 1. For each multi-label data set $\mathcal{D}$, we use $|\mathcal{D}|$, $L(\mathcal{D})$ and $dim(\mathcal{D})$ to represent its number of examples, number of class labels and number of features respectively. In addition, $LCard(\mathcal{D})$ denotes the average number of labels per example, $LDen(\mathcal{D})$ denotes the result of normalization of $LCard(\mathcal{D})$, $DL(\mathcal{D})$ denotes the number of label type. These data sets cover a broad range of scenarios, including music (CAL500, emotions), images (flags, corel5k), audio (birds) and biology (yeast). Meanwhile, these data sets also cover different scale of instances and class labels.

Table 1. Characteristics of the experimental data sets

Data set	$|\mathcal{D}|$	$dim(\mathcal{D})$	$L(\mathcal{D})$	$LCard(\mathcal{D})$	$LDen(\mathcal{D})$	$DL(\mathcal{D})$	Domain
CAL500	502	68	174	26.044	0.150	502	Music
Yeast	2417	103	14	4.237	0.303	198	Biology
Flags	194	19	7	3.392	0.485	54	Images
Corel5k	5000	499	374	3.522	0.009	3175	Images
Emotions	593	72	6	1.869	0.311	27	Music
Birds	645	300	19	1.014	0.053	133	Audio

4.2 Evaluation Metrics

We consider three popular evaluation criteria to evaluate the performance of all methods: Accuracy, Macro-averaged F-Measure (MACRO-F) and Micro-averaged F-Measure (MICRO-F), which evaluate different aspects of classifier performance. The definitions and discussions of them can be found in [9,15,22,25]. For these metrics, the larger the value, the better the performance of the classifier.

4.3   Experiment Settings

In our experiments, LTPML is compared with several state-of-the-art multi-label classification algorithms: BR, CC, ECC, CLR, RAkEL and ML-kNN. As introduced before, these algorithms cover almost all the categories of multi-label classification algorithms. The iteration number p is set to be 1 ($K = K_1 = 2$) in experiments. When p is set to be 1, LTPML-α is same as LTPML-β and are all named as LTPML in experiments. For LTPML-α and LTPML-β, we can tune the iteration number p and the increasing sequence S to achieve better performance.

For fair comparison, SMO provided by Weka [5] is employed as the base classifier for all the comparing algorithms. Parameters for each comparing algorithm are suggested by the corresponding publication. The comparing multi-label algorithms, such as CC, CLR, ECC, RAkEL and ML-kNN, are provided by the MULAN [17]. Five-fold cross-validation is performed three times on each experimental data set.

4.4   Experimental Results

Tables 2, 3 and 4 give the detailed results in terms of different evaluation metrics, and the best result on each data set is marked in bold. The numbers in parentheses denote the rank of the algorithms among the comparing algorithms. ↑ denotes the larger the better. For the reason of memory, we can't get the experimental result of algorithm CLR on corel5k.

We can see that these algorithms (LTPML, ECC, CLR and RAkEL) which consider label correlations in a proper way obtain better performance than BR and ML-kNN which ignore correlations among labels. It indicates that exploiting label correlations reasonably can improve the performance of classifier.

The results in tables also indicate that LTPML possesses highest average rank comparing with other algorithms on all three measures, with (1.5, 1.0, 1.3) respectively. It is worth noting that in terms of MACRO-F, LTPML outperforms all the comparing algorithms on all the data sets. In some cases (e.g. CAL500), LTPML significantly outperforms the comparing algorithms. These results suggest that the real-valued functions $f_j\,(\cdot)\,(1 \leq j \leq q)$ learned by LTPML is capable of yielding reasonable predictive confidence. Although our proposed method runs the best on average, the results on some data sets are not as good as on other data sets. For example, when used on yeast, LTPML is outperformed by ECC and RAkEL in terms of Accuracy and MICRO-F.

Overall, the experimental results demonstrate that LTPML can provide robust and preferable solutions in multi-label scenarios.

4.5   The Iteration Number

The iteration number p which determines the total number of topics participated in the algorithms plays an important role in LTPML. It is easier and clearer to illustrate the effect of p on LTPML-α and LTPM-β with S in which the difference

Table 2. Performance in terms of Accuracy↑.

Data set	Algorithm						
	LTPML	BR	CC	ECC	CLR	RAkEL	ML-kNN
CAL500	**0.2976**[1]	0.2041[4]	0.1864[7]	0.2447[2]	0.1982[5]	0.2097[3]	0.1981[6]
Yeast	0.5178[3]	0.4994[6]	0.4854[7]	**0.5325**[1]	0.5036[5]	0.5272[2]	0.5143[4]
Flags	**0.6152**[1]	0.5995[4]	0.5830[7]	0.5940[5]	0.6066[3]	0.6121[2]	0.5887[6]
Corel5k	**0.1390**[1]	0.0787[5]	0.1018[3]	0.1346[2]	-	0.0799[4]	0.0132[6]
Emotions	**0.5889**[1]	0.5167[7]	0.5311[5]	0.5603[3]	0.5268[6]	0.5744[2]	0.5436[4]
Birds	0.6323[2]	0.6247[5]	0.6282[4]	0.6175[6]	0.6290[3]	**0.6381**[1]	0.5460[7]
AvgRank	1.5	5.2	5.5	3.2	4.4	2.3	5.5

Table 3. Performance in terms of MACRO-F↑.

Data set	Algorithm						
	LTPML	BR	CC	ECC	CLR	RAkEL	ML-kNN
CAL500	**0.1780**[1]	0.0977[5]	0.1149[3]	0.1661[2]	0.0976[6]	0.1016[4]	0.0957[7]
Yeast	**0.4192**[1]	0.3241[7]	0.3508[5]	0.3684[3]	0.3365[6]	0.3630[4]	0.3823[2]
Flags	**0.6794**[1]	0.6328[5]	0.6245[6]	0.6482[4]	0.6490[3]	0.6587[2]	0.5827[7]
Corel5k	**0.2316**[1]	0.2001[5]	0.2070[3]	0.2254[2]	-	0.2012[4]	0.1935[6]
Emotions	**0.6870**[1]	0.6050[6]	0.5973[7]	0.6632[3]	0.6269[5]	0.6661[2]	0.6289[4]
Birds	**0.4317**[1]	0.3973[6]	0.4121[2]	0.4070[4]	0.4059[5]	0.4107[3]	0.1744[7]
AvgRank	1.0	5.7	4.3	3.0	5.0	3.2	5.5

Table 4. Performance in terms of MICRO-F↑.

Data set	Algorithm						
	LTPML	BR	CC	ECC	CLR	RAkEL	ML-kNN
CAL500	**0.4539**[1]	0.3310[4]	0.3073[7]	0.3859[2]	0.3214[6]	0.3387[3]	0.3221[5]
Yeast	0.6521[3]	0.6328[6]	0.6166[7]	0.6523[2]	0.6366[5]	**0.6528**[1]	0.6459[4]
Flags	**0.7501**[1]	0.7403[4]	0.7223[7]	0.7344[5]	0.7414[3]	0.7471[2]	0.7248[6]
Corel5k	**0.2141**[1]	0.1426[5]	0.1557[3]	0.2108[2]	-	0.1436[4]	0.0281[6]
Emotions	**0.7029**[1]	0.6530[6]	0.6469[7]	0.6754[3]	0.6639[5]	0.6902[2]	0.6670[4]
Birds	**0.5174**[1]	0.4968[6]	0.4984[5]	0.5038[4]	0.5042[3]	0.5140[2]	0.3078[7]
AvgRank	1.3	5.2	6.0	3.0	4.4	2.3	5.3

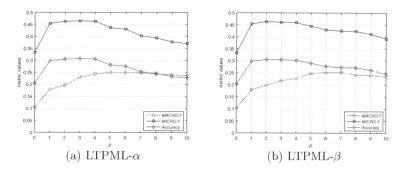

Fig. 1. Results of LTPML-α and LTPML-β under different iteration number p on CAL500

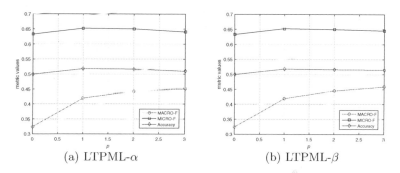

Fig. 2. Results of LTPML-α and LTPML-β under different iteration number p on yeast

of the adjacent elements increases gradually. Here, we set S as a prime number sequence in experiments, i.e., $S = \{2, 3, 5, \ldots\}$.

Due to page limit, the influence of p on LTPML-α and LTPML-β is evaluated by choosing two data sets CAL500 and yeast which have 174 class labels and 14 class labels respectively. Considering the total number of labels of each data set, the iteration number p is set from 1 to 10 on CAL500 and from 1 to 3 on yeast. Experimental results are shown in Figs. 1 and 2 and LTPML degenerates to BR when $p = 0$.

As shown in the figures, with more topics introduced, the performance of LTPML-α and LTPML-β gets better in terms MACRO-F. This is because the more topics introduced, the more detailed information about label correlations introduced. It is worth noting that though the value of MACRO-F is increasing at first, the growth becomes slow gradually and the curves become flat later. The reason is that, as the iteration number or the topic number increases, the useful information about label correlations is exploited more and more completely and it is more difficult to boost the learning performance through the information owned by topics. Figure 1 shows that the value of MACRO-F may reduce in some time. This suggests that topics which contain too sparse information about label correlations will confuse the classifiers in LTPML.

Experimental results show that the value of Accuracy and MICRO-F first increases then decreases as p increases for both LTPML-α and LTPML-β. The reason is that the utilization of the valuable information about label correlations can be maximized when the topic number is suitable for LTPML, which leads to a peak of performance of the corresponding algorithm. When the topic number is small, the more topics, the more useful information about correlations can be mined and used. On the contrary, a large topic number will disperse the information about correlations and make each subsequent or current topic contain little useful information about label correlations. We can observe that both Accuracy and MICRO-F decrease more mildly in LTPML-β than in LTPML-α in the figures. This is because, in LTPML-β, the beneficial information about label correlations carried by former topics can reduce the adverse influence of topics which contain little information about correlations among labels. Specifically, the optimization of the threshold of each label in LTPML which depends on F-measure can alleviate the adverse influence of the increased topics on LTPML in terms of MACRO-F, e.g. compared with Accuracy and MICRO-F, the drop point of MACRO-F seems to be delayed.

In general, LTPML-α and LTPML-β have semblable performance on CAL500 and yeast. Similar trends can be observed on other data sets. With different settings of S, both LTPML-α and LTPML-β can obtain good performance at a relative large p. Compared with LTPML-α, LTPML-β can ease the detrimental effect caused by topics containing little useful information about label correlations while need more expensive computation as p increases. It is a trade-off between computation and performance.

5   Conclusions

In this paper, a simple and efficient multi-label classification framework named LTPML is proposed, which works by leveraging the exploitation of global label correlations. More precisely, several label-topic pairs based multi-class learners are combined to yield the predictive model. Two algorithms named LTPML-α and LTPML-β are derived according to different way of selecting the topics. The empirical experimental results have shown that LTPML can work more effectively than other state-of-the-art multi-label classification algorithms. Furthermore, we also show the influence of the topics as the iteration number changes.

Acknowledgements. This paper is supported by the National Key Research and Development Program of China (Grant No. 2016YFB1001102), the National Natural Science Foundation of China (Grant Nos. 61502227, 61375069), the Collaborative Innovation Center of Novel Software Technology and Industrialization at Nanjing University and the Fundamental Research Funds for the Central Universities (Grant Nos. 020214380036, 020214380038).

References

1. Barutcuoglu, Z., Schapire, R.E., Troyanskaya, O.G.: Hierarchical multi-label prediction of gene function. Bioinformatics **22**(7), 830 (2006)
2. Blei, D.M., Ng, A.Y., Jordan, M I : Latent Dirichlet allocation. J. Mach. Learn. Res. **3**, 993–1022 (2003)
3. Boutell, M.R., Luo, J., Shen, X., Brown, C.M.: Learning multi-label scene classification. Pattern Recogn. **37**(9), 1757–1771 (2004)
4. Brinker, K.: Multilabel classification via calibrated label ranking. Mach. Learn. **73**(2), 133–153 (2008)
5. Hall, M., Frank, E., Holmes, G., Pfahringer, B., Reutemann, P., Witten, I.H.: The weka data mining software: an update. ACM SIGKDD Explor. Newslett. **11**(1), 10–18 (2009)
6. Kim, D., Kim, S., Oh, A.: Dirichlet process with mixed random measures: a non-parametric topic model for labeled data. Computer Science pp. 727–734 (2012)
7. Li, X., Ouyang, J., Zhou, X.: Supervised topic models for multi-label classification. Neurocomputing **149**(PB), 811–819 (2015)
8. Li, X., Ouyang, J., Zhou, X.: Labelset topic model for multi-label document classification. J. Intell. Inf. Syst. **46**(1), 83–97 (2016)
9. Pillai, I., Fumera, G., Roli, F.: Designing multi-label classifiers that maximize F measures: state of the art. Pattern Recogn. **61**, 394–404 (2017)
10. Ramage, D., Hall, D., Nallapati, R., Manning, C.D.: Labeled LDA: a supervised topic model for credit attribution in multi-labeled corpora. In: Conference on Empirical Methods in Natural Language Processing, vol. 1, pp. 248–256 (2009)
11. Read, J., Pfahringer, B., Holmes, G., Frank, E.: Classifier chains for multi-label classification. Mach. Learn. **85**(3), 333 (2011)
12. Rubin, T.N., Chambers, A., Smyth, P., Steyvers, M.: Statistical topic models for multi-label document classification. Mach. Learn. **88**(1–2), 157–208 (2012)
13. Schapire, R.E., Singer, Y.: A boosting-based system for text categorization. Mach. Learn. **39**(2–3), 135–168 (2000)
14. Sun, F., Tang, J., Li, H., Qi, G.J., Huang, T.S.: Multi-label image categorization with sparse factor representation. IEEE Trans. Image Process. **23**(3), 1028–1037 (2014)
15. Tsoumakas, G., Katakis, I., Vlahavas, I.: Mining multi-label data. In: Maimon, O., Rokach, L. (eds.) Data Mining and Knowledge Discovery Handbook, pp. 667–685. Springer, Boston (2009). https://doi.org/10.1007/978-0-387-09823-4_34
16. Tsoumakas, G., Katakis, I., Vlahavas, I.: Random k-labelsets for multilabel classification. IEEE Trans. Knowl. Data Eng. **23**(7), 1079–1089 (2011)
17. Tsoumakas, G., Spyromitros-Xioufis, E., Vilcek, J., Vlahavas, I.: Mulan: a java library for multi-label learning. J. Mach. Learn. Res. **12**(7), 2411–2414 (2011)
18. Wang, X., Sukthankar, G.: Multi-label relational neighbor classification using social context features. In: ACM SIGKDD International Conference on Knowledge Discovery and Data Mining, pp. 464–472 (2013)
19. Wang, X., Li, G.Z.: Multilabel learning via random label selection for protein subcellular multilocations prediction. IEEE/ACM Trans. Comput. Biol. Bioinf. **10**(2), 436–446 (2013)
20. Wieczorkowska, A., Synak, P., Ra, Z.W.: Multi-label classification of emotions in music. In: Kłopotek, M.A., Wierzchoń, S.T., Trojanowski, K. (eds.) Intelligent Information Processing and Web Mining. Advances in Soft Computing, vol. 35, pp. 307–315. Springer, Berlin (2006). https://doi.org/10.1007/3-540-33521-8_30

21. Wu, B., Zhong, E., Horner, A., Yang, Q.: Music emotion recognition by multi-label multi-layer multi-instance multi-view learning. In: ACM International Conference on Multimedia, pp. 117–126 (2014)
22. Wu, X.Z., Zhou, Z.H.: A unified view of multi-label performance measures (2016)
23. Zhang, M.L., Zhang, K.: Multi-label learning by exploiting label dependency. In: ACM SIGKDD International Conference on Knowledge Discovery and Data Mining, pp. 999–1008 (2010)
24. Zhang, M.L., Zhou, Z.H.: ML-KNN: a lazy learning approach to multi-label learning. Pattern Recogn. **40**(7), 2038–2048 (2007)
25. Zhang, M.L., Zhou, Z.H.: A review on multi-label learning algorithms. IEEE Trans. Knowl. Data Eng. **26**(8), 1819–1837 (2014)

Similarity Calculations of Academic Articles Using Topic Events and Domain Knowledge

Ming Liu[✉], Bo Lang, and Zepeng Gu

State Key Laboratory of Software Development Environment,
Beihang University, Beijing, China
{liuming,langbo,guzepeng}@nlsde.buaa.edu.com

Abstract. While studies investigating the semantic similarity among concepts, sentences and short text fragments have been fruitful, the problem of document-level semantic matching remains largely unexplored due to its complexity. In this paper, we explore the document-level semantic similarity issue in the academic literatures using an interpretable method. To integrally describe the semantics of an article, we construct a topic event model that utilizes multiple information facets, such as the study purposes, methodologies and domains. Furthermore, to better understand the documents and achieve a more accurate similarity comparison, we incorporate external knowledge into the topic event construction and similarity calculation. Our approach achieves significant improvements over state-of-the-art methods.

Keywords: Scientific literature analysis
Document semantic similarity · Text understanding · Topic event
Domain ontology

1 Introduction

Long text understanding and similarity calculations in the academic domain have great significant applications, such as plagiarism detection, automatic technical surveys, and citation recommendation. Recent, the semantic similarity of text, including word semantics [1,2] and sentence semantics [3], has received increasing attention. However, studies rarely focus on semantic understanding and similarity calculations at the document-level. Currently, to the best of our knowledge, no public dataset is available. Long documents usually have a sophisticated structure and massive topics; thus, capturing and measuring the semantic similarity among long documents is challenging.

The semantics of long text can be derived from the semantics of combinations of small text units. Many recent studies have used this approach to determine the semantic similarity among larger text units. For example, the similarity in sentence semantics can be determines by examining the semantic similarities

© Springer International Publishing AG, part of Springer Nature 2018
Y. Cai et al. (Eds.): APWeb-WAIM 2018, LNCS 10987, pp. 45–53, 2018.
https://doi.org/10.1007/978-3-319-96890-2_4

between word pairs obtained from two sentences [4]. In addition to lexical semantics, global sentence-level features are also considered in measuring the semantic
similarity [3,5,6]. However, these studies have only focused on short texts, and
their results may not apply to document-level texts.

Studies exploring the semantic similarity among documents are relatively
rare. Existing methods, such as TF-IDF [7] and Latent Dirichlet Allocation
(LDA) [8], focus on information retrieval at the surface level and thus, measure
the document similarity using statistics of words rather than semantics, neglect-
ing the documents structure and meanings of words. For instance, the following
are two snippets of text:

A :"*Jack borrowed a book from the teacher*"

B :"*The teacher borrowed a book from Jack*"

The above-mentioned methods consider texts A and B as equal due to the
limitations of the bag of words assumption, but the two texts actually have opposite meanings. Many researches [9,10] also attempted to add external knowledge
to the document representation. However, these methods still suffer from problems, such as computational complexity and representational opaqueness.

Long documents contain many topic transitions and different focuses; thus
capturing the core semantics in these documents is challenging. In this paper,
we consider the core semantic issue in each document as an event called a topic
event (TE). The TE is a structured summary extracted from each document.
We construct a TE based on the article structure using multiple information
fields. Hence, the semantic similarity among academic papers can be measured by
the TE similarity. Our methods display outstanding performance. In summary,
the main contributions of our study are as follows: (1) We constructed a topic
event model that can integrally represent the semantics of long documents and
proposed a general method for TE similarity calculations. (2) We provided an
ontology-based automatic TE construction method without labeled data. (3) We
introduce a document semantic matching corpus using fine-grained annotations
for the first time.

2   Topic Event

2.1   Definition of Topic Event

Academic articles are used to convey research progress. To obtain a global understanding of each article, we define the structure of a TE in formulation 1.

$$Topic\ Event = \{*Eid\ : ID\ of\ Topic\ Event$$
$$*Style\ : essential\ type\ of\ the\ research\ article$$
$$*Domain\ : the\ domain\ of\ research\ study$$
$$*Target\ : research\ target\ of\ the\ articles$$
$$*Methodology\ : approaches\ used\ in\ the\ research\ study$$
$$*Metadata : \{date,\ author,\ organization,\ citation,\ keywords,\ venue\}$$

Name : *optional name of the developed system or method*

Tools : *optional tools used in the study*

Conclusion : *optional conclusion sentences in the article*

Dataset : *optional datasets used in the study*}   (1)

Items annotated with * are essential, while other items are optional. *Style* indicates the research type, and elements, such as *Domain, Target, Methodology*, and *Keywords*, are terminologies extracted from the papers, while elements such as *Conclusion, Background, Performance* and *Forecast* are key sentences in papers.

2.2   Domain Knowledge

Research Style Ontology. To express the knowledge implicated by types of research styles, we develop the style categories of topic event in Fig. 1(a) using Protégé. The research style includes *Theoretical Origination, Methodology Improvement, System Implementation, Issue Solution, Survey, Analysis*, and *Phenomenon Discovery*. The styles of research studies implicate the variance of researches in terms of difficulties and categories.

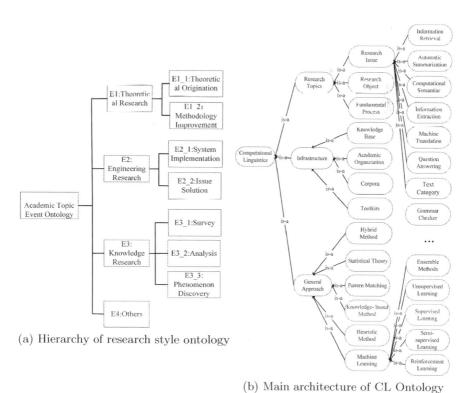

(a) Hierarchy of research style ontology

(b) Main architecture of CL Ontology

Fig. 1. The domain knowledge bases

Domain Ontology. To measure the semantic similarity among terminologies, we manually construct a domain ontology using the same schema used by Word-Net. The concepts extracted from the AAN corpus are used to manually construct the Computational Linguistics Ontology (CL Ontology), and it architecture is shown in Fig. 1(b). Each node of the CL Ontology is a synset. The relationships among the nodes in the ontology are hyponymous relationship.

2.3   Automatic Construction of Topic Events

Current event extraction approaches have mainly focused on newswire sentences. However, in the academic domain, labeled entities and relationships are rarely available for training an extraction model. Many structured annotations in academic articles, such as citations, authors, keywords and journals, are identifiable. Thus, the main goal of extraction is to identify terminologies, such as *target*, and *methodology*, from the article content. We propose an ontology and pattern-based extraction method, which is described in Fig. 2. First, we divide the academic articles into different chunks, and the *Title, Abstract, Introduction* and *Conclusion* sections are considered to include a global description of the research study. Then, we identify the relevant sentences in these sections using trigger words and the best event arguments are chosen after pattern matching. Finally, the extracted event items are delivered to the domain ontology to expand the related event semantic items.

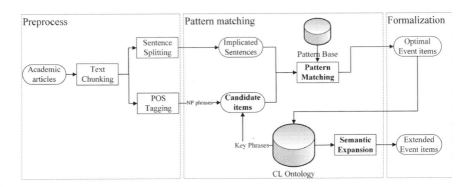

Fig. 2. Process of topic event extraction

Recognition of Semantic Item. Altogether, over 550 patterns are used for the extraction of the *Target* and *Methodology*. Several typical extracting patterns of *Target* and *methodology* are shown in Table 1. For example, when the following sentence is identified using the trigger word *propose* in the *Introduction* section.

  *"In this paper, we **propose** a supervised machine learning **approach for** semantic relation extraction."*

It matches the preceding target pattern, i.e. *approach for*. Thus, the terminology *semantic relation extraction* is chosen as a candidate target. To select the

Table 1. Typical extracting patterns for TE elements

Semantic item	Pre-patterns	Post-patterns
Target	the problem of, the task of, system to, survey on, approach for, framework for	overview, evaluation, track, system, called, process
Methodology	by use of, that using, which employ, takes advantage of, methods of, through an	based framework, method to, algorithm for, techniques to, is applied, performs much better

normative terminology, we derive the domain synset of all the concepts from the CL Ontology, such as the synset of a concept *"Relation Extraction"* as follows:

"Relation Extraction = {Relation Extraction, Semantic Relation Extraction, Semantic Relation Detection, Relationship Extraction, Relation Classification"}.

We calculate the edit distances between the candidate target phrase, i.e., *semantic relation extraction*, and all CL Ontology synset. Thus, the concept with minimum distance, i.e., *'Relation Extraction* is chosen the normal target item.

Ontology-Based Semantic Expansion. Many semantic items are correlated such as the research purpose and domain, and the adopted methodology and toolkit. To extract the target of an academic article, we use CL Ontology to infer its domain. We predefined 12 research domains. The domain concept with the highest semantic similarity to the target concept is chosen as the domain. The details are shown in Algorithm 1.

Algorithm 1. Semantic Inference of *domain* in TE

Require: Domain ontology $KG \vee$ target $t \vee$ Predefined domain $d_1,...,d_n$
Ensure: Output the domain of TE d_{TE}

   $Sim_{max} \leftarrow 0$
   $d_{TE} \leftarrow d_1$
   for all d such that $d \in$ Predefined domain$\{d_1, ..., d_n\}$) **do**
      $Sim\_KG(d,t) \leftarrow$ the semantic similarity of concepts d and t in KG;
      if $Sim\_KG(d,t) > Sim_{max}$ **then**
         $Sim_{max} \leftarrow Sim\_KG(d,t)$;
         $d_{TE} \leftarrow d$;
      end if
   end for

---

3   Topic Event Similarity Calculation

We measure the topic event similarity using the linear combination of several main element similarities, and the topic event similarity can also be extended using other elements. The similarity between topic events E_i and E_j is defined in Eq. (2) as follows:

$$sim(D_i, D_j) = sim(E_i, E_j) = \sum_{k=1}^{N} w_k S_k(L_{ik}, L_{jk}), \tag{2}$$

where w_k is the weight of the kth element in the topic event, and S_k is the similarity function between the kth elements of L_i and L_j. L_i and L_j are two topic events whose elements are defined as $L = \{$ *Target, Domain, Style, Methodology, Keywords, Date* $\}$.

Research Style Similarity. The *Style* similarity between different types of topic events is measured use a method similar to the Wu and Palmer method [1], which is based on the research style ontology shown in Fig. 2, and the formula is expressed in Eq. (3) as follows:

$$sim_{style} = \frac{2\,depth_{LCS}}{depth_{style_i} + depth_{style_j}}, \tag{3}$$

where $Style_i$ and $Style_j$ are the types of the two topic events. LCS is the least common subsumer of two style nodes.

Terminology Similarity. The contents of *Target, Domain, Methodology* and *Keywords* are collections of terminologies that can be measured using the Wu and Palmer method based on the domain ontology. The ontology-based concept semantic similarity is measured by Eq. (4) as follows:

$$sim_{concept} = \frac{2\,depth_{LCS}}{depth_{ec_i} + depth_{ec_j}}, \tag{4}$$

where ec_i and ec_j represent terminologies in the topic event.

Date Similarity. We assume that academic articles with close publication dates are more similar, and that academic articles with longer durations between publication dates are less similar. Thus, the date similarity can be measured using Eq. (5).

$$sim_{date} = \frac{1}{1 + |(year_i + \frac{month_i}{12}) - (year_j + \frac{month_j}{12})}|. \tag{5}$$

4   Experimental Evaluation

Current semantic datasets cannot validate long documents. Hence, we constructed a new document understanding dataset using real-world corpus, i.e.,

ACL venues [11]. The paper pairs were annotated using both 2-level and 5-level annotations, which is available at https://github.com/buaaliuming/DSAP-document-semantics-for-academic-papers/tree/buaaliuming-annotation.

We also manually annotated the golden TEs of academic papers for contrast. We conducted the following different TE-based methods: the TE method based on the golden TEs and lexical semantics in the CL ontology (TE_Onto), the TE method based on the automatically extracted TEs and the lexical semantics in the CL ontology (AutoTE_Onto), the TE method based on the golden TEs and lexical semantics from the LSA (TE_LSA), and the TE method based on the automatically extracted TEs and the lexical semantics in the LSA (AutoTE_LSA). The weights of *Target, Domain, Style, Methodology, Keywords* and *Date* are set to 0.3, 0.25, 0.25, 0.1, 0.05 and 0.05 respectively. We chose 4 state-of-the-art methods ,i.e., the LDA-based method, TF-IDF,and Paragraph Vector, as benchmarks.

4.1   Results and Discussion

We evaluated the performance of TE extraction, which is shown in Table 2. Then, we calculated the Pearson's correlations of predicted similarity scores using the human annotated ground truth. According to the results shown in Table 3, our methods have a distinct advantage over the baseline methods.

Table 2. Precision of TE extraction

TE items	Precision
Target	0.874
Methodology	0.305
Domain	0.695
Date	1.000
Research style	0.821

Table 3. Pearson's correlations

Correlations	5-level annota-tion	2-level annota-tion
AutoTE_Onto	**0.568**	0.456
AutoTE_LSA	0.463	0.327
LDA	0.537	0.250
TF-IDF	0.519	0.245
Paragraph_vector	0.541	0.327

We further set different thresholds θ to predict whether two documents are semantically similar. Paper pairs are considered semantically similar if their similar score is greater than θ. As shown in Fig. 3, the ROC and PR curves of our TE methods have a remarkable advantage over those of the baseline methods. The accuracy of our TE methods was always superior over the accuracy of the baseline methods using different θ. Our best F1-score was 0.639, while the best F1-score of the baseline methods was 0.546.

The Impact of External Knowledge. Our methods obtained the lexical knowledge from the CL Ontology. The Paragraph Vector and LDA obtained the lexical knowledge from a large corpus. The TF-IDF method is the only method

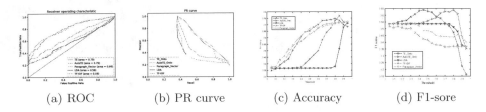

(a) ROC   (b) PR curve   (c) Accuracy   (d) F1-sore

Fig. 3. Comparison of our methods and benchmarks at different thresholds

that does not use external knowledge, and it performed poorly. Thus, methods using external knowledge perform better than the method that does not use external knowledge. Knowledge is indispensable for long text understanding, and our knowledge-intensive approaches are effective in harvesting semantics of long texts.

5   Conclusions

For the first time, we construct a TE model to represent the semantics of a document and measured the semantic similarity among academic articles, and introduce a domain ontology into the construction of TEs and similarity computations. The experimental results shows that our method improve the performance of the semantic similarity analysis in long texts.

Acknowledgments. This research was supported by the Foundation of the State Key Laboratory of Software Development Environment (No. SKLSDE-2017ZX-03).

References

1. Wu, Z., Palmer, M.: Verbs semantics and lexical selection, pp. 133–138. ACL (1994)
2. Mikolov, T., Chen, K., Corrado, G., et al.: Efficient estimation of word representations in vector space. arXiv preprint arXiv:1301.3781 (2013)
3. Agirre, E., Gonzalez-Agirre, A., Lopez-Gazpio, I., et al.: SemEval-2016 task 2: interpretable semantic textual similarity. In: SemEval@ NAACL-HLT, pp. 512–524 (2016)
4. Corley, C., Mihalcea, R.: Measuring the semantic similarity of texts. In: Proceedings of the ACL Workshop on Empirical Modeling of Semantic Equivalence and Entailment, pp. 13–18. Association for Computational Linguistics (2005)
5. Henry, S., Sands, A.: VRep at SemEval-2016 task 1 and task 2: a system for interpretable semantic similarity. In: SemEval, pp. 577–583 (2016)
6. Dolan, B., Quirk, C., Brockett, C.: Unsupervised construction of large paraphrase corpora: exploiting massively parallel news sources, p. 350. ACL (2004)
7. Salton, G., Wong, A., Yang, C.S.: A vector space model for automatic indexing. Commun. ACM **18**(11), 613–620 (1975)
8. Blei, D.M., Ng, A.Y., Jordan, M.I.: Latent Dirichlet allocation. J. Mach. Learn. Res. **3**, 993–1022 (2003)

9. Schuhmacher, M., Ponzetto, S.P.: Knowledge-based graph document modeling. In: WSDM, pp. 543–552 (2014)
10. Zhang, M., Qin, B., Zheng, M., et al.: Encoding distributional semantics into triple-based knowledge ranking for document enrichment, pp. 524–533. ACL (2015)
11. Radev, D.R., Muthukrishnan, P., Qazvinian, V.: The ACL anthology network corpus. In: Proceedings of the 2009 Workshop on Text and Citation Analysis for Scholarly Digital Libraries, pp. 54–61. Association for Computational Linguistics (2009)

Sentiment Classification via Supplementary Information Modeling

Zenan Xu[1], Yetao Fu[1], Xingming Chen[1], Yanghui Rao[1(✉)], Haoran Xie[2],
Fu Lee Wang[3], and Yang Peng[2]

[1] School of Data and Computer Science, Sun Yat-sen University, Guangzhou, China
`raoyangh@mail.sysu.edu.cn`
[2] Department of Mathematics and Information Technology,
The Education University of Hong Kong, Tai Po, Hong Kong
[3] School of Science and Technology, The Open University of Hong Kong,
Kowloon, Hong Kong

Abstract. Traditional methods of annotating the sentiment of a document are based on sentiment lexicons, which have been proven quite efficient. However, such methods ignore the effect of supplementary features (e.g., negation and intensity words), while only consider the counts of positive and negative words, the sum of strengths, or the maximum sentiment score over the whole document primarily. In this paper, we propose to use convolutional neural network (CNN) and long short-term memory network (LSTM) to model the role of negation and intensity words, so as to address the limitations of lexicon-based methods. Results show that our model can not only successfully capture the effect of negation and intensity words, but also achieve significant improvements over state-of-the-art deep neural network baselines without supplementary features.

Keywords: Negation words · Intensity words
Sentiment supplementary information

1 Introduction

Sentiment analysis is a fundamental task of classifying given instances into classes such as positive, neutral, and negative, or fine-grained classes (e.g., very positive, positive, neutral, negative, very negative) in natural language processing. The traditional way of conducting the above task is based on sentiment lexicons [2,7]. Lexicon-based methods mainly exploit features such as the counts

The research has been supported by the National Natural Science Foundation of China (61502545, U1611264, U1711262), Guangdong Science and Technology Program grant (2017A050506025), a grant from the Research Grants Council of the Hong Kong Special Administrative Region, China (UGC/FDS11/E03/16), and the Internal Research Grant (RG 92/2017-2018R) of The Education University of Hong Kong.

© Springer International Publishing AG, part of Springer Nature 2018
Y. Cai et al. (Eds.): APWeb-WAIM 2018, LNCS 10987, pp. 54–62, 2018.
https://doi.org/10.1007/978-3-319-96890-2_5

of positive/negative words, total strengths, and the maximum strength [8]. Although such methods have been shown simple and efficient, they are typically based on bag-of-words models which ignore the semantic composition problem. For sentiment classification, the problem of semantic composition can appear in different ways including negation reversing (e.g., not interesting), negation shifting (e.g., not terrific), and intensification (e.g., very good). Another stream of work focuses on employing machine learning methods, e.g., there are various deep neural networks including CNN [9], recursive autoencoders [12], and LSTM [4], being exploited into sentiment analysis. However, these models also present the above limitation despite their great success.

To address the aforementioned semantic composition problem, we here present a hybrid model for sentiment classification by modelling the supplementary information of negation and intensity words. For example, we change sentence "the movie is not good" to "the movie is bad", and sentence "the movie is very boring" to "the movie is boring + boring". Particularly, we address the issue of semantic composition based on the linguistic role of negation and intensity words. The main contribution of this study is that we develop a backward LSTM to model the reversing effect of negation words and the valence that modified by the intensity words on the following content.

2   Proposed Model

This research aims to tackle the semantic composition issues of traditional lexicon-based methods for sentiment classification. The semantic composition problem can be dealt by modeling the linguistic role of negation and intensity words through a LSTM network. We incorporate the proposed sentiment supplementary information extracted from negation and intensity words into three neural networks, CNN [8], LSTM [6], and CharSCNN [5], and denote these new models as NIS-CNN, NIS-LSTM, and NIS-CharSCNN, where "NIS" means "Negation and Intensity Supplement". In this paper, we mainly introduce the NIS-CNN model, whose architecture is shown in Fig. 1.

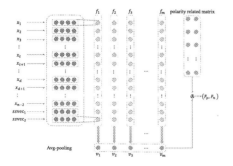

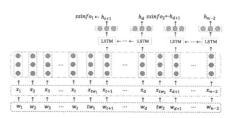

Fig. 1. The architecture of NIS-CNN   **Fig. 2.** Generation of $ssinfo$

2.1 Sentiment Supplementary Vector

We use LSTM to model the effect of negation and intensity words, which is called sentiment supplementary information. The generation of sentiment supplementary information ($ssinfo$) is shown in Fig. 2. A LSTM cell block consists of an input gate I_t, a memory cell C_t, a forget gate F_t, and an output gate O_t to make use of the information from the history $x_1, x_2, \ldots, x_t$ and $h_1, h_2, \ldots, h_{t-1}$ to generate O_t. Formally, O_t is computed as follows:

$$I_t = \sigma(W_i x_t + U_i h_{t-1} + V_i c_{t-1} + b_i), \tag{1}$$

$$F_t = 1.0 - I_t, \tag{2}$$

$$G_t = tanh(W_g x_t + U_g h_{t-1} + b_g), \tag{3}$$

$$C_t = f_t \odot c_{t-1} + i_t \odot g_t, \tag{4}$$

$$O_t = \sigma(W_o x_t + U_o h_{t-1} + V_o c_t + b_o), \tag{5}$$

where x_t is the word embedding of word w_t, σ denotes the sigmoid function, $\odot$ is element-wise multiplication. $\{W_i, U_i, V_i, b_i, W_g, U_g, b_g, W_o, U_o, V_o, b_o\}$ are LSTM parameters.

Now, we denote each negation word and intensity word as target word (tw). We now discuss three situations. Firstly, the model is unchanged if a sentence contains no tw. Secondly, if we have a sentence $S_t = [x_1, \ldots, x_t, tw, x_{t+1}, \ldots, x_n]$, which contains one tw, we use the backward LSTM on words $\{x_{t+1}, x_{t+2} \ldots, x_n\}$ and we get a $ssinfo$. Last but not the least, if we have another sentence $S_d = [x_1, \ldots, x_t, tw_1, x_{t+1}, \ldots, x_d, tw_2, x_{d+1} \ldots, x_n]$, which contains two tw, we use a backward LSTM on words $\{x_{t+1}, \ldots, x_d\}$ and words $\{x_{d+1} \ldots, x_n\}$ to achieve $ssinfo1$ and $ssinfo2$. To preserve the simplicity of the proposed model, we do not consider a sentence contains more than two target words.

After adding the $ssinfo$ into the original sentence and deleting tw, we get a new sentence $\{x_1, x_2, \ldots, x_{t-1}, x_t, \ldots, x_n, \lambda * ssinfo\}$. Here we call the $\lambda * ssinfo$ as a sentiment supplementary vector ($ssvec$). When it comes to intensity and negation words, the value of λ will be initially set to $+1$ and -2 respectively.

2.2 Training and Testing

This task aims to extract the feature map vector (denote as R) of every sentence through a simple CNN, and multiply it by the weight vector to calculate the relevancy between the sentence and the polarity. Finally, we choose the polarity with the largest relevancy as the label for the sentence.

A convolution operation which involves m filters $W \in R_1 * d$ is applied to the words one by one to generate the feature map of the sentence: $V_i = g(W * x_i)$, where "$*$" is a two-dimensional convolution operation and g indicates a non-linear function. The pooling layer is applied to calculate the whole representation of the sentence from the sentiment information extracted by the filters from all words in the text. The average pooling will be used in this case, which aims to capture the average sentiment information so as to apply on the feature vector v. The average pooling is defined as:

$$r_{avg} = \frac{1}{n-h+1} \sum_{j=1}^{n-h+1} v_j.$$ (6)

The model uses two polarity related weight vectors (denoted as C_p and C_n) and feature map vector R obtained by pooling-layer to generate the score under different polarities (denoted as $Score_p i$, $Score_n i$) of the i-th sentence. Here we use $L_i = 1$ and $L_i = 0$ to indicate positive and negative sentiment. For the polarity, we use the softmax to calculate the possibility of being positive and negative as s_i^1 and s_i^0. s_i^1 and s_i^0 are estimated as

$$s_i^1 = \frac{e^{Score_{pi}}}{e^{Score_{pi}} + e^{Score_{pn}}},$$ (7)

$$s_i^0 = \frac{e^{Score_{pn}}}{e^{Score_{pi}} + e^{Score_{pn}}}.$$ (8)

We use cross-entropy to calculate the loss of the model. Assumed that there have N training sentences, the loss function is defined as:

$$L|\theta| = -\sum_{i=1}^{N} log s_i^{L_i} + \frac{\lambda_r}{2}||\theta||^2,$$ (9)

where θ is the set of model parameters, λ_r is a parameter for $L2$ regularization.

3   Experiments

3.1   Dataset

We evaluate the proposed model on three datasets. The first one is Movie Review (MR) [11], in which every sentence is annotated with two classes as positive and negative. The second one is Stanford Sentiment Treebank (SST) [12], where each sentence is classified into five classes, including very negative, negative, neutral, positive, and very positive. The third one is Sentiment Labelled Sentences (SLS) [10], which is collected from reviews of products (Amazon), movies (IMDB), and restaurants (Yelp). Statistics of the three datasets are summarized in Table 1.

Negation and intensity words are derived from Linguistic Inquiry and Word Count (LIWC2007), in which a certain word is labelled according to its characteristic or property. We use all negation words from the Negate part of LIWC2007 and the intensity words manually from the Adverb part by removing some words that are obviously not intensity words.

Table 1. Dataset statistics. S: Number of sentences. L: Average sentence length. V: Vocabulary size. $|N|$: Percentage of documents with negation words. $|I|$: Percentage of documents with intensity words.

Dataset	S	L	V	$|N|$	$|I|$
MR	10662	20	18376	33.7%	53.2%
SST	9613	17	17439	25.8%	49.8%
SLS	3000	12	5170	27.8%	39.0%

3.2   Experiment Design

To evaluate the performance of the proposed NIS-CNN, NIS-LSTM, and NIS-CharSCNN, we implement the following baselines for comparison:

- CNN: which generates sentence representation by a convolutional layer with multiple kernels (i.e., kernels' size of 3, 4, 5 with 100 feature maps each) and pooling operations. Note that dropout operations are added to prevent over-fitting [8].
- LSTM: The whole corpus is processed as a single sequence, and LSTM generates the sentence representation by calculating the means of the whole hidden states of all words. The hidden state size is empirically set to 128 [6].
- CharSCNN: which employs two convolutional layers to extract features from characters to sentences. Following the convolutional layers are two fully-connected layers, the output of the second convolutional layer is passed to them to calculate the sentiment score. Empirically, the context windows of words and characters are set to 1. The convolution state size of the character-level layer and that of the word-level layer are respectively set to 20 and 150 [5].

Our experiments are implemented using the TensorFlow [1] and Keras [3] Python libraries. We use Stochastic Gradient Descent with Adadelta [13] for training. We set the batch size at each iteration to 32 and the size of word embeddings to 300 for all datasets and models. All other parameters are initialized to their default values as specified in the TensorFlow and Keras library. For all datasets, we randomly select 80% samples as the training set, 10% as validation samples, and the remaining 10% for testing.

In our negation and intensity supplement method, LSTM's hidden state sizes d and the dropout rate p are tuned on the validation set for each dataset. The values of d in MR, SST and SLS are 128, 256 and 128 respectively, and the values of p in MR, SST and SLS are 0.5, 0.3 and 0.2 respectively.

3.3   Evaluation Metrics

We use *Accuracy* to evaluate the model performance, as follows:

$$Accuracy = \frac{\sum_{i=1}^{N} tp_i + tn_i}{\sum_{i=1}^{N} tp_i + fp_i + tn_i + fn_i}, \tag{10}$$

where tp_i is 1 if the i-th sentence is positive and the prediction is positive, otherwise, it is 0. tn_i is 1 if the i-th sentence is negative and the prediction is negative, otherwise, it is 0. fp_i is 1 if the i-th sentence is negative and the prediction is positive, otherwise, it is 0. fn_i is 1 if the i-th sentence is positive and the prediction is negative, otherwise, it is 0. N is the number of sentences.

3.4   Results and Analysis

As shown in Table 2, in all datasets, the experimental results of NIS-CNN are superior to those baselines (e.g., CNN, LSTM, and CharSCNN) that do not consider negation and intensity words. We can conclude that the linguistic role of negation and intensity words that our model captured is effective.

Table 2. Accuracy (%) of all models on MR, SST and SLS datasets.

Model	MR	SST	SLS
CNN	75.8	80.2	85.6
IS-CNN	76.7	80.6	85.5
NS-CNN	78.6	82.1	87.4
NIS-CNN	**78.9**	**82.3**	**88.2**
LSTM	75.9	75.8	85.3
IS-LSTM	76.0	76.3	85.0
NS-LSTM	76.7	77.0	86.1
NIS-LSTM	**77.2**	**77.6**	**86.8**
CharSCNN	73.5	81.7	82.0
IS-CharSCNN	73.1	81.2	82.4
NS-CharSCNN	**74.6**	**82.9**	82.8
NIS-CharSCNN	74.5	82.7	**83.3**

We also conduct ablation experiments to evaluate the functional performance of negation words and intensity words respectively, these experiments are conducted on the entire dataset. First of all, we conduct the experiment with no negation and intensity words. Then we remove either negation words or intensity words each time on the basis of our model and execute the NS-CNN and the IS-CNN on the whole dataset respectively. In Table 2, significant improvement can be observed between CNN and NIS-CNN on MR (the accuracy rises from 75.8% to 78.9%), SST (the accuracy rises from 80.2% to 82.3%), SLS (the accuracy rises from 85.6% to 88.2%), which validates the effectiveness of NIS-CNN on modelling the linguistic role of negation and intensity words.

To further validate the effectiveness of the supplementary information, we conduct similar ablation experiments on LSTM and CharSCNN. Improvements can also be seen between LSTM and NIS-LSTM on MR, SST, SLS, as well as between CharSCNN and NIS-CharSCNN on MR, SST, and SLS.

However, we find that methods with negation words only show significant improvement on the accuracy of binary classification compared with methods without negation and intensity words, while methods with intensity words only show a slight improvement and even a little descend. To explore the reason behind such phenomenon, we conduct detailed experiments as follows.

Table 3. Examples about the effect of negation words on MR dataset. *NW*: Negation word. *C*: Content. *Pos*: The probability of predicted Positive (%). *Neg*: The probability of predicted Negative (%).

Sentence	NW	C	CNN		NS-CNN	
			Pos	Neg	Pos	Neg
You cannot help but get caught up	cannot	Positive	38.1	61.9	56.1	43.9
Hollywood wouldn't have the guts to make	not	Positive	41.7	58.3	59.6	40.4
The story is nowhere near gripping enough	nowhere	Negative	69.1	30.9	36.4	63.6

For negation words, we extract all the sentences with negation words in MR dataset and compare the probability under different polarity predicted by CNN and NS-CNN. We can see in Table 3, for those sentences with negation words that were annotated with the false label by CNN, NS-CNN could correct such faults and consequently improved the accuracy. Therefore when we modeled the sentiment reversing effect of negation words and introduce it into CNN, we could correct those sentences that are classified into wrong classes by CNN.

Table 4. Examples about the effect of intensity words on MR dataset. *IW*: Intensity word. *C*: Content. *Pos*: The probability of predicted Positive (%). *Neg*: The probability of predicted Negative (%).

Sentence	IW	C	CNN		IS-CNN	
			Pos	Neg	Pos	Neg
An extremely unpleasant film	extremely	Negative	22.8	77.2	6.60	93.4
Really quite funny	really	Positive	73.5	26.5	85.7	14.3
Too silly to take seriously	too	Negative	19.8	80.2	16.3	83.7
The tenderness of the piece is still intact	still	Positive	52.8	47.2	48.7	51.3

For intensity words, we observe that intensity words just change the sentiment level of the sentence with intensity words but do not change the sentiment polarity. For example, in Table 4, the sentence "An extremely unpleasant film" with the intensity word "extremely" is labelled correctly by CNN. When considering the sentiment shifting effect of intensity words, the probability of negative predicted by IS-CNN is still higher than the probability of positive, while the label keeps negative too. In summarize, when a sentence is annotated with a false label, considering intensity words will not help to correct it. Intensity words should play a more significant role in fine-grained sentiment classification tasks.

4   Conclusion

In this work, we proposed an effective model for sentiment classification. The proposed model addressed the sentiment reversing effect of negation words and the sentiment shifting effect of intensity words. Experimental results validate the effectiveness of our model. In the future, we plan to introduce the attention mechanism to model the valence of every word in the sentence, including the negation and intensity words that change the sentiment of the sentence. Furthermore, we will apply the similar process on negation and intensity words to conjunctions, which may shift the sentiment level of a sentence to some extent.

References

1. Abadi, M., Barham, P., Chen, J.M., Chen, Z.F., Davis, A., Dean, J., Devin, M., Ghemawat, S., Irving, G., Isard, M., Kudlur, M., Levenberg, J., Monga, R., Moore, S., Murray, D.G., Steiner, B., Tucker, P., Vasudevan, V., Warden, P., Wicke, M., Yu, Y., Zheng, X.Q.: Tensorflow: a system for large-scale machine learning. In: OSDI, pp. 265–283 (2016)
2. Baccianella, S., Esuli, A., Sebastiani, F.: SentiWordNet 3.0: an enhanced lexical resource for sentiment analysis and opinion mining. In: LREC, pp. 2200–2204 (2010)
3. Choi, K., Joo, D., Kim, J.: Kapre: On-GPU audio preprocessing layers for a quick implementation of deep neural network models with Keras. CoRR, abs/1706.05781 (2017)
4. Chung, J.Y., Gulcehre, C., Cho, K.H., Bengio, Y.: Empirical evaluation of gated recurrent neural networks on sequence modeling. CoRR, abs/1412.3555 (2014)
5. Guerini, M., Gatti, L., Turchi, M.: Sentiment analysis: How to derive prior polarities from SentiWordNet. In: EMNLP, pp. 1259–1269 (2013)
6. Hochreiter, S., Schmidhuber, J.: Long short-term memory. Neural Comput. **9**(8), 1735–1780 (1997)
7. Hu, M.Q., Liu, B.: Mining and summarizing customer reviews. In: KDD, pp. 168–177 (2004)
8. Kim, S.M., Hovy, E.: Determining the sentiment of opinions. In: COLING, Article no. 1367 (2004)
9. Kim, Y.: Convolutional neural networks for sentence classification. CoRR, abs/1408.5882 (2014)
10. Mikolov, T., Chen, K., Corrado, G., Dean, J.: Efficient estimation of word representations in vector space. CoRR, abs/1301.3781 (2013)

11. Pang, B., Lee, L.: Seeing stars: Exploiting class relationships for sentiment categorization with respect to rating sales. In: ACL, pp. 115–124 (2005)
12. Socher, R., Perelygin, A., Wu, J., Chuang, J., Manning, C.D., Ng, A., Potts, C.: Recursive deep models for semantic compositionality over a sentiment treebank. In: EMNLP, pp. 1631–1642 (2013)
13. Zeiler, M.D.: ADADELTA: an adaptive learning rate method. CoRR, abs/1212.5701 (2012)

Training Set Similarity Based Parameter Selection for Statistical Machine Translation

Xuewen Shi, Heyan Huang, Ping Jian$^{(\boxtimes)}$, and Yi-Kun Tang

Beijing Engineering Research Center of High Volume Language Information
Processing and Cloud Computing Applications, School of Computer Science
and Technology, Beijing Institute of Technology, Beijing 100081,
People's Republic of China
{xwshi,hhy63,pjian,tangyk}@bit.edu.cn

Abstract. Log-linear model based statistical machine translation systems (SMT) are usually composed of multiple feature functions. Each feature function is assigned a weight as a model parameter. In this paper, we consider that different input source sentences may have discrepant needs for model parameters. To adapt the model to different inputs, we propose a model parameters selection method for log-linear model based SMT systems. The method is mainly based on the characteristics of different feature functions themselves without any assumption on unseen test sets. Experimental results on two language pairs (Zh-En and Ug-Zh) show that our method leads to the improvements up to 2.4 and 2.2 BLEU score respectively, and it also shows the good interpretability of our proposed method.

Keywords: Statistical machine translation · Log-linear model
Parameter selection

1 Introduction

Log-linear model based statistical machine translation (SMT) [8] is usually composed of multiple feature functions. Each feature function $h(\cdot)$ is assigned a weight λ as a model parameter. Then, given a sentence f, the posterior probability distribution of a candidate translation e is approximated by:

$$Pr(e|f) = p_\lambda(e|f) = \frac{exp[\sum_{n=1}^{N} \lambda_n h_n(e,f)]}{\sum_{e'} exp[\sum_{n=1}^{N} \lambda_n h_n(e',f)]}, \tag{1}$$

and SMT system will choose the sentence with the highest probability. In this framework, in addition to the estimation of the feature functions, model parameter is also an important factor that affects the translation quality. Obtaining a

This work was supported by the National Key Research and Development Program of China (Grant No. 2017YFB1002103) and the National Natural Science Foundation of China (No. 61732005).

group of appropriate log-linear model parameters $\boldsymbol{\lambda} = \{\lambda_1, \lambda_2, ..., \lambda_N\}$ is often formulated as an optimization problem that finds the parameters minimizing errors of the translations generated by the model on a development corpus D_{dev}:

$$\boldsymbol{\lambda}^* = \arg\min_{\boldsymbol{\lambda}} E(e^*, \boldsymbol{\lambda}, D_{dev}), \tag{2}$$

where $E(e^*, \boldsymbol{\lambda}, D_{dev})$ is an error function. The optimization process is often based on an assumption that both the development set and the test set observe an identical distribution. However, it is not easy to get the suitable development set without the knowledge of test set. Empirical studies [2] show that by using different development corpora to tune the same phrase based SMT systems [3], translation performance can vary obviously.

For development set selection and parameter adaption, most previous researches focus on how to find tuning samples similar to the future test data [6,12,15]. Those approaches focus on the characteristics and domains of data, without considering the nature of different feature functions. In practice, the requirements for model parameters may be very different even for sentences in the same domain, due to the nature of different feature functions. Previous works from the view of the nature of feature functions include [4,5,14]. Those approaches re-tune the model according to the test sentences and the test set needs to be known before parameters construction, which is hard to operate in practice. Zahran and Tawfik [13] adapt parameters pool to avoid re-tuning. They cluster development set by using representations of sentences.

In this paper, we propose a novel model parameter selection method based on training set similarity (TSS). The proposed method mainly has two advantages: (i) the similarity calculation method is designed according to the characteristics of the function. Unlike sentence embedding in [13], the TSS based method does not need a training process in principle, so that it is suitable for low resource language translations; (ii) the parameters construction stage does not need any knowledge about test sets, and it ensures the operable of the method in practice. Experimental results on two language pairs (Zh-En and Ug-Zh) show that our method leads to the improvements up to 2.4 and 2.2 BLEU score respectively.

2   Training Set Similarity Based Parameter Selection

We first construct a parameters pool by tuning on different cluster of development sets respectively, and then select a model parameter group for each specific input sentence during testing step. We propose a novel empirical similarity calculation method, training set similarity (TSS). In the step of parameters pool construction, TSS is used as the basis to divide the development set into different clusters. For parameter selection, we calculate TSS for sentences to be tested and select a suitable parameter group according to the category of its TSS.

2.1   Training Set Similarity

For most development set adaption approaches [4,5,13], similarity is usually used as a basis for dividing and selecting data and it is often calculated between the test data and the development data. Unlike other approaches, TSS measures the similarity between a sentence and the SMT training set. The method is mainly based on two considerations: (i) since the training set contains the whole raw linguistic knowledge of the log-linear model and (ii) using training set as the reference standard to compute similarity does not need any knowledge about unseen data. Formally, given a sentence s and the training set D_{tr}, a TSS is calculated as follows:

$$TSS(s, D_{tr}) = \arg\max_{s_{tr} \in D_{tr}} sim(s, s_{tr}), \tag{3}$$

$$sim(s, s_{tr}) = \frac{2 \times M}{L}, \tag{4}$$

where M is the number of matching words between two sequences, and L is the total number of words in both sentences. In this paper, we use the gestalt pattern matching algorithm [10] to count the matching words between two sentences. Usually, sim lower than 0.5 means the two sequences have very little in common, while the two sequences are similar when the score is higher than 0.8.

2.2   Parameters Pool Construction

In the step of parameters pool construction, we divide the development into different subsets by sentence TSS categories, and obtain a parameters pool by tuning on each subset. We define a categories set $C = \{c | c \in [1, K]\}$ and a classification function $f(\cdot)$ first. The classification function $f(sim)$ is a step function of sim and returns an integer category number c. Unlike the previous works [4,13] using clustering techniques, the mapping tables in this paper are handcrafted. In addition, the division granularity of similarity value ranges is also a noteworthy problem. First, according to the order of magnitude, the size of the data under each similarity value range should reach 10^3. For data with higher TSS scores, we try to divide them with fine granularity.

Given the development set D_{dev} and the training set D_{tr}, the parameters pool P is obtained by the following procedure:

1. For each $s_i = (src_i, tgt_i)$ in D_{dev}, calculate the TSS of src_i using $sim_i = TSS(src_i, D_{tr})$, and then categorize it using $g(s_i) = c_i = f(sim_i)$, we regard $g(s_i)$ as the category of s_i;
2. Divide the development set D_{dev} into K subsets based on the category c_i of each sentence pair s_i in D_{dev}, for each subset $D_{dev}(c) = \{s | s \in D_{dev}, g(s) = c\}$, $c \in [1, K]$;
3. Tune model on each subset $D_{dev}(c)$ and get K groups of model parameters $\lambda(1)$ to $\lambda(K)$, then combine them getting the parameters pool $P = \{\lambda(c) | c \in [1, K]\}$; In this step, we use the MERT algorithm [7] optimized for BLEU [9] metric in tuning.

2.3   Parameter Selection

At the step of testing, we select the model parameters from parameters pool according to the TSS category of the input sentence. Let s_{src} be the source sentence to be translated, the corresponding model parameters $\boldsymbol{\lambda_{src}}$ is selected by the following procedure:

1. Calculate the TSS of s_{src}: $sim_{src} = TSS(s_{src}, D_{tr})$;
2. Categorize sim_{src} and get its category: $c_{src} = f(sim_{src})$;
3. Select the parameters $\boldsymbol{\lambda_{src}}$ from P according to c_{src}: $\boldsymbol{\lambda_{src}} = \boldsymbol{\lambda}(c_{src})$.

3   Experiments

3.1   Datasets and Setups

In experiments, we evaluate our proposed parameter selection method in two translation directions: Zh-En and Ug-Zh. The Ug-Zh translation is used to test the method performance of low resource translations.

Zh-En Datasets: We totally extract 2.6M sentence pairs from LDC corpora[1] and randomly split them into three parts as the training set, the development set and one of the test sets (Test1) in the size of 2.1M, 50K and 5K respectively. We also use NIST2002–2006, 2008 and 2012 datasets as our additional test sets.

Ug-Zh Datasets: We use the training and development datasets from the Uyghur-to-Chinese News Translation Task of CWMT 2017 MT Evaluation Campaign[2] as the whole Ug-Zh Datasets. We randomly split the CWMT 2017 training sets into three parts: a training set, a development set and a test set (Test1) in the size of 0.3M, 30K and 3K respectively, We use the CWMT 2017 development dataset as another test set (Test2).

For each language pair, we built a phrase-based SMT baseline system using Moses [3] with its 14 default feature functions. The Chinese part for all datasets is segmented by the LTP Chinese word segmentor [1]. For the Uyghur side of data, we use subword units [11] as the processing units. All the reported translation performance results in this paper were measured in case-insensitive BLEU [9].

3.2   Results

For all language pairs, the set Test1 is extracted in the same way as development set, and other test sets are in unknown distributions. The above settings of data are used to verify the performances of the proposed method on different test sets in different distributions. Tables 1 and 2 show the contrast between Moses baseline system and our parameter selection method on a macro level. From those tables, it can be seen that by using the same development set, the

---

[1] The corpora include LDC2002E18, LDC2003E07, LDC2003E14, Hansards portion of LDC2004T07, LDC2004T08 and LDC2005T06.

[2] http://ee.dlut.edu.cn/CWMT2017/evaluation_en.html.

Table 1. The comparisons of different development data on Zh-En translation. The underlined <u>BLEU scores</u> are results that tuning on those corresponding test data.

Models	Dev set	Test1	02	03	04	05	06	08	12
Moses	Dev	24.63	29.96	28.70	29.90	29.01	28.48	22.95	21.47
+PS	Dev	**25.28**	**31.77**	**30.89**	32.32	31.04	30.63	23.49	21.95
Moses	06	21.97	31.69	30.82	**32.62**	**31.27**	**<u>33.08</u>**	26.12	24.31
	08	21.52	30.89	30.15	32.01	30.39	32.10	**<u>26.57</u>**	24.27
	12	20.91	30.63	30.00	32.05	30.06	31.79	26.06	**<u>25.27</u>**

proposed parameter selection (PS) method improves the SMT baseline systems up to 2.4 and 2.2 BLEU scores on Zh-En and Ug-Zh translations respectively.

For Zh-En translation experiments, the test sets NIST2002–2006 are in similar domain while NIST2008 and NIST2012 contain some other styles of data. Table 1 shows the comparisons of different development data on Zh-En translation. Except the situation of directly tuning on a specific test set, using NIST2006 as development set gains generally better results, since the development set matches the most domains of the test sets. Using our parameter selection method shows competitive results comparing with Moses baseline system tuning on in-domain data, which is verified the stability of our methods on different domains test data.

For the Ug-Zh translation, Our proposed PS method outperforms all the Moses baseline systems tuning on different development sets, including systems directly tuning on the test set. The results prove the effectiveness of the TSS based data selection strategy on low resources languages.

Furthermore, we analyze the distribution of BLEU on different sentence length as shown in Fig. 1. The performances of our proposed method on long sentences and short sentences are obviously better than baseline systems for both language pairs. This distribution illustrates that for the sentences which are too long or too short, it is more necessary to make special adjustments to the parameters.

Table 2. The comparisons of different development data on Ug-Zh translation.

Models	Dev set	Test1	Test2
Moses	Dev	24.84	41.72
+PS	Dev	**26.06**	**42.74**
Moses	Test1	<u>24.97</u>	41.48
	Test2	23.76	<u>42.66</u>

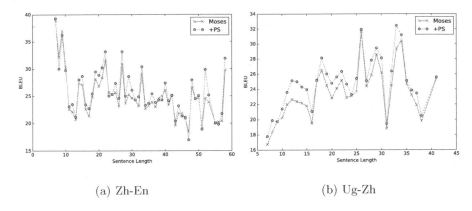

(a) Zh-En   (b) Ug-Zh

Fig. 1. The distribution of BLEU scores on different sentence length.

Table 3. Experimental results of different TSS ranges in Ug-Zh translations. "Sents" is the number of sentences under the TSS range.

Datasets	Models	[0.0, 0.5]	(0.5, 0.6]	(0.6, 0.7]	(0.7, 0.8]	(0.8, 0.99]	(0.99, 1.0]
Dev	Moses	18.61	30.99	37.79	46.92	50.05	58.36
	+PS	**19.03**	**31.68**	**38.43**	**47.63**	**52.91**	**68.07**
Test1	Sents	2164	271	146	87	108	224
	Moses	18.40	30.78	39.53	52.33	47.60	58.66
	+PS	**18.68**	30.78	**39.72**	**52.62**	**51.32**	**69.33**
Test2	Sents	486	39	21	15	65	80
	Moses	34.50	49.58	**52.29**	**53.81**	66.45	66.58
	+PS	**35.56**	**52.84**	51.62	50.06	**67.90**	**67.27**

3.3   Analysis

In order to further verify the effectiveness of the proposed parameter selection method, we take Ug-Zh translation as an example to analyze the method's performances on different TSS ranges as shown in Table 3.

For all the three groups in the table, the improvements are mainly concentrated in the range of TSS 0.8 to 1.0. For the translations of those sentences the translation model plays a more important role than other feature functions. For the source sentences with lower TSS scores, their phrases that can be directly translated are usually short and scattered. As a result, those sentences need a language model to integrate these scattered translated phrases into fluent sentences. With regard to the negative results of Test2, we blame the reason for this part of the test data more sparse. Since the method is tuning on documents, it cannot ensure whether each sentence can be translated better.

We show two example translations in Table 4. For the first example, the TSS is 0.51, which is a lower similarity ratio. In the Moses baseline translation, the

Table 4. Example on Zh-En translations. "Source" is a romanized chinese sentence after segmentation, "Reference" is a gold-standard translation. The underlined words means the main difference, "words" means the better translations, while "words" means the worse translations.

Source	*nuli tigao nongye chanyehua jingying de shuiping, jiakuai tiaozheng nongye he nongcun jingji jiegou, tigao ....*
TSS	$0.51 \in (0.5, 0.6]$
Reference	we must strive to raise the level of industrialization of agriculture, speed up the adjustment of the structure of agriculture and the rural economy, and increase ...
Moses	and make efforts to enhance the level of the industrialized operation of agriculture, speed up the structural adjustment of agriculture and the rural economy, increase ...
+PS	and make efforts to enhance the level of the industrialized operation of agriculture, speed up the adjustment of the structure of agriculture and the rural economy, and increase ...
Source	*ta jiang yu shieryue shiqiri zai fenling caipan fayuan titang*
TSS	$0.91 \in (0.8, 0.99]$
Reference	he will appear in Fanling Magistrates' Courts on December 17
Moses	he will appear in Fanling Magistracy on December 17
+PS	December 17 he will appear in Fanling Magistrates' Courts

language model and direct phrase translation probability score is -81.63 and -21.29 respectively while the score is -74.93 and -26.35 respectively in the translation of our parameter selection method. This indicates that in this case, language model is more effective than translation models and our parameter selection method allocates the appropriate weights for them indeed.

For the second example, the TSS is 0.91, which means that most of the continuous phrases appear in the training data. In this sentence, the phrase *"fenling caipan fayuan"* is a named entity which could be found in phrase table with high probabilities. Our proposed parameter selection method takes advantage of the translation model and successfully translated the named entity *"fenling caipan fayuan"* into "Fanling Magistrates' Courts".

4   Conclusions

In this paper, we present a TSS based parameter selection method for SMT. This method builds a parameters pool for SMT systems based on the characteristics of its feature functions, and allocates suitable parameters to SMT system for different translation inputs when decoding. There is no assumptions on the distribution of unknown test set in the whole process. This paper mainly studies the relationship between language models and translation models at sentence level translation. In the future, we plan to introduce more precise development set categorization strategies in order to balance multiple features functions.

References

1. Che, W., Li, Z., Liu, T.: LTP: a Chinese language technology platform. In: Proceedings of the 23rd International Conference on Computational Linguistics: Demonstrations, pp. 13–16 (2010)
2. Hui, C., Zhao, H., Song, Y., Lu, B.L.: An empirical study on development set selection strategy for machine translation learning. In: Proceedings of the Joint Fifth Workshop on Statistical Machine Translation and MetricsMATR, pp. 67–71 (2010)
3. Koehn, P., Hoang, H., Birch, A., Callison-Burch, C., Federico, M., Bertoldi, N., Cowan, B., Shen, W., Moran, C., Zens, R., et al.: Moses: open source toolkit for statistical machine translation. In: Proceedings of the 45th Annual Meeting of the ACL on Interactive Poster and Demonstration Sessions, pp. 177–180 (2007)
4. Li, M., Zhao, Y., Zhang, D., Zhou, M.: Adaptive development data selection for log-linear model in statistical machine translation. In: Proceedings of the 23rd International Conference on Computational Linguistics, pp. 662–670 (2010)
5. Liu, L., Cao, H., Watanabe, T., Zhao, T., Yu, M., Zhu, C.: Locally training the log-linear model for SMT. In: Proceedings of the 2012 Joint Conference on Empirical Methods in Natural Language Processing and Computational Natural Language Learning, pp. 402–411 (2012)
6. Lü, Y., Huang, J., Liu, Q.: Improving statistical machine translation performance by training data selection and optimization. EMNLP-CoNLL **2007**, 343–350 (2007)
7. Och, F.J.: Minimum error rate training in statistical machine translation. In: Proceedings of the 41st Annual Meeting on Association for Computational Linguistics-Volume 1, pp. 160–167 (2003)
8. Och, F.J., Ney, H.: Discriminative training and maximum entropy models for statistical machine translation. In: Proceedings of the 40th Annual Meeting on Association for Computational Linguistics, pp. 295–302 (2002)
9. Papineni, K., Roukos, S., Ward, T., Zhu, W.J.: BLEU: a method for automatic evaluation of machine translation. In: Proceedings of the 40th Annual Meeting on Association for Computational Linguistics, pp. 311–318 (2002)
10. Ratcliff, J.W., Metzener, D.E.: Pattern-matching-the Gestalt approach. Dr. Dobbs J. **13**(7), 46 (1988)
11. Sennrich, R., Haddow, B., Birch, A.: Neural machine translation of rare words with subword units. In: Proceedings of the 54th Annual Meeting of the Association for Computational Linguistics (Volume 1: Long Papers), vol. 1, pp. 1715–1725 (2016)
12. Song, X., Specia, L., Cohn, T.: Data selection for discriminative training in statistical machine translation. In: 17th Annual Conference of the European Association for Machine Translation, EAMT, pp. 45–53 (2014)
13. Zahran, M.A., Tawfik, A.Y.: Adaptive tuning for statistical machine translation (AdapT). In: Gelbukh, A. (ed.) CICLing 2015. LNCS, vol. 9041, pp. 557–569. Springer, Cham (2015). https://doi.org/10.1007/978-3-319-18111-0_42

14. Zhao, Y., Ji, Y., Xi, N., Huang, S., Chen, J.: Language model weight adaptation based on cross-entropy for statistical machine translation. In: Proceedings of the 25th Pacific Asia Conference on Language, Information and Computation, pp. 20–30 (2011)
15. Zheng, Z., He, Z., Meng, Y., Yu, H.: Domain adaptation for statistical machine translation in development corpus selection. In: 2010 4th International Universal Communication Symposium (IUCS), pp. 2–7 (2010)

Social Networks

Identifying Scholarly Communities from Unstructured Texts

Ming Liu[1(✉)], Yang Chen[2], Bo Lang[1], Li Zhang[3], and Hongting Niu[1]

[1] State Key Laboratory of Software Development Environment,
Beihang University, Beijing, China
{liuming,langbo,niuhongting}@nlsde.buaa.edu.com
[2] National Computer Network Emergency Response Technical Team/Coordination
Center of China, Beijing, China
chenyang@cert.org.cn
[3] School of Government, Beijing Normal University, Beijing, China
zhangli@bnu.edu.cn

Abstract. Scholarly community detection has important applications in various fields. Previous studies have relied heavily on structured scholar networks, which have high computational complexity and are challenging to construct in practice. We propose a novel alternative that can identify scholarly communities directly from large textual corpora. To our knowledge, this is the first study intended to detect communities directly from unstructured texts. Generally, academic articles tend to mention related work and researchers. Researchers that are more closely related to each other are mentioned in a closer grouping in lines of academic text. Based on this correlation, we develop an intuitional method that measures the mutual relatedness of researchers through their textual distance. First, we extract and disambiguate the researcher names from academic articles. Then, we embed each researcher as an implicit vector and measure the relatedness of researchers by their vector distance. Finally, the communities are identified by vector clusters. We implement and evaluate our method on three real-world datasets. The experimental results demonstrate that our method achieves better performance than state-of-the-art methods.

Keywords: Scientific literature analysis · Community detection
Scientific information extraction · Representation learning

1 Introduction

Community detection is attracting substantial research attention. A community refers to a compact "clique" or "cluster" of densely connected nodes in a network. Typically, the entities in a community share similar properties and have more interactions with one another than with the remainder of the network. Communities typically correspond to objects in the real world, such as criminal gangs, interest-based social groups in a social network or units of a cell in a

© Springer International Publishing AG, part of Springer Nature 2018
Y. Cai et al. (Eds.): APWeb-WAIM 2018, LNCS 10987, pp. 75–89, 2018.
https://doi.org/10.1007/978-3-319-96890-2_7

protein-protein interaction network. In the academic domain, the research community typically consists of researchers who share common areas of science [1,2]. The effective detection of scholarly communities can be potentially used in many applications, including reviewer recommendation, bibliography recommendation, researcher impact measurement and research trend analysis. Many algorithms have been devised, which range from straightforward graph partitioning methods, such as hierarchical clustering and the Girvan-Newman [3] algorithm, to more sophisticated optimization techniques based on the maximization of various objective functions [5–14].

Nevertheless, several challenges are encountered in identifying scholarly communities from networks. Bibliographic networks typically refer to collaborative networks or citation networks. The nodes of those networks are authors or documents, and the edges represent collaboration or citation relationships. The formalizations of community detection lead to NP-hard problems, and both the adjacency matrix decomposition of the networks and the traversal on graphs are time-consuming and require substantial computational resources. In addition, bibliographic networks such as collaborative networks and citation networks are coarse-grained. The references of an article are unbiased; however, references are cited for various purposes, such as for comparison, or dismissal, or disproval or are merely mentioned in passing. Thus, network-based community detection approaches may fail to capture this type of fine-grained relatedness. Moreover, due to the limitations and boundaries of academic publishers, it is challenging to exchange scholar metadata, and this incompleteness of the bibliographic network may decrease the results of research community detection.

Structured bibliographic networks are scarce and expensive, while the corpora are ubiquitous. As representation learning techniques improve, textual corpora may provide a cheaper and faster alternative. Previous research is widely cited in scholarly articles, and similar research tends to be presented in a close grouping when authors are writing their manuscripts. For example, in Fig. 1, the researchers who are mentioned in the 3rd line, namely, *"Collings"*, *"Duffy"*, *"Culotta"*, *"Sorensen"*, *"Bunescu"*, and *"Mooney"*, are more closely related to one another than to researcher *"Zhang"*, who is mentioned in line 6.

1   *Recent results mainly rely on kernel-based approaches. Many of them*
2   *focus on using tree kernels to learn parse tree structure related features*
3   *(Collins and Duffy, 2001; Culotta and Sorensen, 2004; Bunescu and Mooney,*
4   *2005). Other researchers study how different approaches can be combined to*
5   *improve the extraction performance. For example, by combining tree kernels*
6   *and convolution string kernels, (Zhang et al., 2006) achieved the state of the*
7   *art performance on ACE (ACE, 2004), which is a benchmark dataset for*
8   *relation extraction.*

Fig. 1. Example of citation sentences in academic articles.

Our approach is based on the intuitive assumption that researchers who are mentioned nearby in lines of academic content are relatively closely related to one another in terms of research area. We call this type of researcher relatedness the mention distance (MD). The more closely two researchers are mentioned in a text, the more closely they are related and the shorter their MD should be. Scholarly communities can be identified by grouping researchers according to MD.

However, several challenges should be addressed when using MD to identify the scholarly communities. A major obstacle is the extraction and disambiguation of researcher names from various academic articles. Most researchers are cited and mentioned with abbreviated names or surnames, and many researchers share the same surnames or even the same full names. It is challenging to identify individual researchers as the basic elements of scholarly communities. Other key challenges that we had to solve are quantifying the degree of MD reasonably and identifying the real scholarly community.

In this study, we first extract each reference from academic articles and recognize all the entities, such as names, venues (journal and conference names), and publication years, in the "References" section of each academic paper. Then, we use those entities to disambiguate the researcher names. Each researcher is embedded as a real-valued vectors through the use of representation learning techniques, and the MDs of researchers are quantified via vector distances. Finally, the researchers are clustered based on their MDs to form the final research communities. In this work, we focus on non-overlapping communities and evaluate the performance of the new approach on real-world datasets. The experimental results demonstrate that our method achieves outstanding performance. In summary, our contributions are as follows:

(1) We propose a framework for discovering hidden communities directly from academic articles without any reliance on bibliographic networks. To our knowledge, this is the first work on detecting communities directly from large unstructured text corpora.
(2) We propose the MD for measuring the mutual relatedness of researchers and define out an embedding-based MD that can accurately capture the fine-grained relatedness of researchers.
(3) Our experimental results demonstrate that our method requires fewer computational resources and is more accurate than current methods, which reveals that an academic corpus is a reliable resource for discovering scholarly communities and is much cheaper than the use of bibliographic networks.

The remainder of this paper is organized as follows. In Sect. 2, we describe related research. In Sect. 3, we describe the general framework for scholarly community detection based on researcher embedding. In Sect. 4, we evaluate our experimental results and we present our conclusions in Sect. 5.

2 Related Work

The detection of clusters and communities has attracted numerous researchers from various disciplines. The general notion of community was first identified by Girvan and Newman [15] to refer to the phenomenon that nodes in many real networks appear to be grouped into subgraphs in which the density of internal connections is larger than the density of connections with the rest of the nodes in the network. Few research works are directly related to ours. Most community detection methods are based on network structures.

Graph partitioning approaches seek to partition a graph into disjoint subgraphs, such as min-cut or the betweenness [3] algorithm. Recently, Attractor was proposed [16] as a distance-dynamic community detection algorithm. Attractor characterizes the target network as an adaptive dynamical system in which the distances among nodes that belong to the same community tend to decrease, while the distances among those in different communities gradually increase. Finally, the communities can be easily obtained by removing the edges with maximal distances.

Many community detection works follow another technique, which is based on the optimization of various objective functions, such as modularity [9], description code length [11], and block model likelihood [17–19] etc. An optimal value of those objective functions represents a good community division. Modularity, the most widely used objective function, relies on comparing the strengths of inter-community and intracommunity connections with those of a null model in which edges are randomly rewired. Optimizing modularity is an NP-hard problem. Many algorithms seek to optimize the modularity scores with heuristic methods, such as tabu [6], external optimization (EO) [7], and the spectral algorithm (SP) [9]. For large-scale networks, the Louvain algorithm [10] starts with each node in a network belonging to its own community, and nodes are moved from one community to another iteratively until there is no further increase in the modularity score. Different from the Louvain method, the smart local moving (SLM) approach [4] starts with the community structure that is obtained in the first iteration of the algorithm. The authors of [11] used the description code length as objective function, which is reputed to be the best method for community detection. In the Infomap technique, each node in the network is encoded with a Huffman code according to its own community. Then, the community detection problem becomes the problem of minimizing the description code length. The authors of [8] presented an effective general search strategy for the joint optimization of various objective functions, including modularity and description code length.

In addition to the above discrimination models, generation models can be used for community detection. For instance, the block model is a generation model [17–19], in which edges are added between its nodes with certain probabilities. Each probability is determined by the probability that two nodes that generate edges belong to the same community. LDA-based methods [20–22] aim to find sets of nodes that correspond to similar "topics" and have links among one another. These topic models are based on the paradigm in which community member nodes generate links and node attributes. However, these methods assume soft community memberships, leading to unrealistic assumptions about the structure of community overlaps. The optimization of modularity maximization models or stochastic models for community detection can be achieved by eigenvalue decomposition, which is equivalent to reconstructing a low-rank modularity matrix. Thus, the approach presented in [23] optimizes the stochastic block model via a nonlinear reconstruction method that uses deep neural networks to best represent and reconstruct the network topology.

In recent years, node embedding has emerged as a promising method for community detection. The node2vec technique [24,25] conducts a biased random walk on a graph to obtain the context for each node and generates continuous feature representations for nodes. Additionally, node content has been used as an important feature for specific community detection [12,13].

3   Methodology

It is a general phenomenon that scholar articles cite previous research. When authors write manuscripts and present the lines of related work, the same categories of techniques are summarized and reviewed together. When the research topic changes, authors may start a new paragraph or sentence. Based on this intuition, we argue that researchers and techniques of similar areas are arranged nearby in the lines of scholar texts. Thus, we can quantify researcher relatedness for detecting hidden communities in an academic corpus.

3.1   Overall Framework

Figure 2 illustrates the framework of our method, which consists of reference entity recognition, researcher relatedness quantification and community clustering.

Researchers are the basic elements of research communities. In the reference entity recognition step, an academic corpus is converted to plain text. We recognize the "Reference" zone of each academic article and extract each reference as an individual process unit. Subsequently, the reference entities, including researcher names, venues, and publication data, are recognized and disambiguated.

In the next step, i.e., the researcher relatedness quantification step, each researcher mentioned in the corpus, particularly in the "Related Work" section, is identified. Then, the key issue is to quantify each researcher with proper representations. The citation linking process helps each researcher find the exact positions where he or she is mentioned in the corpus. The research embedding process represents each researcher using an implicit vector.

Finally, MDs are used to detect researcher communities, and a clustering algorithm is used to identify the researcher communities.

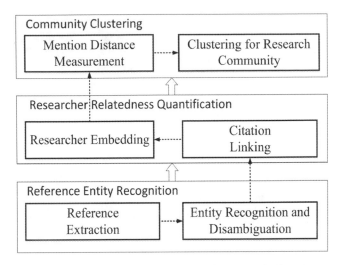

Fig. 2. Overall framework of our method.

3.2   Reference Entity Recognition and Disambiguation

This section presents the details of reference entity recognition and disambiguation.

Reference Segmentation. We design several regex expressions for recognizing each individual reference string from plain texts. Each reference contains the researcher names, publication data and venue. Different types of reference entities are segmented with dots. Time expression is the easiest entity to identify. Thus, the publication year serves as the anchor for distinguishing each individual reference. Features around the anchor are used to identify the references; for example, the first line of a reference typically starts with author names, which start with capital letters, while the final line of a reference contains the venue name, volume or conference abbreviation and ends with a dot.

Reference Entity Recognition (RER). The references in plain texts are divided into individual references after reference segmentation. Reference entities include the researcher names, publication years, titles and venues. We highlight these entities in Fig. 3 with different colors. The blue token combination represents the researcher names, and the green token combination represents the publication years. The article titles are emphasized in red, and the venues in purple.

To recognize these mentioned entities, each token in the reference string should be labeled. Formula 1 expresses our labels, which consist of *"first name"*, *"surname"*, *"title"*, *"venue"*, *"year"* and *"others"*.

$$Y = \{firstname, surname, title, venue, year, others\} \tag{1}$$

A token's label depends not only on its current status but also on the labels of surrounding tokens. The conditional random field (CRF), which is a

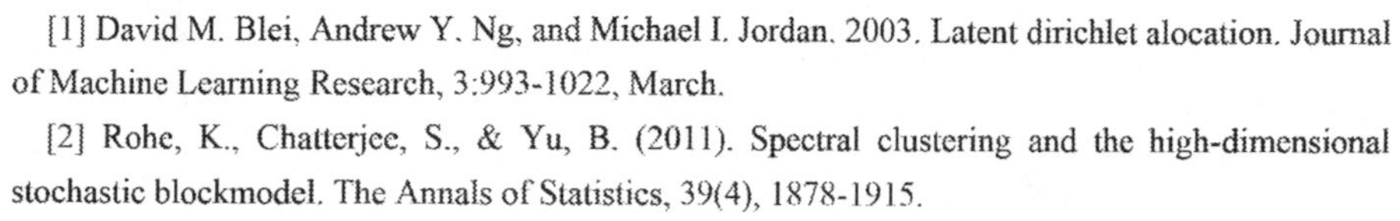

[1] David M. Blei, Andrew Y. Ng, and Michael I. Jordan. 2003. Latent dirichlet alocation. Journal of Machine Learning Research, 3:993-1022, March.

[2] Rohe, K., Chatterjee, S., & Yu, B. (2011). Spectral clustering and the high-dimensional stochastic blockmodel. The Annals of Statistics, 39(4), 1878-1915.

Fig. 3. Examples of raw reference tokens with label information. (Color figure online)

state-of-the-art technique for sequence labeling, is used to model the sequence of reference tokens [26]. The tokens are transformed into vectors of features before being classified using CRF. The joint probability of features of the reference tokens X and their labels Y can be formulated probabilistically shown in the following equation:

$$p(y|x) = \frac{1}{Z(x)} exp[\sum_{i,k} \lambda_k t_k(y_{(i-1)}, y_i, x, i) + \\ \sum_{i,k} \mu_k s_k(y_{(i)}, y_i, x, i)], \tag{2}$$

where $s_k(y_i, x, i)$ denotes the state function of the features of the ith token and the complete observation sequence and $t_k(y_{i-1}, y_i, x, i)$ denotes the transition function of the previous token to the ith token. $Z(x)$ is the normalization factor and λ_k and μ_k are the corresponding weights, which can be estimated using the logarithmic maximum likelihood approach.

The features of each token are character strings, digits or lowercase/uppercase letters depending on whether the token is a specific character (e.g., a dot, square bracket, comma or dash), a specific word in a dictionary of cities, or a specific word in a dictionary of journal titles and conference names.

After each token is labelled using the CRF classifier, the Viterbi algorithm is used to assemble each reference entity. In the case of title or source labels, neighboring tokens with the same label are concatenated.

Researcher Name Disambiguation (RND). In this section, we develop a clustering-based RND method. Due to different writing habits, it is common for one researcher to be mentioned in different forms. For instance, researcher "*David M. Blei*" is often cited with various aliases, such as "*Blei, David M.*", "*Blei, D. M.*", and "*David Blei*". Additionally, a more confusing problem may be "Name Blindness", particularly for Chinese names. Many different persons share the same surname. Therefore, many researchers are cited using the same forms, e.g., both "*Zhang, Jianfeng*" and "*Zhang, Jie*" may be cited as "*Zhang, J.*".

We define the problem as follows: Each researcher R_i is cited with several aliases $\{A_{R_i 1}, A_{R_i 2}, \ldots, A_{R_i n}\}$. Different researchers R_i and R_j may share an alias A_i. Given multiple similar aliases $A = \{A_1, A_2, \ldots, A_i\}$ and their corresponding references $C = \{C_1, C_2, \ldots, C_i\}$, our goal is to determine which aliases belong to the same authors. We regard this as a clustering problem and we group all the aliases in A into different clusters. Aliases in the same cluster refer to the same researcher.

Two principles are used to identify the confusing names before clustering. (1) Abbreviated names are regarded as referring to the same person if they collaborate with at least one common author, e.g., "*Zhang, M.*" and "*Zhang, Ming*" are considered to refer to the same person if they both collaborated with "*Wang, Wei*". (2) The middle name is often omitted in the reference, e.g., "*David Blei*" is often used to refer to "*David M. Blei*". We regard such a name as equal to a fully expressed name if both entries have the same first name and the same surname.

In the clustering process, the only contextual information of A_i is its reference C_i. Our assumption is that the same researcher tends to focus on a similar research area, collaborate with similar coauthors, and publish articles in similar venues. The context feature of each alias A_i is shown in Eq. 3.

$$F(Context_{A_i}) = \{F_{author}, F_{Venue}, F_{Domain}\} \tag{3}$$

In detail, the context feature $F(Context_{A_i})$ is the collection of three features. First, we learn the word embeddings using word2vec from the complete academic corpus. Entities such as *venue* and *author* are viewed as words and assigned unique vector representations.

F_{author} is the feature representation of the authors that occur in each reference C_i. First, we construct an author dictionary from the complete set of references, and F_{author} is filled with the embeddings of authors in reference C_i according to the dictionary order; the empty authors in the dictionary are padded with 0s. Additionally, we filter out the authors whose names appear fewer than 2 times. Thus, we can guarantee that F_{author} is not excessively sparse and that the dictionary can cover all possible coauthors. F_{Venue} is formed directly with the *venue* embedding. F_{Domain} is the domain feature of A_i, and we use *title* as the domain representation. In natural language processing, averaging word vectors is the most common approach for sentence representation. We average the word embeddings in *title* of C_i as F_{Domain}.

Finally, the context vectors $F(Context_{A_i})$ are clustered based on the cosine distance, and the aliases in the same clusters are considered the same researcher. In the end, we rename the confusing researcher names by unified identification.

Citation Linking. Regex expressions are used to recognize each individual quotation string in lines of academic content. First, we scan the sentences in each academic article and use regex expressions to identify all sentences, including citations that contain the family name and publication year. Second, the metadata of each reference are used to identify the correct reference. Finally, the mentioned surnames are replaced by the researcher identification.

3.3   MD and Community Clustering

Inspired by the success of word embedding [27,28], we propose an embedding-based MD. Since each researcher is mentioned in content, the name of each researcher can be viewed as a special word. We leverage the representation learning approach to embed each researcher as a vector representation. A sliding window is used to estimate the relatedness of researchers.

Researcher Embedding. We assume that researchers that co-occur in the same window C are related. When the sliding window scans the whole corpus, we obtain the groups of related researchers. Similar to the continuous bag-of-words (CBOW) model in word2vec, the vectors of researchers around the centroid researcher are averaged as the context vector, and the context vector is used to predict the centroid researcher vector. Given a sequence of related researchers $R_1, R_2, R_3, \ldots, R_T$, the objective is to maximize the average log probability in formula 4 to learn the representations for each researcher R_t. T is the total occurrence of all researchers. C is the sliding-window size, which is a hyper-parameter.

$$\max \frac{1}{T} \sum_{t=1}^{T} \log p(R_t | R_{t-c}^{t+c}), \tag{4}$$

where C is the sliding-window size, R_{t-c}^{t+c} is the set researchers that are covered by the window of size C that is centered at R_t. $p(R_t | R_{t-c}^{t+c})$ is defined with the softmax function as follows:

$$p(R_t | R_{t-c}^{t+c}) = \frac{\exp(e'_{R_t}{}^\top \sum_{-c \leq j \leq c, j \neq 0} e_{R_{t+j}})}{\sum_{\theta=1}^{W} \exp(e'_{R_\theta}{}^\top \sum_{-c \leq j \leq c, j \neq 0} e_{R_{t+j}})}. \tag{5}$$

where e_R is the "input" vector representation of researcher R and e'_R is the "output" vector representations of researcher R. W is the total number of researchers in the training corpus. Stochastic gradient ascent is used to update embeddings. After the training process converges, each researcher is assigned a final vector representation, and related researchers are mapped to similar positions in the vector space.

Mention Distance. The cosine distance reflects the diversity of vectors via vector angles, which can reflect similarity of researchers with different degree of representation values. The Euclidean distance measures the difference between researchers using the absolute distance in the vector space and enables the researcher vectors to become weakly discriminating. Thus, we measure the MD by the cosine distance of vectors after all researchers are embedded into vectors. The MD between researcher i and researcher j is expressed in the following equation:

$$Dis_{mention} = \frac{R_i R_j}{|R_i||R_j|} = \frac{\sum_{k=1}^{n} w_{ik} w_{jk}}{\sqrt{(\sum_{k=1}^{n} w_{ik}^2) \cdot (\sum_{k=1}^{n} w_{jk}^2)}}, \tag{6}$$

where R_i and R_j denote the vector representations of researchers i and j.

Clustering. In the end, we apply the k-means clustering approach to obtain the community memberships for each researcher. The researcher vectors and community number k are input into the clustering algorithm, and the distances of each researcher vector from the clusters are calculated to determine the final memberships.

4   Experimental Evaluation

The experiments are executed on a machine with a 3.4 GHz Intel CPU and 12 GB of internal memory.

4.1   Datasets and Baselines

Current community discovery datasets are mainly graph datasets, which are unsuitable for evaluating our experiments results. We use three real-world datasets with both academic corpora and the corresponding collaboration networks to conduct our experiments. In our collaboration network, each node represents a researcher, and the edges between nodes represent the collaboration relationships.

Relational Classification Dataset (RCD).[1] This classification dataset contains scientific papers from past ACL venues, These papers are manually classified into three research areas: "machine translation", "dependency parsing" and "summarization". All the authors of those papers who belong to the same research areas are regarded as members of the same community.

Publication Classification Dataset (PCD).[2] This classification dataset contains scientific publications from AAN, which are classified into 31 research areas using conference session information. Each conference session is regarded as a ground-truth community in this dataset.

ACL Conference Dataset (ACD).[3] In this dataset, the ground-truth community is labelled automatically. We use the same criteria that are adopted for the DBLP dataset [2] to define the ground-truth community. We choose 10 distinguishing conferences, and the researchers in the same conference are deemed to belong to the same research community.

Table 1. Statistics of three real-world collaboration networks

Dataset	Nodes	Edges
RCD	12995	39201
PCD	971	1293
ACD	11901	37133

The collaboration networks of the above three datasets are expected to contain all corresponding researchers and collaboration relationships of the articles in the RCD, PCD, and ACD corpora. We derive these networks from the

---

[1] http://clair.si.umich.edu/homepage/downloads/aan_relational/.

[2] http://clair.si.umich.edu/homepage/downloads/aan_session/.

[3] http://aclweb.org/anthology/.

AAN collaboration network[4]. Specifically, we traverse the entire AAN collaboration network and extract all edges that contains the authors in the above three datasets. Thus, we obtain the subgraphs that correspond to the three datasets after removing all noise edges and invalid nodes. The statistics of the three real-world collaboration networks are shown in Table 1.

It is highly possible that some researchers are never mentioned in our closed datasets. To use additional quoted sentences and ensure that the researchers are assigned the most reasonable possible vectors, we use the complete ACL Anthology corpus to train our researcher vectors and the k-means algorithm to detect research communities. In the RCD dataset, we set the number of communities k as 3; in the PCD dataset, we set k as 31; and in the ACD dataset, we set k as 10. For fair comparison, we conduct these baseline experiments in the corresponding collaboration networks. We compare our method with state-of-the-art approaches from several categories.

Graph Partitioning. These methods seek to partition a graph into disjoint subgraphs as communities, which include: Attractor [16], the EO algorithm [7], the Louvain algorithm [10], the SLM algorithm [4], and the fast Newman (FN) algorithm [5].

Graph Clustering. These methods embed nodes in graphs as real-valued representations and find communities via clustering, such as SP [9] and node2vec-based approaches [24, 25].

Graph Encoding. Infomap [11] encodes each node in the network using huffman code according to its community for the optimization of the description code length.

Hybrid Method. Combo [8] uses a general search strategy for the joint optimization of modularity and description code length.

4.2   Results and Discussion

Normalized Mutual Information (NMI). We use the NMI as the overall performance measure of various community detection methods. The NMI is more thorough than simply counting the number of misclassified nodes.

$$NMI = \frac{2I(X;Y)}{H(X) + H(Y)}, \qquad (7)$$

where X and Y denote two discrete variables, namely, the ground-truth community labels and the predicted community labels, respectively, $I(X;Y)$ denotes the mutual information of X and Y, and $H(X)$ and $H(Y)$ represent their corresponding entropies.

Table 2 shows that our method outperforms the state-of-the-art community detection approaches. On the PCD dataset, our method achieves approximate results with several state-of-the-art approaches. On the RCD and ACD datasets,

---

[4] http://clair.eecs.umich.edu/aan/index.php.

Table 2. Comparison of NMI

Category	Methods	PCD	RCD	ACD
Graph partitioning	Attractor	0.148	0.218	0.097
	EO	**0.257**	0.151	0.100
	FN	0.247	0.117	0.079
	Louvain	0.250	0.134	0.088
	SLM	0.248	0.122	0.083
Encoding	InfoMap	0.150	0.410	0.082
Hybrid	Combo	0.247	0.003	0.023
Graph clustering	SP	0.187	0.071	0.044
	Node2vec	0.233	0.397	0.124
Our method		0.248	**0.473**	**0.167**

our method achieves a remarkably high NMI. The reason could be that our method captures the fine-grained researcher relatedness in scholarly content, while the network-based methods leverage predominantly the document-level researcher relatedness. Our experiments illustrate that the scholar corpus constitutes another type of reliable knowledge source for research community detection. In the research embedding process, we test various values of sliding-windows size C, ranging from 8 to 20, to obtain different research vectors, and we find that the community detection performances are similar, as shown in Fig. 4(a).

Accuracy. The Como method, the node2vec-based method, and our method can specify the number of predicted communities. We can further calculate the accuracy of their results using the ground-truth communities. We use the Hungarian method to find the optimal assignment between the predicted communities and the ground-truth communities. Subsequently, we calculate the accuracy of the predicted communities. Figure 4(b) illustrates the accuracy comparison of our method and two baseline methods. We achieve relatively high accuracies, i.e., 0.479 and 0.348, on the RCD and ACD datasets, respectively.

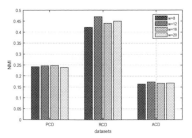

(a) NMI obtained at various parameters

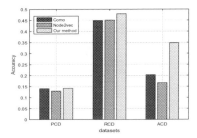

(b) Comparison of Accuracy

Fig. 4. Performance of our methods

Runtime. Our approach is relatively efficient. As shown in Table 4, Louvain and SLM are the fastest approaches. Our approach follows them closely, within a few seconds. In the PCD dataset, which is a relatively small dataset, our method is slower than Louvain, SLM, and InfoMap but is quicker than most of the network-based methods. The embedding-based method is competitive with ours. InfoMap is relatively fast because node encoding does not require a large amount of computing resources. The SP method requires the process of matrix factorization and, thus, consumes more time. Our method is based on the numerical calculation of vectors. Therefore, this method is fast and requires fewer computational resources than most baselines.

Table 3. Comparison of elapsed time

Category	Methods	PCD	RCD	ACD
Graph partitioning	Attractor	>2 h	>20 h	>20 h
	EO	1.9 s	>11 h	>13 h
	FN	2.2 s	>2 h	>2 h
	Louvain	<0.1 s	<0.1 s	<0.1 s
	SLM	<0.1 s	1.0 s	1.0 s
Encoding	InfoMap	<0.1 s	10 s	9.7 s
Hybrid	Combo	6.0 s	362 s	696 s
Graph clustering	SP	11.0 s	769 s	997 s
	Node2vec	0.15 s	0.27 s	7.68 s
Our method		0.34 s	0.28 s	7.46 s

Impact of RND. To evaluate the effectiveness of RND, we manually label the ground-truth data for eight authors. We use the method in Sect. 3.2 and k-means clustering to disambiguate the aliases. The precision, recall and F-1 score are shown in Table 3. We find that the coauthor and venue features are the most distinguishing features. The venue features are less distinguishing due to the challenge of effective similarity measurement of venues. The context information is insufficient due to the limitations of each individual reference. If we could obtain additional contextual information, such as the author affiliation and email address, the name disambiguation performance could be improved. We conduct the comparison experiments with the author vectors with disambiguated researcher names and the author vectors without disambiguated researcher names. According to the results in Fig. 5, RND can improve the overall performance of scholar detection significantly, particularly on the RCD and ACD datasets.

Table 4. Performance of RND

Author names	P	R	F-1
Zhang, Jie	0.674	0.472	0.573
Zhang, Jianfeng	0.739	0.723	0.731
Wang, Wei	0.705	0.831	0.768
Wang, Wen	0.792	0.508	0.650
Wang, Haifeng	0.920	0.712	0.816
Blei, David M	0.921	0.887	0.904
Marcu, Daniel	0.893	0.799	0.846
Moore, Johanna	0.929	0.819	0.874
Average	0.822	0.718	0.770

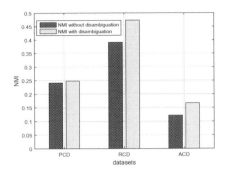

Fig. 5. Impact of RND

5 Conclusions

In this study, we develop a novel method that can detect communities from a vast amount of easily obtained text, and eliminate the dependency on structured networks. We propose and define the Mention Distance (MD), which is used to quantitatively measure the degree of relations between researchers. Our method embeds authors in a vector space using representation learning methods and efficiently detects researcher clusters via the MD. Additionally, we propose an RND method that is based on the contextual information in references, which can facilitate community detection. Our experimental results on three real-world datasets show that the proposed method has the advantages of better performance and faster processing speed than state-of-the-art approaches. Moreover, our experiments illustrate that the academic corpus is another reliable knowledge source for research community detection.

Acknowledgments. This research was supported by the Foundation of the State Key Laboratory of Software Development Environment (No. SKLSDE-2017ZX-03)

References

1. Radicchi, F., Castellano, C., Cecconi, F., et al.: Defining and identifying communities in networks. Proc. Natl. Acad. Sci. USA **101**(9), 2658–2663 (2004)
2. Yang, J., Leskovec, J.: Defining and evaluating network communities based on ground-truth. Knowl. Inf. Syst. **42**(1), 181–213 (2015)
3. Newman, M.E.J., Girvan, M.: Finding and evaluating community structure in networks. Phys. Rev. E **69**, 026113 (2004)
4. Waltman, L., van Eck, N.J.: A smart local moving algorithm for large-scale modularity-based community detection. Eur. Phys. J. B **86**(11), 471 (2013)
5. Newman, M.E.J.: Fast algorithm for detecting community structure in networks. Phys. Rev. E **69**(6), 066133 (2004)
6. Arenas, A., Fernandez, A., Gomez, S.: Analysis of the structure of complex networks at different resolution levels. New J. Phys. **10**(5), 053039 (2008)

7. Duch, J., Arenas, A.: Community detection in complex networks using extremal optimization. Phys. Rev. E **72**(2), 027104 (2005)
8. Sobolevsky, S., Campari, R., Belyi, A., et al.: General optimization technique for high-quality community detection in complex networks. Phys. Rev. E **90**(1), 012811 (2014)
9. Newman, M.E.J.: Modularity and community structure in networks. Proc. Natl. Acad. Sci. **103**(23), 8577–8582 (2006)
10. Blondel, V.D., Guillaume, J.L., Lambiotte, R., Lefebvre, E.: Fast unfolding of communities in large networks. J. Stat. Mech.: Theor. Exp. 2008(10), P10008 (2008)
11. Bohlin, L., Edler, D., Lancichinetti, A., Rosvall, M.: Community detection and visualization of networks with the map equation framework. In: Ding, Y., Rousseau, R., Wolfram, D. (eds.) Measuring Scholarly Impact, pp. 3–34. Springer, Cham (2014). https://doi.org/10.1007/978-3-319-10377-8_1
12. Yang, J., McAuley, J., Leskovec, J.: Community detection in networks with node attributes. In: ICDM (2013)
13. Wang, T., Brede, M., Ianni, A., et al.: Detecting and Characterizing Eating-Disorder Communities on Social Media. In: WSDM, pp. 91–100 (2017)
14. Rohe, K., Chatterjee, S., Yu, B.: Spectral clustering and the high-dimensional stochastic blockmodel. Ann. Stat. **39**(4), 1878–1915 (2011)
15. Girvan, M., Newman, M.: Community structure in social and biological networks. Proc. Natl. Acad. Sci. U.S.A. **99**(12), 7821–7826 (2002)
16. Shao, J., Han, Z., Yang, Q., et al.: Community detection based on distance dynamics. In: KDD 2015, pp. 1075–1084 (2015)
17. Zhang, H., Zhao, T., King, I., et al.: Modeling the homophily effect between links and communities for overlapping community detection. In: IJCAI, pp. 3938–3944 (2016)
18. Han, Y., Tang, J.: Probabilistic community and role model for social networks. In: KDD, pp. 407–416 (2015)
19. Jin, D., Chen, Z., He, D., Zhang, W.: Modeling with node degree preservation can accurately find communities. In: AAAI (2015)
20. Liu, Y., Niculescu-Mizil, A., Gryc, W.: Topic-link LDA: joint models of topic and author community. In: ICML (2009)
21. Balasubramanyan, R., Cohen, W.W.: Block-LDA: jointly modeling entity-annotated text and entity-entity links. In: SDM (2011)
22. Chang, J., Blei, D.M.: Relational topic models for document networks. In: AISTATS (2009)
23. Yang, L., Cao, X., He, D., et al.: Modularity based community detection with deep learning. In: IJCAI, pp. 2252–2258 (2016)
24. Ding, W., Lin, C., Ishwar, P.: Node embedding via word embedding for network community discovery. IEEE Trans. Sig. Inf. Process. Netw. **3**(3), 539–552 (2017)
25. Grover, A., Leskovec, J.: Node2vec: scalable feature learning for networks. In: KDD, pp. 855–864 (2016)
26. Peng, F., McCallum, A.: Information extraction from research papers using conditional random fields. Inf. process. Manag. **42**(4), 963–979 (2006)
27. Mikolov, T., Chen, K., Corrado, G., et al.: Efficient estimation of word representations in vector space. arXiv preprint arXiv:1301.3781 (2013)
28. Wang, D., Zhang, H., Liu, R., Liu, X., Wang, J.: Unsupervised feature selection through gram-schmidt orthogonalization - a word co-occurrence perspective. Neurocomputing **173**, 845–854 (2016)

A Hybrid Spectral Method for Network Community Detection

Jianjun Cheng[1,2], Longjie Li[1], Haijuan Yang[3], Qi Li[1], and Xiaoyun Chen[1(✉)]

[1] School of Information Science and Engineering,
Lanzhou University, Lanzhou 730000, China
`chenxy@lzu.edu.cn`
[2] Gansu Resources and Environmental Science Data Engineering Technology
Research Center, Lanzhou 730000, China
[3] Department of Electronic Information Engineering,
Lanzhou Vocational Technical College, Lanzhou 730070, China

Abstract. Community detection has been paid much attention, and
a large number of community-detection methods have been proposed
in the last decade. Spectral methods are widely used in many applications due to their solid mathematical foundations. In this paper, we
propose a hybrid spectral method to effectively identify communities
from networks. This method begins with a network-sparsification operation, which is expected to remove some between-community edges from
the network to make the community boundaries clearer and sharper,
then it utilizes an iterative spectral bisection algorithm to partition the
network into small communities, and finally some of the small communities are merged to obtain the resulting community structure. We conducted extensive experiments on five real-world networks and two artificial networks, the experimental results show that our proposed method
can extract high-quality community structures from networks effectively.

1 Introduction

Community structure is the most significant structural characteristic of many
complex networks, which means vertices in a network can be partitioned
into some groups naturally, and within-group edges are relatively denser than
between-group ones. Those vertex groups are so-called "communities", which
always correspond to functional modules of networks, such as groups of Web
pages sharing the same topics in WWW networks [15,29], pathways or modules
in metabolic networks and PPI networks [14,19,21], groups of people sharing
interests or activities [13,23,26,31] in real social systems, and so forth.

Hence, we can explore structural properties of networks via detecting communities from them. And to some extent, the functional properties of a network
are determined by the interactions of the communities, so that we can further
predict the functional characteristics of networks through community detection.
Therefore, community detection is of great importance not only in theoretical
studies, but also in practical applications.

© Springer International Publishing AG, part of Springer Nature 2018
Y. Cai et al. (Eds.): APWeb-WAIM 2018, LNCS 10987, pp. 90–104, 2018.
https://doi.org/10.1007/978-3-319-96890-2_8

A great deal of community-detection methods have been proposed in recent years, such as hierarchical methods [8,13,23,26,39], methods based on modularity optimization [2,8,11,23,24,27], random walk based methods [10,32,36,37], spectral methods [1,3–7,9,12,16,18,22,25,27,30,33,34,40], and so on. Among them, spectral methods are a type of ones utilize the eigenspectra of various kinds of matrices associated with networks to detect community structures, and the bisection spectral methods [4,6,25,27,30] are a special scenario, they split the network into two sub-networks according to some information of certain eigenvector's components. Owing to their solid mathematical foundations, the bisection spectral methods can extract high-quality community structures from networks. And the results are more interpretable, more persuasive than those extracted by the methods that depends only on empirical studies. However, they are limited, by and large, to the division of the network into only two sub-networks. For networks containing more than two communities, a naïve idea is to bisect the two sub-networks recursively to get the resulting community structure, but the quality of the results are not guaranteed [25,27].

However, there are some works try to eliminate the limitations for the aforementioned advantages of spectral methods. For instance, Newman [24,27] proposed a spectral bisection method, which first splits the network into two sub-networks according to the signs of elements of the leading eigenvector of the modularity matrix, then the subsequent bisections are based recursively on the signs of elements of the leading eigenvector of the generalized modularity matrices, rather than on those of modularity matrices directly. Cheng et al. [6] proposed a divisive spectral community-detection method, DBSCD, which begins with a network-sparsification algorithm to remove some between-community edges from the network to make the community boundaries clearer and sharper, then iteratively divides the network/sub-network into two sub-networks according to the signs of elements of the eigenvector corresponding to the second largest eigenvalue of the network transition matrix. The experimental results show that these efforts can promote the quality of the resulting community structures effectively.

2   Motivation

The spectral bisection methods mentioned above fall into the category of divisive ones, they detect communities from networks via repeatedly bisecting the networks/sub-networks only. Therefore, once some vertices are misclassified into the wrong communities, there will be no chance that the misclassification be revised. This is a critical shortcoming for traditional spectral bisection method. For example, Fig. 1 illustrates the result extracted by DBSCD from the Santa Fe Institute scientist collaboration network [13] and the ground-truth community structure of that network. The deviation between the extracted community structure and the ground truth is apparently. The six vertices, say, '102', '103', '104', '106', '107', and '108' should be members of the same community with vertex '105' in the ground-truth community structure shown in Fig. 1(a). But as the result of the iterative bisection procedure, they are wrongly assigned to the opposite community as illustrated in Fig. 1(b).

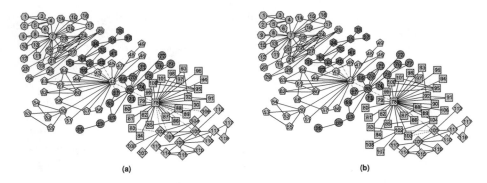

Fig. 1. The deviation between the result of DBSCD and the ground-truth community structure on Santa Fe scientist collaboration network. (a) The ground-truth community structure. (b) The community structure extracted by DBSCD. (The vertices in different communities are plotted in different shapes and shades, and this illustration style applies to the following figures as well.)

But in our experiments, we found that if we continue to divide the sub-networks further, some of the misclassified vertices, as is the aforementioned six-vertex group, might form a separate sub-network as a result of one time of bisection. Even though, the misclassified groups of vertices can not return to the correct communities to which they belong originally, because this method detects communities only using the divisive strategy.

The proposed method in this paper is motivated by these findings, in which we first continue the bisection process according to the signs of elements of the eigenvector corresponding to the second largest eigenvalue of the network/sub-network transition matrix repeatedly, until every sub-network can not be divided further, and then merge some of the sub-networks to obtain the final community structures. The intention of this merge process is to make the groups of misclassified vertices to return to their correct communities. That is to say, the proposed method detects communities not by dividing networks into sub-networks top down only, but also agglomerating small communities into larger ones bottom up. Therefore, it is a "hybrid" community-detection method.

3   The Proposed Method

The literature [6] shows that the spectral methods can benefit from the sparsification of networks, we employ the same sparsification algorithm in this paper to pre-process the network as well. Therefore, the proposed community-detection method consists of three phases in logical, the pre-processing phase to sparsify the input network, the division phase to split the sparsified network into small intermediate communities by repeatedly bisecting the network/sub-networks according to the signs of elements of the eigenvector corresponding to the second largest eigenvalue of the network transition matrix, and the merge phase to get the resulting community structure. The pseudo-code outlining the framework of the proposed method is listed in Algorithm 1.

Algorithm 1. The framework of the proposed community-detection method

---

Input: $G(V, E)$, network; θ, similarity threshold; K, the number of communities
Output: CS, the extracted community structure

1 $G \leftarrow$ network_sparsification(G, θ)
2 $CS \leftarrow$ division(G)
3 $CS \leftarrow$ merge(G, CS, K)

4 **return** CS

The networks involved in this paper are the undirected and unweighted simple graphs, which can be represented as $G = (V, E)$, where V and E stand for the vertex set and edge set, respectively; the numbers of vertices and edges are denoted as $n = |V|$, and $m = |E|$, respectively. Algorithm 1 begins with function "network_sparsification()", which is corresponding to the pre-processing operation. Its basic idea is to calculate similarities between end vertices of edges, and edges which are incident to vertices whose similarities are smaller than the given threshold, θ, are removed from networks, but some cases need to be considered specially. The logic of the sparsification is presented as Algorithm 2.

Algorithm 2. The network-sparsification algorithm: network_sparsification(G, θ)

Input: $G(V, E)$, network; θ, similarity threshold
Output: G, the sparsified network

1 **foreach** $(u, v) \in E$ **do**
2 $\quad$ $d_{min} \leftarrow$ min$\{d_u, d_v\}$
3 $\quad$ **if** $d_{min} \leqslant 2$ **then**
4 $\quad\quad$ continue   /* C programming language style continue */

5 $\quad$ **if** $d_{min} = 3$ **then**
6 $\quad\quad$ $x \leftarrow arg\,min_w \{d_w | w \in \{u, v\}\}$
7 $\quad\quad$ **if** $max\{d_w | w \in N(x)\} \leqslant d_{min}$ **then**
8 $\quad\quad\quad$ continue   /* do not remove the edge from the network */

9 $\quad$ **if** $(sim(u, v) < \theta)$ *and* $(sim(v, u) < \theta)$ **then**
10 $\quad\quad$ G.remove_edge(u, v)

11 **return** G

In Algorithm 2, d_u, d_v and d_w are the degrees of vertices u, v and w, respectively; $sim(u,v)$ is the similarity between vertices u and v, which is calculated as $sim(u,v) = \frac{|N(u) \cap N(v)|}{|N(u)|}$, where, $N(u)$ and $N(v)$ are the neighbor-vertex sets of u and v, respectively. Steps 3 through 8 are the special considerations, which ignore edges connecting too small degree vertices, and keep them in the network untouched. The intention of these operations is to keep the network from being partitioned into too small pieces at the pre-processing phase.

Then, the function "division()" in Algorithm 1 is responsible for splitting the sparsified network into small sub-networks. This task is accomplished by repeatedly bisecting the network/sub-networks according to the signs of elements of eigenvector corresponding to the second largest eigenvalue of the network transition matrix. The rationale behind this procedure is as below.

In the literature [25], Newman derived the formulas to present the mechanism of his spectral bisection method starting from optimizing the modularity, and achieved the formula

$$Q = \frac{\lambda}{2}, \tag{1}$$

where λ is the eigenvalue of the generalized eigenvector equation with constraints

$$\boldsymbol{As} = \lambda \boldsymbol{Ds}, \tag{2}$$

$$s.t. \ \boldsymbol{M}^T \boldsymbol{s} = 0,$$

where $\boldsymbol{A}$ is the network adjacency matrix; $\boldsymbol{D}$ is a diagonal matrix with element $D_{ii} = \sum_{j=1}^{n} A_{ij}$; $\boldsymbol{M}$ is a vector with element $M_i = D_{ii}$; and $\boldsymbol{s}$ is the solution vector whose element $s_i = \pm 1$ indicates the community membership of the corresponding vertex, i.e., $s_i = +1$ is to classify vertex i into group 1, and $s_i = -1$, into group 2.

For our proposed method, we should take the largest eigenvalue of Eq. (2) to be the value of λ in principle, so that the modularity Q can be maximized. But it is not allowed, since $\boldsymbol{s} = (1, 1, \cdots, 1)^T$ is obviously a solution of Eq. (2), and according to the Perron-Frobenius theorem, it is the eigenvector corresponding to the largest eigenvalue. But, this vector does not satisfy the constraint, $\boldsymbol{M}^T \boldsymbol{s} = 0$. Therefore, we have to take the second best choice to have the second largest eigenvalue to be the value of λ, and the corresponding eigenvector to be the solution vector. The elements of the eigenvector are float numbers, but $s_i = \pm 1$. Thus, we round the elements of the eigenvector to $+1$ or -1, this is equivalent to checking the signs of the eigenvector elements. If it is positive, we classify the corresponding vertex to group 1, otherwise, to group 2. This strategy divides the network into two sub-networks, we call it repeatedly to divide the sparsified network into indivisible small sub-networks. In the iterative procedure, we can control the sub-networks needed to be bisected to be connected, so that $\boldsymbol{D}$ is invertible. Therefore, we can rearrange the terms in Eq. (2) as

$$\boldsymbol{D}^{-1} \boldsymbol{As} = \lambda \boldsymbol{s}. \tag{3}$$

The matrix $T = D^{-1}A$ is the transition matrix of the network. Obviously, the division procedure to split the sparsified network into small sub-networks is based on it, and the splitting procedure is listed as the pseudo-code in Algorithm 3. After sparsification, the network might be disconnected. Therefore, we take each connected component as an initial community, which forms the current community structure CS. We construct a sub-network sg_i for each community $C_i \in CS$, calculate the second largest eigenvalue and the corresponding eigenvector of the sub-network transition matrix. If the sub-network is divisible, we partition it into two communities C_{i1} and C_{i2} according to the signs of the eigenvector components, substitute C_i with C_{i1} and C_{i2} by inserting them into CS and removing C_i from CS. This bisection is repeated until no community can be divided further, and CS is returned as the intermediate community structure.

Algorithm 3. The iterative spectral bisection strategy: division(G)

Input: $G(V, E)$, networks
Output: CS, the intermediate community structure containing small
        communities

1 $CS \leftarrow G$.connected_components()
2 **repeat**
3 | bSplit $\leftarrow false$
4 | **foreach** $C_i \in CS$ **do**
5 | | $sg_i \leftarrow G$.subgraph(C_i)
6 | | $T \leftarrow G$.transition_matrix(sg_i)
7 | | $(\lambda_2, s) \leftarrow$ second_largest_eigenval_eigenvec(T)
8 | | **if** *IsDivisible*(sg_i, λ_2, s) **then**
9 | | | bSplit $\leftarrow true$
10 | | | $C_{i1} \leftarrow \{v|s[v] > 0, v \in C_i\}$
11 | | | $C_{i2} \leftarrow \{v|s[v] < 0, v \in C_i\}$
12 | | | $CS \leftarrow CS \cup \{C_{i1}, C_{i2}\}\backslash\{C_i\}$

13 **until** (bSplit $= false$)
14 **return** CS

The function "IsDivisible()" in Algorithm 3 determines whether a sub-network sg_i is divisible or not. If two principles are reached [6], we must stop the bisection. When all of the transition matrix eigenvalues except the largest one are negative, the corresponding sub-network should not be divided further for violating the objective of modularity optimization [6]. And if there are some zero-value components or near-zero-value components in the eigenvector corresponding to the second eigenvalue, further bisection is not needed either, for further division might result in more than two disconnected sub-graphs. We integrate them when implementing the method in experiments.

At last, the function "merge()" in Algorithm 1 merges some of the intermediate communities to obtain the resulting community structure. In pursuit of efficiency, we take the strategy similar to those of FastQ [8, 23] to merge pairs of small communities iteratively. In each iteration, we calculate the modularity increment led to by merging a pair of communities, and join those two communities whose merge can obtain the largest modularity increment. This merge process is repeated until the community number reaches, K, the number of communities given by user. The pseudo-code of the merge phase is listed in Algorithm 4.

Algorithm 4. The community merge algorithm: merge(G, CS, K)

Input: $G = (V, E)$, networks; CS: the intermediate community structure
        consisted of small sub-networks; K, the number of resulting
        communities
Output: CS, the resulting community structure

1 calculate matrix e and vector a
2 **while** $|CS| > K$ **do**
3   **for** $\forall C_i, C_j \in CS$ **do**
4          $\Delta Q_{ij} \leftarrow 2(e_{ij} - a_i a_j)$
5      $(r, t) \leftarrow arg\, max_{i,j}\{\Delta Q_{ij} | i, j = 1, 2, \cdots, |CS|, i \neq j\}$
6      $C_{rt} \leftarrow C_r \cup C_t$
7      $CS \leftarrow CS \cup C_{rt} \backslash \{C_r, C_t\}$
8   update matrix e and vector a
9 **return** CS

In order to determine which pair of communities should be joined, we need to calculate quickly the modularity increment led to by joining two communities. We denote the number of communities contained in the intermediate community structure as k, the modularity corresponding to the community structure is defined as $Q = \sum_{i=1}^{k}(e_{ii} - a_i^2)$, where e_{ii} is the diagonal element of a $k \times k$ matrix e, whose element e_{ij} is the ratio of edges associating vertices in communities i with vertices in community j; a_i is element of a $k-$dimensional vector a, $a_i = \sum_{j=1}^{k} e_{ij}$ represents the proportion of edges connecting to vertices in community i. Based on this, the modularity increment led to by merging communities i and j into one can be deduced as $\Delta Q_{ij} = 2(e_{ij} - a_i a_j)$ [23], which can be calculated quickly. And in Algorithm 4, we calculate matrix e and vector a at the first step, and update them after joining the selected pair of communities in each iteration.

The entire merge process is similar to algorithm FastQ, but differs from the latter. The merge process here gets started from the intermediate community structure, rather than from the status that each community contains one vertex as its sole member in FastQ. Moreover, we terminate the merge process when the community number reaches the number of communities given by user, rather than all vertices are merged into one community in FastQ. In this way, the merge process here works with a higher efficiency.

4   Experiments and Analysis

We tested the performance of the proposed community-detection method on five real-world networks and two artificial networks generated using the LFR benchmark network generator software [17]. The statistical information about them is as listed in Table 1.

Table 1. The statistical information about the seven network datasets

Network	$|V|$	$|E|$	Community#	μ	Reference
Karate club	34	78	2	–	[13,26,38]
Dolphin	62	159	2	–	[20]
Risk map	42	83	6	–	[35]
Scientists' collaboration	118	197	6	–	[13]
Football game schedule	115	613	12	–	[13]
Artificial_1000	1000	15135	16	0.5	–
Artificial_10000	10000	98103	407	0.6	–

We have conducted experiments on these networks to testify the proposed method, and compared the results with those of several other popular algorithms, namely spectral clustering algorithm [28], Newman2006 [27], PPC [37], Newman2013 [25], FastQ [8,23], and DBSCD [6]. As Newman2013 can only be applied to two-community networks, we performed the experiments about it only on the karate club network and the bottlenose dolphin social network. Our proposed method has two parameters, the similarity threshold θ and the number of communities K. We adopted the parameter settings as what was used by DBSCD [6] in the experiments, i.e, for the similarity threshold θ, we set $\theta = 0.15$ for the five real-world networks, and for the other networks, the mode of similarity values in $[0.05, 0.2]$ was taken as the value of θ empirically; for the number of communities K, we set its value for each network as the number listed in the 'Community#' column in Table 1.

We evaluate the quality of the identified community structure quantitatively in terms of three evaluation metrics, modularity (Q), accuracy (A) and normalized mutual information (NMI), and the experimental results are recorded in Table 2. We report the rank for each method or algorithm on each metric per network (numbers in parentheses), and calculate a score by averaging the numbers of every method (numbers in the 'score' column in Table 2). The final rank of each algorithm is listed in the last column in Table 2, and the highest score and rank on each network are typed in bold. Below we analyze the seven networks in detail.

The Karate Club Network. This is a well-known network, which contains 34 vertices standing for members of a karate club and 78 edges representing relationships between the karate club members, respectively. A dispute between

Table 2. The experimental results of the comparison algorithms and the proposed method. (Q: modularity, A: accuracy, NMI: normalized mutual information.)

Network	Algorithm	Q	A	NMI	Score	Rank
Karate club	Ground truth	0.371	1.00	1.00		
	Spectral clustering	0.313(7)	0.912(4)	0.646(7)	6.000	7
	Newman2006	0.393(2)	0.618(7)	0.677(6)	5.000	6
	PPC	0.42(1)	0.676(6)	0.687(5)	4.000	3
	Newman2013	0.360(6)	0.971(3)	0.836(3)	4.000	3
	FastQ	0.381(3)	0.735(5)	0.692(4)	4.000	3
	DBSCD	0.371(4)	1.00(1)	1.00(1)	**2.000**	1
	Proposal	0.371(4)	1.00(1)	1.00(1)	**2.000**	1
Dolphin	Ground truth	0.373	1.00	1.00		
	Spectral	0.379(6)	0.984(2)	0.889(2)	3.333	2
	Newman2006	0.491(3)	0.484(7)	0.449(7)	5.667	7
	PPC	0.519(1)	0.597(6)	0.496(6)	4.333	6
	Newman2013	0.385(4)	0.968(3)	0.814(3)	3.333	2
	FastQ	0.495(2)	0.694(5)	0.573(5)	4.000	5
	DBSCD	0.385(4)	0.968(3)	0.814(3)	3.333	2
	Proposal	0.373(7)	1.00(1)	1.00(1)	**3.000**	1
Risk map	Ground truth	0.621	1.00	1.00		
	Spectral clustering	0.589(5)	0.833(4)	0.818(4)	4.333	4
	Newman2006	0.547(6)	0.762(6)	0.723(6)	6.000	6
	PPC	0.621(4)	0.81(5)	0.803(5)	4.667	5
	FastQ	0.625(3)	0.929(3)	0.894(3)	3.000	3
	DBSCD	0.631(1)	0.976(1)	0.956(1)	**1.000**	1
	Proposal	0.631(1)	0.976(1)	0.956(1)	**1.000**	1
Scientists' collaboration	Ground truth	0.739	1.00	1.00		
	Spectral clustering	0.695(6)	0.703(6)	0.772(6)	6.000	6
	Newman2006	0.708(5)	0.831(4)	0.834(5)	4.667	5
	PPC	0.751(1)	0.847(3)	0.877(3)	2.333	2
	FastQ	0.749(2)	0.831(4)	0.867(4)	2.333	2
	DBSCD	0.740(3)	0.949(2)	0.936(2)	2.333	2
	Proposal	0.739(4)	0.983(1)	0.968(1)	**2.000**	1
Football game schedule	Ground truth	0.601	1.00	1.00		
	Spectral clustering	0.538(5)	0.791(4)	0.908(4)	4.333	4
	Newman2006	0.493(6)	0.652(5)	0.758(5)	5.333	5
	PPC	0.588(3)	0.887(3)	0.92(3)	3.000	3
	FastQ	0.550(4)	0.583(6)	0.751(6)	5.333	5
	DBSCD	0.601(1)	1.000(1)	1.000(1)	**1.000**	1
	Proposal	0.601(1)	1.000(1)	1.000(1)	**1.000**	1
Artificial_1000	Ground truth	0.43	1.00	1.00		
	Spectral clustering	0.375(4)	0.715(4)	0.886(4)	4.000	4
	Newman2006	0.224(6)	0.439(5)	0.464(6)	5.667	6
	PPC	0.384(3)	0.927(3)	0.894(3)	3.000	3
	FastQ	0.356(5)	0.395(6)	0.671(5)	5.333	5
	DBSCD	0.422(2)	0.985(2)	0.972(2)	2.000	2
	Proposal	0.428(1)	0.996(1)	0.993(1)	**1.000**	1
Artificial_10000	Ground truth	0.397	1.00	1.00		
	Spectral clustering	0.179(6)	0.288(3)	0.555(4)	4.333	4
	Newman2006	0.187(5)	0.03(6)	0.21(6)	5.667	6
	PPC	0.301(4)	0.161(4)	0.561(3)	3.667	3
	FastQ	0.311(3)	0.032(5)	0.368(5)	4.333	4
	DBSCD	0.387(2)	0.933(1)	0.977(2)	1.667	2
	Proposal	0.397(1)	0.882(2)	0.979(1)	**1.333**	1

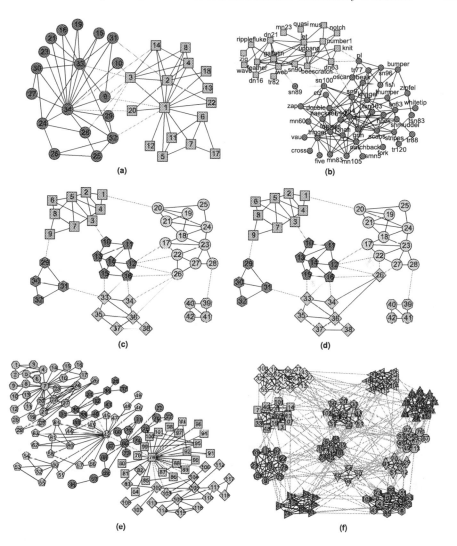

Fig. 2. The community structures detected by our proposed method from the real-world networks. (a) The extracted result from the karate club network. (b) The detected result from the dolphin social network. (c) The ground-truth community structure of the Risk map network. (d) The uncovered result from Risk map network. (e) The identified community structure from the scientists' collaboration network. (f) The extracted result from the football game schedule network. The extracted results from the karate club network, the dolphin social network and the football game schedule network are identical to the ground-truth community structures of these three networks.

the president and the instructor led to the club members to split into two fractions. Undoubtedly, the proposed method can detect the community structure from this network effectively. Figure 2(a) illustrates the result extracted by our proposed method, which is identical to the two fractions. From the perspective of evaluation metrics, the modularity of the community structure is not the largest, but both accuracy and NMI achieve their maxima, say, 1, as the identified result is identical to the ground-truth community structure. Therefore, the combination index of the three metrics, score, of the result is the highest.

The Dolphin Social Network. This is also a famous network depicting 62 dolphins living in Doubtful Sound, New Zealand. This network is constructed by taking dolphins as vertices, connecting edges between pairs of vertices who are observed being occurring together frequently, there are 62 vertices and 159 edges totally. This network is believed to break into two communities clearly. Figure 2(b) shows the extracted result by our proposed method, which is also exactly the same as the ground-truth community structure of this network. The comparison of the metrics on this network yields the similar result with those of the previous network: the modularity is not the largest, but both accuracy and NMI are 1. Therefore, the score is the highest.

Risk Map Network. This network is a political map loaded in the board game, Risk. The map is divided into 42 territories and organized as six continents. Therefore, it contains 42 vertices and 83 edges, and the 42 vertices can be grouped into six communities naturally, as illustrated in Fig. 2(c). Running the proposed method on this network, we obtained the result presented in Fig. 2(d), all but one of vertices are classified into the correct communities. The only one vertex being misclassified, vertex '26', connects to vertices located in three different communities with two edges each community, it is reasonable that vertex '26' be classified into any one of the three communities. Considering quantitatively, all of the modularity, accuracy and NMI of the community structure found by the proposed method are the largest among the comparison algorithms. Hence, our proposed method ranks the first on this network naturally.

The Scientists' Collaboration Network. This is a network describing the co-author relationships between 118 scientists working at the Santa Fe Institute, New Mexico, it consists of 118 vertices and 197 edges. According to scientists' specialities, the vertices can be divided into 6 communities, as illustrated in Fig. 1(a). This network is one of those motivated the proposed method. As mentioned previously, DBSCD misclassified 6 vertices into the incorrect communities, and to revise those misclassifications to make them return to their correct community is the original intention of the proposed method. The proposed method does work as expected, it merges the 6 misclassified vertices into the correct community, Fig. 2(e) shows the detected result. Considering quantitatively from evaluation metric perspective, both accuracy and NMI are much larger than those of comparison algorithms. This result manifests to some extent that the proposed method overcomes the shortcoming of the traditional spectral bisection methods with a high degree of success, it outperforms the comparison algorithms significantly on this network.

The Football Game Schedule Network. This network is corresponding to the Division I American college football game schedule played by 115 teams during year 2000 season, in which vertices represent teams, and edges stand for games between the two teams they associate. There are 115 vertices and 613 edges in this network, the teams is divided into 12 conferences, and games are more frequent between teams belonging to the same conference than teams of different conferences. When feeding this network as the input, the proposed method identified the result as illustrated in Fig. 2(f), which is also identical to the ground-truth community structure of this network, all values of the modularity, accuracy and NMI are the largest. Consequently, our proposed method achieves the top rank on this network too.

The Artificial Networks. The two artificial networks are both synthesized using the LFR benchmark generator [17]. This software works with some parameters tuning the characteristics of the generated networks. In this paper, we take the same parameter settings as those adopted by DBSCD, i.e., for the first network: the vertex number is 1000; the average and maximum of vertex degrees are 30 and 50; the exponent of vertex degree and community size distributions are -2 and -1; the minimum and maximum communities contain about 30 and 100 vertices, respectively; the mixing parameter μ is 0.5. For the second one, it contains 10000 vertices, the average and maximum of degrees are 20 and 50, the degree and community size distribution exponents are also -2 and -1, the minimum and maximum communities contain about 10 and 50 vertices, separately, and $\mu = 0.6$.

For the LFR network, μ is the most vital parameter, which tunes the ratio of between-community edges to total edges in networks. Obviously, $\mu = 0.5$ is a transition point, when $\mu > 0.5$, communities in networks will be fuzzy. Consequently, the above parameter settings yielded two networks whose community boundaries are not so clear. Taking them as inputs, our proposed method extracted the effective community structures from them successfully. Figures 3 and 4 illustrate the evaluation metrics corresponding to the identified community structures of the two networks, respectively.

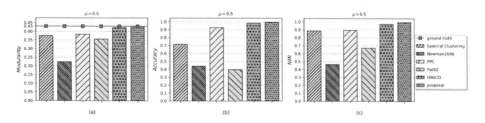

Fig. 3. The three metrics obtained from the first LFR networks. (a) Modularity (Q). (b) Accuracy (A). (c) NMI.

Among the comparison algorithms, only DBSCD can achieve good results from these two artificial networks. The results extracted by DBSCD are much better than those of other comparison algorithms: all of the modularities, accuracies and NMI's of DBSCD are much larger than those of other comparison algorithms. The advantages are more apparent especially on the second artificial network, which can be seen obviously from Fig. 4. Because the mixing parameter

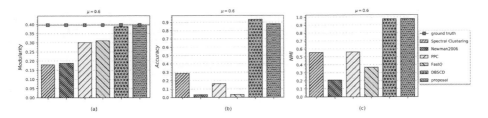

Fig. 4. The three metrics obtained from the second LFR networks. (a) Modularity (Q). (b) Accuracy (A). (c) NMI.

$\mu = 0.6$ for this network, the community structure of this network tends to be obscure. Thus, the other comparison algorithms can almost not detect the effective community structure from it. The proposed method performs better than DBSCD on both of these two artificial networks. For the first one, the superiority of the proposed method can be intuitively observed from Fig. 3 visually. For the second one, although the accuracy of the proposed method is smaller than that of DBSCD, but the modularity and NMI of the proposed method are both larger than those of the DBSCD conversely. Therefore, the combination index, score, of the proposed method is the largest once again.

5   Conclusion

In this paper, we propose a hybrid spectral method to detect communities from networks. This method overcomes the shortcoming of the traditional spectral bisection methods that they detect communities using divisive strategy only. First, it utilizes a network-sparsification algorithm to remove some between-community edges to make the community structure more prominent. Then, it divides the sparsified network into small communities using an iterative spectral bisection method, which bisects each network/sub-network according to the signs of elements of the eigenvectors corresponding to the second largest eigenvalue of the network transition matrix. At last, this method merges some of the small communities to obtain the final result.

In effect, the sparsification operation works as a pre-processing of the network, and the merge procedure takes the role of post-processing of the intermediate community structure. And just because of the existence of the pre-processing operation and the post-processing procedure, the proposed method overcomes the shortcoming of traditional spectral community-detection methods highly successfully. The experimental results show that the proposed method can identify high-quality community structures from networks effectively.

Acknowledgements. This work was partially supported by the Fundamental Research Funds for the Central Universities, China (grant ID: lzujbky-2017-sp24, lzujbky-2017-192), program of Hui-Chun Chin and Tsung-Dao Lee Chinese Undergraduate Research Endowment, CURE (grant ID: LZU-JZH1923), and the Young Scientists Fund of the National Natural Science Foundation of China (grant ID: 61602225).

References

1. Arenas, A., Díaz Guilera, A., Pérez Vicente, C.J.: Synchronization reveals topological scales in complex networks. Phys. Rev. Lett. **96**(11), 114102 (2006)
2. Blondel, V.D., Guillaume, J.L., Lambiotte, R., Lefebvre, E.: Fast unfolding of communities in large networks. J. Stat. Mech: Theory Exp. **2008**(10), P10008 (2008)
3. Capocci, A., Servedio, V.D.P., Caldarelli, G., Colaiori, F.: Detecting communities in large networks. Phys. A: Stat. Theor. Phys. **352**(2–4), 669–676 (2005)
4. Chan, T.F., Ciarlet Jr., P., Szeto, W.K.: On the optimality of the median cut spectral bisection graph partitioning method. SIAM J. Sci. Comput. **18**(3), 943–948 (1997)
5. Chauhan, S., Girvan, M., Ott, E.: Spectral properties of networks with community structure. Phys. Rev. E **80**, 056114 (2009)
6. Cheng, J., Li, L., Leng, M., Lu, W., Yao, Y., Chen, X.: A divisive spectral method for network community detection. J. Stat. Mech: Theory Exp. **2016**(3), 033403 (2016)
7. Cheng, X., Shen, H.: Uncovering the community structure associated with the diffusion dynamics on networks. J. Stat. Mech: Theory Exp. **2010**(04), P04024 (2010)
8. Clauset, A., Newman, M.E.J., Moore, C.: Finding community structure in very large networks. Phys. Rev. E **70**, 066111 (2004)
9. Donetti, L., Muñoz, M.A.: Detecting network communities: a new systematic and efficient algorithm. J. Stat. Mech: Theory Exp. **10**, P10012 (2004)
10. van Dongen, S.: Graph clustering by flow simulation. Ph.D. thesis, University of Utrecht (2000)
11. Duch, J., Arenas, A.: Community detection in complex networks using extremal optimization. Phys. Rev. E **72**, 027104 (2005)
12. van Gennip, Y., Hunter, B., Ahn, R., Elliott, P., Luh, K., Halvorson, M., Reid, S., Valasik, M., Wo, J., Tita, G.E., Bertozzi, A.L., Brantingham, P.J.: Community detection using spectral clustering on sparse geosocial data. SIAM J. Appl. Math. **73**(1), 67–83 (2013)
13. Girvan, M., Newman, M.E.J.: Community structure in social and biological networks. Proc. Nat. Acad. Sci. **99**(12), 7821–7826 (2002)
14. Guimera, R., Nunes Amaral, L.A.: Functional cartography of complex metabolic networks. Nature **433**(7028), 895–900 (2005)
15. Kleinberg, J., Lawrence, S.: The structure of the web. Science **294**, 1849–1850 (2001)
16. Krzakala, F., Moore, C., Mossel, E., Neeman, J., Sly, A., Zdeborov, L., Zhang, P.: Spectral redemption in clustering sparse networks. Proc. Nat. Acad. Sci. **110**(52), 20935–20940 (2013)
17. Lancichinetti, A., Fortunato, S., Radicchi, F.: Benchmark graphs for testing community detection algorithms. Phys. Rev. E **78**(4), 046110 (2008)
18. de Lange, S., de Reus, M., Van Den Heuvel, M.: The Laplacian spectrum of neural networks. Front. Comput. Neurosci. **7**, 189 (2014). https://doi.org/10.3389/fncom.2013.00189
19. Lewis, A., Jones, N., Porter, M., Deane, C.: The function of communities in protein interaction networks at multiple scales. BMC Syst. Biol. **4**(1), 100 (2010)
20. Lusseau, D., Schneider, K., Boisseau, O., Haase, P., Slooten, E., Dawson, S.: The bottlenose dolphin community of doubtful sound features a large proportion of long-lasting associations. Behav. Ecol. Sociobiol. **54**(4), 396–405 (2003)

21. Menche, J., Sharma, A., Kitsak, M., Ghiassian, S.D., Vidal, M., Loscalzo, J., Barabási, A.-L.: Uncovering disease-disease relationships through the incomplete interactome. Science **347**(6224) (2015). https://doi.org/10.1126/science.1257601
22. Nascimento, M.C.V.: Community detection in networks via a spectral heuristic based on the clustering coefficient. Discret. Appl. Math. **176**, 89–99 (2014)
23. Newman, M.E.J.: Fast algorithm for detecting community structure in networks. Phys. Rev. E **69**, 066133 (2004)
24. Newman, M.E.J.: Modularity and community structure in networks. Proc. Nat. Acad. Sci. **103**, 8577–8582 (2006)
25. Newman, M.E.J.: Spectral methods for community detection and graph partitioning. Phys. Rev. E **88**, 042822 (2013)
26. Newman, M.E.J., Girvan, M.: Finding and evaluating community structure in networks. Phys. Rev. E **69**, 026113 (2004)
27. Newman, M.E.: Finding community structure in networks using the eigenvectors of matrices. Phys. Rev. E **74**(3), 036104 (2006)
28. Ng, A.Y., Jordan, M.I., Weiss, Y.: On spectral clustering: analysis and an algorithm. In: Dietterich, T., Becker, S., Ghahramani, Z. (eds.) Advances in Neural Information Processing Systems 14, pp. 849–856. MIT Press (2002)
29. Pan, Y., Li, D.H., Liu, J.G., Liang, J.Z.: Detecting community structure in complex networks via node similarity. Physica A **389**(14), 2849–2857 (2010)
30. Pothen, A., Simon, H.D., Liou, K.P.: Partitioning sparse matrices with eigenvectors of graphs. SIAM J. Matrix Anal. Appl. **11**(3), 430–452 (1990)
31. Qin, H., Liu, T., Ma, Y.: Mining user's real social circle in microblog. In: 2012 IEEE/ACM International Conference on Advances in Social Networks Analysis and Mining (ASONAM), pp. 348–352, August 2012
32. Rosvall, M., Bergstrom, C.T.: Multilevel compression of random walks on networks reveals hierarchical organization in large integrated systems. PLOS ONE **6**(4), 1–10 (2011)
33. Shen, H., Cheng, X., Fang, B.: Covariance, correlation matrix, and the multiscale community structure of networks. Phys. Rev. E **82**, 016114 (2010)
34. Shi, P., He, K., Bindel, D., Hopcroft, J.E.: Local lanczos spectral approximation for community detection. In: Ceci, M., Hollmén, J., Todorovski, L., Vens, C., Džeroski, S. (eds.) ECML PKDD 2017. LNCS (LNAI), vol. 10534, pp. 651–667. Springer, Cham (2017). https://doi.org/10.1007/978-3-319-71249-9_39
35. Steinhaeuser, K., Chawla, N.V.: Identifying and evaluating community structure in complex networks. Pattern Recogn. Lett. **31**(5), 413–421 (2010)
36. Su, Y., Wang, B., Zhang, X.: A seed-expanding method based on random walks for community detection in networks with ambiguous community structures. Sci. Rep. **7**, 41830 (2017). https://doi.org/10.1038/srep41830
37. Tabrizi, S.A., Shakery, A., Asadpour, M., Abbasi, M., Tavallaie, M.A.: Personalized pagerank clustering: a graph clustering algorithm based on random walks. Phys. A **392**(22), 5772–5785 (2013)
38. Zachary, W.: An information flow model for conflict and fission in small groups. J. Anthropol. Res. **33**, 452–473 (1977)
39. Zarandi, F.D., Rafsanjani, M.K.: Community detection in complex networks using structural similarity. Phys. A: Stat. Mech. Appl. **503**, 882–891 (2018). http://www.sciencedirect.com/science/article/pii/S0378437118303066
40. Zhang, X., Nadakuditi, R.R., Newman, M.E.J.: Spectra of random graphs with community structure and arbitrary degrees. Phys. Rev. E **89**, 042816 (2014)

Personalized Top-*n* Influential Community Search over Large Social Networks

Jian Xu[1]([⊠]), Xiaoyi Fu[2], Liming Tu[1], Ming Luo[1], Ming Xu[1], and Ning Zheng[1]

[1] Hangzhou Dianzi University, Hangzhou, China
`jian.xu@hdu.edu.cn`
[2] Hong Kong Baptist University, Hongkong, China

Abstract. User-centered analysis is one of the aims of online community search. In this paper, we study personalized top-*n* influential community search that has a practical application. Given an evolving social network, where every edge has a propagation probability, we propose a maximal *pk*-Clique community model, that uses a new cohesive criterion. The criterion requires that the propagation probability of each edge or each maximal influence path between two vertices that is considered as an edge, is greater than *p*. The maximal clique problem is an NP-hard problem, and the introduction of this cohesive criterion makes things worse, as it may add new edges to existing networks. To conduct personalized top-*n* influential community search efficiently in such networks, we first introduce a search space refinement method. We then present pruning based and heuristic based search approaches. The proposed algorithms more than double the efficiency of the search performance for basic solutions. The effectiveness and efficiency of our algorithms have been verified using four real datasets.

Keywords: Community search · Online · Pruning · Heuristic search

1 Introduction

Online community analysis aims to find communities that have certain relationships with a query node in an online manner. As the communities for different vertices residing in a network may have very different characteristics, the ability for personalized community detection, which online community search provides, is more meaningful.

In this paper, we study modeling and querying of the top-*n* influential communities to a specific query node, termed as personalized top-*n* influential community search. As the influential communities around a user represent the social contexts for that user, top-*n* influential community search provides a useful tool for other analytical tasks, such as influential social community discovery and accurate community influence modeling. The following is a example.

© Springer International Publishing AG, part of Springer Nature 2018
Y. Cai et al. (Eds.): APWeb-WAIM 2018, LNCS 10987, pp. 105–120, 2018.
https://doi.org/10.1007/978-3-319-96890-2_9

Example (User-centered influential community discovery). Users in social networks are usually surrounded by many communities. Suppose that Mr. Spike is a Twitter user. Though he does not belong to any music community yet, he is surrounded by many music fans. So, to learn which are influential music communities for Spike we conduct a top-n influential community search.

But we face challenges in this study. The first challenge is how to identify all communities around a user in evolving and dynamic social networks. The second is how to calculate the influence of all these communities efficiently.

To address these two challenges, we model a social network as a graph G with vertices representing individuals and edges representing connections or relationships between any two individuals. While influence is propagated in the network according to a stochastic cascade model, such as Independent Cascade (IC) model [5], each edge (u, v) in the network is associated with a propagation probability $w(u, v)$. With cliques [1], we propose the concept of a maximal pk-Clique community. A k-clique is a complete subgraph that includes at least k vertices, and is not contained in any other complete subgraph. A maximal pk-Clique community is a community in a social network that has edges between nodes u and v in the graph that can either be original edges in the social network or a path from u to v, and its corresponding maximum influence probability is greater than p. We argue that, with propagation probability of every edge greater than a specific value p, users are well-connected with each other in such a community. A maximal pk-Clique ensures that a discovered community is connected and cohesive. We also develop pruning based and heuristic search based algorithms to efficiently find top-n influential communities with the support of an auxiliary data structure. Our contributions can be summarized as follows:

- We propose a novel cohesive criterion to define an explicit community model in evolving and dynamic social networks;
- A search space that contains all communities surrounding a user is identified. We also propose two search approaches that reduce time complexity by more than two times, while preserving the robustness of the search approaches;
- The experimental results from four real datasets confirm that our algorithms are correct and show that the proposed algorithms significantly outperform the baseline algorithm (Table 1).

Table 1. Notations

Symbol	Description	Symbol	Description
$G(V, E)$	A graph	C_{pk}, C	A pk-Clique
	Node set V, edge set E	$D(C_i, C_j)$	Diversity of clique C_i, C_j
$w(u, v)$	Propagation probability	$\mathbb{C}$	A collection of pk-Cliques
	Edge (u, v)	$Pr()$	Aggregated influence of a node set
P_{uv}	A path between u and v	$\Gamma(v)$	The adjacent nodes of v
MIP	Maximum Influence Path	ε	A threshold
s	A querying node	V_{in}	Nodes with influence $\geq \varepsilon$
p	Propagation probability	V_{out}	Nodes with influence $< \varepsilon$

2   Preliminaries

Consider a directed graph $G = (V, E)$ with an edge labeled $w : E \rightarrow (0, 1]$, where V is a set of vertices representing users of a social network and E is a set of edges between vertices representing user to-user connections. For every edge $(u, v) \in E$, weight $w(u, v)$ denotes the propagation probability of the edge, which is the probability that v is influenced by u through the edge (u, v).

To model the influence process, we adopt the IC model [5] in this paper. For a path $P = \langle v_0, v_1, \ldots, v_k \rangle$ in G, we define the propagation probability of the path as the product of the weights of its constituent edges:

$$w(P) = \prod_{i=0}^{k-1} w(v_i, v_{i+1}) \tag{1}$$

There may be more than one path between one vertex and the other, and different paths may have a different propagation probability. We use the Maximum Influence Path (MIP) to approximate the real influence from one vertex to another within the social network. A maximum influence path from vertex u to vertex v is defined as any path P with a maximal weight $w(P)$.

Definition 1. *(pk-Clique Community)*
    Given a graph G, and two parameters p and k, a pk-Clique community is an induced subgraph $C_{pk} = (V_{pk}, E_{pk})$ of G that satisfies the following requirements:

- *Any two vertices in C_{pk} can be reached via an MIP between them in the subgraph C_{pk}, and the weight of every MIP is greater than p. And there are at least k vertices in C_{pk}.*
- *A pk-Clique community is maximal if C_{pk} is not a subgraph of any other pk-Clique community.*

The difference in this study from previous community models [7,12,14] is that, for the first time, we model the relationship between users in a community with an edge as a dynamic propagation probability. We argue this is an appropriate measurement of cohesiveness in a community.

Existing maximal clique enumeration algorithms suffer from exploring a huge search space [18]. Though the cohesiveness measurement used in pk-Clique is reasonable, the introduction of pk-Clique will add more new edges to a network, thus leading to an increased number and size of cliques. These cliques usually have high similarity, as many cliques share a large portion of vertices. In order to obtain comprehensive knowledge of communities around a query node, it is inappropriate to report all these cliques for this is redundant. We define the concept of l-diversity for returned cliques.

Definition 2. *(l-diversified and l-similar)*
    Given two maximal cliques C_i, C_j in graph G and a parameter $l(0 \le l \le 1)$. We define the diversity of two maximal cliques C_i and C_j as

$$D(C_i, C_j) = \frac{|C_i \cap C_j|}{|C_i \cup C_j|} \tag{2}$$

If $D(C_i, C_j) \geq l$, then cliques C_i and C_j are l-diversified, otherwise cliques C_i and C_j are l-similar.

Given a vertex s and a pk-Clique community in the graph, we will use aggregated influence, which is the weight union of the MIPs from nodes in a community to s, to estimate the influence from the community to s. We also use a threshold ε to filter out vertices with little influence.

To calculate the influence of a community, the problem is that a vertex can be influenced by another vertex through different paths. Many works [12] adopt a model known as an "expectation model" to simplify the paths by supposing a dependence between paths, and mostly the MIP is the path that is adopted. In this work we are mainly concerned with the relationship between two vertices. We always assume that the paths between a different pair of vertices are independent. That is, even if two paths that share a sub-path lead to the same destination. We also assume that there are virtually two independent paths from two sources to the destination. This approximation simplifies the calculation of the aggregated influence of a community.

Definition 3. *(Aggregated Influence of Community)*
Aggregated influence of a maximal pk-Clique community to a vertex s, is defined as the influence probability that s is influenced by any vertices in this community, and the weight of every maximum influence path from these vertices to vertex s is greater than ε. Denote aggregated influence as $Pr(v \mid V(C))$ by the IC model, it is calculated as

$$Pr(v \mid V(C)) = 1 - \prod_{v \in V(c)} (1 - w(P_{v \to s})))$$

(3)

Definition 4. *(Top-n Influential Community Search)*
Suppose $\mathbb{C}$ is the set of influential maximal pk-Clique communities around s. The top-n influential community search can be defined as a query to find n maximal pk-Cliques in $\mathbb{C}$,
Top-n(s) $= \langle C_1, C_2, \ldots, C_n \rangle$ where $Pr(C_1)\, Pr(C_2) > \ldots > Pr(C_n)$ if $\mid \mathbb{C} \mid > n$.

In this work, we explore top-n influential community search, and we also require that the returned communities are l-diversified.

Example 3 (Top-3 maximal pk-Cliques in Fig. 1). Figure 1 illustrates a subgraph surrounding a query node s. Suppose the vertices residing in the dark grey area have influence over s greater than ε (say, 0.2). To simplify the illustration, we set all their influence to s at 0.3; When calculating aggregated influence of a clique, we ignore the vertices residing in the light grey area with an influence of less than ε. But when we check whether a group of vertices is a pk-Clique, these vertices make sense, since a maximal pk-Clique may include the vertices where the influence over s is less than ε. Having done the search, we present all maximal pk-Cliques and their aggregated influence in Table 2. With the setting l-diversified ($l = 0.5$), we obtain cliques $\{2, 3, 5\}$, $\{6, 8, 5\}$, and $\{6, 8, 9, 7\}$.

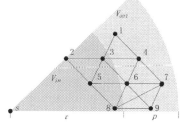

Fig. 1. A query example of node s.

Table 2. Maximal pk-cliques (example 3)

Clique ($k = 3$)	Pr
$\{2, 3, 5\}$	0.657
$\{6, 8, 5\}, \{6, 3, 5\}$	0.51
$\{6, 8, 9, 7\}, \{6, 3, 4\}, \{1, 3, 4\}$	0.3
$\{6, 4, 7\}$	0

3   Basic Solution

3.1   Search Space

In contrast to community detection that aims to find all communities in a graph, personalized search of top-n influential communities only considers communities "near" a given vertex in the social graph. We take the word "near" to mean a vertex that can be reached by a specific community with a probability above a fixed threshold. So we only examine communities that include at least one vertex such that the MIP between this vertex and the query node is above the threshold ε.

We adopt single-source shortest-path algorithms such as Dijkstra algorithm to find all vertices around a query node where the weight of their MIPs is above c. A modified version of Dijkstra algorithm is used that will stop at a vertex when the propagation probability of the MIP between this vertex and the query node is less than ε.

We denote the vertices where the propagation probability to a query node is above ε as V_{in}. From previous discussions we are also aware that a maximal pk-Clique community C, said to be influential to a query node, must include at least one vertex that is included in V_{in}. If we denote this vertex as v, to determine whether C is a maximal pk-Clique, we must examine every vertex where the propagation probability of the MIP between these vertices to v is greater than p.

We derive the following lemma.

Lemma 1. *(Given a maximal pk-Clique that has influence on s, the minimum propagation probability of an MIP between a vertex in the clique and the query node s is $p\varepsilon$.)*

Given a graph $G = (V, E)$ with influence function $w : E \rightarrow (0, 1]$, probability p, ε, and a query node s, for any vertex u, if the MIP between vertex u and node s is less than $p\varepsilon$, it will not be included by any maximal pk-Clique that has influence on s.

Proof. Suppose u is a vertex with $w(P_{us}) < p\varepsilon$. And $P_{us} = \langle u, v_1, v_2, \ldots, v_k, s \rangle$ is a maximal influential path from u to s. There will be a node $v_i (1 \leq i \leq k)$

that is the last node along path P with $w(P_{v_i s}) \geq \varepsilon$. In other words, node v_i is the last node in set V_{in} along path P.

Suppose node v_i is included in a maximal pk-Clique denoted by C and path P can be decomposed into P_{uv_i} and $P_{v_i s}$. We know that sub-paths of a maximal influential path are maximum influence paths. Then, paths P_{uv_i} and $P_{v_i s}$ are maximum influence paths. Because with $w(P_{v_i s}) \geq \varepsilon$, we have $w(P_{uv_i}) < p$. According to Definition 1, vertex u will not be included in C.

With Lemma 1 and Definition 1, we propose Algorithm 1, that returns a candidate subgraph G', including all vertices around a query node s with influence to s greater than $p\varepsilon$.

Algorithm 1. Search Space

Input:
    A graph $G = (V, E)$, vertex s and probability ε, p
Output:
    A graph $G' = (V', E')$

1: $V\_in = \{v \mid v \in V \ \& \ w(P_{vs}) > \varepsilon\}$; Set influence attribute of nodes with $w(P_{vs})$;
2: $V\_out = \{v \mid v \in V \ \& \ \varepsilon > w(P_{vs}) > p\varepsilon\}$; Set influence attribute of nodes with 0;
3: Copy $G' = (V', E')$ from G;
4: **for** each edge $(u, v) \in E'$ **do**
5:     **if** $w(u, v) < p$ **then**
6:         Delete (u, v) from G' ;
7: **for** each vertex $u \in V'$ **do**
8:     **for** each vertex $v \in \{v \mid v \in V' \ \& \ w(P_{uv}) > p\}$ **do**
9:         **if** $(u, v) \notin E'$ **then**
10:            Add (u, v) to G' ;
11: **return** G';

Algorithm 1 obtains all the vertices consisting of V_{in} and V_{out}, and any edge between these vertices remains unchanged in graph G'. Since we use a modified version of Dijkstra's algorithm to retrieve influential vertices at lines 1–2, the time complexity of identifying V_{in} and V_{out} is $O(| E' | \, lg \, | V' |)$.

3.2   Basic Algorithm

In this and following sections we will use the terms maximal pk-Clique and maximal clique interchangeably. Using social networks related to a query node s, the first step of a top-n Influential Community Search (ICS) is to identify the subgraph, including all candidate cliques, by executing Algorithm 1. Then the ICS enumerates all maximal cliques in this area. Among all these returned maximal cliques, we filter out those cliques where the size is less than k, calculate the integrated influence of each community, and then push it into a heap.

Algorithm 2. Basic Top-n ICS

Input:
   A graph $G = (V, E)$, parameter n, k, l
Output:
   A sorted list of cliques

1: Run Algorithm 1;
2: $heap = \emptyset$; $list = \emptyset$;
3: $\mathbb{C} \leftarrow \text{MCE}(V, \emptyset, \emptyset)$;
4: **for** each clique $C \in \mathbb{C}$ **do**
5:    **if** $|C| \geq k$ **then**
6:        Push($heap, C$);
7: **while** $(list.length() < n \mid heap \neq \emptyset)$ **do**
8:    $C = \text{Pop}(heap)$;
9:    **if** C l-diversified to $list[0 : (list.length() - 1)]$ **then**
10:        $list.append(C)$;
11: **return** $list$;

When all the maximal cliques have been put in the heap, the algorithm selects l-diversified top-n maximal cliques in the heap and returns the result list.

The procedure of the top-n influential community search is presented in Algorithm 2. Algorithm 2 uses the procedure, Maximal Cliques Enumerate (MCE), to generate all maximal cliques in a graph. MCE is an implementation of maximal clique enumeration method proposed by Bron-Kerbosch [1]. Its performance is discussed in [17]. MCE takes G' as input. Where G' is originally obtained from Algorithm 1. But before being fed into Algorithm 2, it is transformed into an undirected graph. After collecting all maximal cliques at line 3, Algorithm 2 pushes every maximal clique into a data structure $heap$. As its name implies, $heap$ uses the heap to manage maximal cliques during the insertion. The sort key is the aggregated influence of each clique. Lines 5–6 finish this operation. After we have the ordered cliques, the next step is to select l-diversified results. Lines 8–12 check each clique to see whether it is l-diversified with the previous appended cliques in the $list$, if it is, it will also be appended to the list.

Algorithm 2 achieves the worst-case time complexity of $O(3^{|V|/3})$ for a $|V|$-vertex graph [17]. Using d-degenerate order in the first-round iteration, the complexity is reduced to $O(|V|3^{d/3})$ [4].

4   Efficient Influential Community Search

Algorithm 2 is a functionally correct procedure, but we are unlikely to be satisfied with its performance, particularly with the large social networks that now exist.

4.1   Search Space Refinement

The first possible method for accelerating the entire work is to remove all vertices where the degree is less than k. From the perspective of each vertex in the graph, if a vertex's degree is less than k, then it absolutely will not be a member of any maximal pk-Clique. We make the following observation.

Observation 1: A vertex with a degree less than k will not be a member of any maximal pk-Clique.

Many core decomposition algorithms use a bottom-up approach to determine a vertex's class. That is, what kind of maximal clique dose this vertex belong to? We adopt such a bottom-up approach to remove all vertices, so they will not be included in any maximal pk-Clique, and conduct search space refinement. We use Algorithm 3 for this purpose.

With binary sort to order vertices, this refinement can be done in $O(|V'| + |E'|)$ time complexity.

Algorithm 3. Search Space Refinement

Input:
    A graph $G = (V, E)$, vertex s and probability ε, p
Output:
    A graph $G' = (V', E')$

 1: Same with Algorithm 1 lines 1-12;
 2: Sort vertices in G in ascending order of their degree;
 3: **while** $(G' \neq \emptyset)$ **do**
 4:     $d = $ the minimum vertex degree in G';
 5:     **if** $d < k$ **then**
 6:         **for** each $v \in \{v \mid v \in V'$ & $v.degree \leq d\}$ **do**
 7:             Delete v and all edges incident to v from G';
 8:             Re-sort the remaining vertices in G';
 9:     **else**
10:         Break;
11: **return** G';

4.2   Pruning Based Algorithm

Recalling Definition 3, the aggregated influence of the community is defined as the probability that s is influenced by any vertex in a maximal pk-Clique community. Algorithm 2 calculates the aggregated influence of each identified community. Is it possible to derive an upper bound of aggregated influence of to-be-found communities in advance? With this bound, can we prune unnecessary search branches in the graph? To this end, we propose Definition 5, following Observation 2.

Definition 5. *(Aggregated Influence of vertex set)*
    The aggregated influence of a set of vertices to a vertex s, is defined as the probability that s is influenced by any vertex in this set. The calculation is the same as that in Definition 3.

Suppose a node set V_s and all maximal pk-Clique communities it includes. Denote all these communities as $\mathbb{C}$, and for any $C \in \mathbb{C}$, we have

$$
\begin{aligned}
Pr(v \mid v \in V_s) &- Pr(v \mid v \in C\,) \\
&= 1 - \prod_{v \in V_s} (1 - w(P_{v \to s})) - (1 - \prod_{v \in C} (1 - w(P_{v \to s}))) \\
&= \prod_{v \in C} (1 - w(P_{v \to s}))(1 - \prod_{v \in V_s - C} (1 - w(P_{v \to s}))) \\
&> 0
\end{aligned}
\tag{4}
$$

With the above inequation, we obtain Observation 2.

Observation 2: The aggregated influence of a vertex set is bigger than any maximal clique residing in it.

With Observation 2, we have the upper bound of the aggregated influence of maximal cliques in a subgraph. This upper bound will help us make the right decision before exploring the branch of a search tree. The remaining question is how to calculate this upper bound.

In the previous subsection, Algorithm 3 removes the vertices where the degree was less than k. It is important to notice that it decreases the number of vertices, leading to a reduced search space, but it does not ensure that there are no maximal cliques left with a size less than k. Thus we also have a lower size bound of expected maximal cliques, that is k. Now we have Algorithm 4.

Algorithm 4. Pruning Based Top-n ICS

Input:
    A graph $G = (V, E)$, parameter n, k, l
Output:
    A sorted list of cliques

```
 1: Run Refined Searching Space;
 2: list = ∅;  cliques_new = 0;   i = 0;
 3: repeat
 4:     cliques_old = cliques_new;
 5:     queue = [];  //length fixed
 6:     MCP(V_in ∪ V_out, ∅, ∅, queue, n2^i, k);
 7:     cliques_new =clique number in queue;
 8:     while (list.length() < n | queue == ∅) do
 9:         C = pop(queue);
10:         if  C l-diversified to list[0 : list.length() − 1]  then
11:             list.append(C);
12:     i = i + 1;
13: until (list.length() == n|cliques_old == cliques_new)
14: return list;
```

Algorithm 5. MCP(P, X, C, q, n, k)

Input:
Vertex set P, X, C,parameter q, n, k
Output:
maximal cliques

1: **if** $P \cup X == \emptyset$ **then**
2:    Insert C into q;
3: Choose a pivot $u \in P \cup X$;
4: **for** each vertex $v \in P \setminus \Gamma(u)$ **do**
5:    $T = C \cup \{v\} \cup ((P \cup X) \cap \Gamma(v))$;
6:    **if** $Pr(T) \leq Pr(q[n])$ & $|T| < k$ **then**
7:      **return**
8:    MCP$(P \cap \Gamma(v), X \cap \Gamma(v), C \cup \{v\}, q, n, k)$;
9:    $P = P \setminus \{v\}$;
10:    $X = X \cup \{v\}$;

Algorithm 4, Pruning Based Top-n ICS, calls Algorithm 5, MCP (Maximal Cliques with Pruning), to complete the top-n influential community search. Then it selects l-diversified n maximal cliques. If the number of previous returned top-n maximal cliques is less than n after the selection, Algorithm 5 calls MCP once again, but this time it will enlarge the result set to twice its previous one. That is, it will set parameter n to $2n$ and pass to procedure MCP. Until we have arrived at n maximal pk-Cliques which are l-diversified in the returned *list*.

Algorithm 5 MCP takes more inputs than Algorithm 3 MCE. Parameter q keeps track of currently found top-n influential communities. MCP uses an efficient global priority queue to manage the cliques found while running it.

The second main difference between MCP and MCE is lines 5–7. Vertex set T denotes a union of vertex sets, including C, $\{v\}$, and $(P \cup X) \cap \Gamma(v)$. Set C includes the vertices that are part of maximal cliques to be found. $\{v\}$ is currently an expanding vertex. And $(P \cup X) \cap \Gamma(v)$ includes the vertices to be extended. According to Observation 2, the aggregate influence of set T is the upper bound of any pk-Cliques to be retrieved. Thus, if the aggregate influence of set T is less than the minimum aggregate influential pk-Cliques in the queue, for instance, clique $q[n]$, there is no need to undergo further searches along this branch. For the same reason, if the size of T is less than k, there is no need to conduct further searches either. This is the pruning based top-n influential community search. When a maximal clique is found, it is inserted into q. This operation is executed with line 2. At the end of MCP, we have a queue with n maximal pk-Cliques.

Theorem 1. *(Correctness of algorithm Pruning Based Top-n ICS)*
Given a graph $G = (V, E)(V \neq \emptyset)$ and a query node s, the algorithm Pruning Based Top-n ICS generates top-n l-diversified maximal pk-cliques without duplication.

Proof. It has been proofed in [17], that MCE generates all, and only, maximal cliques without duplication that contain all vertices in C, some vertices in P, and no vertices in X, without duplication. MCP follows MCE, and just skips over some branches that do not lead to top-n maximal cliques. Thus MCP returns the top-n maximal cliques with input vertex sets P, X, C as well.

With the statements at lines 10–14, the returned cliques in $list$ are l-diversified.

Now we come to analyze its time complexity. In the best case, MCP finds top-n maximal pk-Cliques first. Thus the time complexity is $O(n)$. But in the worst case, MCP finds top-n maximal pk-Cliques at the last moment. If the returned top-n maximal pk-Cliques are not l-diversified, it will call MCP once again, until n is greater than $|\mathbb{C}|$. The time complexity in this case is $O(3^{|V|/3}log_2^{|\mathbb{C}|})$.

4.3   Heuristic Based Algorithm

According to Eq. 3, if the probability distribution of each node's influence in V_{in} is uniform, then the more nodes that reside in V_{in} for a clique, the bigger the aggregated influence it will acquire. This leads to Observation 3.

Observation 3: It is generally expected that a vertex $u \in V$ such that $V = V_{in} \cup V_{out}$ and $u = max\{|V_{in} \cap \Gamma(u)|\}$ has a high probability of being included in an influential maximum clique.

With Observation 3, we propose Algorithm 6, Heuristic MCP(HMCP). To minimize $P \setminus \Gamma(u)$, MCP chooses a pivot $u \in P \cup X$ that maximizes $|P \cap \Gamma(u)|$ at line 3 of Algorithm 5. In contrast to MCP, HMCP chooses a pivot that maximizes $|V_{in} \cap P \cap \Gamma(u)|$. That is, expending of a subtree that has a high probability of leading to influential maximum cliques becomes a priority.

The heuristic based top-n ICS maintains an additional list which keeps the sets of all vertices that are in V_{in} and adjacent to v in $G' = (V', E')$. Note that the rather time-consuming calculation at this step is carried out only at the beginning of the main program and not in HMCP. Therefore, the total time required to select the pivot is very small.

Algorithm 6. HMCP(P, X, C, q, n, k)

Input:
    Vertex set P, X, C,parameter q, n, k
Output:
    Maximal cliques

1: **if** $P \cup X == \emptyset$ **then**
2:    Insert C into q;
3: Choose a pivot $u \in P \cup X$
           where $u = max\{|V_{in} \cap P \cap \Gamma(u)|\}$
4: Same with Algorithm 5 lines 4-10;

Example 4 (An execution of a heuristic search). With the settings left the same as in Example 3, it is easy to find a heuristic based search that will perform the outermost recursive calls in the order of $\{5, 7, 1\}$, instead of $\{6, 2, 1\}$ in MCP.

5   Experimental Study

In this section, we study the performance of the proposed algorithms over four real datasets. All the algorithms are implemented with Python2.7 and run on a CentOS server (Intel i7-7700 3.6 GHz CPU and 32 GB RAM).

A. Data sets

We used four real social network datasets available at https://snap.stanford. edu. For each dataset, we selected one of the vertices with the most neighbors, that is, vertex 2070 in Facebook, vertex 349932090 in Twitter, vertex 104***590 in Google+, and vertex 1 in Epinions as query nodes. For those with no propagation probability with each edge in the original datasets, we set a random value which uniformly distributed $(0, 1)$ to each edge and took this value as their propagation probability. Experiments in this work focused on the scenario when $\varepsilon = 0.2$, $p = 0.6$, and $n = 3$.

B. Evaluation of the maximal pk-Clique model

The Label Propagation Algorithm (LPA) is considered to be an accurate model in detecting communities in social networks [19]. Wang et al. [19] also proposed that Normalized Mutual Information (NMI) [3] is a popular criterion for evaluating the accuracy of community detection models. The score of NMI stands for the agreement of two results. We compute the score of NMI for the results of our maximal pk-Clique model and the LPA model in this experiment. The scores we achieved were about 83% for Twttier, 58% for Google+, 67% for Facebook, and 63% for Epinions.

C. The necessity to l-diversify

Figure 2 presents the redundancy over different maximal pk-Clique sizes. It indicates that similarity increases as the size of maximal clique increases.

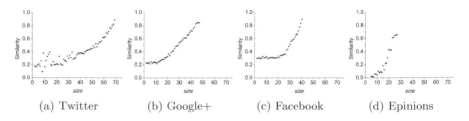

(a) Twitter   (b) Google+   (c) Facebook   (d) Epinions

Fig. 2. Similarity between communities

D. Evaluation of search space refinement

Given a social network and a parameter k, early removal of vertices where the degree is less than k will lead to a smaller search space. The results presented in Table 3 show that the refinement approach works in different networks, but the effectiveness varies and depends on the topology of the graph in an actual situation.

Table 3. Vertices removed in a refined search space

Data	Total	*k*	5	10	15	20	25	30	35	40	45	50	55	60	65	70	75
Twitter	208		0	0	2	2	1	7	8	1	5	4	4	1	1	5	3
Google+	491		29	55	55	37	35	34	24	44	50	-	-	-	-	-	-
Facebook	709		14	52	46	55	70	9	9	3	-	-	-	-	-	-	-
Epinions	49112		2.0E4	7936	2855	1491	1071	721	620	507	-	-	-	-	-	-	-

E. Efficiency of proposed algorithms

We implemented a search space refinement in all three algorithms for fair comparison between results. Tables 4, 5, 6 and 7 illustrate the results (time in seconds) obtained from different datasets, where PS denotes Pruning based top-*n* ICS, HS denotes Heuristic based top-*n* ICS and BS denotes Basic Search algorithm.

Table 4. Efficiency over Twitter

k	5	15	25	35	45	55
PS	10.85	10.44	10.21	9.29	8.21	8.07
HS	10.55	10.30	10.26	9.28	8.13	8.01
BS	131	130	130	132	130	129

Table 5. Efficiency over Google+

k	5	15	25	35	45	55
PS	0.79	0.78	0.53	0.20	0.09	-
HS	0.78	0.77	0.52	0.20	0.04	-
BS	1.16	1.17	1.17	1.18	1.21	-

It is clear that proposed pruning based PS and heuristic based HS algorithms appear far more efficient than the basic solution, both in time and space costs. It also shows that the HS performs better than the PS in most cases. We notice that time cost does not decrease dramatically as vertices are progressively removed.

Table 6. Efficiency over Facebook

k	4	8	12	16	20	24
PS	2.94	2.93	2.79	2.70	2.50	2.16
HS	2.75	2.70	2.72	2.11	2.06	1.82
BS	15.24	14.92	15.17	15.30	14.94	14.73

Table 7. Efficiency over Epinions

k	4	8	12	16	20	24
PS	9.06	8.92	8.83	8.84	8.53	8.42
HS	9.00	8.79	8.68	8.75	8.45	8.37
BS	20.41	19.94	19.37	19.36	19.36	18.76

Tables 8, 9, 10 and 11 compare the space cost (in kbytes) of different algorithms. The results show that the PS and the HS require less memory than the BS in each case.

Table 8. Space cost over Twitter

k	5	15	25	35	45	55
PS	725	721	712	696	406	<10
HS	720	704	672	634	303	<10
BS	2.0E6	2.0E6	2.0E6	2.0E6	34271	315

Table 9. Space cost over Google+

k	5	15	25	35	45	55
PS	602	<10	<10	<10	<10	-
HS	613	<10	<10	<10	<10	-
BS	11280	885	573	500	<10	-

Table 10. Space cost over Facebook

k	4	8	12	16	20	24
PS	704	705	18	32	<10	<10
HS	646	610	16	<10	<10	<10
BS	2.1E5	1.9E5	6935	4739	5722	5673

Table 11. Space cost over Epinions

k	4	8	12	16	20	24
PS	18165	<10	<10	<10	<10	<10
HS	13656	<10	<10	<10	<10	<10
BS	1.0E5	213	<10	<10	<10	<10

F. Evaluation of scalability

To further test the performance of proposed algorithms, we enlarged the search space over Facebook and Epinions datasets. The results are presented in Tables 12 and 13. The cost increases as the number of vertices increases in both cases.

Table 12. Scalability over Facebook

Vertices	97	147	616	744
PS	0.0027	0.005	4.61	1189.98
HS	0.0023	0.004	4.54	1179.98
BS	0.0048	0.108	31.75	24583.70

Table 13. Scalability over Epinions

Vertices	4104	20151	43892	65169
PS	6.22	9.00	9.24	9.54
HS	6.19	8.86	9.11	9.27
BS	14.84	20.22	20.63	21.25

G. Case Study

We built a co-author network from the DBLP (DataBase systems and Logic Programming) data set for this case study. The diameter of the network is 6. A vertex represents an author and an edge is added between two authors if they are co-authors. The propagation probability is calculated according the frequency of co-authorships. We performed a top-3 ICS ($\varepsilon = 0.2, p = 0.8, k = 5$) query for "Alexanderm T", the results are shown in Table 14.

Table 14. Top-3 influential communities for Alexanderm T

n	Pr	Members
1	0.92	Chad D, Djoerd H, Ivan K, Jaap K, Julia K, ller, Lucas B
2	0.83	Chi W, Clare R.V, Fangbo T, Heng J, Jialu L, Jiawei H, Lance M.K, Xiang R
3	0.79	Bo Z, Jiawei H, Jing G, Lu S, Qi L, Wei F, Yaliang L

6   Related Work

The detection of community, which is defined as natural divisions of network nodes into densely connected subgroups [15], has been widely studied in biological networks and social networks [8,15,16,20].

Concepts relating to graph properties like k-clique, k-core and so on have been extensively studied in random graphs. Recently, implicit community models k-core [14], kr-Clique [12] have been proposed to discover communities.

An online community search, which finds communities around a query vertex online, recently attracted a lot of attentions, for example, [2].

In considering models for the spread of influence through a social network, a dynamic cascade model called the Independent Cascade (IC) model, was first investigated by Goldenberg et al. [5]. Granovetter et al. [6] were among the first to propose another Linear Threshold (LT) model. Kempe et al. [9] discussed maximizing the spread of influence problem with these models in social network analysis. Based on these works, there appears to be a lots of proposals focusing on the spread of influence problem, for example, [10,11,13].

7   Conclusions

In this paper, we study the personalized top-n influential community search problem in social networks. In particular, we propose a novel community model based on the maximal pk-Clique concept which uses a new cohesive criterion. We first present search space refinement. Then we introduce a pruning approach and a heuristic search approach. Extensive experiments over real social networks verify the effectiveness and efficiency of our search algorithms.

Acknowledgment. This work is supported by the National Natural Science Foundation of China (No. 61572165), the Natural Science Foundation of Zhejiang Province (No. LZ15F 020003). Xiaoyi Fu's work is supported by Hong Kong Research Grants Council (No. 12200817, 12201615 and 12258116).

References

1. Bron, C., Kerbosch, J.: Algorithm 457: finding all cliques of an undirected graph. Commun. ACM **16**(9), 575–577 (1973)
2. Cui, W., Xiao, Y., Wang, H., Lu, Y., Wang, W.: Online search of overlapping communities. In: Proceedings of the ACM SIGMOD, pp. 277–288 (2013)
3. Danon, L., Dazguilera, A., Duch, J., Arenas, A.: Comparing community structure identification. J. Stat. Mech. Theory Exp. **2005**(09) (2005)
4. Eppstein, D., Maarten, L., Strash, D.: Listing all maximal cliques in sparse graphs in near-optimal time. Comput. Sci. **6506**, 403–414 (2010)
5. Goldenberg, J., Libai, B., Muller, E.: Talk of the network: a complex systems look at the underlying process of word-of-mouth. Mark. Lett. **12**(3), 211–223 (2001)
6. Granovetter, M.: Threshold models of collective behavior. **83**, 1420–1443 (1978)

7. Huang, X., Cheng, H., Qin, L., Tian, W., Yu, J.X.: Querying k-truss community in large and dynamic graphs. In: Proceedings of the ACM SIGMOD, pp. 1311–1322 (2014)
8. Huang, X., Lakshmanan, L.V.S., Xu, J.: Community search over big graphs: models, algorithms, and opportunities. In: IEEE International Conference on Data Engineering, pp. 1451–1454 (2017)
9. Kempe, D., Kleinberg, J., Tardos, E.: Maximizing the spread of influence through a social network. In: Proceedings of the ACM SIGKDD, pp. 137–146 (2003)
10. Lee, J.R., Chung, C.W.: A query approach for influence maximization on specific users in social networks. IEEE Trans. Knowl. Data Eng. **27**(2), 340–353 (2015)
11. Li, H.P., Hu, H., Xu, J.: Nearby friend alert: location anonymity in mobile geosocial networks. IEEE Pervasive Comput. **12**(4), 62–70 (2013)
12. Li, J., Wang, X., Deng, K., Yang, X., Sellis, T., Yu, J.X.: Most influential community search over large social networks. In: Proceedings of the ICDE, pp. 871–882 (2017)
13. Li, R.H., Qin, L., Yu, J.X., Mao, R.: Finding influential communities in massive networks. VLDB J. **2**, 1–26 (2017)
14. Li, R.H., Yu, J.X., Mao, R.: Efficient core maintenance in large dynamic graphs. IEEE Trans. Knowl. Data Eng. **26**(10), 2453–2465 (2014)
15. Newman, M.E., Girvan, M.: Finding and evaluating community structure in networks. Phys. Rev. E **69**(2), 026113 (2004)
16. Ruan, J., Zhang, W.: An efficient spectral algorithm for network community discovery and its applications to biological and social networks. In: Seventh IEEE International Conference on Data Mining, pp. 643–648 (2007)
17. Tomita, E., Tanaka, A., Takahashi, H.: The worst-case time complexity for generating all maximal cliques and computational experiments. Theor. Comput. Sci. **363**(1), 28–42 (2006)
18. Wang, J., Cheng, J., Fu, W.C.: Redundancy-aware maximal cliques. In: Proceedings of the ACM SIGKDD, pp. 122–130 (2013)
19. Wang, M., Wang, C., Yu, J.X., Zhang, J.: Community detection in social networks: an in-depth benchmarking study with a procedure-oriented framework. Proc. VLDB Endow. **8**(10), 998–1009 (2015)
20. Zhu, Q., Hu, H., Xu, C., Xu, J., Lee, W.C.: Geo-social group queries with minimum acquaintance constraints. VLDB J. **26**(5), 1–19 (2014)

Matrix Factorization Meets Social Network Embedding for Rating Prediction

Menghao Zhang, Binbin Hu, Chuan Shi(✉), Bin Wu, and Bai Wang

Beijing Key Lab of Intelligent Telecommunications Software and Multimedia,
Beijing University of Posts and Telecommunications, Beijing 100876, China
{Jack,hubinbin,shichuan,wubin,wangbai}@bupt.edu.cn

Abstract. Social recommendation becomes a current research focus,
which leverages social relations among users to alleviate data sparsity
and cold-start problems in recommender systems. The social recommendation methods usually employ simple similarity information as social
regularization on users. Unfortunately, the widely used social regularization cannot make a good analysis of the users' social relation characteristics. In order to overcome the shortcomings of social recommendations,
we propose a new framework for which combines network embedding and
probabilistic matrix factorization. We make use of social relation features
extracted from social networks, on top of which we learn an additional
layer that uncovers the social dimensions that explain the variation in
people's feedback. Furthermore, the influence of different social network
embedding strategies on our framework are compared. Experiments on
three real datasets validate the effectiveness of the proposed solution.

Keywords: Network embedding · Matrix factorization
Social recommendation

1 Introduction

With the continuous development of the e-commerce, and the rapid growth in the
number and variety of goods, customers need to spend much time to find what
they want. The process of visiting a large number of irrelevant information will
be certainly drowned in information overload problem and it will undoubtedly
continue to lose customers. To confront these challenges, personalized recommender systems emerged. Obviously, in recommender systems, the recommendation methods are the most critical component. Collaborative filtering is one
of the most important technology of recommender systems [1]. It recommends
items, which have been evaluated positively by another similar user or by a set
of such users. Probabilistic Matrix Factorization (PMF) [2] is one of the most
successful collaborative filtering methods. And the goal of the PMF is to decompose user-item rating matrix into user factor matrix and item factor matrix. This

© Springer International Publishing AG, part of Springer Nature 2018
Y. Cai et al. (Eds.): APWeb-WAIM 2018, LNCS 10987, pp. 121–129, 2018.
https://doi.org/10.1007/978-3-319-96890-2_10

matrix factorization model performs well on the sparse, large, and imbalanced datasets and scales linearly with the number of observations.

Traditional matrix factorization methods always ignore social relationships among users. However, in our real life, when we seek advice from our friends on restaurants or books, we are actually requesting verbal social recommendations. Hence, in order to provide more personalized recommendations to improve recommender systems, we need to incorporate users' social network information. A few social recommendation methods have been proposed [3,4] on the basis of intuition that users' social relations can be employed to strengthen traditional recommender systems. And these social recommendation methods usually use user similarity and regularization constraints.

Recently, the deep learning model and social network embedding (SNE) methods have been widely studied [5–7]. Some researches apply the deep learning methods to the recommender systems [8,9]. These methods provide novel approaches for recommendation systems. However, they all have a high time complexity and space complexity and they do not make full use of social relations to strengthen the rating prediction.

In this paper, we focus on the impact of SNE on social recommender systems. We will study how to combine social network embedding methods with probabilistic matrix factorization, and propose a framework named as matrix factorization meets social network embedding for rating prediction (called MERP). The framework takes care both the accuracy and efficiency of social recommender systems. The basic idea of the method is to generate user social network features by SNE and user latent feature with PMF, then combine these two features to predict ratings. First of all, we utilize SNE in social information network to generate user social network features. Next, we put these features in optimized matrix factorization model which can learn user-item latent feature and social feature simultaneously. Finally, we combine latent feature and social feature to predict user-item rating. Our experiments show that the combination of SNE and PMF is more accurate than other rating prediction method in the social recommendation model.

2   The MERP Methods

In this section, we build our framework (MERP) which combines user social network embedding and matrix factorization method. Firstly, we describe our social network embedding model. And then we systematically interpret how to integrate social network information with rating information.

2.1   Social Network Embedding

Recent progress in representational learning for network embedding opened new ways for feature learning of discrete objects. Particularly, the Skip-gram model [10] aims to learn continuous words' feature representations by optimizing a neighborhood preserving likelihood objective. And some network embedding (also called network representation learning (NRL)) methods use Skip-gram model to learn the nodes latent factors. DeepWalk [5], LINE [7] and Node2vec [6] are three representative social network embedding methods (SNE).

One of the original methods in network embedding methods is DeepWalk. The basic steps of the algorithm are as follows: it scans over the nodes of a network, and for every node it aims to embed it such that the node's latent features can predict the nearby nodes (i.e., nodes inside some slide window). The node feature representations are learned by optimizing the likelihood objective using Skip-gram and Hierarchical Softmax. The Skip-gram objective is based on the distributional hypothesis. It states that nodes in similar places tend to have similar meanings. In other words, similar nodes tend to appear in similar node neighborhoods.

2.2 MERP

In recommender systems, an efficient and effective approach is to factorize the user-item rating matrix. Its basic formulation assumes the following model to predict the preference of a user u toward an item i:

$$\hat{x}_{u,i} = \mu + b_u + b_i + \gamma_u^T \gamma_i \tag{1}$$

where μ is global offset, b_u and b_i are user and item bias terms, and γ_u and γ_i are vectors describing latent factors of user u and item i.

We firstly introduce DeepWalk network embedding method to implement our model. In recommendation system, social relations provide an independent source for recommendation. We think of our user social relations as a network, then we utilize DeepWalk to learn a latent space representation of social interactions θ_u. Our extended predictor takes the form

$$\hat{x}_{u,i} = \mu + b_u + b_i + \gamma_u^T \gamma_i + \theta_u^T \theta_i \tag{2}$$

where μ, b_u, b_i, γ_u, γ_i are as in Eq. 1. θ_u, θ_i are social factors whose inner product models the social interaction between u and i.

One naive way to implement the above model would be to directly use social network embedding features f_u of user u as θ_u in the above equation. However, this would present issues due to the high dimensionality of the features in question. Therefore, we propose to learn an embedding kernel which linearly transforms such high-dimensional features into a much lower-dimensional (say 16 or so) 'social rating' space:

$$\theta_u^T = f_u^T E^T \tag{3}$$

Here, E is a matrix embedding DeepWalk feature space into social space and f_i is the original social feature vector for user u. The numerical values of the projected dimensions can then be interpreted as the extent to which a user exhibits a particular social rating factor. This embedding is efficient in the sense

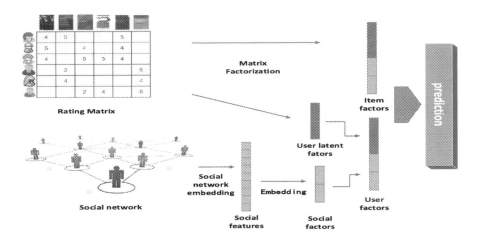

Fig. 1. Diagram of the preference predictor of MERP.

that all users share the same embedding matrix which significantly reduces the number of parameters to learn.

Next, we introduce a visual bias term b' whose inner product with f_u models items' overall property toward the social appearance of a given user. In summary, our final model is shown in Fig. 1 and our final prediction formula is:

$$\hat{x}_{u,i} = \mu + b_u + b_i + \gamma_u^T \gamma_i + (f_u^T E^T)\theta_i + f_u^T b' \tag{4}$$

2.3   Model Learning and Discussion

We blend the users' social network embeddings in to matrix factorization framework for learning the parameters of our model. And we adopt SGD to optimize the following objective:

$$\mathcal{L} = min \sum_{u,i \in U, I} (\hat{x}_{ui} - x_{ui})^2 + \lambda(\|\gamma_u\|_F^2 + \|\gamma_i\|_F^2 \tag{5}$$
$$+ b_u^2 + b_i^2) + \beta_\theta \|\theta_i\|_F^2 + \beta_E \|E\|_F^2 + \beta_b {b'}^2$$

As mentioned above, we select three different methods in social network embedding: DeepWalk, LINE, Node2vec. These three methods have their own advantages. LINE applies to large-scale information network. DeepWalk adopt random walk strategy in the network node search, while Node2vec provides two strategies: depth-first search (DFS) and breadth-first search (BFS). These two strategies make Node2vec learn node representations obeying two principles: the ability to learn representations that embed nodes from the same network community closely together, as well as to learn representations where nodes with similar structural roles get embedded together. In social information network, we find that BFS outperforms DFS which illustrates that embed nodes from the same network community closely is more suitable in social recommendation.

3   Experiments

In this section, we perform experiments on multiple real-world datasets and present the result analysis.

3.1   Datasets and Metrics

We use three popular datasets to validate the effectiveness of our model. The Douban dataset [11] includes 1000 users and 5000 movies with 176308 movie ratings ranging from 1 to 5. As for the Yelp dataset [11], it includes 9581 users and 14037 items with 171109 ratings ranging from 1 to 5. And the Epinions dataset is a larger dataset, consisting 40163 users and 139738 items with 664827 ratings ranging from 1 to 5. We use two common metrics to evaluate the performance of different methods, namely Mean Absolute Error (MAE) and Root Mean Square Error (RMSE) [11].

3.2   Compared Method

We consider the following baselines to validate the effectiveness of MERP. (1) PMF [2]/NMF [12]: They are typical matrix factorization methods. (2) $UserMean/ItemMean$: It employs a user/item's mean rating to predict the missing ratings directly. (3) $BPMF$ [13]: It is a matrix factorization method with Bayesian framework. (4) $SoMF$ [3]: It is the MF based recommendation method with social regularization.

After obtaining the social network embeddings, we could calculates users' similarity with them directly and makes predictions with weighted average method, which is called ERP. Instead of the native idea, our model MERP is a more flexible model to leverage the social network embeddings.

3.3   Effectiveness Experiments

This section will validate the effectiveness of MERP through comparing its different variations to baselines. For a fair comparison of PMF, UserMean, Item-Mean, NMF, BPMF, SoMF, we use the same parameters in both methods. For all the experiments in this experiments, the latent factor number is fixed to 10. In MERP, we set social latent factors $D = 64$, social embedding features $F = 10$, the regularization coefficient λ, β_θ, β_E, β_b are set to trivial values 0.05, 0.001, 0.001, 0.1. In this experiment, we choose DeepWalk method as our network embedding method, we have fixed the window size (5) and the walk length (10) to emphasize local structure.

For these three datasets, we use different ratios (80%, 60%, 40%, 20%) of data as training data. For example, the training data 80% means that we select 80% of the ratings from user-item rating matrix as the training data to predict the remaining 20% of ratings. The random selection was carried out 10 times independently in all the experiments. We report the average results on three

Table 1. Effectiveness experimental results on three datasets (the improvement is based on PMF)

Dataset	Training	Metrics	PMF	UserMean	ItemMean	NMF	BPMF	SoMF	ERP	MERP
Douban	80%	MAE	0.6272	0.6922	0.6600	0.6265	0.5986	0.6047	0.6724	**0.5848**
		Improve		−10.36%	−5.23%	0.32%	4.56%	3.59%	−7.21%	6.76%
		RMSE	0.7870	0.8668	0.8283	0.8015	0.7651	0.7625	0.8478	**0.7396**
		Improve		−10.14%	−5.25%	−1.84%	2.78%	3.11%	−7.73%	6.02%
	60%	MAE	0.6251	0.6896	0.6619	0.6314	0.6066	0.6129	0.6790	**0.5861**
		Improve		−9.89%	−5.89%	−1.01%	2.96%	1.95%	−8.62%	6.24%
		RMSE	0.7825	0.8621	0.8298	0.8063	0.7712	0.7745	0.8611	**0.7416**
		Improve		−10.17%	−6.04%	−3.04%	1.44%	1.02%	−10.04%	5.23%
	40%	MAE	0.6342	0.6951	0.6712	0.6657	0.6303	0.6399	0.6873	**0.5888**
		Improve		−9.60%	−5.83%	−4.97%	0.61%	−0.90%	−8.37%	7.16%
		RMSE	0.7950	0.8683	0.8737	0.8596	0.8032	0.8029	0.8820	**0.7474**
		Improve		−9.22%	−9.90%	−8.13%	−1.03%	−1.00%	−10.94%	5.99%
	20%	MAE	0.6600	0.6979	0.6848	0.7243	0.6895	0.7132	0.7382	**0.5957**
		Improve		−5.74%	−3.76%	−9.74%	−4.47%	−8.06%	−11.85%	9.74%
		RMSE	0.8276	0.8744	0.8764	0.9396	0.8848	0.8891	0.9890	**0.7538**
		Improve		−5.65%	−5.90%	−13.53%	−6.91%	−7.43%	−19.50%	8.92%
Yelp	80%	MAE	0.8155	0.8500	0.8177	0.8208	0.8889	0.8390	0.8922	**0.7847**
		Improve		−4.33%	−0.27%	−0.65%	−9.00%	−2.88%	−9.41%	3.78%
		RMSE	1.0420	1.0894	1.0618	1.0474	1.1664	1.0733	1.2282	**1.0032**
		Improve		−4.55%	−1.90%	−0.52%	−11.94%	−3.00%	−17.87%	3.72%
	60%	MAE	0.8281	0.8550	0.8286	0.8307	0.9295	0.8549	0.9229	**0.7969**
		Improve		−3.25%	−0.06%	−0.31%	−12.24%	−3.24%	−11.45%	3.77%
		RMSE	1.0560	1.0942	1.0771	1.0619	1.2192	1.0884	1.2877	**1.0152**
		Improve		−3.62%	−2.00%	−0.56%	−15.45%	−3.07%	−21.94%	3.86%
	40%	MAE	0.8470	0.8691	0.8468	0.8535	1.0091	0.8666	0.9871	**0.8091**
		Improve		−2.61%	0.02%	−0.77%	−19.14%	−2.29%	−16.54%	4.47%
		RMSE	1.0832	1.1133	1.0990	1.0905	1.3240	1.0976	1.3973	**1.0258**
		Improve		−2.78%	−1.46%	−0.67%	−22.23%	−1.33%	−29.00%	5.30%
	20%	MAE	0.8894	0.8973	0.8828	0.8935	1.0810	0.8711	1.1419	**0.8359**
		Improve		−0.89%	0.74%	−0.46%	−21.54%	2.06%	−28.39%	6.02%
		RMSE	1.1332	1.1525	1.1434	1.1404	1.4067	1.1005	1.6341	**1.0519**
		Improve		−1.70%	−0.90%	−0.64%	−24.14%	2.89%	−44.20%	7.17%
Epinions	80%	MAE	0.8730	0.9385	0.8981	0.8520	0.9473	0.8615	0.9129	**0.8293**
		Improve		−7.50%	−2.88%	2.41%	−8.51%	1.32%	−4.57%	5.01%
		RMSE	1.1167	1.2115	1.1628	1.1061	1.2427	1.1083	1.1846	**1.0638**
		Improve		−8.49%	−4.13%	0.95%	−11.28%	0.75%	−6.08%	4.74%
	60%	MAE	0.8951	0.9467	0.9132	0.8665	0.9935	0.8768	0.9349	**0.8429**
		Improve		−5.79%	−2.02%	3.20%	−10.99%	2.04%	−4.45%	5.83%
		RMSE	1.1454	1.2217	1.1791	1.1258	1.3035	1.1264	1.2105	**1.0757**
		Improve		−6.66%	−2.31%	1.71%	−13.80%	1.66%	−5.68%	6.09%
	40%	MAE	0.9253	0.9626	0.9444	0.8887	1.0731	0.9016	0.9666	**0.8643**
		Improve		−4.03%	−2.06%	3.96%	−15.97%	2.56%	−4.46%	6.59%
		RMSE	1.1844	1.2477	1.2145	1.1345	1.4111	1.1636	1.2438	**1.0968**
		Improve		−5.34%	−2.54%	4.21%	−19.14%	1.76%	−5.02%	7.40%
	20%	MAE	0.9599	1.0033	0.9983	0.9290	1.2144	0.9198	1.0278	**0.9080**
		Improve		−4.52%	−4.00%	3.22%	−26.51%	4.18%	−7.07%	5.41%
		RMSE	1.2282	1.3002	1.2718	1.2068	1.5947	1.2029	1.3077	**1.1374**
		Improve		−5.86%	−3.55%	1.74%	−29.84%	2.06%	−6.47%	7.39%

different datasets and also record the improvement of all methods compared to the baseline PMF.

The performance of all the methods are shown in Table 1. And we can get the following conclusions. MERP always perform better than the original methods on each dataset and all ratios. To some extent, MERP is considered to be a combination of two methods PMF and ERP. MERP always performs better than PMF and ERAP which illustrates the effectiveness of the combination of these two methods. By comparing the three datasets, we can find that on more sparse datasets the MERP can performs better.

3.4 Impact of Different Strategies

Experiments in this section will validate the sensitivity of different network embedding strategies in MERP. Here we compare two metods in MERP: Deep-Walk and Node2vec. In Node2vec, there are two different strategies: breadth first search (BFS) and depth first search (DFS). So we compare a total of three strategies: DeepWalk, Node2vec-BFS, Node2vec-DFS with the same experiment setting as Scct. 3.3.

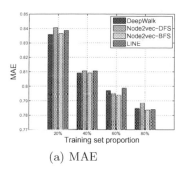
(a) MAE

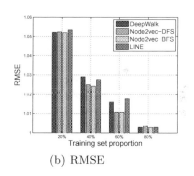
(b) RMSE

Fig. 2. The comparison of different strategies in MERP.

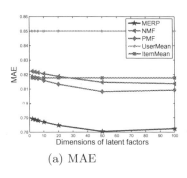
(a) MAE

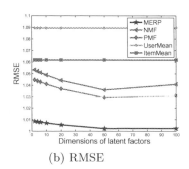
(b) RMSE

Fig. 3. The comparison on different latent dimensions.

As shown in Fig. 2(a), all the methods improve as training set proportion increases. However, when we try different strategies, we find that the Node2vec-BFS performs better then other network embedding strategies. The result illustrates that breadth first search strategy is more suitable in social recommendation.

3.5   Parameter Study

For matrix factorization based methods, the latent dimension is an important parameter to tune. And our model also involves such a parameter. We vary it from 0 to 100 with a step of 10, and examine how the performance changes with regard to the latent dimension. As shown in Fig. 3(a), using 50 latent dimension yields the best performance and MERP performs better in all latent dimensions.

4   Conclusion

In this paper, we propose a framework which combines social network embedding and matrix factorization to predict the unknown user-item ratings in the rating matrix. In order to fully utilize the social information network to solve data sparseness, scalability, and predictive quality in social recommendation problem, we introduce social network embedding method into matrix factorization. MERP makes a better performance on rating prediction accuracy. We analyze the performance of MERP in terms of its dependency on training set size, different network search strategies and latent dimension number, our method performs well in all cases.

Acknowledgements. This work is supported by the National Natural Science Foundation of China (No. 61772082, 61375058), the National Key Research and Development Program of China (2017YFB0803304), and the Beijing Municipal Natural Science Foundation (4182043).

References

1. Zhang, M., Hu, B., Shi, C., Wang, B.: Local low-rank matrix approximation with preference selection of anchor points. In: International Conference on World Wide Web Companion, pp. 1395–1403 (2017)
2. Mnih, A., Salakhutdinov, R.R.: Probabilistic matrix factorization. In: NIPS, pp. 1257–1264 (2008)
3. Ma, H., Zhou, D., Liu, C., Lyu, M.R., King, I.: Recommender systems with social regularization. In: WSDM, pp. 287–296 (2011)
4. Tang, J., Hu, X., Liu, H.: Social recommendation: a review. Soc. Netw. Anal. Min. **3**(4), 1113–1133 (2013)
5. Perozzi, B., Al-Rfou, R., Skiena, S.: Deepwalk: online learning of social representations. In: SIGKDD, pp. 701–710 (2014)
6. Grover, A., Leskovec, J.: node2vec: scalable feature learning for networks. In: SIGKDD, pp. 855–864 (2016)

7. Tang, J., Qu, M., Wang, M., Zhang, M., Yan, J., Mei, Q.: Line: large-scale information network embedding. In: WWW, pp. 1067–1077 (2015)
8. Wang, H., Wang, N., Yeung, D.Y.: Collaborative deep learning for recommender systems. In: SIGKDD, pp. 1235–1244 (2015)
9. Liang, D., Altosaar, J., Charlin, L., Blei, D.M.: Factorization meets the item embedding: regularizing matrix factorization with item co-occurrence. In: RecSys, pp. 59–66 (2016)
10. Mikolov, T., Chen, K., Corrado, G., Dean, J.: Efficient estimation of word representations in vector space. In: ICLR Workshop (2013)
11. Shi, C., Hu, B., Zhao, W.X., Yu, P.S.: Heterogeneous information network embedding for recommendation. arXiv preprint arXiv:1711.10730 (2017)
12. Lee, D.D., Seung, H.S.: Learning the parts of objects by non-negative matrix factorization. Nature **401**(6755), 788–791 (1999)
13. Salakhutdinov, R., Mnih, A.: Bayesian probabilistic matrix factorization using Markov chain Monte Carlo. In: ICML, pp. 880–887 (2008)

An Estimation Framework of Node Contribution Based on Diffusion Information

Zhijian Zhang[1,2], Ling Liu[3], Kun Yue[1(✉)], and Weiyi Liu[1]

[1] School of Information Science and Engineering,
Yunnan University, Kunming, China
kyue@ynu.edu.cn
[2] College of Science, Kunming University of Science and Technology,
Kunming, China
[3] Yunnan Police Officer Academy, Kunming, China

Abstract. As a key problem in social network studies, identifying important nodes is useful for spreading prediction and restraint. Nowadays, many researches focus on identifying important nodes based on network structure, which is completely necessary. However, the role of a node in diffusion is simultaneously determined by network structure and diffusion characteristics. In this paper, we aim to find the contributive nodes that play important roles in influence spreading without network structure information. First, we formulize the concept of node contribution to influence diffusion to describe the importance of nodes in the spreading processes. Then, we propose an estimation framework and give the method to estimate node contribution based on diffusion samples. Accordingly, the Contribution Estimation algorithm is proposed upon the framework. Finally, we implement our algorithm and test the efficiency on two weighted social networks.

Keywords: Social network · Influence diffusion · Node contribution
Contribution Estimation

1 Introduction

The explosive growth of online social networks changes the way how the information spread among people. Due to the properties of online social networks, news, advertisements, headlines and even rumors can spread to persons around the world in few seconds cross the online social networks. This process is called influence diffusion in social network research [1]. It is significant to get a thorough understanding of this phenomenon and clarify the keys to the diffusion process. However, these two problems are both arduous due to dynamic changes, large scale and the complexity of the social networks.

Since Kempe et al. [2] proposed two classical diffusion model to describe the influence spreading process and formalized the influence maximization problem, influence diffusion in social networks has been paid much attention. Besides the diffusion model and the influence maximization problem, some of recent efforts estimate the status of social network by sampling based on statistical theory [3]. In those researches, how to identify important nodes is a crucial problem, especially for

© Springer International Publishing AG, part of Springer Nature 2018
Y. Cai et al. (Eds.): APWeb-WAIM 2018, LNCS 10987, pp. 130–137, 2018.
https://doi.org/10.1007/978-3-319-96890-2_11

predicting and restraining the influence diffusion in social networks. The important node makes many nodes activated directly or indirectly in the diffusion, in other words, the influence spreading effect will be weakened without these nodes.

There are many concepts are proposed to identifying the important nodes in social networks. Such as degree centrality [4], betweenness centrality [5], second order centrality [6] and structural holes [7]. Degree centrality [4] is an obvious indicator to describe node importance, but it cannot reflect the entire network structure information which affects the diffusion. Furthermore, betweenness centrality [5, 8] describes the probability that information pass the node, called importance in network structure. Although, Newman [8] proposed a random work based method to estimate node betweenness centrality, this method is time consuming in large scale social networks. Researchers also proposed some other indicator to measure node importance, such as closeness centrality [9], structural holes [7] and super-mediators [10]. Unfortunately, all of these methods are topology based and cannot reflect the relationship between nodes and the characteristics of influence diffusion.

Considering the characteristics of diffusion process, Chen [11] propose a local ranking algorithm named ClusterRank, which considers not only the number of neighbors and the neighbors' influences. Yang and Leskovec [12] developed the linear influence model and estimate node influence according to the user activity information. Tan et al. [13] proposed a noise tolerant time-varying factor graph model (NTT-FGM) for modeling and predicting social actions. In these method, user attributes or action information are necessary.

Many researchers are committed to finding important nodes [15–18]. However, the node importance of diffusion is determined by not only network topology but also the characteristics related to influence spreading such as influence power and activate probability. Unfortunately, little attempt has been given to identify node importance efficiently based on diffusion characteristics.

In this paper, we focus on solving the following problem: how to identify important nodes for influence diffusion in a dynamic and large scale online social network? As we know, it is almost impossible to obtain the complete information about the social network structure due to the large scale and the dynamic characteristics of the network. Besides, some nodes important to network structure may have little contribution to influence diffusion. In other words, we are to identify the important nodes contributive to influence diffusion based on diffusion information such as activation information in spreading processes and the activate probability in the diffusion model. Thus, the main challenges of our work is how to describe the importance of a node in influence diffusion and how to identify important nodes in large scale dynamic social networks.

We first clarify the definition of the contribution of node v in graph G within radius r, which reflects the importance of node v for the influence diffusion process rather than the importance for the structure of a graph. Then, we propose a method to estimate the node contribution based on the influence spreading information instead of network structure. Finally, we implement the influence spreading model and our Contribution Estimation algorithm and identify the contributive nodes in two real weighted social networks. By comparison, the experiment results show that our method and algorithm can identify the key nodes of diffusion process in two real social networks. Without these contributive nodes, the average number of nodes which an influence diffusion expects to activate is significantly decreased.

2   Contribution Estimation Framework

For the diffusion process, we consider the contribution of a node to the spreading process. First, we mathematically model the online social network as a directed network $G = <V, E>$, where V is the set of nodes and E is the set of edges between nodes. A node is inactive or active, and a node is inactive at the beginning of diffusion. The diffusion process starts from a set of initialized active nodes, called as seeds, and the process unfolds in discrete time steps. If a node is activated by active nodes, it becomes active and stays active. The diffusion process terminates if there is no more nodes activated.

We employ $C_v(G)$ to describe the contribution of node v in influence diffusion via social networks G. In an influence diffusion process, we consider estimating node contribution based on diffusion observation information rather than network structure. More specifically, the contribution of node v to the diffusion d_i is estimated by the change of the number of activated nodes after v is activated which is observed or estimated by sampling.

First, we present the definition of diffusion distance. If active node v makes node w activated in r steps/time interval, we say diffusion distance from v to w is r. For instance, if v activates w directly, then $r = 1$. Thus, $C_v^r(G)$ is called as the contribution of node v in graph G within radius r, denoted as follows

$$C_v^r(G) = E\big(A_r^v(G)\big) \tag{1}$$

Where $A_r^v(G)$ denotes the number of activated nodes at the end of a specific diffusion starting from seed nodes set s in graph G in radius r. And $E\big(A_r^v(G)\big)$ denotes the expected number of $A_r^v(G)$.

In order to find the node contribution, we propose a method free from both network structure and diffusion model. This method estimates the contribution based on the observation of the number of activated nodes in the diffusion process. So, we define the contribution of node v to the diffusion d_i within radius r, described as follows:

$$\varphi_v^r(d_i, G) = A_{t+r}(d_i, G) - A_t(d_i, G) \tag{2}$$

where $A_t(d_i, G)$ is the number of activated nodes at time t in diffusion d_i in graph G, and $A_{t+r}(d_i, G)$ is the number of activated nodes at time $t + r$.

As $\varphi_v^r(d_i, G)$ is the summary contribution of the nodes activated at time t to specific diffusion d_i, we utilize the expected number of $\varphi_v^r(G)$ to represent the contribution of node v to any diffusion in graph G.

$$\Phi_v^r(G) = E\big(\varphi_v^r(G)\big) \tag{3}$$

This means that $\Phi_v^r(G)$ indicates the expected value of increase number of activated nodes within radius r after node v is activated for any diffusion in graph G.

Therefore, we can estimate $\Phi_v^r(G)$ based on samples of diffusion processes. Let $D = \{d_1, d_2, \ldots, d_n\}$ denote a set of different diffusion samples in graph G and v is activated in these diffusion, where d_i is the i-th diffusion. Then, $\varphi_v^r(d_i, G)$ can be calculated. Let $\hat{\varphi}_v^r(G)$ present the average of $\varphi_v^r(d_i, G)$, and we have

$$\hat{\varphi}_v^r(G) = \frac{\sum_{i=1}^{|D|} (A_{t+r}(d_i, G) - A_t(d_i, G))}{|D|} \qquad (4)$$

Thus, $\hat{\varphi}_v^r(G)$ is regarded as an estimator of $C_v^r(G)$.

3 Contribution Estimation

According to the above discussion, we give an algorithm to estimate node contribution for diffusion, where we calculate $\Phi_v^r(G)$ in different diffusion trails $D = \{d_1, d_2, \ldots, d_n\}$ to estimate the contribution of node v in radius r for the diffusion in graph G.

For any $d_i \in D$, we first compute the number of nodes $\varphi_v^r(d_i, G)$, inactivated during the period between time t_v and time t_{v+r} according to Formula (2). Then, we obtain the summary of $\varphi_v^r(d_i, G)$ in diffusion samples D. The average of $\varphi_v^r(d_i, G)$ is regarded as the contribution of node v in graph G. The contribution of concerned nodes can be obtained by the same method. Finally, we can rank the nodes by the result and find the most contributive nodes.

Algorithm 1: Contribution Estimation

Input: Node v, radius r and diffusion samples $D = \{d_1, d_2, \ldots, d_n\}$. In each diffusion d_1, the time t_v node v is activated and the number of activated nodes at time t_v and t_{v+r} is required.

Output: contribution of node v in radius r $\Phi_v^r(G)$

Steps:

    Step 1: $\Phi_v^r(G) \leftarrow 0$

    Step 2: For each $d_i \in D$ **Do**

$$\Phi_v^r(G) \leftarrow \Phi_v^r(G) + \left(A_{t+r}(d_i, G) - A_t(d_i, G)\right)$$

    End For

    Step 3: $\Phi_v^r(G) \leftarrow \dfrac{\Phi_v^r(G)}{|D|}$

Output: $\Phi_v^r(G)$

The running time of Algorithm 1 is mainly determined by the number of diffusion samples and the number of nodes which we focus on. The complexity of Algorithm 1 is $O(n)$ for the situation with one node contribution computing, where $n = |D|$ is the number of diffusion samples. In a particular network, we compute the contribution of some latent nodes instead of the whole nodes set.

4 Experimental Results

4.1 Data Sets and Experiment Setup

In order to evaluate our algorithm, we employed two datasets[1] of real social networks, which are both weighted networks [14]. The networks are who-trusts-whom network of people who trade using Bitcoin on the platforms BitcoinOTC and BitcoinAlpha. The BitcoinOTC network has 5881 nodes and 35592 edges, and the BitcoinAlpha network has 3783 nodes and 24186 edges. In these networks, a node represents a user in the social network and edges indicate the bitcoin trade between users.

Since we consider the node contribution for influence diffusion, we employ Independent Cascade model [2] to simulate the information diffusion process. The IC model is a classic diffusion model in which there is an activate probability at every edge. If a node is activated, it tries to activate it's inactivate neighbor nodes with the activate probability once. In our experiments, the activate probability is determined by the edge weights in BitcoinOTC and BitcoinAlpha networks.

4.2 Effectiveness Evaluation

Before computing the node contribution, we implemented IC model and simulated the influence diffusion process on the above two networks with randomly selected seed nodes. We record the numbers of activated nodes at different time and compute the node contribution based on the Contribution Estimation algorithm.

Then, we implement our Contribution Estimation algorithm and compute the node contribution in the two social networks. We computed the node contribution with different radius ($r = 1$, $r = 2$ and $r = 3$), and sorted the nodes by contribution value.

In order to evaluate the effectiveness of our Contribution Estimation algorithm, we deleted the contributive nodes in networks, simulated the influence spreading and compared the average number of corresponding activated nodes in original networks. The result are shown in Figs. 1 and 2 respectively.

Figure 1(a)–(c) show the average number of activated nodes in original bitcoinotc network in which different number of contributive nodes are deleted when radius equals to 1, 2 and 3. Figure 2(a)–(c) show the similar results in bitcoinalpha network. We can see that the average number of activated nodes significantly decreases when the contributive nodes were deleted. This means that our Contribution Estimation algorithm is effective for identifying the important nodes in influence diffusion. Without these contributive nodes, the mount of nodes that an influence spreading can activate will decrease sharply. In another word, these nodes are critical for the diffusion capability of a social network.

Figures 1(d) and 2(d) show the comparison results of $r = 1$, $r = 2$ and $r = 3$, where we deleted top 50 contributive nodes in bitcoinotc and bitcoinalpha. It can be seen that the contributive node with greater radius is more pivotal than those with smaller radius, which is in line with the reality. Since some contributive nodes with lower radius are

---

[1] The datasets are public and available at http://snap.stanford.edu/data/index.html.

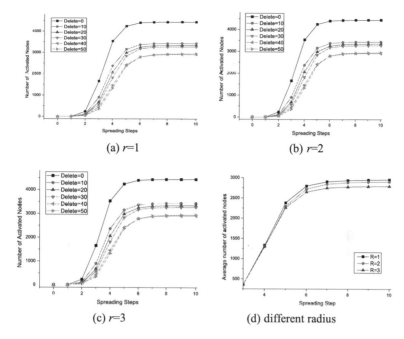

Fig. 1. Spreading comparison in Bitcoinotc

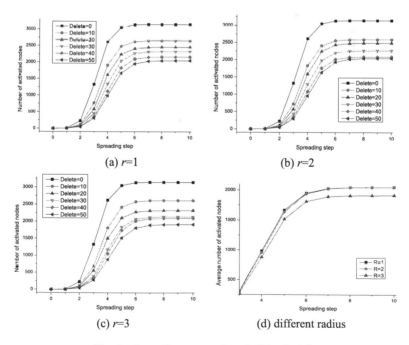

Fig. 2. Spreading comparison in Bitcoinalpha

also contributive when radius is large, the difference of average number of activated nodes among $r = 1$, $r = 2$ and $r = 3$ is not very obvious.

5   Conclusion and Future Work

In this paper, we focus on identifying the contributive nodes for influence diffusion in social networks. First, we clarify the definition of node contribution and propose the estimation framework of node contribution based on diffusion information. In our method, the network structure information is not necessary. Then, we give the Contribution Estimation algorithm correspondingly. Experiment results show that our method find the contributive nodes efficiently in two real social networks. Without these contributive nodes, the influence spreading effect is significant weakened.

During the experiments, we note that the weight of edges or the activate probability is crucial for influence diffusion studies especially in some unweight cases. Thus, our immediate future work is to improve our contribution estimation method and work on identifying the relationship between nodes.

Acknowledgement. This paper was supported by the Research Foundation of Educational Department of Yunnan Province (2017ZZX133), Applied Basic Research Project of Yunnan Province (Youth Program) (2015FD037), National Natural Science Foundation of China (61472345) and Research Foundation of Yunnan University (2017YDJQ06).

References

1. Leskovec, J., Krause, A., Guestrin, C., et al.: Cost-effective outbreak detection in networks. In: KDD, pp. 420–429 (2007)
2. Kempe, D., Kleinberg, J., Tardos, É.: Maximizing the spread of influence through a social network. In: KDD, pp. 137–146 (2003)
3. Hardiman, S.J., Katzir, L.: Estimating clustering coefficients and size of social networks via random walk. In: International Conference on World Wide Web, vol. 9, pp. 539–550. ACM (2013)
4. Bonacich, P.: Factoring and weighting approaches to status scores and clique identification. J. Math. Sociol. **2**(1), 113–120 (1972)
5. Freeman, L.C.: A set of measures of centrality based on betweenness. Sociometry **40**(1), 35–41 (1977)
6. Kermarrec, A.M., Merrer, E.L., Sericola, B., Trédan, G.: Second order centrality: distributed assessment of nodes criticity in complex networks. Comput. Commun. **34**(5), 619–628 (2011)
7. Burt, R.S.: Structural Holes: The Social Structure of Competition, pp. 53–58. Harvard University Press, London (2009)
8. Newman, M.E.J.: A measure of betweenness centrality based on random walks. Soc. Netw. **27**(1), 39–54 (2005)
9. Sabidussi, G.: The centrality index of a graph. Psychometrika **31**(4), 581–603 (1966)

10. Saito, K., Kimura, M., Ohara, K., Motoda, H.: Discovery of super-mediators of information diffusion in social networks. In: Pfahringer, B., Holmes, G., Hoffmann, A. (eds.) DS 2010. LNCS (LNAI), vol. 6332, pp. 144–158. Springer, Heidelberg (2010). https://doi.org/10. 1007/978-3-642-16184-1_11
11. Chen, D.D., Gao, H., Lü, L., Zhou, T.: Identifying influential nodes in large-scale directed networks: the role of clustering. PLoS One 8(10), e77455 (2013)
12. Yang, J., Leskovec, J.: Modeling information diffusion in implicit networks. In: IEEE International Conference on Data Mining, pp. 599–608. IEEE (2011)
13. Tan, C., Tang, J., Sun, J., Lin, Q., Wang, F.: Social action tracking via noise tolerant time-varying factor graphs. In: ACM SIGKDD International Conference on Knowledge Discovery and Data Mining, pp. 1049–1058. ACM (2010)
14. Kumar, S., Spezzano, F., Subrahmanian, V.S., Faloutsos, C.: Edge weight prediction in weighted signed networks. In: IEEE International Conference on Data Mining, pp. 221–230. IEEE (2017)
15. Yang, Y., Gang, X.: Mining important nodes in directed weighted complex networks. Discret. Dyn. Nat. Soc. 2017(5), 1–7. (2017)
16. Sarkar, A., Chattopadhyay, S., Dey, P., Roy, S.: The importance of seed nodes in spreading information in social networks: a case study. In: International Conference on Communication Systems and Networks, pp. 395–396. IEEE (2017)
17. Zhou, J., Zhang, Y., Cheng, J.: Preference-based mining of top-K influential nodes in social networks. Future Gener. Comput. Syst. 31(1), 40–47 (2014)
18. Muhuri, S., Chakraborty, S., Setua, S.K.: An edge contribution-based approach to identify influential nodes from online social networks. In: IEEE International Symposium on Nanoelectronic and Information Systems, pp. 155–160. IEEE (2017)

Multivariate Time Series Clustering via Multi-relational Community Detection in Networks

Guowang Du, Lihua Zhou[✉], Lizhen Wang, and Hongmei Chen

School of Information, Yunnan University, Kunming 650500, China
bingwei2642@qq.com,
{lhzhou, lzhwang, hmchen}@ynu.edu.cn

Abstract. Clustering multivariate time series is a challenging problem with numerous applications. The presence of complex relations amongst individual series poses difficulties with respect to traditional modelling, computation and statistical theory. In this paper, we propose a method for clustering multivariate time series by using multi-relational community detection in complex networks. Firstly, a set of multivariate time series is transformed into a multi-relational network. Then, an algorithm for multi-relational community detection based on multiple nonnegative matrices factorization (MNMF) is proposed and is applied to identify time series clusters. The transformation of time series from time-space domain to topological domain benefits from the ability of networks to characterize both local and global relationship amongst nodes (representing data samples), while the use of MNMF can give full play to complex relations amongst individual series and preserve the multi-way nature of multivariate information. Preliminary experiment indicates promising results of our proposed approach.

Keywords: Multivariate time series · Clustering · Multi-relational network
Community detection · Matrix factorization

1 Introduction

With the rapid growth of digital sources of information, enormous amounts of time series data are being continually generated and collected. Mining these data is helpful to discover the hidden knowledge and information, thus it is receiving increasing attention from researchers in the recent years.

Clustering time series, i.e. dividing a set of time series into groups such that similar ones are put in the same group [1], is a fundamental task of time series data mining (TSDM). This task has been extensively studied and a great number of approaches have been proposed. Ferreiraa and Zhao [2] thought that time series clustering requires not only local information, but also global knowledge to capture the pattern formation of given time series, but in general only the local relationship amongst neighbour data samples can be easily identified, while long distance global relationship remains unknown in the original form of time series. To use the global knowledge for clustering time series, Ferreiraa and Zhao [2] proposed a network-based approach. This approach

© Springer International Publishing AG, part of Springer Nature 2018
Y. Cai et al. (Eds.): APWeb-WAIM 2018, LNCS 10987, pp. 138–145, 2018.
https://doi.org/10.1007/978-3-319-96890-2_12

uses distance functions to transform a set of time series into a network (represented by a graph), where each time series is represented by a node and the most similar ones are connected, and then apply community detection algorithms to identify groups of strongly connected nodes and, consequently, identifies time series clusters. This network-based clustering technique can capture arbitrary cluster shapes.

In this paper, we extend the network-based approach [2] to cluster multivariate time series data. Firstly, we use a distance function to transform multivariate time series into a multi-relational network, which is modelled as multiple single-relational graphs, each reflecting similarity amongst time series of a variable. A multi-relational network means that series are related through various kinds of heterogeneous relationship types, which jointly affect attributes of series. Then, we use MNMF to jointly factorize multiple relation matrices to obtain the clusters.

In summary, the specific contributions of this paper are highlighted as follows:

(1) The idea of using community detection in complex networks for univariate time series clustering is extended to multivariate time series clustering. The extension benefits from the ability of networks to characterize both local and global relationship amongst nodes (representing data samples), thus the local relationship and long distance global relationship amongst neighbour data samples can be used.

(2) A MNMF-based algorithm for multi-relational community detection is proposed. The proposed algorithm jointly factorizes multiple relation matrices, thus the complex relations amongst individual series can be effectively incorporate into the process of community detection to improve the clustering performance.

(3) Numerical study in real data has been conducted, and the effectiveness of our proposed approach is validated by the preliminary experimental results.

The remaining sections of this paper are organized as follows: a brief overview of related work about clustering time series, detecting multi-relational communities and NMF is given in Sect. 2. Section 3 introduces our proposed method for multivariate time series clustering. Experiments and results are presented in Sect. 4. Finally, we conclude the paper in Sect. 5.

2 Related Work

One way to clustering time series is feature-based approach [3, 4], which first extracts some features from input time series, and then applies traditional clustering algorithms to cluster extracted features. Another way to cluster time series data is transformation-based approach [2, 5], which transform time series data into other models and then the clustering are carried out on the transformed models.

Community detection [6] is one of the important tasks in social network analysis. The algorithms for multi-relational networks must take the existence of an edge in multiple graphs into account in order to obtain meaningful results rather than just concern with the existence of an edge in a graph. One way to detect communities in a multi-relational network is to transform a multi-relational network into a single-relational network, and then use single-relational network algorithms to detect communities, such as methods of Wu et al. [7], Tang et al. [8] and Ströele et al. [9]. Zhou et al. [10] proposed a coalition formation game theory [11]-based approach to detect communities in multi-relational social networks. This method handles multi-relational social networks directly rather than transforms them. In addition, Lin et al. [12], Zhang et al. [13] and Li et al. [14] also studied community detection problem in multi-relational networks, but in their researches, a multi-relational network not only contains multiple typed relations but also contains more than one typed objects. The multi-relational networks in this study are limited to contain multiple typed relations but just one typed objects.

NMF [15] aims at learning the representation parts of the original data by approximating the target matrix into the product of two low-rank matrices. Specifically, given an $m \times n$ nonnegative matrix A where all the elements are nonnegative, NMF decomposes A into two nonnegative matrices $P_{m \times c}$ and $Q_{n \times c}$ such that $A \approx PQ^T$, s.t. $P \geq 0, Q \geq 0$. The optimal $P_{m \times c}$ and $Q_{n \times c}$ are learned by minimizing a particular loss function, such as Euclidean distance $D(P, Q) = ||A - PQ^T||_2$. In general, c is much smaller than $\min\{m, n\}$.

In recent years, NMF has been extensively investigated and extended for the problem of community detection [16, 17]. Single nonnegative matrix factorization can tackle a data set with only one type of feature or relationship. Gupta et al. [18] and Huang et al. [4] employed jointly multiple matrices factorization to analyse data sets with multiple types of features, aiming to derive a solution which uncovers the common latent structure shared by multiple features.

3 Multivariate Time Series Clustering via Multi-relational Community Detection in Networks

Given a set of multivariate time series, the problem of clustering such data is concerned with the discovering of inherent groupings of the data according to how similar or dissimilar the time series are to each other. In this section, we begin with the data representation, and then introduce the network construction, and finally present MNMF for detecting communities.

3.1 Data Representation

A univariate time series is an ordered sequence of n real values observed at time points $t = 1, \ldots, n$, while a multivariate time series consists of $r \times n$ observations corresponding to r variables. Let $X = \{X^1, X^2, \ldots, X^m\}$ represent a collection of m multivariate time series. The series $X^i \in X$ can be written as $X^i = \{x_{ijt}\}, j = 1, \ldots, r; t = 1, \ldots, n$, indicating $r \times n$ observations in total. Let $X_j^i = \{x_{ijt}\} t = 1, \ldots, n$ represents the time series corresponding to the j-th variable of $X^i, X_j = \{X_j^1, X_j^2, \ldots, X_j^m\}$ be the set of m time series corresponding to the j-th variable of $X = \{X^1, X^2, \ldots, X^m\}$.

A single-relational network is often represented as a graph $G_j = (V_j, E_j)$ composed by a set of nodes $V_j = \{V_j^1, \ldots, V_j^m\}$ and a set of edges $E_j = \{(V_j^i, V_j^k)|V_j^i, V_j^k \in V_j\}$, where the nodes represent objects and edges indicate interactions amongst objects.

A multi-relational network can be represented as multiple single-relational graphs, each reflecting interactions amongst entities in one kind of relation type. In this paper, one variable of a multivariate time series corresponds to one relation type, thus $X_j = \{X_j^1, X_j^2, \ldots, X_j^m\}$ can be transformed into a single-relational network $G_j = (V, E_j)$, reflecting similarity amongst time series $X_j^i \in X_j$, $i = 1, \ldots, m$, where $V = \{v_1, \ldots, v_m\}$ represents m multivariate time series to be clustered, and $(v_i, v_k) \in E_j$ if X_j^i is similar to X_j^k. Let A_j be an adjacency matrix of G_j with $A_j(v_i, v_k) = 1$ if $(v_i, v_k) \in E_j$ for any pair of nodes $v_i, v_k \in V$ and 0 otherwise. Let $G = \{G_1, G_2, \ldots, G_r\}$ be a multi-relational network transformed from multivariate time series $X = \{X^1, X^2, \ldots, X^m\}$ with r variables, where $G_i = (V, E_i)$, $i = 1, \ldots, r$ corresponds to $X_i = \{X_i^1, X_i^2, \ldots, X_i^m\}$, the adjacency matrix of G_i is A_i. $G = \{G_1, G_2, \ldots, G_r\}$ can also be denoted as $G = (V, E)$, where $E = \bigcup_{i=1}^r E_i$.

In the following, variable and relation are interchangeable, and node and multivariate time series are interchangeable. Likewise, cluster and community are interchangeable.

3.2 The Network Construction

To construct a network from a set of time series, a distance function is first used to measure the similarity amongst time series, and then each time series is represented as a node and it is connected to its k most similar ones (k-NN) or is connected to ones of which similarities with the node are higher than a threshold value ε (ε-NN). In the case of multivariable time series, the above process is repeated many times and each time deals with a set of time series with respect to a variable.

3.3 Multiple Nonnegative Matrices Factorization (MNMF)

Given a set of adjacency matrices $A = \{A_1, A_2, \ldots, A_r\}$ of a multi-relational network, the aim of MNMF is to find a nonnegative matrix P and matrices $\{Q_i\}_{i=1,\ldots,r}$. Matrix P is a $m \times c$ membership matrix, representing the possibility of each node belonging to a community. Matrix $\{Q_i\}_{i=1,\ldots,r}$ is also a $m \times c$ matrix, representing the connectivity within each community amongst m nodes with respect to the i-th relation. Then,

$$A_i \approx P \times Q_i^T \tag{1}$$

In the following, we introduce the objective function of our proposed method and the updating rules for learning the optimal parameters. To integrate multiple types of relations simultaneously in the process of detecting communities, we represent a multi-relational network by multiple adjacent matrices and factorize these matrices simultaneously. Therefore, the objective function is defined as follow:

$$J(P, Q_1, \ldots, Q_r) = \frac{1}{2}\left(\sum_{i=1}^r ||A_i - PQ_i^T||_F^2\right) \tag{2}$$

Subject to $P > 0$ and $Q_i > 0$, $\forall i = 1, \ldots, r$.

This objective function aims to minimize the sum of the distances between the adjacent matrix $A_i(i = 1, \ldots, r)$ and the multiplications of the membership matrix P by the connectivity matrix $Q_i(i = 1, \ldots, r)$. $\| \cdot \|_F^2$ denotes the Frobenius norm, and Q_i^T is the transpose of Q_i.

Now, the task is to learn optimal parameters $P^*, Q_i^*(i = 1, \ldots, r)$ that can minimize objective function in Eq. (2). We use stochastic gradient descent method to solve this task. The updating rules for P and $Q_i(i = 1, \ldots, r)$ are shown in Theorems 1 and 2 respectively. The proof of Theorems 1 and 2 are omitted due to the limitation of space.

Theorem 1. If fixed $P, Q_1, \ldots, Q_{s-1}, Q_{s+1}, \ldots Q_r$, the objective function $J(P, Q_1, \ldots, Q_r)$ is nonincreasing under the updating rule:

$$
(Q_s)_{ij} = (Q_s)_{ij} \frac{(A_s^T P)_{ij}}{(Q_s P^T P)_{ij}} \tag{3}
$$

Theorem 2. If fixed $Q_1, \ldots, Q_r$, the objective function $J(P, Q_1, \ldots, Q_r)$ is nonincreasing under the updating rule:

$$
P_{ij} = P_{ij} \frac{\left(\sum_{s=1}^{r} A_s Q_s \right)_{ij}}{\left(\sum_{s=1}^{r} P Q_s^T Q_s \right)_{ij}} \tag{4}
$$

Algorithm 1 describes the overall procedure of MNMF. It is able to guarantee that the objective function converges to local minimum based on the Theorems 1 and 2. After obtaining the membership matrix P, the community structure of the multi-relational network G can be discovered.

```
Algorithm 1: MNMF
Input: Adjacent matrices A_1,..., A_r , the number of communi-
ties c
Output: membership matrix P and connectivity matrix
Q_i(i = 1,...,r)
Initialize: Randomly choose an initial P,Q_1,...,Q_r .
Repeat
   For s = 1 to r
      Fixed P,Q_1,...,Q_{s-1},Q_{s+1},...Q_r , update Q_s with Equation (3).
   End for
   Fixed Q_1,...,Q_r , update P with Equation (4).
Until convergence.
```

Membership matrix P contains the membership information of every node. After normalizing P by row, the value in this matrix is the possibility of the node belonging to the community. In our experiment, we assign the node to the community corresponding to the max possibility when we evaluate the clustering results.

4   Experimental Evaluation

In the experiments of this paper, we use Robot Execution Failures (LP4: failures in approach to ungrasp position) time series dataset taken from the UCI repository [19]. This dataset contains 117 6-variables time series. Each variable represents a force or a torque which is measured at 15 regular time intervals starting immediately after failure detection. The total observation window for each variable was of 315 ms, thus there is a total of 90 observations, where Fx1 … Fx15 is the evolution of force Fx, the same for Fy, Fz and the torques Tx, Ty and Tz. All 117 series are classified three classes, where 21% is normal, 62% is collision and 18% is obstruction.

We use three metrics including *rand index* (RI) [20], *adjusted rand index* (*ARI*) [21], *normalized mutual information* (*NMI*) [21] and *Purity* [22] to evaluate the performance of our proposed algorithm. We use the ε-NN construction method with the Dynamic Time Warping (DTW) distance function to conduct our experiments. Since the final results of the MNMF depend on the initial values, we run the algorithms MNMF 20 times and obtain the clustering results which have the minimal value of the objective function. The convergence condition for MNMF is $\Delta J(P, Q_1, \ldots, Q_r) < \delta$, where $\Delta J(P, Q_1, \ldots, Q_r)$ is the absolute increment of the objective function values in adjacent iterations, δ is a threshold value. The best performance metrics of MNMF on LP4 data set and corresponding parameters are shown in Table 1.

Table 1. The best parameters and performance metrics of MNMF on LP4 data set

ε	δ	RI	ARI	NMI	Purity
0.45	1.00E−6	0.93	0.85	0.77	0.91

From Table 1, we can see that the results of our approach are promising. It indicates that it is effective to cluster multivariate time series clustering via multi-relational community detection in networks. In 117 series, there are two series of collision are classified into normal, and eight series of obstruction are classified into normal.

5   Conclusion

In this paper, an approach for clustering multivariate time series by using multi-relational community detection in complex networks is proposed. The proposed approach first transforms a set of multivariate time series into a multi-relational network, and then uses MNMF to identify communities representing multivariate time series clusters. The transformation of time series from time-space domain to topological domain benefits from the ability of networks to characterize both local and global relationship amongst data samples, while the use of MNMF can give full play to complex relations amongst individual series and preserve the multi-way nature of multivariate information.

Acknowledgement. This research was supported by the National Natural Science Foundation of China (61762090, 61262069, 61472346, and 61662086), The Natural Science Foundation of Yunnan Province (2016FA026, 2015FB114), the Project of Innovative Research Team of Yunnan Province, and Program for Innovation Research Team (in Science and Technology) in University of Yunnan Province (IRTSTYN).

References

1. Esling, P., Agon, C.: Time-series data mining. ACM Comput. Surv. **45**(1), 1–34 (2012)
2. Ferreira, L.N., Zhao, L.: Time series clustering via community detection in networks. Inf. Sci. **326**, 227–242 (2016)
3. Maharaj, E.A., D'Urso, P.: Fuzzy clustering of time series in the frequency domain. Inf. Sci. **181**(7), 1187–1211 (2011)
4. Huang, X.H., Ye, Y.M., Xiong, L.Y., Lau, R.Y.K., Jiang, N., Wang, S.K.: Time series k-means: a new k-means type smooth subspace clustering for time series data. Inf. Sci. **367–368**(1), 1–13 (2016)
5. Deng, W., Wang, G., Xu, J.: Piecewise two-dimensional normal cloud representation for time-series data mining. Inf. Sci. **374**(2016), 32–50 (2016)
6. Fortunato, S.: Community detection in graphs. Phys. Rep. **486**, 75–174 (2010)
7. Wu, Z., Yin, W., Cao, J., Xu, G., Cuzzocrea, A.: Community detection in multi-relational social networks. In: Lin, X., Manolopoulos, Y., Srivastava, D., Huang, G. (eds.) WISE 2013. LNCS, vol. 8181, pp. 43–56. Springer, Heidelberg (2013). https://doi.org/10.1007/978-3-642-41154-0_4
8. Tang, L., Wang, X., Liu, H.: Community detection via heterogeneous interaction analysis. Data Min. Knowl. Discov. **25**(1), 1–33 (2012)
9. Ströele, V., Zimbrão, G., Souza, J.M.: Group and link analysis of multi-relational scientific social networks. J. Syst. Softw. **86**(7), 1819–1830 (2013)
10. Zhou, L., Yang, P., Lü, K., Zhang, Z., Chen, H.: A coalition formation game theory-based approach for detecting communities in multi-relational networks. In: Dong, X., Yu, X., Li, J., Sun, Y. (eds.) WAIM 2015. LNCS, vol. 9098, pp. 30–41. Springer, Cham (2015). https://doi.org/10.1007/978-3-319-21042-1_3
11. Saad, W., Han, Z., Debbah, M., Hjørungnes, A., Basar, T.: Coalitional game theory for communication networks: a tutorial. IEEE Signal Process. Mag. **26**(5), 77–97 (2009)
12. Lin, Y.-R., Choudhury, M.D., Sundaram, H., Kelliher, A.: Discovering multi-relational structure in social media streams. ACM Trans. Multimed. Comput. Commun. Appl. **8**(1), 1–28 (2012)
13. Zhang, Z., Li, Q., Zeng, D., Gao, H.: User community discovery from multi-relational networks. Decis. Support Syst. **54**(2), 870–879 (2013)
14. Li, X.T., Ng, M.K., Ye, Y.M.: Multicomm: finding community structure in multi-dimensional networks. IEEE Trans. Knowl. Data Eng. **26**(4), 929–941 (2014)
15. Lee, D.D., Seung, H.S.: Learning the parts of objects by non-negative matrix factorization. Nature **401**(6755), 788–791 (1999)

16. Wang, F., Li, T., Wang, X., Zhu, S., Ding, C.: Community discovery using nonnegative matrix factorization. Data Min. Knowl. Discov. **22**, 493–521 (2010)
17. Ma, X., Dong, D.: Evolutionary nonnegative matrix factorization algorithms for community detection in dynamic networks. IEEE Trans. Knowl. Data Eng. **29**(5), 1045–1058 (2017)
18. Gupta, S.K., Phung, D., Adams, B., Venkatesh, S.: A matrix factorization framework for jointly analyzing multiple nonnegative data sources. In: Yada, K. (ed.) Data Mining for Service. SBD, vol. 3, pp. 151–170. Springer, Heidelberg (2014). https://doi.org/10.1007/978-3-642-45252-9_10
19. https://archive.ics.uci.edu/ml/datasets/Robot+Execution+Failures
20. Halkidi, M., Batistakis, Y., Vazirgiannis, M.: On clustering validation techniques. J. Intell. Inf. Syst. **17**(2–3), 107–145 (2001)
21. Huang, X., Ye, Y., Guo, H., Cai, Y., Zhang, H., Li, Y.: DSKmeans: a new kmeans-type approach to discriminative subspace clustering. Knowl. Based Syst. **70**(2014), 293–300 (2014)
22. Solomonoff, A., Mielke, A., Schmidt, M., Gish, H.: Clustering speakers by their voices. In: IEEE International Conference on Acoustics, Speech and Signal Processing, vol. 2, pp. 757–760 (1998)

Recommender Systems

NSPD: An N-stage Purchase Decision Model for E-commerce Recommendation

Cairong Yan, Yan Huang, Qinglong Zhang$^{(\boxtimes)}$, and Yan Wan

School of Computer Science and Technology,
Donghua University, Shanghai 201620, China
wofmanaf@gmail.com

Abstract. In this paper, we proposed a scalable framework W&D (wide & deep framework) plus to capture users' personal interest for e-commerce recommender systems by combining the advantage of W&D and Residual Units. To better model users' actual purchase processes, we build an NSPD (N-stage Purchase Decision Model) based on W&D plus by splitting the shopping process into n stages. According to a real scenario, we maximize each stage probability and multiply them as the final probability. Besides, we capture users' evolving sequential preference to recommend the right product at the right time period. Experimental results on IJCAI 2015 from Tmall and Amazon Clothing, Shoes and Jewelry demonstrate that NSPD can outperform existing state-of-the-art models significantly.

Keywords: E-commerce recommender systems
Wide & Deep framework · Residual Units

1 Introduction

E-commerce RSs (recommendation systems) is a personalized recommendation tool to enhance the overall marketing performance of e-commerce platforms. By establishing user-centric personalized marketing strategy, e-commerce RSs can help platforms support the most needed information to the users at the most appropriate time, provide a more comfortable shopping experience, and enhance customer loyalty. The accuracy of the recommendation results is a key factor in determining the success or failure of the RSs. If the product recommended by the RSs is not required by the user, the user will lose confidence in the RSs and use the recommended information as Spam. Therefore, it is crucial to improving the accuracy of RSs.

Traditional works, including CF (collaborative filtering), user-item graph models, regression-based models and MF (matrix factorization) [1] have been proposed and applied to increase the accuracy of the recommendation results.

Supported by National Natural Science Foundation of China (grant No. 61402100, 61472075).

150   C. Yan et al.

These methods mainly made use of features such as implicit feedback [2,3], temporal dynamics [4,5], social influence [6,7], or the sequence of historical action records [8,9] to model users' purchase habits. Other works, such as VBPR (visual Bayesian personalized ranking) [10] and TVBPR (temporal dynamics VBPR) [11] further make use of visual features extracted from product images and demonstrated in some domains like fashion, visual features play an important role of the RSs.

Recently, DNN have quickly evolved and yielded great successes in several fields such as speech recognition, computer vision, and natural language processing. [12] introduced the concept of Fashion DNA, a mapping of fashion products to vectors using deep neural networks that are optimized to forecast purchases across a large group of customers. Based on item properties, [12] can circumvent the cold start problem and provide products recommendations even in the absence of prior purchase records. [13] presented a general framework named NCF by replacing the inner product of matrix factorization with a neural architecture that can learn an arbitrary function from data, and shows better recommendation performance of using deeper layers of neural networks. [14] presented us the W&D learning framework, which combines the strength of wide linear models and deep neural networks, and can effectively memorize sparse feature interactions using cross-product feature transformations, as well as generalize to previously unseen feature interactions through low-dimensional embeddings.

Yet, there is still no model that can analyze users' actual purchase processes (e.g., if a user wants to buy a headphone in Amazon, he may query the headphone category, then select the headphone style, the brand, the feature, etc., and finally select a product.). The whole process is shown in Fig. 1.

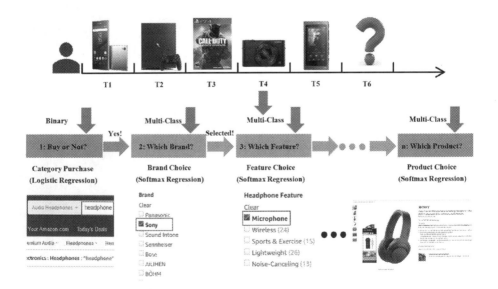

Fig. 1. General workflow of the purchase decision.

In this paper, our goal is to recommend the right products at the right period to the right user for e-commerce platforms by constructing an n-stage purchase decision model. We believe good RSs should maximally imitate the real scenario of users' normal life. Methodologically, we split users' shopping process into n stages based on real scenarios, then we compute the probability of each stage and select the product which gains the highest probability (here, we only concentrate on the accuracy of RSs. For diversity, we can simply select the top-k probability of each stage), we multiply the selected probability and the result is the interest of final selected product under all conditions.

To better capture more features that may affect user' purchase decision, we proposed a deep neural network based scalable framework W&D by combining the advantage of W&D [14] and Residual Units [16]. We modify W&D by employing RCNN [15] to capture text information, CNN to capture visual information and other engineering techniques to capture more features. Then we add a concatenated layer to merge all these features. To increase network multiplicity and make neural network deeper, we further add a Residual layer before the final scoring layer. In addition, we seek to capture user's evolving sequential preference to make the recommended result more reasonable (i.e., if a user has purchased some game related products such as Play Station®4, Play Station®VR, headphone, etc., within a short time in coming, he may buy VR games rather than other electronic equipment). Existing approaches of e-commerce RSs can be seen as a particular case of our model with only one stage.

Note that stages in our model are independent, so is the data used in each stage. Thus, it is easy to implement NSPD into popular distributed machine learning frameworks such as Theano, MXNET, and TensorFlow. Experimental results show that our model exhibits significant performance improvements on two real-world datasets like IJCAI 2015, Amazon Clothing, Shoes and Jewelry. Specifically, our main contributions are listed as follows:

(1) We build large scalable stage models to capture users' personal interest for e-commerce RSs. Methodologically, we split user's shopping process into n stages according to a real scenario, maximize each stage probability, and multiply them as the final probability.
(2) We proposed a deep neural network based scalable framework W&D plus to capture users' personal interest for e-commerce RSs by combining the advantage of W&D and Residual Units.
(3) We captured users' evolving sequential preference to recommend the right products at the right time period.
(4) Experimental results on IJCAI 2015 from Tmall and Amazon Clothing, Shoes and Jewelry demonstrate that our proposed models can outperform state-of-the-art models significantly.

In addition, NSPD can be easily adapted to coping with sequential and dynamics prediction in other domains by re-split the stage of action process. Particularly, by controlling the start stage and end stage, we can apply the model to different scenarios. E.g., if the stage ends with the merchant select, the final probability is the preference of a merchant, so consider user and merchant as a pair, then the merchant can identify which user can be converted to regular loyal buyers and then target them to reduce promotion cost and increase the ROI (return on investment). Similarly, if the stage ends with the brand selected, the final probability is the brand loyalty, so it is reasonable to recommend new products for the brand followers. Thus, the model can solve the cold start of totally new products perfectly. The overall view of our model is presented in Fig. 2.

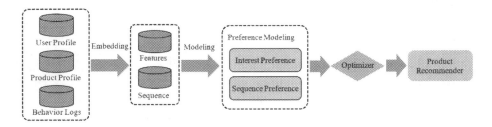

Fig. 2. An n-stage purchase decision model for e-commerce recommendation.

2   Related Work

The most closely related works to ours are (1) e-commerce recommendations that model user preferences or item similarities; (2) works that deal with temporal dynamics and (3) those that address the sequential prediction task we are interested in.

E-commerce Recommendations. The most important goal for e-commerce platform is to turn visitors into paying customers. Therefore, it is essential to investigate users' preferences and purchase behaviors. Much of e-commerce platforms such as Amazon, Tmall, JD.COM etc. have built personalized recommendation engines to increase sales revenues. Among all these engines, the most widely used method is CF. Approaches such as user-based CF, item-based CF and MF have gained great success. The most successful example of item-based CF perhaps is Amazon, who used users' history actions (i.e., browse information, purchase information, context information, etc.) and other explicit and implicit information, as well as item-based collaborative filtering algorithm to recommend the products that users might interested in. [13] present a neural network architecture to model latent features of users and items for collaborative filtering and shown promising results, other works such as [14] which jointly trained wide linear models and deep neural networks to combine the benefits of memorization and generalization for recommender systems. [17] described a deep neural network architecture for recommending YouTube videos, which transform special representations of categorical and continuous features with embeddings and quantile normalization, respectively. [18] presented a dynamic recurrent basket model based on RNN for next basket recommendation, merge users' current interests and global sequential features into users' recurrent and dynamic representation. In this paper, we combine the strength of [13, 14, 18] in the computing of stage probabilities.

Temporal Dynamics. There have been some works in the RSs community taking dynamics into account, mostly based on MF techniques such as [1, 11, 19]. [1] decompose ratings into distinct terms, so the RSs can tackle different temporal aspect separately. Specifically, [1] considered three terms varying over time: item biases, user biases, and user preferences. [19] proposed a new matrix factorization model GBPMF and integrated a temporal factor in the process of the Gibbs sampling algorithm. [11] presented a flexible temporal model that is capable of accounting for multiple sources of visual temporal dynamics, such as dynamics at the user or community level, the introduction of new products, or sales promotions that impact the choices people make in the short term.

Sequential Recommendation. Recent works have shown that sequential patterns can be utilized to improve personalized recommendations at the right time. For example, FPMC [20] combines the strength of Markov chains at modeling the smoothness of subsequent actions and the power of Matrix factorization at modeling personal preferences for the sequential recommendation. Combining FPMC and VBPR, [7] can capture individual users' preferences towards visual art styles, as well as the tendency to interact with items that are 'visually consistent' during a browsing session. Other works like [21] proposed an opportunity model to estimate the follow-up purchase probability of a user at a specific time. To recommend the best next product to each target user, Yap et al. [22] learned user-specific sequential knowledge through personalized sequential pattern mining. [23] presented an architecture for an adaptive user interface with an RNN that performs sequential recommendation of content and control elements, which employs GRU to map user interaction histories to vectors in a Euclidean space shared with user and target item vectors. Moreover, the time interval information between purchase transactions was used to improve the performance of next product recommendations [24].

Similarly, e-commerce recommendation is also a time-sensitive task, since people's interests in products are drifting over time. Therefore, it is necessary to capture the users' evolving sequential preferences. To make recommendations at the right time, this paper focuses on e-commerce recommendations for a specific period by considering users' evolving sequential preferences.

3   Model Description

3.1   Feature Processing

We begin to describe our model from feature processing. We divide the original individual features into four types: textual feature, categorical feature, continuous feature, and visual feature.

Each feature is represented as a vector. For textual data, such as product descriptions (e.g., "This adorable basic ballerina tutu is perfect for dance recitals. Fairy Princess Dress up, costume, play and much."), we use RCNN to convert the string into a real valued vector. We first apply a bi-directional recurrent structure to capture the contextual information, then employ a max-pooling layer to judge which feature plays a key role in the text and converts texts with various length into a fixed-length vector. We then take the output of the max pooling layer as the text feature vector.

Categorical inputs such as brand are represented by a one-hot vector. Consider the example in Fig. 1, brands have eight values: Panasonic, Sennheiser, Sony, Sound Intone, Bose, Beats, Audio-Technica, and Ausdom, then we can use a vector with length 8 to represent brands, and the values in the vector correspond to each brand (e.g., Panasonic: (1, 0, 0, 0, 0, 0, 0, 0), Sennheiser: (0, 1, 0, 0, 0, 0, 0, 0), etc.).

For the continuous feature, we use z-score normalization to speed up the gradient calculation. The normalized value d_i of row D in the i-th column is calculated as Normalized$(d) = (d_i - \bar{D})/std(D)$ We save the parameters (mean, variance) of train set and use them to normalize the test set data.

In addition, we extract visual features f_i as a 4096-dimensional feature vector from product image by taking the output of the second fully-connected layer (i.e., FC7) of AlexNet [25], we then use a weight matrix $\boldsymbol{w}_f$ to embedding original feature space f_i into a low-rank visual space (i.e., the final visual feature is $\boldsymbol{w}_f^T f_i$, here w_f is a $4096 \times F$ matrix, then the dimension of $\boldsymbol{w}_f^T f_i$ should be F, $F \ll 4096$).

However, simply using raw features rarely provides optimal results. Therefore, in both industry and academia, there exists a large body of work on feature engineering to transform the raw features. One major type of transformation is to construct functions based on a combination of cross features, and use their outputs as the input to a learner. Cross-product transformations employed in this paper is like [14]. Given individual features $x_i \in \Re^{n_i}$ and $x_j \in \Re^{n_j}$, cross feature $x_{i,j}$ is defined in $\Re^{n_i} \times \Re^{n_j}$. Here we use AND (age=20, gender=male) to transform raw features into cross features, whose value is 1 if the user's age range is 20 and his gender is male. The overall view of feature processing is presented in Fig. 3.

3.2   N-stage Purchase Decision Model

Interest Preference. Given m_1 users and m_2 items, we denote the order list of N observed events as $E = \{e_j = (u_j, i_j, t_j, q_j, d_j)\}_{j=1}^N$ on time window $[0, T]$, where $u_j \in \{1, \cdots, m_1\}$, $i_j \in \{1, \cdots, m_2\}$, $0 \leq t_1 \leq t_2 \leq \cdots \leq T$. This represents the interaction between user u_j, item i_j at time t_j with the interaction context q_j after $k-1$ stages chosen of d_j. Here q_j can be a high dimension vector such as the text review, user profile or simply some embedding of static user/item features. $d_j \in \{d_j^1, \cdots, d_j^{n-1}\}$ represents the $n-1$ stages (e.g., category chose, brand chose, etc., in Fig. 1) we set. Then user's preference can be represented by the joint probability of stage decisions.

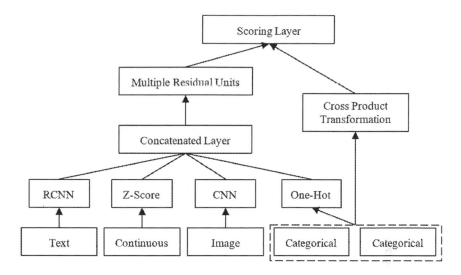

Fig. 3. Wide & Deep plus framework for n-stage purchase decision model.

$$P(e_j) = P(d_j^1) \times P(d_j^2|d_j^1) \times \cdots \times P(i_j|d_j^{n-1}) \times P(t_j|i_j) \tag{1}$$

Here $P(t_j|i_j)$ represents user's evolving sequential preference, which is the likelihood of the prediction of buying i_j at time t_j. To simplify, we set $t_0^u = 0$ for all users.

Stage probability can be turned into a binary classification problem or a multi-classification problem. Consider the example in Fig. 1, whether a user will make a purchase from a category is a binary classification problem. If so, we model which brand the user will choose from this category as a multi-classification problem.

In this paper, we model the decision stage based on W&D framework [14], which is flexible enough to include a range of features and to fit different prediction scenarios (for different stages of purchase behavior). The prediction of stage k (here we assume stage k is a binary classification problem or multi-classification problem) is

$$P(d^k|d^{k-1}) = f(\boldsymbol{w}_{\text{wide}}^{\text{T}}[x_k, \phi(x_k)] + \boldsymbol{w}_{\text{deep}}^{\text{T}}\alpha_k^{(lf)} + b) \tag{2}$$

where $f(*)$ is the sigmoid function for binary classification or softmax function for multi-classification, $\phi(x_k)$ is the cross-product transformation of the k-th stage related individual features x_k, and b is the bias term. Parameter $\boldsymbol{w}_{\text{wide}}$ is the vector of all wide model weights and $\boldsymbol{w}_{\text{deep}}$ is the weight applied on the final activations $\alpha^{(lf)}$.

Recently, ResNet [16] has achieved remarkable success, both in image recognition and text classification. In this paper, we also add residual layers to the W&D plus framework. The multiple Residual Units we used here is similar to [26], which is slightly modified original Residual Units since we do not use convolution kernels. We called the modified W&D framework W&D plus, which is presented in Fig. 3. The unique property of Residual Unit is to add back the original input feature after passing it through two layers of ReLU transformations. Specifically:

$$x^o = F(x^i) + x^i \tag{3}$$

Where $F(*)$ denotes the function that maps the input x^i of the Residual Unit to the output x^o. The Residual Units is presented in Fig. 4. To avoid over fitting, in each Unit, we use batch normalization [27] and dropout [28] (in NSPD, neurons in the training part is kept with a probability of 0.5).

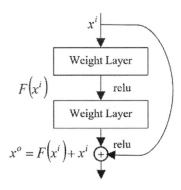

Fig. 4. Residual Units

Sequence Preference. For each user, we formulate the corresponding embedding $f_u(t)$ after u's l-th event $e_l^u = (i_l^u, t_l^u, q_l^u, d_l^u)$ as follow:

$$f_u(t) = \sigma\Big(\underbrace{\boldsymbol{w}_1(t_l^u - t_{l-1}^u)}_{\text{temporal drift}} + \underbrace{\boldsymbol{w}_2 q_l^u}_{\text{interaction feature}} \Big) \tag{4}$$

Here, i_l^u is the l-th action record item ordered by the corresponding even time t_l^u. $\boldsymbol{w}_1$ and $\boldsymbol{w}_2$ are weight matrices. $\sigma(*)$ is a sigmoid function. Then the conditional probability of observing an event in a small window $[t', t]$ can be represented as

$$P(t|i) = f_u(t)\exp\Big(-\int_{t'}^{t} f_u(\tau)d\tau \Big) \tag{5}$$

where $t > t'$ and t' is the last time point when user u have an action with items. Particularly, all weight matrix used in this paper is initialized by He initialization. [29] points out that if DNN uses ReLu as activation function, He initialization can gain a better result.

Log-Likelihood Maximization. We infer each stage independently. Assume Θ_k be the parameter sets of stage k, y_k be the result, c_k be the corresponding

labels, then the parameters can be learned by maximizing the log-likelihood function as follows:
For binary classification:

$$J_k = \sum_{\Theta_k} \Big[c_k \log y_k + (1 - c_k) \log(1 - y_k) \Big] \tag{6}$$

For multi-classification or sequence preference:

$$J_k = \sum_{\Theta_k} c_k \log y_k \tag{7}$$

Here, we use BPR [28] to optimize NSPD. Note that not purchasing an item does not necessarily indicate a user dislike it, so optimize a criterion that means purchased item is more preferred than non-purchased ones is more reasonable. NSPD can be evaluated with the widely-used AUC measure:

$$\text{AUC} = \frac{1}{N} \sum_{u,t} \frac{1}{|E_{u,t}^+||E_{u,t}^-|} \sum_{i \in E_{u,t}^+, i' \in E_{u,t}^-} \sigma\Big(P(e^i) > P(e^{i'}) \Big) \tag{8}$$

where $E_{u,t}^+$ is the set of items selected by user u at time t and $E_{u,t}^-$ include items (to simply, we only sample 10 items) that were not selected. $\delta(*)$ is the indicator function ($\delta(x) = 1$ if x is true, otherwise $\delta(x) = 0$). In practice, we maximize the following log-likelihood function:

$$\arg\max_{\Theta} \sum c_n \sum_{i' \neq i} \log\Big(P(e^i) > P(e^{i'}) \Big) - \frac{\lambda}{2} \|\Theta\|^2 \tag{9}$$

where Θ is the set of all parameters and λ is the L2-regularized parameter.

3.3   Model Optimization

The optimization problem can be learned by SGD. However, since each stage of our model is independent, so is the data used in each stage. So, our model can be trained in a distributed framework. Here, we use ASGD [31] to train NSPD.

4   Experiments

4.1   Experimental Setup

Datasets. To evaluate the efficacy of NSPD, we performed experiments on real datasets IJCAI 2015 from Tmall and Amazon Clothing, Shoes and Jewelry. IJCAI 2015 datasets are the sales data of the "Double 11" day of 2014 at Tmall, which include a number of merchants and their new buyers acquired during the event, and six months of user activity log data before the event. The detailed information about IJCAI 2015 datasets is shown in Table 1.

The second group of datasets Amazon Clothing, Shoes and Jewelry contains 5 large categories: Women's, Men's, Girls', Boys' and Baby's Clothing& Accessories, of which the images are available with high quality (typically centered on a white background) and have previously been shown to be effective for recommendation tasks. We process each data set by taking user's review histories as implicit feedback and extracting visual features from one image of each item. We discard users who have performed fewer than 5 actions. Statistics of Clothing, Shoes, and Jewelry are shown in Table 2. Price in Table 2 means the max price of the datasets and description means the average length of description text (characters such as basic punctuation, tabs, and newlines are filtered out).

Table 1. IJCAI 2015 statistics.

Data	User	Merchants	Pairs	Positive pairs	Positive %
Train	212,062	1,993	260,864	15,952	6.12%
Test	212,108	1,993	261,477	16,037	6.13%

Table 2. Amazon datasets statistics.

Dataset	Users	ASIN	Feedback	Cate	Brands	Price	Des	Rate	Time
Baby	89868	32507	123985	80	287	500	58	5	Mar 2003–Jul 2014
Boys	163120	40604	200207	138	629	180	61	5	Mar 2003–Jul 2014
Girls	113852	41826	142219	147	517	489	87	5	Mar 2003–Jul 2014
Men	1179387	267857	1675610	195	2781	1000	80	5	Mar 2003–Jul 2014
Women	1630803	603277	2855258	237	3441	1000	69	5	Mar 2003–Jul 2014

Baselines. MF based methods are currently state-of-the-art for modeling implicit feedbacks datasets. Therefore, we mainly compare against state-of-the-art MF methods it this area, both in the field of item recommendation and sequential data prediction. In addition, to demonstrate our proposed W&D plus framework works, we also compare it with the original one.

(1) BPR-MF: Introduced by [30], is a state-of-the-art method for personalized ranking on implicit feedback datasets. It uses standard MF as the underlying predictor.
(2) BPR-TMF: This model extends BPR-MF by making use of taxonomies and temporal dynamics, i.e., it adds a temporal category bias as well as a temporal item bias in the standard MF predictor.
(3) VBPR: This method models raw visual signals for recommendation using the BPR framework [10], but does not capture any temporal dynamics as we do in this work.

(4) TVBPR: This method models visual dimensions and captures visual temporal dynamics, but only relies on explicit timestamps [11].

(5) W&D: This method combines the strength of both wide linear models and deep neural networks. It can effectively memorize sparse feature interactions as well as generalize to previously unseen feature interactions through low-dimensional embedding.

(6) W&D plus: This method extends W&D by making use of Residual Units and embedding more features such as visual features and text features, presented in Fig. 3.

(7) NSPD: A method proposed in this paper in Eq. (8). Different stage split will be experimented with and compared to other methods. One-stage NSPD without sequential preference with temporal interval assessment is W&D plus model.

Ultimately, these methods are designed to evaluate (1) the performance of the current state-of-the-art methods on our task; (2) the importance of modeling visual features and sequentially dynamics; and (3) the strength of our proposed NSPD.

Platform. All experiments are conducted on 3 Linux servers, each of them has 8 physical cores on Intel®Core^{TM} i7-7700 CPU @ 3.60 GHz processors and 65.86 GB memory.

Implementation. We implement W&D, W&D plus and NSPD in Python, BPR-MF, BPR-TMF, VBPR, and TVBPR is implemented in C++. For dataset IJCAI 2015, we split NSPD into three stages, i.e., category purchase, brand choice, and product choice. Since most users of IJCAI 2015 dataset are one-time buyers, so we do not consider sequence preference. For Amazon Clothing, Shoes and Jewelry datasets, we also split NSPD into three stages like IJCAI 2015. For names appeared in Sect. 4.2, we make some explanations: 2-stage NSPD means the purchase process is split into two stages, i.e., category select, product purchase. 3-stage NSPD means the purchase process is split into three stages, i.e., category select, brand select, product purchase. 4-stage NSPD means sequential preference is added beyond 3-stage NSPD. The initial learning rate of all models is 0.001. For lacking sufficient computation resource, each Residual Unit only consists of 3 layers and each stage only have 2 Residual Units. The output size of the second Residual Unit is half of the first one (in fact, we set the output size of Residual Units to be [128, 64]). Parameters of RCNN and AlexNet is same as [15, 25].

4.2   Performance

Demonstrate of Residual Layer and Stage Decision. To speed up, we only select 20% data in this part. First, we set an experiment on Amazon Clothing, Shoes and Jewelry to test W&D plus. We use cross-entropy (the small the better) and AUC (the higher the better) as Evaluation Methodology. The best result is shown in Table 3.

From Table 3 we can see, W&D plus beats W&D by making use of Residual Units and visual features. The average improvement is 20.6% on loss aspect and 10.0% on AUC aspect.

Since IJCAI 2015 has no information of the image, here we compare W&D, W&D plus and NSPD only with non-visual features. The best result is shown in Table 4.

From Table 4 we can see that W&D plus gains 2.8% improvement over W&D by making use of Residual Units. Considering stage split, NSPD gains 2.2% improvement with 2-stages and 5.5% improvement with 3-stage over W&D plus.

Table 3. Performance of W&D, W&D plus on Amazon Clothing, Shoes and Jewelry.

Models	Evaluation	Baby	Boys	Girls	Men	Women
W&D	loss	0.525	0.463	0.474	0.492	0.399
	AUC	0.680	0.664	0.685	0.638	0.719
W&D plus	loss	0.376	0.413	0.407	0.671	0.382
	AUC	0.824	0.709	0.715	0.67	0.806

Table 4. Performance of W&D, W&D plus and NSPD on IJCAI 2015.

Models	W& D	W& D plus	2-stage NSPD	3-stage NSPD
AUC	0.562	0.578	0.591	0.610

Performance Comparison of All Models. We evaluate all models on Amazon Clothing, Shoes and Jewelry with two settings: All Items and Cold Start. All Items measures the overall ranking accuracy, including both warm start and cold start scenarios. Cold Start uses the subset comprising only cold items with interactions fewer than 5, which is especially significant in the domains like fashion where new items are constantly added to the system and the data is incredibly long-tailed. Table 5 compares the performance of different models with their best result.

We make comparisons to better explain and understand our finding as follows:

(1) BPR-TMF beats the state-of-the-art method BPR-MF for personalized ranking from implicit feedback by modeling temporal dynamics. This also been true when we compare VBPR with TVBPR, which lead to 7.0%, 8.7%, 13.6%, 9.3%, and 9.6% improvement respectively.

(2) TVBPR gain significant improvements over BPR-TMF by making use of additional visual signals. This lead to 3.3% improvement on Baby' Clothing, 8.9% on Boys' Clothing, 9.4% on Girls' Clothing, 6.2% on Men's Clothing and 5.6% on Women's Clothing on all items setting.

Table 5. Performance of all models.

Data	Set	BPR-MF	BPR-TMF	VBPR	TVBPR	W&D++	2-NSPD	3-NSPD	4-NSPD	Impv
Baby	All	0.5330	0.5475	0.5390	0.5658	0.5757	0.6479	0.6519	0.7159	26.5%
	Cold	0.4805	0.4925	0.4857	0.5085	0.5552	0.6005	0.6212	0.6534	28.5%
Boys	All	0.5180	0.5309	0.5321	0.5784	0.5741	0.6459	0.6758	0.7267	25.6%
	Cold	0.4973	0.5060	0.5083	0.5585	0.5516	0.5851	0.6105	0.6747	20.8%
Girls	All	0.4991	0.5156	0.5066	0.5640	0.5838	0.6915	0.7231	0.7545	33.8%
	Cold	0.4937	0.5100	0.4882	0.5435	0.5759	0.6306	0.6580	0.6854	26.1%
Men	All	0.6419	0.6612	0.6525	0.7025	0.6762	0.6914	0.7185	0.7361	4.8%
	Cold	0.6824	0.5107	0.4865	0.5913	0.5566	0.6629	0.6364	0.6398	8.2%
Women	All	0.6322	0.6803	0.6551	0.7183	0.7085	0.7290	0.7364	0.7576	5.5%
	Cold	0.4889	0.5198	0.4814	0.7179	0.6727	0.6913	0.7143	0.7267	1.2%

(3) W&D plus shows almost the same performance with TVBPR, which gain slightly improvement on Baby' Clothing, Boys' Clothing, and Girls' Clothing, but lost to TVBPR on Men's Clothing and on Women's Clothing. The reason leads to this may be W&D plus do not consider temporal dynamics like TVBPR do.

(4) NSPD split users purchase behaviors into three stages and gain promising performance on all the five datasets. The average improvement over TVBPR is 12.0% on all items setting and 14.8% on cold start setting.

(5) NSPD obtains more improvements when considering sequential preference (4-stages), both in all items setting and cold start setting. The average improvement over TVBPR is 19.1% on all items setting and 16.3% on cold start setting.

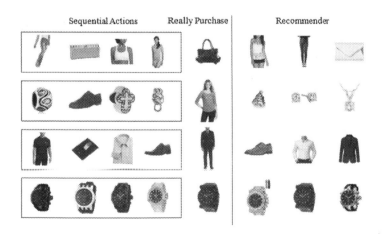

Fig. 5. Demonstration of sequential consistency

Sequential Prediction. User's habit differs greatly, especially in a big dataset like ours. To some people, the long-term interaction is at play, as demonstrated in the above experiments. However, to others, short-term interactions are more important (i.e., people are more likely to click or buy items similar to what they have just bought or reviewed). In this part, we randomly sample 2 users for each of the five Amazon Clothing, Shoes and Jewelry datasets. We generate a top-3 recommendation list for the selected users with NSPD (4-stage) to test the sequential effects on RSs. The result is shown in Fig. 5. From Fig. 5 we can see that baby, boys, girls, and men show strong sequential consistency. For women, historical actions sequence seems to have limited effect on future actions.

5   Conclusion and Discussion

In this paper, we proposed a scalable framework W&D plus to capture user's personal interest in e-commerce RSs by combining the advantage of W&D and Residual Units. To better model user's actual purchase processes, we build NSPD based on W&D plus by splitting the shopping process into n stages according to a real scenario, we maximize each stage probability and multiply them as the final probability. Besides, we capture user's evolving sequential preference with temporal interval assessment to recommend the right products at the right time period. We further joint user's personal interest and sequence preference to combine the benefits of personal models and sequential models. In addition, we implemented NSPD into distributed machine learning frameworks to cope with the continuous increasing data challenge. Experiments results on IJCAI 2015 from Tmall and Amazon Clothing, Shoes and Jewelry demonstrate that NSPD can outperform state-of-the-art models significantly.

However, there are still problems to be solved. One of them is how to generate a top-k recommendation list for any target user (in this paper, we simply recommended products for several sampled users) since the number of products is typically very large (e.g., millions) and materializing the entire rating prediction list for all users is practically infeasible.

In the future, we plan to study the retrieval phase of RSs (i.e., recommended phase), we will focus on reducing top-k retrieval cost.

References

1. Koren, Y., Bell, R., Volinsky, C.: Matrix factorization techniques for recommender systems. Computer **42**(8), 99–110 (2009)
2. Koren, Y., Bell, R.: Advances in collaborative filtering. In: Ricci, F., Rokach, L., Shapira, B. (eds.) Recommender Systems Handbook, pp. 77–118. Springer, Boston, MA (2015). https://doi.org/10.1007/978-1-4899-7637-6_3
3. Li, H., et al.: FEXIPRO: fast and exact inner product retrieval in recommender systems. In: Proceedings of the 2017 ICDM, pp. 835–850. ACM (2017)
4. Koren, Y.: Collaborative filtering with temporal dynamics. Commun. ACM **53**(4), 89–97 (2010)

5. Zhang, Y., et al.: Daily-aware personalized recommendation based on feature-level time series analysis. In: Proceedings of the 24th International Conference on World Wide Web, pp. 1373–1383. International World Wide Web Conferences Steering Committee (2015)

6. Ma, H., et al.: SoRec: social recommendation using probabilistic matrix factorization. In: Proceedings of the 17th ACM Conference on Information and Knowledge Management, pp. 931–940. ACM (2008)

7. He, R., et al.: Vista: a visually, socially, and temporally-aware model for artistic recommendation. In: Proceedings of the 10th ACM Conference on Recommender Systems, pp. 309–316. ACM (2016)

8. Mobasher, B., et al.: Using sequential and non-sequential patterns in predictive web usage mining tasks. In: Proceedings of the 2nd International Conference on Data Mining, pp. 669–672. IEEE (2002)

9. He, R., McAuley, J.: Fusing similarity models with Markov chains for sparse sequential recommendation. In: Proceedings of the 16th ICDM, pp. 191–2000. IEEE (2016)

10. He, R., McAuley, J.: VBPR: visual Bayesian personalized ranking from implicit feedback. In: Proceedings of the 30th AAAI Conference on Artificial Intelligence, pp. 144–150. AAAI (2016)

11. He, R., McAuley, J.: Ups and downs: modeling the visual evolution of fashion trends with one-class collaborative filtering. In: Proceedings of the 25th International Conference on World Wide Web, pp. 507–517. International World Wide Web Conferences Steering Committee (2016)

12. Bracher, C., Heinz, S., Vollgraf, R.: Fashion DNA: merging content and sales data for recommendation and article mapping. arXiv preprint arXiv:1609.02489 (2016)

13. He, X., et al.: Neural collaborative filtering. In: Proceedings of the 26th International Conference on World Wide Web, pp. 173–182. International World Wide Web Conferences Steering Committee (2017)

14. Cheng, H.T., et al.: Wide & deep learning for recommender systems. In: Proceedings of the 1st Workshop on Deep Learning for Recommender Systems, pp. 7–10. ACM (2016)

15. Lai, S., et al.: Recurrent convolutional neural networks for text classification. In: Proceedings of the 29th AAAI, pp. 2267–2273. AAAI (2015)

16. He, K., et al.: Deep residual learning for image recognition. In: Proceedings of the 29th CVPR, pp. 770–778. IEEE (2016)

17. Covington, P., Adams, J., Sargin, E.: Deep neural networks for YouTube recommendations. In: Proceedings of the 10th ACM Conference on Recommender Systems, pp. 191–198. ACM (2016)

18. Yu, F., et al.: A dynamic recurrent model for next basket recommendation. In: Proceedings of the 39th International ACM SIGIR, pp. 729–732. ACM (2016)

19. Yan, C., et al.: A method of Bayesian probabilistic matrix factorization based on generalized gaussian distribution. J. Comput. Res. Dev. 53(12), 010–018 (2016)

20. Rendle, S., Freudenthaler, C., Schmidt-Thieme, L.: Factorizing personalized Markov chains for next-basket recommendation. In: Proceedings of the 19th International Conference on World Wide Web, pp. 811–820. ACM (2010)

21. Wang, J., Zhang, Y.: Opportunity model for e-commerce recommendation: right product; right time. In: Proceedings of the 36th International ACM SIGIR, pp. 303–312. ACM (2013)

22. Yap, G.-E., Li, X.-L., Yu, P.S.: Effective next-items recommendation via personalized sequential pattern mining. In: Lee, S., Peng, Z., Zhou, X., Moon, Y.-S., Unland, R., Yoo, J. (eds.) DASFAA 2012. LNCS, vol. 7239, pp. 48–64. Springer, Heidelberg (2012). https://doi.org/10.1007/978-3-642-29035-0_4

23. Soh, H., et al.: Deep sequential recommendation for personalized adaptive user interfaces. In: Proceedings of the 22nd International Conference on Intelligent User Interfaces, pp. 589–593. ACM (2017)

24. Zhao, G., et al.: Increasing temporal diversity with purchase intervals. In: Proceedings of the 35th International ACM SIGIR, pp. 165–174. ACM (2012)

25. Krizhevsky, A., Sutskever, I., Hinton, G.E.: ImageNet classification with deep convolutional neural networks. In: Advances in Neural Information Processing Systems, pp. 1097–1105 (2012)

26. Shan, Y., et al.: Deep crossing: web-scale modeling without manually crafted combinatorial features. In: Proceedings of the 22nd SIGKDD, pp. 255–262. ACM (2016)

27. Ioffe, S., Szegedy, C.: Batch normalization: accelerating deep network training by reducing internal covariate shift. arXiv preprint arXiv:1502.03167 (2016)

28. Srivastava, N., et al.: Dropout: a simple way to prevent neural networks from overfitting. J. Mach. Learn. Res. **15**(1), 1929–1958 (2014)

29. He, K., et al.: Delving deep into rectifiers: surpassing human-level performance on ImageNet classification. In: Proceedings of the 28th CVPR, pp. 1026–1034. IEEE (2015)

30. Rendle, S., et al.: BPR: Bayesian personalized ranking from implicit feedback. In: Proceedings of the 25th AUAI, pp. 452–461. AUAI (2009)

31. Zhang, S., et al.: Asynchronous stochastic gradient descent for DNN training. In: Proceedings of 2013 IEEE International Conference on Acoustics, Speech and Signal Processing, pp. 6660–6663. IEEE (2013)

Social Image Recommendation Based on Path Relevance

Zhang Chuanyan[⊠], Hong Xiaoguang, and Peng Zhaohui

Shandong University, Jinan 250101, China
chuanyan_zhang@sina.cn, {hxg, pzh}@sdu.edu.cn

Abstract. With the incredibly growing amounts of images shared on the social network, it's necessary to ease the uses' burden on the information overload through recommendation systems. For social image recommendation, heterogeneous information, such as image content, tags, user relationships, in addition to the user-image preferences, is extremely valuable for making effective recommendations. However, most existing social image recommendation methods mainly focus on the user-image topological structure, but largely ignore the context information. In this paper, we explore a novel algorithm for social image recommendation based on path relevance(PR). Firstly, we model both the user behaviors and image properties on a heterogeneous network and PR, a unified distance measure, is given. Then, top-k images are recommended for each personalized user through PR between user and image. Our methods can tackle the challenges of highly spares representation problem in the social network scenario. Further, an approximate method is proposed to adaptively learn the weights of different semantic edges on the heterogeneous network. We evaluate our approach with a newly collected 100,000 social image data set from Flickr. The experimental results demonstrate that our method leads to more effective recommendations, with a significant performance gain over the state-of-the-art alternatives.

Keywords: Recommendation systems · Social image recommendation
Information network · Relevance measure · Sparse representation

1 Introduction

With the advent of social media networks, such as YouTube, Twitter, Facebook, and Flickr, tremendous multimedia data, likely images and videos are growing in an unprecedented speed. Information recommendation is considered as the key technology to ease the data overload on social network.

To facilitate information recommendation, many methods have been proposed. Generally, the traditional recommendation methods can be classified into 2 categories: (1) Content-based filtering (CBF) method and (2) Collaborative filtering (CF) method. CBF recommends items with the similar contents as the ones of user's interests, where user's interests and item contents are modeled as vectors with the same dimensions. Thus, the key problem of CBF is to discover series of attributes to represent item and user' preferences with high accuracy and coverage [1]. It's suggested that the linkage in

© Springer International Publishing AG, part of Springer Nature 2018
Y. Cai et al. (Eds.): APWeb-WAIM 2018, LNCS 10987, pp. 165–180, 2018.
https://doi.org/10.1007/978-3-319-96890-2_14

social network represents much more information than the human defined feature vector about the items, especially for the web-scale data. The linkage, on behalf of user behavior, i.e., the "human signal", could be binary ratings "Like" on Facebook, retweets on Twitter, numeric ratings on IMDb and so on. CF uses the users' preference matrix to list every user's signal about all the items in social network where similarity can be measured between row vectors or column vectors. Comparing with CBF, CF can efficiently bridge the semantic gap between multimedia and user preference [2]. To improve the efficiency of CF, Matrix factorization (MF) is adopted which factorize the user's preference matrix into a product of two low rank matrices U and V and conducts CF by exploiting U and V [3].

Even MF based CF has been proved to be one of the most successful approaches to recommendation, it still has some limitations. Since the users' preference matrix is difficult to express more information, the MF based CF can't handle content information images, such as social relationships, tags and image content, which are valuable for making effective recommendations [4]. Further researches study the context-aware recommendation (CAR). General MF methods for CAR tend to introduce many contextual factors to extends the classical two-dimensional matrix factorization problem into an n-dimensional version of the same problem [5, 6]. Obviously, a huge number of parameters must be learned using the training data. In fact, it is shown that the number of model parameters grow exponentially with the number of contextual factors. For social image recommendation, there are thousands of dimensions for images and inappropriate handling of them will cause curse of dimensionality.

Moreover, the sparsity of the users' preference matrix poses a great challenge on CF-based algorithms. When the matrix becomes sparse, MF-based CF will be extremely non-robust and easily disturbed by noise. Compared with the music and movie recommendation scenario in which there are intensive feedbacks provided by users and limited items, the user's preference matrix becomes more highly sparsely in the scenario of social image recommendation. To avoid the sparsity problem, community recommendation is proposed [7, 8]. Instead of personalized interest inference, community recommendation aggregates information from similar users or items to increase the information density. Thus, it leads to lose of personalization.

In conclusion, both structural and context information are important to make effective image recommendation, however current CF-based CAF methods hardly handle the high dimensions of images and the sparsity of social network. To tackle these challenges, we propose a novel method in this paper termed Path Relevance Recommendation (PRR) for social image recommendation. PRR incorporates the user's behaviors, i.e., the structure information and the context information (e.g., user relationship, tag, image content) in a heterogeneous network. Then, Path relevance is defined to measure the distance between user and image in the heterogeneous network and top-k images are recommended to each user. Further, to weight the contribution of different semantic edges, an iterative method is proposed to adaptively learn the weights. The proposed PRR is of merits in its generalization capability since other valuable information can be easily incorporated into the network without influencing the inner complexity. Even the network is still sparse, but it doesn't affect the result of the distance measure. The main contributions of this paper are summarized below:

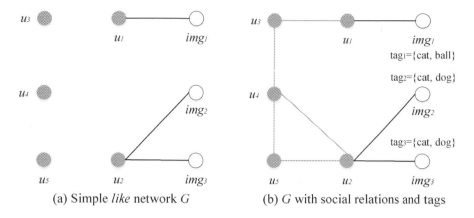

(a) Simple *like* network G   (b) *G* with social relations and tags

Fig. 1. Heterogenous network about user-image

1. We study the problem of image recommendation on social network and propose PRR method. The intuition is that the context information could be equally critical as the structure information. And motivated by the success of relevance search, we incorporated both the 2 kinds of information in a heterogeneous network and facilitate the personalized recommendation through relevance between user and item.
2. A neighborhood random walk model is used to measure the user-image relevance on a unified attributed network which models both the structure and context information. Theoretical analysis is provided to proof the efficiency of Path relevance to represent the users' preferences.
3. We propose an approximate weight self-adjustment method to learn the degree of contribution of different semantic edges in PR.
4. A real-world social image data set is collected from Flickr. The experiments demonstrate that our method can recommend high-quality images for each user even in sparse network.

The rest of this paper is organized as follows. Section 2 introduces the preliminary concepts for social image recommendation problem. Section 3 presents a unified framework to integrate user behaviors and context information. We propose an iterative recommendation algorithm for social image in Sect. 4. Section 5 presents experimental results, followed by related works in Sect. 6. Finally, Sect. 7 concludes the paper.

2   Preliminary

Heterogeneous Network can be denoted as a graph $G = (V, E)$, where V is the set of vertices, E is the set of edges. If there are more than one types of vertices or edges in G, G is a Heterogeneous Network, otherwise, it's a Homogeneous Network. Taking Fig. 1 as an example, graph (a) and (b) are both heterogeneous network. Further, if the vertices in a heterogeneous network G are associated with a set of attributes, G is an **Attributed Network**, e.g., image nodes have tags in graph (b) of Fig. 1.

An attributed network is denoted as $G = (V, E, \Lambda)$ and $\Lambda = \{a_1, \ldots, a_m\}$ is the set of attributes associated with vertices in V. Thus, for each vertex v_j of a specific class, if the vertex class has an attribute a_i, the value of v_j on attribute a_i can be denoted as $a_i(v_j)$. Explained specially, $a_i(v_j)$ can be a set of attribute values. In other words, for a single attribute of a vertex, there may be more than one values.

Definition 1. Given an attributed network $G = (V, E, \Lambda)$, where V is the set of users and images, each edge in E represents a relationship between vertex, and Λ denote the attribute values of the vertex, *social image recommendation* is to select *top-k* images which are not linked to the user and should be most relevant to him.

Table 1 lists the main symbols and their definitions.

Table 1. Definition of main symbols

Symbol	Definition
G	The attributed network
G_u	The unified attributed network
$V, U, M, T,$ Att	Vertex set, V contains all the vertex in G_u; U is the user set; M is the image set; T is the tag set and Att is the set of attributed vertex.
a_i	The ith attribute of image
E	The edge set in G_u
ET, et	ET contains all the edge types in G_u, $et_i \in ET$
A	The transition probability matrix
p_i	The path relevance vector from vertex i to all the vertex in G_u
W	The edge type weight vector
$Top\text{-}k$	$Top\text{-}k$ images are selected to recommend

The original goal of recommendation systems is find out the potential interested items for users. In this paper, we use relevance to express the degree of user' interest. Therefore, in social image recommendation, there are 2 issues: a relevance measure to representing user's preference and a recommendation algorithm.

3 Relevance on Attributed Network

3.1 Unified Attributed Network

In an attribute network G, each vertex of a specific class is associated with a set of attributes, such as tag attribute for image vertex. Considering the conditions of social image recommendation, it's easy to define the relevance measure according to structural edges or vertex properties in G. For example, we can use intersect or *cos* function to measure the distance between vertex according to attributes; meanwhile, the structural relevance can be measure by path distance on G. However, if we combine structural and attribute relevance as the final relevance measure, it become confused and difficult to interpret. Instead of modeling structural information and attribute information separately, we propose a unified attributed network to transform the vertex attribute information into structural node and edge in graph.

Definition 2. Given an attributed network $G = (V, E, \Lambda)$, the **Unified Attributed Network** is denoted as G_u, $G_u = (V \cup Att, E \cup E_a)$. The attribute vertex Att is the set of all attribute values. Assume $v_i \in V, v_j \in Att$, the edge $(v_i, v_j) \in E_a$ if and only if v_i has an attribute x, $a_x(v_i) = v_j$, and E_a is called attributed edge.

To build G_u based on G, the information is modeled as follow:

- Structural information: It's the users' behavior, i.e., users *like* images in social network. In G_u, the information is modeled as edges between users and images.
- Context information: The social relationship can be represented through edge. The content information, i.e., the attributes in attributed network G, includes tags and image content features. Attribute nodes and edges are built to transform G to G_u.

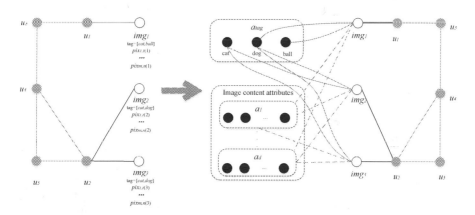

Fig. 2. From G to G_u

Figure 2 shows the unified attributed network G_u which is transformed from an attributed network G. The image node in G has two kinds of attributes: tag and image content. We collect all the tag values as an attribute a_{tag}, and add one attributed vertex to G for each different value. Thus, in Fig. 2, the tag attribute nodes including ball, cat, and dog. Then, we add the attributed edges for G_a between images and their attributed nodes. For example, because the img_1 has the tag values {*ball, cat*}, we should add 2 attributed edges: $(img_1, ball)$ and (img_1, cat).

For the image representation, taking all the pixels of each image as the image content attributes is unreasonable. In this paper, we use the state-of-art 4,096-dim DeCAF [9] feature to represent image content. Therefore, for every image node in G_u, there are 4096 attributed nodes linking to it. Comparing with discrete values of the tag attribute, the values of content attribute are continuous numbers. If we create an attributed node for each number, the continuous distance is lost. For example, assume the domain of a content attribute a_i is {10, 11, 50, 100}. Comparing with the difference between 10 and 100, 10 and 11 are too close to combine them as a single node in G_u. Thus, for the attributed nodes with continuous number domain, we combine the nearest attributed node pair as one until the node number is limited to d. Firstly, sort numerical attributed nodes by values. Then, compare the distance between the neighborhood nodes v_i and v_{i+1}, and combine the nearest pair as $v_i(j) = [v_i(j), v_{i+1}(j)]$. Thirdly, continue to compare the neighborhood pairs by the range boundary and combine them through revise the boundary value. Finally, we will get $4096*d$ attributed nodes in G_u.

3.2   Path Relevance

Given G_u, for social image recommendation, the following work is to design a relevance measure on G_u that can express the degree of user' interest to a specific item. In this paper, random walk with restart [10] that gives the sum of all the path probabilities for a given node pair in G is adopted.

Definition 3. Given a network $G = (V, E)$, let A be the $N \times N$ transition probability matrix of G, c is the restart probability, based on random walk with restart on graph, for any $v_i \in V$, the **Path Relevance** from v_i to all the nodes in G is defined as

$$p_i = (1 - c)Ap_i + cq_i \tag{1}$$

where p_i and q_i are all column vector, and $p_i(j)$ is the path relevance from v_i to v_j. q_i is a zero vector with the i-th element set to 1.

The original goal of recommendation systems is find out the potential interested items for users. In this paper, we use path relevance to express the degree of user' interest. If and only if the path relevance gives higher values to the items that user directly links than the items without directly edges in G, the relevance measure can represent user's preference.

Theorem 1. For social image recommendation, given a unified attributed network G_u and any node v_i in G_u, path relevance can provide a stable probability vector p_i. Without loss of generality, let v_j is lined to v_i, while v_k is not. Then, there must be a constant $c \in (0, 1)$ for Eq. 1 that make $p_i(i) > p_i(k)$.

Proof. Since q_i is a zero vector with the i-th element set to 1, the derivative function of $p_i(i)$ about c is

$$\frac{\partial p_i(i)}{\partial c} = 1 - A(i, :)p_i \tag{2}$$

In random walk, $A(i, :)p_i < 1$, thus, $p_i(i)$ is an increasing function about c. Considering the extreme situation, when $c = 1$, $p_i = q_i$. It means $p_i(i) \in (0, 1)$ and the limitation of $p_i(i)$ is 1 with c approaching 1. For any node x in each iteration,

$$p_i(x) = (1 - c)A(x, :)p_i = (1 - c)\{\sum_{y \neq x} A(x, y)p_i(y) + A(x, i)p_i(i)\} \tag{3}$$

If the node x is lined to $i, A(x, i) > 0$, otherwise, $A(x, i) = 0$. Since $\sum_y p_i(y) = 1$, with the increasing of pi(i), $\sum_{y \neq x} A(x, y) p_i(y)$, denoted as ε, will decrease to a minimum value. Obviously, $\varepsilon + A(j, i) p_i(i) > \varepsilon$. Therefore, there must be a constant $c \in (0, 1)$ for Eq. 1 that make $p_i(j) > p_i(k)$.

Based on Theorem 1, the path relevance measure can be used to represent the user's preference. Compared with homogeneous network or bipartite network, the unified attributed network G_u has more types of vertices and edges. Obviously, different type of edge means different semantic relationship between two objects. Therefore, they may have different degree of contributions in the random walk. For example, given a random walk started from u_1 in Fig. 2, the next state may be user node u_3 or image node img_1, which respectively represents the friendship and "*like*" behavior. For example, we could consider that the images posted by a user means more preference information than his friends. Thus, it's necessary to quantize the importance of different edge types by different weights of edges.

Without loss of generality, given G_u and the transition probability matrix A, any edge $e(i, j) \in E$ with a type $et(i, j), et(i, j) = et_k \in ET$. We assume that each type et_k has a weight w_k, $\sum_{k=1}^{|ET|} w_k = |ET|$. Considering the edge weights, A is rewritten as $\bar{A}$,

$$\bar{A}(i, j) = \frac{A(i, j) w_k}{\sum_j A(i, j) w_k} \tag{4}$$

4 Social Image Recommendation

4.1 Recommendation Objective Function

We propose an iterative adaptive method compute the weights. Firstly, we create an objective function about W. Then, we compute W through optimizing the function. We use path relevance on G_u to select top-k relevant images for given user. In other words, the set of top-k images have the most similarity to the set of the i-th user's neighborhood nodes which are taken as the observed interested images. Moreover, the method should ensure that similar users have the similar rating to the same items. Given 2 users u_i and u_j in G_u, if they are similar, the recommended images should also be similar. In reverse, if u_i and u_j have signally different preferences, their recommended images should be different as well. Therefore, they are positive correlative and we will maximize the correlation to design our objective function.

Let s_{ij} denote the similarity between u_i and $u_j, u_i, u_j \in U \subseteq V$. Based on the path relevance on $G_u, s_{ij} = p_i(j)$. The recommended top-k image set for u_i is denoted as V_i. The vertex set similarity between V_i and V_j is

$$s(V_i, V_j) = \sum_{v_x \in V_i, v_y \in V_j} \frac{d(v_x, v_y)}{|V_i| \times |V_j|} \tag{5}$$

where $d(v_x, v_y) = p_x(y)$. We compose all the s_{ij} as the User Similarity Vector X. When $(i - 1) \times |U| + j = k$, $x_k = s_{ij}$. In the same way, we get the Vertex Set Similarity Vector Y through composing all the $s(V_i, V_j)$. Then, we can get the Pearson Correlation Coefficient about X and Y by

$$\rho_{X,Y} = \frac{\sum (x_i - \bar{x})(y_i - \bar{y})}{\sqrt{\sum (x_i - \bar{x})^2}\sqrt{\sum (y_i - \bar{y})^2}} \tag{6}$$

where $\bar{x}$ and $\bar{y}$ are the average values of X and Y.

Given $G_u = (V, E)$, the weight $W = \{w_1, w_2, \ldots, w_m\}$ about different edge types, and the recommended number top-k, the goal of social image recommendation is to find top-k images V_i for each user $u_i \in U \subseteq V$ and $(u_i, V_i(j)) \notin E$, so that the following objective function is maximized

$$O(\{Vi\}_{i=1}^{|U|}, W) = \rho_{X,Y} \tag{7}$$

subject to $\sum_{i=1}^{|ET|} w_i = |ET|$ and $w_i > 0$.

4.2   Approximate Weight Self-Adjustment

We design an objective function for social image recommendation based on the similarity transitivity on G_u. However, the Eq. (7) are too complicated and even can't promise a unique solution. Thus, we simplify the weight estimation problem. Considering the weight W of different edge types, we can get the transition probability of every edge in G_u through Eq. (4). However, Eq. (1) is not a polynomial function of multi-variable $w_1, w_2, \ldots, w_{|W|}$. That's because there may be multi-edge of the same type linked to one nodes in G_u. Here, we divide the edge types in G_u into 2 categories:

- Multi-Value Type: There may be multi-edges of this type linked to one nodes. In G_u, "*Friendship*", "*like*" and "*Tag*" are all multi-value edge types.
- Single-Value Type: All the image-attribute relations belong to single-value type because there is one and only one value for a content attribute of an image.

Let w_0 denote the edge weight of multi-value type, $W = \{w_1, w_2, \ldots, w_m\}$ denote the edge weights of single-value type. The transition probability of e(u, v) of multi-value type is

$$p_{u,v} = \begin{cases} \frac{w_0}{N(u)w_0} & if(u, v) \in E, u \in U \cup T \\ \frac{w_0}{N(u)w_0 + m} & if(u, v) \in E, u \in M \end{cases} \tag{8}$$

where $N(u)$ is the degree of vertex u without attribute edges.

The transition probability of $e(u, v)$ of single-value type is

$$p_{u,v} = \begin{cases} \frac{w_i}{N(u)w_0 + m} & if\,(u,v) \in E, u \in M, v \in Att_i \\ \frac{1}{d(u)} & if\,(u,v), \in E, u \in Att_i, v \in M \end{cases} \tag{9}$$

where $d(u)$ is the degree of attributed vertex u.

Combine Eqs. (8)–(10), the path relevance turns to be a polynomial function of multi-variable $w_1, w_2, \ldots, w_m$. Further, the objective function is simplified as

$$O(\{V_i\}_{i=1}^{|U|}, W) = \sum_{\substack{v_i, v_j \in U \\ i \neq j}} [p_i(j) - \frac{2\sum p_i(j)}{|U|(|U|-1)}][s(V_i, V_j) - \frac{2\sum s(V_i, V_j)}{|U|(|U|-1)}] \tag{10}$$

Theorem 2. Given a recommendation result $\{V_i\}_{i=1}^{|U|}$ of G_u based on path relevance, there exists a unique solution $W^* = \{w_1^*, \ldots, w_m^*\}$ to maximizes the objective function in Eq. (10).

Proof. Taking Eq. (5) into the objective function, we can infer that Eq. (10) is a polynomial function of multi-variable W with non-negative real coefficients. Assume that the polynomial expression of w_i in Eq. (10) is denoted as $f_i(w_i)$. Then the Lagrange form of Eq. (10) is

$$O'(\{V_i\}_{i=1}^{|U|}, W, \lambda) = \sum_{i-1}^{m} f_i(w_i) + \lambda(\sum_i w_i - m) \tag{11}$$

The partial derivative functions of Eq. (11) are

$$\frac{\partial O'}{\partial w_i} = \frac{\partial f_i(w_i)}{\partial w_i} + \lambda = g_i(w_i) + \lambda + b = 0, (1 \leq i \leq m) \tag{12}$$

where $\frac{\partial f_i(w_i)}{\partial w_i} = g_i(w_i) + b$ with a constant b.

Based on Gauss' Fundamental Theorem of Algebra, Eq. (12) has at least one complex root. Assume $\lambda = \lambda^*$. When $\lambda^* \geq -c$, there are no real number solution to Eq. (12). Considering the condition $\lambda^* < -c$, there must be at least one positive number δ to make Eq. (12) is greater than 0 when $w_i^* = \delta$. Since Eq. (12) is continuous in the real interval $(0, \delta)$, there must be at least one solution in $(0, \delta)$. Therefore, there also is a unique positive root which maximizes the objective function in Eq. (10).

According to Theorem 2, we proof that the objective function Eq. (11) has a unique positive root $W^* - \{w_1^*, \ldots, w_m^*\}$. With constrains of W, the process of weight estimation is a linear programming problem, and we propose an adaptive weight adjustment approach to iteratively improve the objective function.

With $w_0 = 1.0$ fixed, let $W^t = \{w_1^t, \ldots, w_m^t\}$ be the weights of single-value edge type in the ith iteration. Firstly, we initialize $w_i^0 = 1.0$, $1 \leq i \leq m$. Then, the weight w_i in $(i+1)^{th}$ iteration is

$$w_i^{t+1} = \frac{1}{2}(w_i^t + \Delta w_i^t) \qquad (13)$$

To determine the weight increment Δw_i^t, we design a majority vote method: if two images linked into the same user u_i in G_u, the image content attribute that share same value by the two images has direct contribution to represent the user's preference. Increasing the weight of the corresponding edge type will benefit the Path Relevance. The similar user of u_i should have high relevance to these images and the adjustment will increase the relevance between the similar user of u_i and these two images. Therefore, it optimizes the objective function which is designed based on correlation. The *vote* function for w_i is

$$vote_i(v_p, v_q) = \begin{cases} 1 & \exists u \in U, (u, v_p), (u, v_q) \in E \text{ and } v_p(i) = v_q(i) \\ 0 & otherwise \end{cases} \qquad (14)$$

Through counting all the vertex $v_j \in Att_i$, the weight increment Δw_i^t is

$$\Delta w_i^t = \frac{\sum\limits_{v_j \in Att_i} \sum\limits_{v_p, v_q \in M} vote_i(v_p, v_q)}{\frac{1}{m}\sum\limits_{i} \sum\limits_{v_j \in Att_i} \sum\limits_{v_p, v_q \in M} vote_i(v_p, v_q)} \qquad (15)$$

We can get the w_i^{t+1} through Eqs. (13) and (15). The adjustment of weights may be increasing, unchanged or decreasing. Firstly, Eq. (15) ensure that the adjusted weights still satisfy the constraint. Secondly, the *vote* method adjusts the weights towards to the direction of increasing the subjective function in Eq. (10).

4.3   Recommendation Algorithm

By assembling different parts, the approach to social image recommendation based on path relevance is presented in Algorithm 1 (Table 2).

There are double-tier iterations in PRR. The inner loop is to computer the relevance based on random walk with restart and it's time complexity is $O(n^3 t_1)$ where n is the vertex number of G_u and t_1 is the iteration steps. With the outer loop t_2, the time complexity of *PRR* is $O(n^3 t_1 t_2)$. There are many fast-approximate approaches for RWR. Especially, for top-k search based on RWR, time complexity can be reduced to $O(n+m)$ where m is the number of edges and $m \ll n^2$ [10]. Thus, the final approximate time complexity can be $O(n(n+m)t_2)$.

Table 2. The path relevance recommendation algorithm

Algorithm 1. Path relevance recommendation on G_u

Inputes: G_u, top-k, c

Outputs: $\{V_i\}_{i=1}^{|U|}$

1. Build the adjective matrix Adj based on G_u;
2. Normalize the columus of Adj to get A;
3. Initialize W^0

       for all the multi-value edge types, fix a weight $w_0=1.0$;

       others, $w_i^0=1.0$, $1\leq i\leq m$;

Repeat step 4-8 until the objective function converage

4. Based on Eq. (8) and (9), compute $\bar{A}$;
5. Based on Eq. (9), compute q_i, $1\leq i\leq |V|$ through $\bar{A}$;
6. For each user i, select V_i based on q_i;
7. Compute the objective function in Eq. (10);
8. Update the weight $W=\{w_1,...,w_m\}$ based on Eq. (13)-(15);

End

9. Return $\{V_i\}_{i=1}^{|U|}$.

---

5   Experiments and Analysis

5.1   Experiment Setup

Dataset. To our best knowledge, there is no publicly available social image or other media dataset that is image-centric in large scale. Therefore, for the research described in this paper, we design the FlickrImageNetwork (FIN) data set.

Table 3. The statistical details of FlickrIN

Image number	100,000
User number	6,125
Tag number	10,000
Average number of ratings per image	18.24
Density of preference	0.18%
Dimension of visual features	4096

Typically, FIN contains 3 types of information, including user behavior, user relationships and tags. In FIN, we collect 100,000 images from Flickr at random, which are uploaded by 6,125 users. And each image has at least one user-defined tag. In detail, we use randomly breadth first search method to build FIN. Given a set with only one initial user, we randomly choose another user that has friendship with any user in the set and add it into set. Meanwhile, we collect all the images and their tags posted by the user in the set until we have enough images. In our dataset, more than 350,000 unique tags are collected and use the TF-IDF measure to remove the stop works. Further, a tag dictionary sized 10,000 is designed. In FIN dataset, we use the "like" feedback as a binary rating. Since each image is rated by limited users and users usually tend to rate a few of images, the rating matrix is extremely sparse. Table 3 shows the statistical details of our FIN.

Evaluation Metrics. In the experiment, we randomly select 20% "like" feedback information as the test dataset. Most recommendation methods use Root Mean Square Error (RMSE) as their evaluation metric. We also use RMSE as a basic metric in this paper. However, for social image recommendation, the rating feedback is binary rather than scores and RMSE may leads to bias for social image recommendation. Therefore, we design an additional evaluation metric through hit rate. In detail, given a recommendation method, we rank the output rating of each user to the new images and compare with the test data. For a user u_i, assume that u_i has m_i images which he has exhibited a "like" in the test data. Select *top-k* images based on the output of the recommendation method. If there are n_i images that belong to m_i, we can get a hit percentage for u_i. The Hit Rate Score (HRS) is defined as the expectation of all users' hit percentage, $HRS = E(n_i/m_i)$.

Baseline. We compared the performance of PRR with the following baseline methods:

- Content-based filtering (**CBF**) [11]: Image content are modeled as the state-of-art 4096d DeCAF features and user preference is designed through the statistics of his "*like*" images.
- Item-based collaborative filtering (**ICF**) [1]: It computes the similarity between images through user-image matrix. Then, directly give the recommendation through similarity of images.
- Probabilistic Matrix Factorization (**PMF**) [12]: It's a state-of-the arts latent factor based on CF. In this method, we only use the user-image matrix and ignore the context information. 30-dim latent features is used.
- **SoRec** [13]: It's an extended version of PMF that integrate the user's relationship into user-image matrix. In this experiment, the feature vectors of user, factor and image are all 30-dim.
- **STM** [14]: It's also an extended version of PMF that integrate the image tags and content. Image content is modeled as 42,496-dim through HG [15]. In this experiment, the topic space is 256 and the parameters for R, U, V are 1.90, 0.35 and 0.60 respectively.

5.2   Algorithm Evaluation

Information Sensitivity. We study the influence of different information to our method PRR. In G_u, there are 4 types of information: user feedback (M), social relationships (S), tags of image (T) and image content (C). Firstly, we build G_u based on M and C and test PRR. Then, we integrate S and T into G_u to test the influence of these information to the PRR performance. Set *top-k* = 30 (top-k will reduce the performance only if it's smaller than the number of liked images by someone in test dataset), and the test results are shown in Table 4.

Table 4. Information sensitivity of PRR

Information in G_u	Metric	
	RMSE	HRS
M, C	2.8795	0.7524
S, M, C	1.9143	0.8317
T, M, C	2.2854	0.8149
S, T, M, C	1.2583	0.8936

Table 4 shows that with more information in G_u, the values of RMSE decrease and the values of HRS increase. That's proof PRR is sensitive to the additional data. Our method can use the valuable information to improve the performance of IR. Moreover, when social relationships are added in G_u, RMSE decreases by 0.97; while RMSE decreases by 0.59 with the tags added. The improvement on HRS is similar. Thus, we conclude that social relationship is more important than image tag for social image recommendation.

Convergence. We study the convergence in terms of objective function Eq. (11) on top-k and scale of data and Fig. 3(a) and (b) show the trend of the values respectively. Both the figures show that the objective function increases with the iteration times and converges very quickly, usually in 7 times. Figure 3(a) shows that when we the value of top-k increases, the objective function decreases. A reasonable explanation is that, too many images recommended to users reduce the similarity between the recommended sets and further influence the correlation. To test the convergence on the scale of dataset, we randomly select users and images in FIN with 3 different percentages: 0.25, 0.5, 0.75 to build 3 new dataset Sub(0.25), Sub(0.5) and Sub(0.75). Then, the heterogenous networks are built through creating edges if there are any relationships between the selected objects, likely social relation and *like*. Figure 3(b) shows the result on these 3 and the full dataset. That demonstrates the convergence of PRR is robust on data scale.

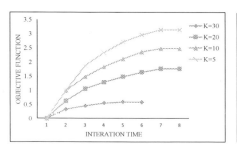
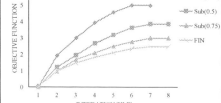

(a) Test on Top-k   (b) Test on Data Scale

Fig. 3. Convergence of PRR

5.3   Method Comparisons

To test the performance of PRR on different sparse datasets, we built another 3 datasets based on FIN. Firstly, we sort all image nodes by *like* degree in G. Then, select top 25, 50 and 75 percent image nodes and their interested user nodes to form 3 datasets: D_1, D_2 and D_3 and the corresponding density respectively are: 8.12%, 2.51% and 1.15%. In the following experiments, we compare our method PRR with other 5 classical methods on D_1, D_2, D_3 and FIN.

Figure 4 shows the result. With the density of dataset decreasing, the dataset become more and more sparse. From D_1 to FIN, both RMSE and HRS show that the PRR methods keep high and stable performance. While the performances of PMF and its extended versions decrease distinctly. Only in D1, SoRec and STM are better than PRR. It's demonstrates that PRR can tackle the challenges of highly sparse representation problem in the social network recommendation.

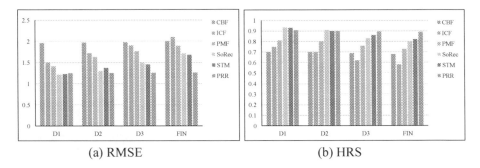

(a) RMSE   (b) HRS

Fig. 4. Performance comparison

6   Related Work

In this paper, we mainly focus on the social image recommendation problem. There are 2 issues in this problem: context-aware and sparse. Few works adopt CBF to recommend images through comparison between the image content and users' profile [1]. Over specification is the main problem for CBF [16]. Most social image recommendation systems are based on CF. However, when the rating matrix become sparse, CF can't achieve satisfactory performance. To alleviate the sparsity problem, many matrix factorization methods have been proposed, such as SVD [17], WMF [18], PMF [12] and CTR [19]. These models seek the resultant high-level features from user image matrix. SoRec [13] extends the MF by incorporating the social links. Considering the text information in social network, some researches extend the CF following the basic idea of topic models [20]. STM [4] is proposed by jointly considering of image content analysis with the user' preferences based on sparse representation. All these extended versions of MF mainly rely on the rating matrix. However, when the rating matrix is very sparse, the performance can't be promised. Moreover, with the model parameter increasing, these models will face curse of dimensionality.

Our work inspired by the success of relevance search, is different from the above methods. The intuition is that the item context information is equally critical as the structure information, and we incorporated the 2 kinds of information in a unified attributed network and facilitate the personalized recommendation thought relevance between user and item on the network.

7 Conclusions

In this paper, we explore a novel algorithm termed PRR for social image recommendation. PRR considers both the user behaviors and item properties through path relevance on the unified attributed network and it can tackle the challenges of highly sparse representation problem. The experimental results demonstrate that PRR leads to more effective recommendations. Future work is to study the different level of user relationships, like friendships, follow and classmates, etc.

References

1. Saveski, M., Mantrach, A.: Item cold-start recommendations: learning local collective embeddings, pp. 89–96 (2014)
2. Su, X., Khoshgoftaar, T.M.: A Survey of Collaborative Filtering Techniques. Hindawi Publishing Corp., Cairo (2009)
3. Salakhutdinov, R., Mnih, A.: Probabilistic matrix factorization. In: International Conference on Neural Information Processing Systems, pp. 1257–1264. Curran Associates Inc. (2007)
4. Li, J., Lu, K., Huang, Z., et al.: Two birds one stone: on both cold-start and long-tail recommendation. In: ACM on Multimedia Conference, pp. 898–906 (2017)
5. Pham, T.A.N., Li, X., Cong, G., Zhang, Z.: A general recommendation model for heterogeneous networks. IEEE Trans. Knowl. Data Eng. 28(12), 3140–3153 (2016)
6. Symeonidis, P., Nanopoulos, A., Manolopoulos, Y.: A unified framework for providing recommendations in social tagging systems based on ternary semantic analysis. IEEE Trans. Knowl. Data Eng. 22(2), 179–192 (2010)
7. Zhang, D., Zhang, D., Chang, E.Y.: Combinational collaborative filtering for personalized community recommendation. In: ACM SIGKDD, pp. 115–123 (2008)
8. Wang, J., Zhao, Z., Zhou, J., et al.: Recommending Flickr groups with social topic model. Inf. Retrieval 15(3–4), 278–295 (2012)
9. Donahue, J., Jia, Y., Vinyals, O., Hoffman, J., Zhang, N., Tzeng, E., Darrell, T.: DeCAF: a deep convolutional activation feature for generic visual recognition. In: ICML (2014)
10. Fujiwara, Y., Nakatsuji, M., Onizuka, M., et al.: Fast and exact top k search for random walk with restart. Proc. VLDB Endow. 5(5), 442–453 (2012)
11. Pazzani, M.J., Billsus, D.: Content-based recommendation systems. In: Brusilovsky, P., Kobsa, A., Nejdl, W. (eds.) The Adaptive Web. LNCS, vol. 4321, pp. 325–341. Springer, Heidelberg (2007). https://doi.org/10.1007/978-3-540-72079-9_10
12. Dieleman, S., Schrauwen, B.: Deep content-based music recommendation. In: International Conference on Neural Information Processing Systems, pp. 2643–2651. Curran Associates Inc. (2013)

13. Hao, M., Yang, H., et al.: SoRec: social recommendation using probabilistic matrix factorization. Comput. Intell. **28**(3), 931–940 (2008)
14. Liu, X., Tsai, M.-H., Huang, T.: Analyzing user preference for social image recommendation. arXiv preprint arXiv:1604.07044 (2016)
15. Zhou, X., Cui, N., Li, Z., et al.: Hierarchical gaussianization for image classification. In: IEEE, International Conference on Computer Vision. IEEE Xplore, pp. 1971–1977 (2009)
16. Adomavicius, G., Tuzhilin, A.: Toward the next generation of recommender systems: a survey of the state-of-the-art and possible extensions. IEEE Trans. Knowl. Data Eng. **17**, 734–749 (2005)
17. Lowe, D.G.: Distinctive image features from scale-invariant keypoints. Int. J. Comput. Vis. **60**(2), 91–110 (2004)
18. Hu, Y., Koren, Y., Volinsky, C.: Collaborative filtering for implicit feedback datasets. In: Eighth IEEE International Conference on Data Mining, pp. 263–272. IEEE (2009)
19. Wang, C., Blei, D.M.: Collaborative topic modeling for recommending scientific articles. In: ACM SIGKDD, pp. 448–456 (2011)
20. Barnard, K., Duygulu, P., Forsyth, D., et al.: Matching words and pictures. J. Mach. Learn. Res. **3**(2), 1107–1135 (2003)

Representation Learning with Depth and Breadth for Recommendation Using Multi-view Data

Xiaotian Han[1], Chuan Shi[1(✉)], Lei Zheng[2], Philip S. Yu[2], Jianxin Li[3], and Yuanfu Lu[1]

[1] Beijing Key Lab of Intelligent Telecommunications Software and Multimedia,
Beijing University of Posts and Telecommunications, Beijing, China
hanxiaotian.h@gmail.com, shichuan@bupt.edu.cn, luyfroot@gmail.com
[2] University of Illinois at Chicago, Chicago, USA
{lzheng21,psyu}@uic.edu
[3] The University of Western Australia, Perth, Australia
jianxin.li@uwa.edu.au

Abstract. Recommender system has been well investigated in the past years. However, the typical representative *CF*-like models often give recommendation with low accuracy when the interaction information between users and items are sparse. To address the practical issue, in this paper we develop a novel **R**epresentation **L**earning with **D**epth and **B**readth (*RLDB*) model for better recommendation Specifically, we design a heterogeneous network embedding method and convolutional neural network based method to learn feature representations of users and items from user-item interaction structure and review texts, respectively. Furthermore, an end-to-end breadth learning model is proposed through employing multi-view machine technique to learn features and fuse these diverse types of features in a uniform framework. Extensive experiments clearly demonstrates that our model outperforms all the other methods in these datasets.

Keywords: Recommender system · Rating prediction
Multi-view machine · Heterogeneous information network embedding

1 Introduction

In the past decade, the rapid growth of e-commerce is changing people's daily life such as online shopping, reading articles, and watching movies. Therefore, effective recommender systems are highly desirable to help the customers by presenting the products or services that are likely of interest to them. *Collaborative Filtering* (*CF*) is the most representative model which assumes that people who share similar preferences in the past tend to have similar choices in the future. *CF* model based recommendations is often lack of sufficient interaction between users and items.

© Springer International Publishing AG, part of Springer Nature 2018
Y. Cai et al. (Eds.): APWeb-WAIM 2018, LNCS 10987, pp. 181–188, 2018.
https://doi.org/10.1007/978-3-319-96890-2_15

Fortunately, there are lots of additional available data information that can help to alleviate the issue of insufficient interaction information between users and items, e.g., the review text. There are some studies [5,6,9] that have used review text to improve the recommendation results. McAuley and Leskovec in [6] and Ling et al. in [5] still mainly used the structure information to construct user-item matrix and then utilized the review text as the auxiliary information to slightly adjust the user-item weights during matrix factorization. In contrast, Zheng et al. [9] only used the review text to learn user behavior and item properties, which are then used to learn the hidden latent features for users and items jointly. But they ignored the effect of structure information in use. In addition, in most cases we need to treat the different views of data to be equally important as much as possible while the importance of noisy features should be discriminated automatically and effectively.

To do this, in this paper, we design a novel *Representation Learning* model with *Depth* and *Breadth* (called *RLDB*) to effectively utilize two views of data (i.e., structure and text information) through two deep neural networks extracting feature representation and a breadth learning model fusing features.

The contributions of this paper are summarized as follows:

- We propose a novel HIN embedding method to extract structure information of users and items for recommendation. As well as we design a CNN and meta-path based heterogeneous network embedding method to effectively learn the representation of the structure information.
- We propose a end-to-end based breadth learning method through employing multi-view machine to effectively combine the structure representations and text representation together.
- We conduct extensive experiments on three real-world datasets to evaluate the effectiveness of RLDB.

2   Proposed Model

In order to integrate different views of data for improving recommendation, we develop a novel model - **R**epresentation **L**earning with **D**epth and **B**readth (*RLDB*), to model the representations of users, the representations of items, and their combinations.

2.1   Model Architecture

The overall architecture of the RLDB model is shown in Fig. 1. The proposed model can learn representations from different views via deep neural networks and effectively couple them with a multi-view machine for recommendation. In this paper, we take the structure information of rating network and text information in reviews of users and items as the four views. The details of RLDB are described below.

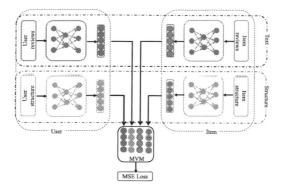

Fig. 1. The architecture of the proposed model

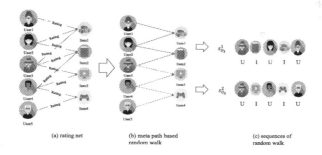

Fig. 2. The generation process of random walk sequences

2.2   Structure Representation

To take users' consumption habits into account, we construct an rating network as a type of heterogeneous information networks [8]. Then we model the representation of the interaction network structure of user/item via a neural network.

Meta-path Based Random Walk. In our model, we use meta-path based random walk to model the representation of the users' structure of rating network as well as the items'. In order to illustrate the meta-path based random walk, we show a toy example of the rating network to illustrate structure of the rating network in Fig. 2(a).

In order to illustrate the generation process, Fig. 2(b) shows the sequence generation process of the user U_3. We set L as 5. The nodes connected by the solid line make up the first sequence $s_{u_3}^1$ on behalf of the user U_3, which contains $U_3I_2U_2I_1U_1$. The dotted line connects the nodes to form the anther sequence $s_{u_3}^2$ containing $U_3I_3U_4I_4U_5$. Figure 2(c) shows the result of the meta-path based random walk. The sequence $s_{U_j}^k$ represents the kth random walk starting with the node s_{U_j} on behalf of the user U_j. The $s_{U_1}^1$ in Fig. 2 is based on the mate-path $UIUIU$. After obtaining the sequence, we concatenate all the random walk sequences starting with s_{U_j} to represent the rating network structure of the user s_{U_j}.

$$S_{U_i} = s_{U_i}^1 \oplus s_{U_i}^2 \oplus s_{U_i}^3 \oplus \cdots \oplus s_{U_i}^n \tag{1}$$

Similar to the user, the concatenated sequence S_{I_i} on behalf of the item I_i's structure.

Learning the Representation of Structure. After getting the concatenated sequences of random walks on behalf of the users or items, we learn the representation of the users and items rating network structure. First, we embed the users and items into a uniform vector space and update the embedding vector via the training process. The nodes are represented as a vector, so we represent this sequence as a two-dimensional matrix and feed it to a convolution neural network with several filters.

This feature map of the kth filter within the convolutional layer with the is as follow:

$$\hat{r}_{U_i}^k = \left[r_k^1, r_k^2, \cdots, r_k^3, \cdots, r_k^{n-h+1} \right] \tag{2}$$

where k represents the kth filter in the convolution layer.

We concatenate multiple features obtained by multiple filters (with varying window sizes) to generate the representation $R_{U_i}^s$ of the rating network structure on behalf of user U_i:

$$R_{U_i}^s = r_{U_i}^1 \oplus r_{U_i}^2 \oplus r_{U_i}^3 \oplus \cdots \oplus r_{U_i}^n \tag{3}$$

We can obtain the item I_i's representation $R_{I_i}^s$ of rating network structure in the same way.

As discussed before, the semantics underneath different paths are different. For example, in the conventional layer, the window size of 3 can model the semantics of the mate-path UIU while window size of 5 can capture the semantics of the mate-path $UIUIU$. Obviously, the distinct semantics under different paths will lead to different information.

2.3   Text Representation

In the part of text information, we utilize CNN to facilitate the deeper understanding of users' review texts. In the first layer, all the review texts for users or items are represented as matrices of word embedding [7] to capture the semantic information. The next layers are the common layers used in CNN based models to discover multiple features for users and items with the multiple features of different filters. The top layer is a max pooling layer which can find the most important features.

Consider the user review texts as an example. Let us denote all the reviews of a user U_1 consisting of n reviews as $d_{U_1 I_1}, d_{U_1 I_2}, d_{U_1 I_3}, \cdots, d_{U_1 I_i}, \cdots, d_{U_1 I_n}$.

$$D_{U_i} = d_{U_i I_1} \oplus d_{U_i I_2} \oplus d_{U_i I_3} \oplus \cdots \oplus d_{U_i I_n} \tag{4}$$

where $d_{U_i I_j}$ indicates the review from user U_i to I_j and $\oplus$ is the concatenation operator.

This feature map of the kth filter within the convolutional layer with the is as follow:

$$\hat{r}_{U_i}^k = \left[r_k^1, r_k^2, \cdots, r_k^3, \cdots, r_k^{n-h+1} \right] \tag{5}$$

To capture the most important feature, we feed the output of a convolutional filter to a max pooling layer. Denote the $r_{U_i}^k = max\{\hat{r}_{U_i}^k\}$ as the feature corresponding to the kth filter. We concatenate multiple features obtained by multiple filters (with varying window sizes) to generate the representation $R_{U_i}^t$ of the review texts on behalf of user U_i:

$$R_{U_i}^t = r_{U_i}^1 \oplus r_{U_i}^2 \oplus r_{U_i}^3 \oplus \cdots \oplus r_{U_i}^n \tag{6}$$

We can obtain the item I_i representation $R_{I_i}^t$ of review texts in the same way.

2.4   Fusion with Multi-view Machine

We obtain four representations of the rating network structure and review texts for users and items respectively.

Inorder to combine the four representations effectively, We further explore all vectors of these representation interactions up to the mth order between inputs from m-view data by the multi-view machine [2]. Equation 7 shows the principle of multi-view machine that interacts all features to the mth-order between inputs from m views.

$$\dot{y} = \sum_{i_1=1}^{I_1+1} \cdots \sum_{i_m=1}^{I_m+1} w_{i_1,\ldots,i_m} \left(\prod_{v=1}^{m} z_{i_v}^{(v)} \right) \tag{7}$$

where the $\mathbf{z}^{(v)^T} = \left(\mathbf{x}^{(v)^T}, 1 \right) \in \mathbb{R}^{I_v+1}, \forall v = 1, 2, \cdots, m$, the $\mathbf{v}$ is the vector of the views, the m is the number of the views.

In our model, the m is 4 because we have four representations of rating network structure and review texts for users and items totally. In our experiments, we use a simple expression shown in [1] that is equivalent to Multi-view Machine as discussed above.

3   Experiments

In this section, we do extensive experiments to validate the effectiveness of the proposed RLDB on three real datasets, compared to the state-of-the-arts.

3.1 Dataset

The experiments are conducted on three representative datasets.

- Amazon[1]: Amazon review datasets [6] contains product reviews and meta-data from Amazon website. In our experiment, we select three of the largest categories including Instant Video, Electronics and Home, called Amazon_V, Amazon_E, and Amazon_H.
- Yelp[2]: It is a large-scale datasets consisting of restaurant reviews.
- TripAdvisor[3]: TripAdvisor is an American travel website company providing hotels booking as well as reviews of travel-related content.

In our experiments, we adopt the widely used criterion of Mean Squared Error (MSE) [6] to evaluate the performance of our proposed model.

$$MSE = \frac{1}{N} \sum_{n=1}^{N} (r_n - \hat{r}_n)^2, \tag{8}$$

where r_n is the nth gold rating score, rn is the nth predicted rating score and N is the total number of ratings.

3.2 Compared Methods

Compared methods involved in experiments are summarized as follows.

- MF [4]: Matrix Factorization is the important popular CF-based recommendation method. It only uses rating matrix as input and estimates two low-rank matrices to predict ratings.
- HFT [6]: Hidden Factor as Topic proposed in employs topic distributions to learn latent factors from user or item reviews.
- ConvMF [3]: ConvMF integrates CNN into probabilistic matrix factorization which captures contextual information of documents.
- DeepCoNN [9]: DeepCoNN adopts two parallel CNN to model user behaviors and item properties from all the review texts.

3.3 Effectiveness Experiments

The performances of RLDB and the compared methods are reported in Table 1 with the best performance shown in bold. From the Table 1, we have the following observations.

The proposed model RLDB outperforms other approaches on all the datasets. Since the CF-based method MF only use the rating score information, it has poor performance on the rating prediction. Although DeepCoNN [9] adopt deep

---

[1] https://snap.stanford.edu/data/web-Amazon.html.
[2] https://www.yelp.com/dataset-challenge.
[3] https://www.tripadvisor.com/.

Table 1. MSE Comparison with other methods. Best results are indicated in bold.

Dataset	MF	HFT-10	ConvMF	DeepCoNN	RLDB
Amazon_V	1.5319	1.5470	1.4159	1.1349	**1.0755**
Amazon_E	1.7918	1.6137	1.9550	1.6331	**1.5139**
Amazon_H	1.4737	1.5103	1.5040	1.4199	**1.3202**
Yelp	1.6129	1.6064	1.9570	1.3982	**1.3160**
TripAdvisor	0.5722	0.4071	0.4895	0.4113	**0.3962**

learning model, its performance is still worse than our RLDB due to the use of single-view data (review information). The ConvMF and RLDB both utilize two views of data. The better performance of RLDB shows the more effective combination of multi-view data in RLDB. In all, the good performance of RLDB attribute to the effective feature extraction with representation learning from multi-view data and fusion mechanism with multi-view machine in RLDB.

3.4   Cold Start Experiment

In this section, we evaluate the MSE score on five types of cold start users with different numbers of rated items (e.g., users with the number of rated items no more than 5). As a representative of CF, MF is included to compare their performance. The results are shown in Fig. 3. We can observe that the RLDB always performs better than MF. More importantly, the superiority of RLDB is more significant for more cold users (i.e., users with less rating interactions). It shows that RLDB has the potential to alleviate the cold start problem. We think it attributes to multi-view data and the delicate design of RLDB: representation learning model with depth and breadth.

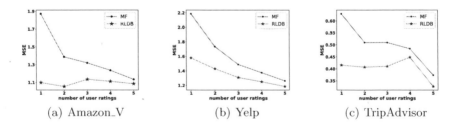

(a) Amazon_V   (b) Yelp   (c) TripAdvisor

Fig. 3. MSE of RLDB compared to MF for users with different number of training reviews

4   Conclusion

In order to make full use of rich multi-view data in recommender system, we propose a novel model RLDB based on representation learning. The RLDB designs a heterogeneous information network embedding method and CNN to learn representation of users and items from structure and text information, respectively. Furthermore, a multi-view machine is employed to effectively fuse these features. Extensive experiments on real-world datasets show RLDB can boost the recommendation performance and benefit for cold start problem.

Acknowledgments. This work is supported by the National Natural Science Foundation of China (No. 61772082, 61375058), the National Key Research and Development Program of China (2017YFB0803304), and the Beijing Municipal Natural Science Foundation (4182043).

References

1. Cao, B., Zheng, L., Zhang, C., Yu, P.S., Piscitello, A., Zulueta, J., Ajilore, O., Ryan, K., Leow, A.D.: DeepMood: modeling mobile phone typing dynamics for mood detection. In: SIGKDD (2017)
2. Cao, B., Zhou, H., Li, G., Yu, P.S.: Multi-view machines. In: WSDM, pp. 427–436. ACM (2016)
3. Kim, D., Park, C., Oh, J., Lee, S., Yu, H.: Convolutional matrix factorization for document context-aware recommendation. In: Proceedings of the 10th ACM Conference on Recommender Systems, pp. 233–240. ACM (2016)
4. Koren, Y., Bell, R., Volinsky, C.: Matrix factorization techniques for recommender systems. Computer, **42**(8) (2009)
5. Ling, G., Lyu, M.R., King, I.: Ratings meet reviews, a combined approach to recommend. In: Proceedings of the 8th ACM Conference on Recommender Systems, pp. 105–112. ACM (2014)
6. McAuley, J., Leskovec, J.: Hidden factors and hidden topics: understanding rating dimensions with review text. In: Proceedings of the 7th ACM Conference on Recommender Systems, pp. 165–172. ACM (2013)
7. Mikolov, T., Sutskever, I., Chen, K., Corrado, G.S., Dean, J.: Distributed representations of words and phrases and their compositionality. In: NIPS, pp. 3111–3119 (2013)
8. Shi, C., Li, Y., Zhang, J., Sun, Y., Philip, S.Y.: A survey of heterogeneous information network analysis. IEEE Trans. Knowl. Data Eng. **29**(1), 17–37 (2017)
9. Zheng, L., Noroozi, V., Yu, P.S.: Joint deep modeling of users and items using reviews for recommendation. In: WSDM, pp. 425–434. ACM (2017)

Attentive and Collaborative Deep Learning for Recommendation

Feifei Li[1], Hongyan Liu[2], Jun He[1(✉)], and Xiaoyong Du[1]

[1] Key Laboratory of DEKE (MOE), School of Information,
Renmin University of China, Beijing, China
{pangxiaobi,hejun,duyong}@ruc.edu.cn
[2] School of Economics and Management, Tsinghua University, Beijing, China
hyliu@tsinghua.edu.cn

Abstract. Deep learning has been successfully introduced to collaborative filtering for recommendation systems in recent years. In these studies, autoencoder models are usually used to extract latent features of items or users, and related researches facilitate the learning techniques using item and user latent factors in matrix factorization models. Inputs of autoencoder models are usually side information of items or profile information of users. However, in many real world applications, we do not have users' profile information. Moreover, results of both matrix factorization and deep learning models are difficult to interpret. To solve these issues, in this paper, we propose a deep collaborative filtering model with attention mechanism. With this model, learning of latent factors of users and items can be facilitated by deep processing of items' tag information. Furthermore, user preferences learned are interpretable. Experiments conducted on a real world dataset demonstrate that our model can significantly outperform the state-of-the-art deep collaborative filtering models.

1 Introduction

Nowadays, recommendation systems become more and more important for many business applications. For instance, recommendation contributes 60% of video clicks on YouTube, and 80% on Netflix. Traditional methods such as collaborative filtering (CF) [5] succeed in recommending items to users under some circumstances. However, they suffer from many problems such as data sparsity and cold start. Therefore, in recent years, many researches have been done to combine collaborative filtering and content-based recommendation [7] to overcome these issues.

Deep learning models have shown their good performance in many domains [2]. They are also introduced to recommendation systems to further improve the recommendation performance [12].

Composite models take advantage of both deep learning and traditional recommendation models and usually achieve better performance than independent deep learning models. For example, CDL [9] and DCF [3] integrate SDAE with

© Springer International Publishing AG, part of Springer Nature 2018
Y. Cai et al. (Eds.): APWeb-WAIM 2018, LNCS 10987, pp. 189–197, 2018.
https://doi.org/10.1007/978-3-319-96890-2_16

PMF, and CVAE [4] integrates VAE [1] with PMF. These models are all shown to have state-of-the-art performance. In these models, different autoencoder models are applied to extract latent features of items, which is helpful to alleviate the data sparsity and cold start problems in traditional recommendation models. Models like CDL and CVAE only extract deep features for items. DCF extracts deep features for both items and users. These models take side information of items and users as inputs to extract latent features. Side information of items is usually their description, which is usually not hard to get. But users' side information includes personal information such as gender and occupation, which is usually difficult to get due to privacy policies. Besides, it cannot describe user preferences directly.

In our proposed model ACR, tag information is used as inputs of deep learning model to extract deep features of both items and users. We only use items' tag information, as items' tagging behavior is common in many domains and there are also many methods to do item tagging. To solve the scarcity of user information, we introduce an attention mechanism to the deep learning model. The attention mechanism extracts user preferences with the help of items' tag information. In mobile app (the real world dataset we use) recommendation, tags of users' downloaded apps can better represent user preferences than user profiles. Meanwhile, we visually showcase the relationship between user preferences and their actual behaviors. And the user preferences drawn by our proposed attention mechanism seem to be pretty understandable.

2   Related Work

2.1   Deep Learning Based Recommendation

In recent years, various deep learning models such as multi-layer perception network (MLP), CNN, RNN, DSSM and autoencoders are introduced to recommendation systems, among which composite models integrating deep learning models with traditional models attract the attention of many researchers.

State-of-the-art collaborative deep recommendation models either only learn deep features of items or use user profiles to infer user preferences. User profiles are usually difficult to obtain and are not in the same semantic space as items. Our model uses tags of items to extract deep features of both users and items. The data can be obtained much more easily and can illustrate the user preferences in the meantime.

2.2   Attention Mechanism

Attention mechanism has shown its effectiveness in many learning tasks, such as machine translation, text classification as well as recommendation [10,11]. Its remarkable effectiveness mainly thanks to the intuition that human beings usually pay attention to a specific area instead of its entire area when recognizing things. The key idea of attention applied in deep learning usually is giving different attention weight to different hidden states.

However, attentive models for recommendation tasks usually add attention to items, indicating that users have different preferences to different items or different components of an item. While we use the attention model to learn users' deep features, illustrating that user preferences can be described by different tags.

3   Our Recommendation Model

In this section we define the task and introduce our ACR model. The intuition behind our model is that deep representation of both items and users can be learned from tag information of items. Tags of items selected (e.g. purchased or downloaded) by each user can reflect user preferences to some extent. Tags of items can usually illustrate item characteristics and functions well, and they are easily to obtain.

3.1   Task Definition and Symbols

Assume that there are N users, M items and L tags and we have implicit feedback of users over items represented by a rating matrix, where value 1 means positive feedback and 0 means missing value. Each item has some tags to describe its characteristics and functions. In this paper, K is the dimension of latent factors and m is the number of tags of each user. Let R be the rating matrix, U be user latent factor matrix, U^+ be collaborative user latent factor matrix, V be item latent factor matrix, V^+ be collaborative item latent factor matrix, P be user preference matrix, Z be latent feature matrix of item, X be the item tag matrix and T be the user tag matrix. Small letters are vectors of corresponding matrixs. λ_* are prior parameters of relative variables in the subscripts. W and b are weight matrix and bias respectively in the deep learning network.

Our task is to predict the items that each user may be interested in with their historial behaviors.

3.2   Attentive and Collaborative Recommendation Model

The graphic model of ACR is illustrated in Fig. 1. In model ACR, VAE is used to extract latent deep features for items, and a forward network with an attention model is used to extract latent deep features to describe user preferences. PMF model learns the interactive relationship between users and items based on rating information and latent deep features learned by deep learning models.

Similar to PMF, we assume that the prior distributions of collaborative latent factor of each user i and each item j are normal distribution.

We assume that the extracted latent feature representation (we call it latent content variable for convenience) for item j is drawn from unit normal distribution.

The content information x_j of item j is generated from latent content variable z_j, and the generation process is described in Algorithm 1.

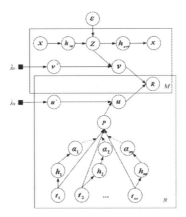

Fig. 1. Graphic model of ACR. The two boxes stand for networks to extract latent factors for items and users. M is the number of items and N is the number of users.

Algorithm 1. Generation network of VAE

1: **for** each layer l in the inference network **do**
2: 　　For each column n of the weight matrix $\boldsymbol{W}_l$, draw $\boldsymbol{W}_{l,*n} \sim N(0, \lambda_w^{-1} \boldsymbol{I}_{K_l})$
3: 　　Draw the bias vector $\boldsymbol{b}_l \sim N(0, \lambda_w^{-1} \boldsymbol{I}_{K_l})$
4: 　　For each row j of $\boldsymbol{h}_l$, draw $\boldsymbol{h}_{l,j*} \sim N(\sigma(\boldsymbol{h}_{l-1,j*}\boldsymbol{W}_l + \boldsymbol{b}_l), \lambda_s^{-1} \boldsymbol{I}_K)$
5: **end for**
6: draw $\boldsymbol{x}_j \sim N(\boldsymbol{h}_L \boldsymbol{W}_{L+1} + \boldsymbol{b}_{L+1}, \lambda_x^{-1} \boldsymbol{I})$

Algorithm 2. Inference network of VAE

1: **for** each layer l in the inference network **do**
2: 　　For each column n of the weight matrix $\boldsymbol{W}_l$, draw $\boldsymbol{W}_{l,*n} \sim N(0, \lambda_w^{-1} \boldsymbol{I}_{K_l})$
3: 　　Draw the bias vector $\boldsymbol{b}_l \sim N(0, \lambda_w^{-1} \boldsymbol{I}_{K_l})$
4: 　　For each row j of $\boldsymbol{h}_l$, draw $\boldsymbol{h}_{l,j*} \sim N(\sigma(\boldsymbol{h}_{l-1,j*}\boldsymbol{W}_l + \boldsymbol{b}_l), \lambda_s^{-1} \boldsymbol{I}_K)$
5: **end for**
6: **for** each item j **do**
7: 　　Draw $\boldsymbol{\mu}_j \sim N(\boldsymbol{h}_L \boldsymbol{W}_\mu + \boldsymbol{b}_\mu, \lambda_s^{-1} \boldsymbol{I}_K), \log \boldsymbol{\sigma}_j^2 \sim N(\boldsymbol{h}_L \boldsymbol{W}_\sigma + \boldsymbol{b}_\sigma, \lambda_s^{-1} \boldsymbol{I}_K)$
8: 　　Draw $\boldsymbol{z}_j \sim N(\boldsymbol{\mu}_j, diag(\boldsymbol{\sigma}_j^2))$
9: **end for**

The inference network parameterized by ϕ is described in Algorithm 2.

Following the reparameterization trick described in [1], we draw samples from $\epsilon \sim N(0, \boldsymbol{I})$ and $\boldsymbol{z}_j = \boldsymbol{\mu}_j + \boldsymbol{\sigma}_j \odot \epsilon$. $\boldsymbol{\mu}_j$ and $\boldsymbol{\sigma}_j$ are generated through a MLP.

Then the item latent factor can be represented by $\boldsymbol{v}_j = \boldsymbol{v}_j^+ + \boldsymbol{z}_j$.

Similarly, the latent user factor can be represented by $\boldsymbol{u}_i = \boldsymbol{u}_i^+ + \boldsymbol{p}_i$.

We use $\boldsymbol{t}_{il}$ to refer to the l-th tag of user i. For each tag, we count the number of apps that labeled by it in a user's download history. And then we choose the top-m tags sorted by the frequency. Each tag is represented by an embedding vector $\boldsymbol{t}_{il} \in \mathbb{R}^K$, which is pre-trained by word2vec model [6]. User preference vector $\boldsymbol{p}_i$ is learned through an attentive deep learning model described as follows.

First, for user content information, embeddings of tags are processed by a hidden layer with weight matrix $\boldsymbol{W}$, $\boldsymbol{W} \in \mathbb{R}^{K_h \times K}$, and bias $\boldsymbol{b} \in \mathbb{R}^{K_h}$, through activation function tanh as shown in Eq. (1), where K_h is the dimension of hidden state $\boldsymbol{h}_{ij}$:

$$\boldsymbol{h}_{il} = tanh(\boldsymbol{W}\mathbf{t}_{il} + \boldsymbol{b}) \tag{1}$$

Then a context vector $\boldsymbol{S} \in \mathbb{R}^{K_h}$ is learned as a feature selector, assigning different weights to the hidden states as shown in Eq. (2):

$$\hat{\alpha}_{il} = \boldsymbol{h}_{il}^T \boldsymbol{S} \tag{2}$$

The attention weight of the lth tag of user i is computed through a softmax function $\alpha_{il} = \frac{\exp \hat{\alpha}_{il}}{\sum_l \exp \hat{\alpha}_{il}}$.

The preference vector of user i, $\boldsymbol{p}_i$, is calculated by Eq. (3):

$$\boldsymbol{p}_i = \sum_l \alpha_{il}\boldsymbol{t}_{il} \tag{3}$$

Then, rating can be predicted according to normal distribution as shown in Eq. (4):

$$\hat{r}_{ij} \sim N((\boldsymbol{u}_i^+ + \boldsymbol{p}_i)^T (\boldsymbol{v}_j^+ + \boldsymbol{z}_j), C_{ij}^{-1}) \tag{4}$$

where C_{ij} is a precision parameter, serving as the confidence for r_{ij} in a similar way used in model CTR [8]. When $r_{ij} = 1$, $C_{ij} = a$, and $C_{ij} = b$ otherwise. a and b are hyper-parameters satisfying $a > b > 0$. Because if the rating is 1, it means that the user likes that item, otherwise it means the user dislikes it or hasn't noticed it before.

To estimate the values of all variables, we use the following loss function based on maximum a posteriori estimation (MAP):

$$
\begin{aligned}
L(U, V, \theta, \phi) = &-\sum_{i,j} \frac{C_{ij}}{2}(r_{ij} - \boldsymbol{u}_i^T \mathbf{v}_j)^2 - \frac{\lambda_u}{2}\sum_i ||\boldsymbol{u}_i - \boldsymbol{p}_i||_2^2 \\
&- \frac{\lambda_v}{2}\sum_j \mathbb{E}_{q_\phi(\boldsymbol{Z}|\boldsymbol{X})}||\boldsymbol{v}_j - \boldsymbol{z}_j||_2^2 - \frac{\lambda_0}{2}(||\boldsymbol{W}||_2^2 + ||\boldsymbol{b}||_2^2) \\
&+ \mathbb{E}_{q_\phi(\boldsymbol{Z}|\boldsymbol{X})}\log p(\boldsymbol{X}|\boldsymbol{Z}) - \mathbb{KL}(q_\phi(\boldsymbol{Z}|\boldsymbol{X})||p(\boldsymbol{Z})) \\
&- \frac{\lambda_w}{2}\sum_l (||\boldsymbol{W}_l||_F^2 + ||\boldsymbol{b}_l||_2^2)
\end{aligned}
\tag{5}
$$

where the first term is prediction error between real value and the predicted rating value and other terms can be regarded as regularizations. The last three terms are related to VAE model, where $\mathbb{KL}$ means KL-divergence between $q_\phi(\boldsymbol{z}_j|\boldsymbol{x}_j)$ and $p(\boldsymbol{Z})$.

3.3   Optimization

The optimization process to learn various variables for our model ACR is described in Algorithm 3. This process mainly contains two steps. First, item latent content variable is generated by VAE model. And then variables $\boldsymbol{S}$, $\boldsymbol{W}$ and $\boldsymbol{b}$ together with $\boldsymbol{U}^+$ and $\boldsymbol{V}^+$ are updated by stochastic gradient descent.

Algorithm 3. Learning process of ACR

1: Initialize W, b, and S randomly
2: Draw $U^+ \sim N(0, \lambda_u^{-1} I_K)$
3: Draw $V^+ \sim N(0, \lambda_v^{-1} I_K)$
4: **for** n in epoches **do**
5:     Generate Z and update variables of VAE as described in Algorithms 1 and 2
6:     Compute P according to Equations (1), (2) and (3)
7:     Sample $\hat{R}$ according to Equation (4)
8:     Compute gradients to update W, b, U^+, V^+ and S respectively
9: **end for**

---

4 Experiments

4.1 Dataset

We conducted experiments on a real world mobile app recommendation dataset obtained from a popular mobile app store. Each app corresponds to an item, with its names and tags as content information. The tags of apps describe their functions such as *Game*, *Camera*, *Chat*, etc. We use app download behaviors to infer implicit feedback information. That is to say, for each download, the corresponding rating in the item-user rating matrix is 1. We aim to predict rating values for all of other apps not yet downloaded. In the download history of a user, for each tag, we count the number of apps that labeled by it. And then we choose the top 20 tags as the content information of the user.

To test the performance of the proposed model on datasets with different degrees of sparsity, we constructed two datasets, a sparse dataset and a dense dataset. For the sparse dataset, all the users have at least 5 download records per month, and for the dense dataset, all the users have at least 15 download records per month. The details of the datasets are shown in Table 1.

4.2 Experiment Setup

Pre-training of Tag Embeddings. Each App in our dataset are described by a set of Tags. Those co-occurred tags usually have similar semantic. Therefore, we treat each tag as a word and the tag list of an app as a sequence. Then, we use skip-gram model [6], window-size is 5, min-count is 1 and embedding dimension is 200 to train embeddings of each tag, which are used to extract latent user features in our proposed model ACR.

Baselines and Experimental Settings. Experiments are conducted on a Tesla K40m GPU. We use models CDL, CVAE, PMF, and bi-CVAE as our comparison baselines. By bi-CVAE we mean that we modify CVAE, and add a VAE model to extract users' deep features. Same as used in our model, top 20 tags of downloaded apps is regarded as users' content information. Other parameters and training process are the same as CVAE.

In our ACR model, top 20 tags of each user are used. The training epoch is 30 for dense dataset and 20 for sparse dataset. Learning rate is 0.001. We also pretrain VAE as CVAE does.

Table 1. Dataset statistics

Dataset	Users	Apps	Tags	Ratings in training set	Ratings in testing set	Sparseness
Sparse	77724	57177	6456	4824451	778485	0.13%
Dense	8558	42417	5344	1046042	177961	0.34%

Table 2. F1-scores of different algorithms (%)

Dataset	Algorithms	F1@5	F1@10	F1@20
Mobile app sparse	PMF	10.40	12.92	13.66
	CDL	10.33	13.79	15.89
	CVAE	10.35	13.83	15.94
	bi-CVAE	11.53	15.41	16.88
	ACR	**13.13**	**16.92**	**17.41**
Mobile app dense	PMF	14.59	19.63	22.59
	CDL	15.59	20.80	24.67
	CVAE	15.85	20.81	24.75
	bi-CVAE	15.72	21.47	25.84
	ACR	**19.94**	**25.44**	**28.30**

4.3 Experimental Results

We use precision@k, recall@k and F1-score@k (k = 5, 10, 20) to evaluate the results. Due to the limited space, we only demonstrate the result of F1-score@k in Table 2. Figures in bold are the best among all the competitive algorithms.

We can see from the results that our ACR model performs better than other competitive models on all the three measures, especially for the top 5 and top 10 apps. Therefore, our model can significantly improve recommendation accuracy. Our model improves more on the dense dataset than on the sparse dataset. As users have more behavior records in the dense dataset, tags of apps they have downloaded are more sufficient to represent user interest.

For other competitive models, bi-CVAE performs better than CVAE and CDL, which indicates that adding content information to the users really works, while an autoencoder cannot achieve the same accuracy as our attention model. CVAE and CDL have nearly the same performance and are better than PMF overall.

4.4 Interpretation of User Preferences

By introducing attention mechanism into our model, we can assign different attention weights to different tags to represent user preferences. We can find that nearly all the apps downloaded by the user contain user preference tags.

That is to say, preference tags can represent users' interest and can further guide their behavior effectively. We omit the demonstration of case study due to the limited space.

5   Conclusions

In this paper we study how to extract latent user features to facilitate latent user factor learning and how to improve the interpretability of deep collaborative recommendation models. To this end, we proposed an attentive and collaborative model, which learns deep representation of both items and users, and implicit relationship between items and users simultaneously. Meanwhile, an attention mechanism is proposed to obtain user preferences represented by tags. Experiments conducted on a real world dataset show that our model achieves higher accuracy compared with state-of-the-art models. Case study further demonstrates that tags with high attention weights can represent the user interest.

Acknowledgment. This work was supported in part by National Natural Science Foundation of China under grant No. U1711262, 71771131, 71272029, 71490724 and 61472426.

References

1. Kingma, D.P., Welling, M.: Auto-encoding variational bayes. arXiv preprint arXiv:1312.6114 (2013)
2. LeCun, Y., Bengio, Y., Hinton, G.: Deep learning. Nature **521**(7553), 436–444 (2015)
3. Li, S., Kawale, J., Fu, Y.: Deep collaborative filtering via marginalized denoising auto-encoder. In: Proceedings of the 24th ACM International on Conference on Information and Knowledge Management, pp. 811–820. ACM (2015)
4. Li, X., She, J.: Collaborative variational autoencoder for recommender systems. In: Proceedings of the 23rd ACM SIGKDD International Conference on Knowledge Discovery and Data Mining, pp. 305–314. ACM (2017)
5. Linden, G., Smith, B., York, J.: Amazon. com recommendations: item-to-item collaborative filtering. IEEE Internet Comput. **7**(1), 76–80 (2003)
6. Mikolov, T., Sutskever, I., Chen, K., Corrado, G.S., Dean, J.: Distributed representations of words and phrases and their compositionality. In: Burges, C.J.C., Bottou, L., Welling, M., Ghahramani, Z., Weinberger, K.Q. (eds.) Advances in Neural Information Processing Systems 26, pp. 3111–3119. Curran Associates, Inc. (2013)
7. Pazzani, M.J., Billsus, D.: Content-based recommendation systems. In: Brusilovsky, P., Kobsa, A., Nejdl, W. (eds.) The Adaptive Web. LNCS, vol. 4321, pp. 325–341. Springer, Heidelberg (2007). https://doi.org/10.1007/978-3-540-72079-9_10
8. Wang, C., Blei, D.M.: Collaborative topic modeling for recommending scientific articles. In: Proceedings of the 17th ACM SIGKDD International Conference on Knowledge Discovery and Data Mining, pp. 448–456. ACM (2011)
9. Wang, H., Wang, N., Yeung, D.Y.: Collaborative deep learning for recommender systems. In: Proceedings of the 21th ACM SIGKDD International Conference on Knowledge Discovery and Data Mining, pp. 1235–1244. ACM (2015)

10. Wang, X., Yu, L., Ren, K., Tao, G., Zhang, W., Yu, Y., Wang, J.: Dynamic attention deep model for article recommendation by learning human editors' demonstration. In: Proceedings of the 23rd ACM SIGKDD International Conference on Knowledge Discovery and Data Mining, pp. 2051–2059. ACM (2017)
11. Zhang, Q., Wang, J., Huang, H., Huang, X., Gong, Y.: Hashtag recommendation for multimodal microblog using co-attention network. In: IJCAI, pp. 3420–3426 (2017)
12. Zhang, S., Yao, L., Sun, A.: Deep learning based recommender system: a survey and new perspectives. arXiv preprint arXiv:1707.07435 (2017)

Collaborative Probability Metric Learning

Hongzhi Liu$^{(\boxtimes)}$, Yingpeng Du, and Zhonghai Wu

School of Software and Microelectronics, Peking University,
Beijing, People's Republic of China
{liuhz,dyp1993,wuzh}@pku.edu.cn

Abstract. Matrix factorization is a widely used collaborative filtering technique. However, the inner-product it relies on is not a proper distance metric because it does not satisfy the triangle inequality. Therefore, it cannot reliably capture similarities of neither item-item pairs nor user-user pairs, which will lead to suboptimal performance and limited interpretability. To solve these problems, we propose a novel collaborative filtering method based on metric learning, which can simultaneously capture the similarities of item-item pairs and user-user pairs besides the users' preferences on items. Different from previous metric learning methods which always only use either global structure information or local neighborhood information, the proposed method integrates both of these two kinds of information within a probability framework. Experimental results confirm the effectiveness of the proposed method.

Keywords: Metric learning · Personalized recommendation
Collaborative filtering · Implicit feedback · Probability framework

1 Introduction

To deal with the information overload problem on the Internet, recommender systems have been used to provide users with personalized recommendations of products or services in various domains [1]. Collaborative filtering (CF) is a popular technique used by recommender systems. It makes predictions and recommendations based on the behavior of other users in the system [2]. Matrix factorization (MF) is a widely used CF technique. It maps both users and items into a joint latent factor space, such that user-item interactions are modeled as inner-products in that space [3]. However, this kind of latent factor representation is not a metric representation because it does not satisfy the triangle inequality. Therefore, it cannot reliably capture similarities of neither item-item pairs nor user-user pairs, which will lead to suboptimal performance and limited interpretability of resulting model [4].

Recently, Hsieh et al. [4] proposed a collaborative filtering method based on metric learning, called Collaborative Metric Learning (CML). It learns a distance metric by pulling the user-item pairs with observed positive feedback closer and pushing the other user-item pairs relative further apart. This idea is inspired by Large Margin Nearest Neighbor (LMNN) [5], which is a local distance metric learning algorithm [6]. It tries to propagate the observed positive user-item relationships to get the global optimum. However, such kind of local propagation is easy to get into a local optimum and difficult to reach the global optimum.

Extensive research about personalized recommendation has been focused on explicit feedback [7,8]. Compared with explicit feedback data, implicit feedback data is more widespread and easier to be collected. However, implicit feedback data is also more difficult to be analyzed, because there is only positive feedback. Most MF methods assume all the unobserved feedback in implicit feedback data as negative examples, which may be caused by unseen instead of dislike [9].

To solve these problems, we propose a novel collaborative filtering method based on metric learning. Different from the local distance metric learning used in CML [4], we propose a hybrid metric learning method which can capture both global structure and local neighborhood information. Therefore, we can get better performance which will be shown in Sect. 5. In addition, we design a probability framework to distinguish between users' preferences and feedback. Based on this framework, we present an optimization criterion derived from a Bayesian analysis of the problem and employ bound optimization to learn it from training data.

2   Problem Definition

Let U and V denote the set of users and items, respectively. With $u_i \subset U$ and $v_j \in V$ denoting a user and an item, respectively, we define the implicit user feedback matrix $\mathbf{R} = (r_{ij})_{m \times n}$ as follows:

$$
r_{ij} = \begin{cases} 1, & \text{if user } u_i \text{ has given positive feedback to item } v_j, \\ 0, & \text{otherwise.} \end{cases}
$$

The set $P_{u_i} \subseteq V$ denotes the items to which user u_i has given positive feedback, i.e. $P_{u_i} = \{v_j | r_{ij} = 1\}$. The rest of items is denoted by $N_{u_i} = V \backslash P_{u_i}$.

Our goal is to give each user u_i a personalized ranking list of items from N_{u_i} using only implicit feedback data, i.e. assuming we only know the positive feedback P_{u_i} for each user u_i.

3   Probability Analysis Framework for Implicit Feedback

Implicit feedback data is collected by recording the feedback action (e.g. click, read, or buy) of users on given lists of items. The given lists are always very short compared with the whole item set. Therefore, most of the items are unaware for each user. The items without observed feedback are not all disliked by a user, but also contain some ones that the user likes. The goal of a recommender system is to find out these items and recommend them to the user.

To explore users' real preference, we simulate the process that users interact with the recommender system. We assume that even a user likes an item, there is only a probability of β that he/she will give feedback to this item. In addition, we assume dislikes will lead to missing values, i.e. users will not give (positive) feedback to items they dislike. We define $\mathbf{Y} = (y_{ij})_{m \times n}$ as the real preferences of users on items, i.e. $y_{ij} = 1$ if user u_i really likes item v_j and $y_{ij} = 0$ otherwise. Therefore, we get the conditional probability of feedback observation as follows:

$$P(r_{ij} = 1|\Theta) = \beta P(y_{ij} = 1|\Theta)$$

where Θ is the parameter vector which describes the recommender system.

Based on these assumptions, we derive a general optimization criterion for personalized recommendation from implicit feedback through a Bayesian analysis of the problem. The goal is to maximize the following posterior probability given the implicit user feedback matrix $\mathbf{R}$.

$$P(\Theta|\mathbf{R}) \propto P(\mathbf{R}|\Theta)P(\Theta), \tag{1}$$

where Θ is the parameter vector of an arbitrary model. We assume the prior density $P(\Theta)$ is a normal distribution with zero mean and covariance matrix $\Sigma_\Theta = \lambda I$, where I is an identity matrix.

Assuming the independence of each pair of user-item, we get the likelihood function $P(\mathbf{R}|\Theta)$:

$$P(\mathbf{R}|\Theta) = \prod_{(u_i,v_j) \in U \times V} P(r_{ij} = 1|\Theta)^{\delta(r_{ij}=1)} P(r_{ij} = 0|\Theta)^{\delta(r_{ij}=0)}$$

where $\delta(x)$ is the indicator function, which returns 1 if x is true and 0 otherwise.

Our goal is to learn the parameters Θ to maximize the posterior probability in Eq. (1). Given the monotonicity of the logarithm function, the model parameters that maximize Eq. (1) are equivalent to those maximize $\ln P(\Theta|R)$. Specifically,

$$\ln P(\Theta|\mathbf{R}) \propto \ln P(\mathbf{R}|\Theta)P(\Theta) = \ln P(\mathbf{R}|\Theta) + \ln P(\Theta)$$

$$= \sum_{u_i \in U} \left(\sum_{v_j \in P_u} \ln \beta P(y_{ij} = 1|\Theta) + \sum_{v_j \in N_u} \ln(1 - \beta P(y_{ij} = 1|\Theta)) \right) + \ln P(\Theta). \tag{2}$$

To simplify the representation in the following of this paper, we set

$$\sum_{U,P_u} \triangleq \sum_{u_i \in U} \sum_{v_j \in P_u} \quad \text{and} \quad \sum_{U,N_u} \triangleq \sum_{u_i \in U} \sum_{v_j \in N_u}.$$

To simplify the optimization process of this problem, we present a lower bound for the posterior probability in Eq. (2) by inequality scaling based on Jensen inequality and the convex property of $\ln(\cdot)$ function.

$$\ln P(\Theta|\mathbf{R}) \propto \ln P(\mathbf{R}|\Theta)P(\Theta)$$

$$\geq \sum_{U,P_u} \ln \beta P(y_{ij} = 1|\Theta) + (1-\beta) \sum_{U,N_u} \ln 1 + \beta \sum_{U,N_u} \ln P(y_{ij} = 0|\Theta) + \ln P(\Theta)$$

$$= \sum_{U,P_u} \ln \beta + \sum_{U,P_u} \ln P(y_{ij} = 1|\Theta) + \beta \sum_{U,N_u} \ln P(y_{ij} - 0|\Theta) - \lambda ||O||^2$$

$$(3)$$

As the constant $\sum_{U,P_u} \ln \beta$ is irrelevant with the optimal setting of parameters Θ, we will drop it for simplification. The key to optimize Eq. (3) is to estimate the conditional probability of $P(y_{ij}|\Theta)$.

4   Collaborative Probability Metric Learning

Based on the above probability analysis framework, we design a metric learning method for collaborative filtering, called Collaborative Probability Metric Learning (CPML). We estimate the conditional probability of $P(y_{ij}|\Theta)$ by metric learning.

4.1   Objective Function

We assume the closer between a user and an item, the higher probability that the user would like the item. In contrast, farther user-item distance means a lower probability that the user would like the item. Based on this assumption, we define the conditional probability of $P(y_{ij}|\Theta)$ as follows:

$$
\begin{cases}
P_G(y_{ij} = 1|\Theta) = 1/\left(1 + e^{(||\bm{u_i}-\bm{v_j}||^2 - \mu)}\right) \\
P_G(y_{ij} = 0|\Theta) = 1/\left(1 + c^{(\mu-||\bm{u_i}-\bm{v_j}||^2)}\right)
\end{cases}
\tag{4}
$$

where $\bm{u_i} \in R^d$ is the latent vector representation of the ith user, $\bm{v_j} \in R^d$ is the latent vector representation of the jth item, and d is the dimensionality of latent factor space.

Therefore, we get an objective function:

$$F_G(\Theta, R) = \sum_{U,P_u} \ln \frac{1}{1 + e^{(||\bm{u_i}-\bm{v_j}||^2 - \mu)}} + \beta \sum_{U,N_u} \ln \frac{1}{1 + e^{(\mu-||\bm{u_i}-\bm{v_j}||^2)}} - \lambda ||\Theta||^2.$$

$$(5)$$

The goal of Eq. (5) is to directly optimize the distance of all example pairs. It is good at capturing the global structure of training data. However, the optimization of Eq. (5) will sacrifice some local relative relationships. In addition, to make personalized recommendation for each user, we only need to take the top-k items (i.e. the k-nearest items) into consideration rather than all items.

Taking into consideration the above reasons, we propose to integrate some local information into the estimation of $P(y_{ij} = 1|\Theta)$. Let n_{u_i} be the nearest item of user u_i without observed feedback, i.e.

$$n_{u_i} = \{v_k| \min_k ||\bm{u_i} - \bm{v_k}||^2 \text{ where } r_{ik} = 0\}.$$

If an item v_j is closer to user u_i than n_{u_i}, then the probability $P(y_{ij} = 1|\Theta)$ should be higher. Otherwise, if an item v_j is farther than n_{u_i}, then the probability $P(y_{ij} = 1|\Theta)$ should be lower. Specifically, we define $P_L(y_{ij} = 1|\Theta)$ as follows:

$$P_L(y_{ij} = 1|\Theta) = 1 / \left(1 + e^{(||\boldsymbol{u}_i - \boldsymbol{v}_j||^2 - ||\boldsymbol{u}_i - \boldsymbol{n}_{u_i}||^2)}\right). \tag{6}$$

By integrating Eqs. (4) and (6), we get a hybrid probability estimation as follows,

$$P(y_{ij} = 1|\Theta) = \alpha P_G(y_{ij} = 1|\Theta) + (1 - \alpha)P_L(y_{ij} = 1|\Theta), \tag{7}$$

where $0 \leq \alpha \leq 1$ is a trade-off weight. Therefore, we get a new objective function:

$$
\begin{aligned}
F(\Theta, \mathbf{R}) &\overset{Eq.\,3}{=} \sum_{U,P_u} \ln P(y_{ij} = 1|\Theta) + \beta \sum_{U,N_u} \ln P_G(y_{ij} = 0|\Theta) - \lambda||\Theta||^2 \\
&= \sum_{U,P_u} \ln\left(\alpha P_G(y_{ij} = 1|\Theta) + (1 - \alpha)P_L(y_{ij} = 1|\Theta)\right) + \beta \sum_{U,N_u} \ln P_G(y_{ij} = 0|\Theta) - \lambda||\Theta||^2 \\
&\geq \alpha \sum_{U,P_u} \ln(P_G(y_{ij} = 1|\Theta)) + (1 - \alpha)\sum_{U,P_u} \ln(P_L(y_{ij} = 1|\Theta)) + \beta \sum_{U,N_u} \ln P_G(y_{ij} = 0|\Theta) - \lambda||\Theta||^2
\end{aligned}
\tag{8}
$$

4.2   Model Learning

We employ the widely used stochastic gradient descent (SGD) method to optimize the objective function in Eq. (8). We randomly sample a user u_i at each iteration. The gradients of objective function $F(\Theta, \mathbf{R})$ with respect to user vector $\boldsymbol{u}_i$ and item vectors $\boldsymbol{v}_j$ are calculated as follows,

$$
\begin{aligned}
\frac{\partial F}{\partial \boldsymbol{u}_i} = &-\sum_{v_j \in P_{u_i}} \frac{2\alpha(\boldsymbol{u}_i - \boldsymbol{v}_j)}{1 + e^{(\mu - ||\boldsymbol{u}_i - \boldsymbol{v}_j||^2)}} + \sum_{v_j \in P_{u_i}} \frac{2(1 - \alpha)(\boldsymbol{v}_j - \boldsymbol{n}_{u_i})}{1 + e^{(||\boldsymbol{u}_i - \boldsymbol{n}_{u_i}||^2 - ||\boldsymbol{u}_i - \boldsymbol{v}_j||^2)}} \\
&+ \sum_{v_j \in N_{u_i}} \frac{2\beta(\boldsymbol{u}_i - \boldsymbol{v}_j)}{1 + e^{(||\boldsymbol{u}_i - \boldsymbol{v}_j||^2 - \mu)}} - \lambda\boldsymbol{u}_i
\end{aligned}
\tag{9}
$$

$$
\frac{\partial F}{\partial \boldsymbol{v}_j} =
\begin{cases}
\frac{2\alpha(\boldsymbol{u}_i - \boldsymbol{v}_j)}{1 + e^{(\mu - ||\boldsymbol{u}_i - \boldsymbol{n}_{u_i}||^2)}} + \frac{2(1 - \alpha)(\boldsymbol{u}_i - \boldsymbol{v}_j)}{1 + e^{(||\boldsymbol{u}_i - \boldsymbol{v}_j||^2 - ||\boldsymbol{u}_i - \boldsymbol{n}_{u_i}||^2)}} - \lambda\boldsymbol{v}_j, & v_j \in P_{u_i} \\
-\frac{2\beta(\boldsymbol{u}_i - \boldsymbol{v}_j)}{1 + e^{(||\boldsymbol{u}_i - \boldsymbol{n}_{u_i}||^2 - \mu)}} - \lambda\boldsymbol{v}_j, & \text{otherwise}
\end{cases}
\tag{10}
$$

With the above gradients, the model parameters could be updated as follows,

$$\boldsymbol{u}_i = \boldsymbol{u}_i + \eta\frac{\partial F}{\partial \boldsymbol{u}_i}, \quad \boldsymbol{v}_j = \boldsymbol{v}_j + \eta\frac{\partial F}{\partial \boldsymbol{v}_j}, \tag{11}$$

where η is the learning rate.

The pseudocode of CPML algorithm is shown in Algorithm 1. It consists of an initialization step (Line 1) and an iterative training process (Line 2–9). The initialization step involves randomly setting the model parameters. The training process adopts the SGD method. In each iteration, we randomly sample a user u_i. Instead of directly using the whole set of items without observed feedback, i.e. N_{u_i}, which will lead to computation impracticable for large datasets, we use a sampling technique. We randomly sample q items from N_{u_i}, denoted as N_u', and use it to estimate N_{u_i}. With these information, we calculate the gradients of latent vectors of user u_i and all items $v_i \in P_{u_i} \bigcup N_u'$ according to Eqs. (9) and (10). Then we update these vectors using the gradients according to Eq. (11). At last, to avoid the curse of dimensionality and ensure the robustness of the learned metric, we bound all the user and item vectors within a hyper unit sphere as in [4].

Algorithm 1. Collaborative Probability Metric Learning (CPML)

Input: Implicit feedback matrix $\mathbf{R}$, dimensionality of latent space d, threshold μ,
    number of sampling items without feedback q, weights α and β
Output: Latent factor matrix of users $\mathbf{U}$ and items $\mathbf{V}$.
1: Randomly initialize the parameters $\mathbf{U} = R^{m \times d}$ and $\mathbf{V} = R^{n \times d}$;
2: **for** $t = 0; t < T; t = t + 1$ **do**
3:     Sample a user u_i from U;
4:     Sample q items from N_{u_i}, denoted as N_u', to estimate N_{u_i};
5:     $n_{v_u} = \min\limits_{v_j \in N_u'} ||v_j - u_i||^2$;
6:     Calculate the gradient $\frac{\partial F}{\partial u_i}$ and $\frac{\partial F}{\partial v_j}$ according to Eq.(9) and Eq.(10);
7:     Update the latent vectors of u_i and $v_i \in P_{u_i} \bigcup N_u'$ according to Eq.(11);
8:     Normalize the latent vectors: $u_i = \frac{u_i}{\max(||u_i||, 1)}$, $v_j = \frac{v_j}{\max(||v_j||, 1)}$;
9: **end for**

---

5   Experiments

5.1   Experimental Setup

Three real-world data sets are used as experimental data, including Movie-Lens100k, Netflix5k5k, and CiteUlike. The same as in [11], we consider the ratings higher than 3 as positive feedback in MovieLens100k and Netflix5k5k. In addition, we remove the users and items which have less than five interactions.

    To evaluate the empirical performance of CPML and the baseline methods, we adopt three widely used evaluation metrics for recommendation, including precision at top k, recall at top k, and AUC. Two-fold cross-validation is adopted and experimental results are recorded as the average of the two runs.

    Four recommendation algorithms are used as baseline methods, including User-kNN, WRMF [10], BPR [11], and CML [4]. User-KNN is a memory-based collaborative filtering method. WRMF [10] is a model-based collaborative filtering method. BPR [11] is a pairwise learning to rank method. CML [4] is a metric learning method for collaborative filtering. For all baseline methods, their hyper parameters are tuned to their best. For CPML, we set the two weights $\alpha = 0.5$ and $\beta = 0.1$, the threshold $\mu = 0.1$, the regularization parameter $\lambda = 0.1$, and the dimensionality of latent space $d = 50$.

5.2   Experimental Results

Table 1 shows the performance of CPML and the baseline algorithms according to different evaluation metrics. CPML outperforms the baselines on all the three data sets. Compared with the best performance of baseline algorithms (denoted as italic in Table 1), the improvements of CPML are up to 13.35%, 12.67%, 6.65% for Recall@50, Precision@10 and AUC, respectively.

Different from MF based methods, CPML and CML learn the latent representation by metric learning. The triangle inequality ensured by CPML and CML is an important characterize for the latent space representation. This is why CPML and CML perform relatively better than the other comparison methods. In addition, metric learning provides an intuitive way to describe the relationship between users and items [4]. Therefore, the models learned by CPML and CML have better interpretability compared with MF based methods.

Table 1. Performance of CPML and baseline algorithms

Dataset	Method	Recall@50	Precision@10	AUC
MovieLens	User-kNN	0.4126	0.4602	0.8652
	WRMF	0.3883	0.4354	0.8340
	BPR	0.4175	0.4368	*0.8911*
	CML	*0.4222*	*0.4807*	0.8704
	CPML	**0.4646**	**0.4964**	**0.9151**
	Improve.	**10.03%**	**3.27%**	**2.70%**
Netflix5k5k	User-kNN	0.3506	0.1994	0.7637
	WRMF	0.3327	0.1998	0.7741
	BPR	0.4219	0.2215	0.8969
	CML	*0.4258*	*0.2388*	*0.8809*
	CPML	**0.4565**	**0.2526**	**0.9154**
	Improve.	**7.22%**	**5.78%**	**2.07%**
CiteULike	User-kNN	0.1861	0.0684	0.6236
	WRFM	0.2196	0.0686	0.7799
	BPR	0.2457	0.0772	0.8218
	CML	*0.2690*	*0.0868*	*0.8457*
	CPML	**0.3049**	**0.0978**	**0.9019**
	Improve.	**13.35%**	**12.67%**	**6.65%**

Figure 1 shows the performance, measured by AUC, of CPML and the baselines on both testing data and training data with varying dimensionality d of latent space. As the dimensionality d increases, the performances of all the methods on the training data increase, except User-kNN which is independent with this parameter. However, their performances on the testing data begin to decease as the dimensionality d increases enough, except CPML. This demonstrates the capacity of CPML for preventing over-fitting.

CML is a local metric learning method which tries to reach the global optimum by propagating local neighborhood relationship. Therefore, it is easy to get into a local optimum. CPML overcomes this problem by integrating local neighborhood information and global structure information.

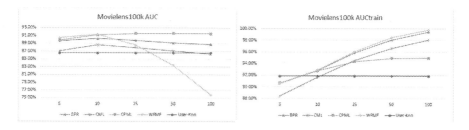

Fig. 1. Performance of different methods with various dimensionality d of latent space, the left is on the testing data and the right is on the training data

6   Conclusion

In this paper, we have studied personalized recommendation from implicit feedback, and proposed a novel algorithm called collaborative probability metric learning (CPML). Different from the inner-product used by MF methods, which does not satisfy the triangle inequality, CPML tries to embed all the users and items into a Euclidean space by metric learning. Therefore, we can get better performance and interpretability. Different from CML [4] which focuses on local information and is easy to get into a local optimum, CPML integrates both local neighborhood information and global structure information. Therefore, CPML has better capability to prevent over-fitting. Extensive experiments demonstrate the promise of our approach in comparison with the traditional collaborative filtering methods and the state-of-the-art metric learning algorithm for collaborative filtering.

Acknowledgements. This work was sponsored by National Key R&D Program of China (Grant No. 2017YFB1002002).

References

1. Lu, J., Wu, D., Mao, M., et al.: Recommender system application developments: a survey. Decis. Support Syst. **74**, 12–32 (2015)
2. Ekstrand, M.D., Riedl, J.T., Konstan, J.A.: Collaborative filtering recommender systems. Found. Trends Hum.-Comput. Interact. **4**(2), 81–173 (2011)
3. Koren, Y., Bell, R., Volinsky, C.: Matrix factorization techniques for recommender systems. Computer **42**(8), 30–37 (2009)

4. Hsieh, C.K., Yang, L., Cui, Y., et al.: Collaborative metric learning. In: International Conference on World Wide Web, WWW 2017, pp. 193–201 (2017)
5. Weinberger, K.Q., Saul, L.K.: Distance metric learning for large margin nearest neighbor classification. J. Mach. Learn. Res. **10**, 207–244 (2009)
6. Kulis, B.: Metric learning: a survey. Found. Trends Mach. Learn. **5**(4), 287–364 (2013)
7. Ricci, F., Rokach, L., Shapira, B.: Recommender systems: introduction and challenges. In: Ricci, F., Rokach, L., Shapira, B. (eds.) Recommender Systems Handbook, pp. 1–34. Springer, Boston (2015). https://doi.org/10.1007/978-1-4899-7637-6_1
8. Satzger, B., Endres, M., Kießling, W.: A preference-based recommender system. In: Bauknecht, K., Pröll, B., Werthner, H. (eds.) EC-Web 2006. LNCS, vol. 4082, pp. 31–40. Springer, Heidelberg (2006). https://doi.org/10.1007/11823865_4
9. Liu, H., Wu, Z., Zhang, X.: CPLR: collaborative pairwise learning to rank for personalized recommendation. Knowl.-Based Syst. **148**, 31–40 (2018)
10. Pan, R., Zhou, Y., Cao, B., et al.: One-class collaborative filtering. In: IEEE International Conference on Data Mining, ICDM 2008, pp. 502–511 (2008)
11. Rendle, S., Freudenthaler, C., Gantner, Z., et al.: BPR: Bayesian personalized ranking from implicit feedback. In: Uncertainty in Artificial Intelligence, UAI 2009, pp. 452–461 (2009)

UIContextListRank: A Listwise Recommendation Model with Social Contextual Information

Zhenhua Huang[1,2(✉)], Chang Yu[2], Jiujun Cheng[2],
and Zhixiao Wang[3]

[1] School of Computer Science, South China Normal University,
Guangzhou 510631, China
jukiehuang@163.com
[2] Department of Computer Science, Tongji University, Shanghai 201804, China
[3] School of Computer Science and Technology,
China University of Mining and Technology, Xuzhou 221116, China

Abstract. With the explosive growth of social network, the exploitation of social information in recommendation models has become increasingly significant. However, most existing models only made use of users' social information, ignoring the value of items' social information. Based on the above fact, we present a listwise learning to rank recommendation model, UIContextListRank, which generates a ranked list of items for individual users directly. We employ matrix factorization to construct a listwise objective function that measures the difference between the predicted lists and the real ones. Furthermore, we express users' social contextual information as their trust friends and items' social contextual information as their concurrent items, and incorporate the social contextual information of both users' and items' into the listwise model to improve recommendation quality. Moreover, we implement our proposed model in a distributed environment to tackle the challenge of overwhelming data. Experiments have been conducted on two real-world datasets to evaluate the proposed model. And the experimental results prove the model's effectiveness and efficiency.

Keywords: Recommender system · Learning to rank
Social contextual information · Spark

1 Introduction

Recommender systems exist everywhere in our life, helping individuals explore and seek their interested items in all aspects. Conventional recommender systems provide recommendation results depending only on the predicted rating scores that users might give to items, while mere ratings can hardly reflect users' preference accurately in actual life. To improve the recommendation quality, researchers have integrated learning to rank [1] techniques to recommendation models. Learning to rank recommendation models focus on directly generating ranked item lists that users may like instead of predicting certain rating, which compensates the disadvantage of conventional models effectively.

© Springer International Publishing AG, part of Springer Nature 2018
Y. Cai et al. (Eds.): APWeb-WAIM 2018, LNCS 10987, pp. 207–215, 2018.
https://doi.org/10.1007/978-3-319-96890-2_18

Learning to rank has three levels [1]: pointwise, pairwise and listwise. Pointwise [2] methods learn a rating function through training data to rate items and the items are then ranked according to the predicted rating scores. Pairwise [3] methods concern about the order of item pairs and get the ranked list through the order between pairs. Listwise [4] methods optimize the recommendation item list directly, which is the most natural way for recommendation. There are two basic approaches to build a listwise recommendation model currently: (1) optimizing the evaluation metrics directly [5, 6], like MAP(mean average precision), NDCG(Normalized Discounted Cumulative Gain). (2) constructing an objective function for optimization [7].

With the popularization of Internet, social networks like Facebook have occupied a significant part in people's everyday life. Users' behaviors and relationships in social networks reflect their personalized features to some extent. Therefore, taking social information into account while recommending has become a beneficial approach to enhance recommendation quality. However, most models [8, 9] only focus on users' social information, the exploitation of items' social information in recommendation hasn't received enough attention. In the meantime, for the explosive growth of data, the efficiency of recommendation models when facing overwhelming volume of data should be guaranteed.

In this paper, we take the above issues into consideration and propose our solution. The main contributions of our work are as follows:

- We propose a listwise learning to rank recommendation model UIContextListRank, taking both users' and items' social contextual information into consideration.
- We extend our model to a paralleled version to tackle the problem of overwhelming data and implement the paralleled model in Spark.
- We evaluate our model on two real-world datasets, and the experimental results show the effectiveness and efficiency of our model in comparison with state-of-the-art baselines.

2   UIContextListRank

2.1   Definition and Notation

We assume there are m users and n items, each user can give an item a rating score. We refer to user as u, item as i, the rating score that u gives to i as r_{ui}. $U = \{u_1, \cdots, u_m\}$ represents the set of users, $I = \{i_1, \cdots, i_n\}$ represents the set of items, $V = \{1, \cdots, v\}$ represents the range of r_{ui}, $F(u) = \{u_j, u_k, \cdots\}$ represents the set of user u's friends $(j, k < m)$. A list of items that user u has rated can be achieved from user u's rating history. Here we use π_u to refer to the list, and $\pi_u = \{i_k | \forall i_k \in I, r_{ui} > 0\}$.

2.2   Social Contextual Information

In the proposed model, we take the social features of both users' and items' into consideration. As in the previous work [10], for a user u who has rated item i and has a social friend set $F(u)$, we can indicate u's social contextual information as

$N^1(u, i) = \{x | x \in F(u), r_{ui} > 0, r_{xi} > 0\}$, which represents the set of u's friends who have also rated item i. And we use $N^1_{u,i} = |N^1(u, 1)|$ to represent the number of friends in $N^1(u, i)$.

In the meantime, we define the social contextual information of items as the count of their concurrent items in all users' rating lists. For instance, if u has rated both i_1 and i_2, then for item i_1, its concurrent item set is $C(i_1) = \{i_2 | r_{ui} > 0, r_{ui_2} > 0, |r_{ui_1} - r_{ui_2}| \leq 1, u \in U\}$. We restrict the difference between r_{ui_1} and r_{ui_2} to 1 to grantee that u's preferences for the concurrent items are similar. Hence, we define i's social contextual information as $N^2(u, i) = \{s | s \in C(i), r_{ui} > 0\}$, and the number of $N^2(u, i)$ is $N^2_{u,i} = |N^2(u, i)|$.

2.3   UIContextListRank Model

According to the social contextual information defined above and the inspiration from [7], we propose a listwise learning to rank recommendation model, UIContextListRank as Eqs. (1) and (2).

$$min_{U^*, V^*} \xi = \sum_{u-1}^{m} \left\{ -\sum_{i=1}^{n} P_{\pi_u}(r_{ui}) log P_{\pi_u}[g(\hat{r}_{ui})] \right\} + \lambda \left(\|U\|_F^2 + \|V\|_F^2 \right) \quad (1)$$

$$\hat{r}_{ui} = (1 - \alpha - \beta) \cdot U_u^T \cdot V_i + \alpha \cdot \frac{1}{N^1_{u,i}} \sum_{x \in N^1(u,i)} U_x^T \cdot V_i + \beta \cdot \frac{1}{N^2_{u,i}} \sum_{y \in N^2(u,i)} U_u^T \cdot V_y$$

$$(2)$$

where ξ denotes the loss function, $\hat{r}_{ui}$ denotes the predicted scores that user u gives to item i. U denotes the 'user-latent factor' matrix, V denotes the 'item-latent factor' matrix. The social contextual information of users' and items' are adopted here to extend U and V. Note that $\hat{r}_{ui}$ is just used to rank items, representing the degree that how users prefer the items. $g(\cdot)$ represents the logistic sigmoid function. λ is the regularization factor that helps prevent overfitting. During training, we use cross entropy to evaluate the similarity between the probability of the item ranked list produced by the model and that of the corresponding actual one.

We use gradient descent algorithms to minimize the model loss. We first substitute Eq. (2) into (1), then use the calculated gradients Eqs. (3) and (4) to update matrix U and V. Note that $g(\hat{r}_{ui})$ represents the derivative of logistic sigmoid function and $\hat{r}'_{ui}$ means that Eq. (2) takes derivative of U_u and V_i separately.

$$\frac{\partial \xi}{\partial U_u} = \sum_{i-1}^{n} \left(\frac{exp(g(\hat{r}_{ui}))}{\sum_{k=1}^{n} exp(g(\hat{r}_{uk}))} - \frac{exp(r_{ui})}{\sum_{k=1}^{n} exp(r_{uk})} \right) g'(\hat{r}_{ui})\hat{r}'_{ui} + 2\lambda U_u \quad (3)$$

$$\frac{\partial \xi}{\partial V_i} = \sum\nolimits_{u=1}^{m} \left(\frac{exp(g(\hat{r}_{ui}))}{\sum_{k=1}^{n} exp(g(\hat{r}_{uk}))} - \frac{exp(r_{ui})}{\sum_{k=1}^{n} exp(r_{uk})} \right) g'(\hat{r}_{ui})\hat{r}'_{ui} + 2\lambda V_i \qquad (4)$$

3   Paralleled UIContextListRank in Spark

The processing of massive scale of data can bring great difficulty to recommender systems, causing problems like long computation time and slow interaction with users. Therefore, in this section, we propose the paralleled UIContextListRank model and implement it in Spark to tackle the challenge of big data.

We have introduced the UIContextListRank model as Eqs. (1) and (2), and gradients of the 'user-latent factor' matrix U_u and 'item-latent factor' matrix V_i as Eqs. (3) and (4) in Sect. 2. To construct a paralleled model, we must eliminate the dependency between user feature matrix and item feature matrix in the iterative process, and assure the current user or item feature matrix is only associated with the state of last user or item feature matrix. Therefore, we use alternating stochastic gradient descent method to optimize the 'user-latent factor' matrix and the 'item-latent factor' matrix separately, dividing the training into two parts to train user feature matrix and item feature matrix separately.

We can deduce that the expressions of U_u^k and V_i^k after one iteration are:

$$U_u^k = U_u^{k-1} - \eta \left[\sum\nolimits_{i=1}^{n} \left(\frac{exp\left(g(\hat{r}_{ui})^{k-1}\right)}{\sum_{j=1}^{n} exp\left(g(\hat{r}_{uj})^{k-1}\right)} - \frac{exp(r_{ui})}{\sum_{j=1}^{n} exp(r_{uj})} \right) g'(\hat{r}_{ui})^{k-1}(\hat{r}'_{ui})^{k-1} \right.$$

$$\left. + 2\lambda Uuk - 1. \qquad (5) \right.$$

$$(\hat{r}'_{ui})^{k-1} = (1 - \alpha - \beta) \cdot (U_u^{k-1})^T \cdot V_i^{h_k} + \alpha \cdot \frac{1}{N_{u,i}^1} \sum\nolimits_{x \in N^1(u,i)} (U_x^{h_k})^T \cdot V_i^{h_k}$$

$$+ \beta \cdot \frac{1}{N_{u,i}^2} \sum\nolimits_{y \in N^2(u,i)} (U_u^{k-1})^T \cdot V_y^{h_k} \qquad (6)$$

$$V_i^k = V_i^{k-1} - \eta \left[\sum\nolimits_{u=1}^{m} \left(\frac{exp\left(g(\hat{r}_{ui})^{k-1}\right)}{\sum_{j=1}^{n} exp\left(g(\hat{r}_{uj})^{k-1}\right)} - \frac{exp(r_{ui})}{\sum_{j=1}^{n} exp(r_{uj})} \right) g'(\hat{r}_{ui})^{k-1}(\hat{r}'_{ui})^{k-1} \right.$$

$$\left. + 2\lambda Vik - 1 \right.$$

$$(7)$$

where η is the learning rate, h_k represents the matrix's temporary state in the k-th iteration.

4 Experiments

4.1 Datasets and Evaluation Metric

We conduct our experiments on two real-world datasets, Epinions[1] and Flixster[2]. Epinions is a website that provides customers' reviews about items for reference. And customers can add the users they trust as friends. It contains 40,163 users, 139,738 items, 487,183 relationships and 664,824 ratings. Flixster is movie website that offers movie lovers to rate movies and make friends. It contains 147,612 users, 48,794 items, 11,794,648 relationships and 8,196,077 ratings.

We use NDCG (Normalized Discounted Cumulative Gain) as the evaluation metric in the following groups of experiments.

4.2 Baselines

We compare our proposed model with three listwise learning to rank recommendation algorithms, which are CofiRank [5], ListRank [7] and SoRank [8].

CofiRank: It directly optimizes NDCG by Maximum Margin Matrix Factorization, instead of constructing an objective function.

ListRank: It uses the TOP probability model and cross-entropy to construct a listwise loss function and optimizes the loss function using stochastic gradient descent.

SoRank: It integrates the information of users' friends into the objective function on the base of ListRank, also optimized by stochastic gradient descent.

4.3 Experimental Results

Experiment 1: NDCG@k

In this experiment, we measure the four models' recommendation performance with NDCG@k. We choose 3000 users from two datasets separately, 80% as training set and 20% as testing set. We set the regularization factor λ in Eq. (1) as 0.01, the dimension d of matrix U and V as 10, the parameters α and β in Eq. (2) as 0.3 and 0.3. Then we compute the value of NDCG@k when $k = 5, 10, 15, 20$.

Figures 1 and 2 illustrate the NDCG@k results of the four algorithms. We can see that UIContextListRank gets the highest values of NDCG in both datasets, which proves that the introduction of both users' and items' social context information to listwise recommendation can significantly improve the accuracy of recommendation. Meanwhile, the four algorithms' values of NDCG increase with the increasing of k, which tells us that the recommendation works better when the result list is longer.

---

[1] http://www.epinions.com
[2] https://www.flixster.com

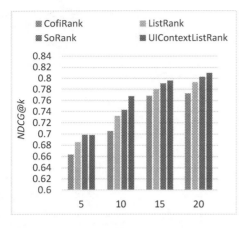

Fig. 1. NDCG@k on Epinions   **Fig. 2.** NDCG@k on Flixster

Experiment 2: The Dimension of Latent Factors

In this experiment, we analyze the impact of the dimension of latent factors on recommendation performance. We use the same datasets and parameter setting as **Experiment 1**, and calculate NDCG@10 as the dimension is set to 10, 20, 30, 50 separately.

Figures 3 and 4 demonstrate that for all the four models, the NDCG@10 increase with the increase of the dimension of latent factors. The improvements of recommendation performance can be naturally explained that the higher the dimension is, the more detailed the features of users and items are. We can also find that UIContextListRank outperforms other models under all circumstance, which strongly proves its effectiveness. As the complexity and running time of experiments can both increase greatly with the increase of dimension, we set the dimension as 10 in other groups of experiments.

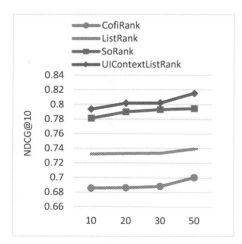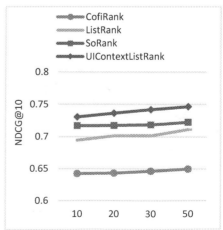

Fig. 3. NDCG@10 on Epinions   **Fig. 4.** NDCG@10 on Flixster

Experiment 3: Number of Users

In this experiment, we are to study the models' performance under different number of users. We randomly choose 3000, 5000 and 10000 users from Epinions and Flixster separately to construct six datasets. We use the same parameter setting as **Experiment 1**, calculating NDCG@10 on the six constructed datasets.

Tables 1 and 2 shows that the number of users does not affect the recommendation accuracy significantly. And our proposed model can always achieve the best performance under different numbers of user, which proves the model's stability and effectiveness.

Table 1. NDCG@10 of different number of users on Epinions

Number of Users	3000	5000	10000
CofiRank	0.6855	0.6821	0.6836
ListRank	0.7321	0.7329	0.7318
SoRank	0.7810	0.7894	0.7857
UIContectListRank	0.7934	0.7969	0.7922

Table 2. NDCG@10 of different number of users on Flixster

Number of Users	3000	5000	10000
CofiRank	0.6420	0.6418	0.6451
ListRank	0.6943	0.6985	0.6972
SoRank	0.7169	0.7170	0.7202
UIContectListRank	0.7305	0.7330	0.7345

Experiment 4: Paralleled UIContextListRank

This experiment is conducted on a cluster consists of six homogeneous compute nodes which is a 64-bit server with 32-core 2.0 GHz CPU, 256 GB of memory and CentOS 6.7 operating system. The evaluation metric is running time. And the datasets and parameter settings are the same as **Experiment 3**.

Figures 5 and 6 indicate that the paralleled model runs much faster than the normal version and the difference increases with the user number increasing. The running time of paralleled model is almost 1/3 of that of the normal model in Epinions, 1/4 of that of normal model in Flixster. The Experimental results strongly show that the implementation in a distributed environment can significantly improve recommendation efficiency when dealing with large amount of training data.

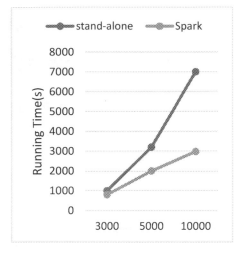

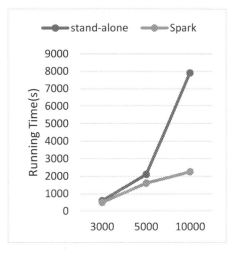

Fig. 5. Running time on Epinions

Fig. 6. Running time on Flixster

5 Conclusions

In this paper, we propose a listwise personalized learning to rank recommendation model, UIContextListRank, integrating the social contextual information of both users' and items'. And we also implement the model in Spark to tackle with the challenge of overwhelming data. The thorough experiments conducted on two different real-world datasets show the effectiveness and efficiency of our model.

In the future, we will try to extract more kinds of social contextual features for recommendation. Also, we want to incorporate other machine learning methods like deep neural networks into recommendation models to improve recommendation quality.

Acknowledgment. This work was supported by the National Natural Science Foundation of China (No. 61772366) and the Natural Science Foundation of Shanghai (No. 17ZR1445900).

References

1. Li, H.: Learning to rank for information retrieval and natural language processing. Synth. Lect. Hum. Lang. Technol. **7**(3), 1–121 (2014)
2. Liu, T.Y.: Learning to rank for information retrieval. Found. Trends® Inf. Retr. **3**(3), 225–331 (2009)
3. Hüllermeier, E., Fürnkranz, J., Cheng, W., Brinker, K.: Label ranking by learning pairwise preferences. Artif. Intell. **172**(16), 1897–1916 (2008)
4. Hofmann, K., Whiteson, S., de Rijke, M.: Balancing exploration and exploitation in listwise and pairwise online learning to rank for information retrieval. Inf. Retr. **16**(1), 63–90 (2013)

5. Weimer, M., Karatzoglou, A., Le, Q.V., Smola, A.J.: Cofirank-maximum margin matrix factorization for collaborative ranking. In: Advances in Neural Information Processing Systems, pp. 1593–1600 (2008)
6. Wu, B.-X., Xiao, J., Zhu, J., Ding, C.: An adaptive kNN using listwise approach for implicit feedback. In: Li, F., Shim, K., Zheng, K., Liu, G. (eds.) APWeb 2016. LNCS, vol. 9931, pp. 519–530. Springer, Cham (2016). https://doi.org/10.1007/978-3-319-45814-4_42
7. Shi, Y., Larson, M., Hanjalic, A.: List-wise learning to rank with matrix factorization for collaborative filtering. In: Proceedings of the Fourth ACM Conference on Recommender Systems, pp. 269–272. ACM (2010)
8. Yao, W., He, J., Huang, G., Zhang, Y.: SoRank: incorporating social information into learning to rank models for recommendation. In: Proceedings of the 23rd International Conference on World Wide Web, pp. 409–410. ACM (2014)
9. Ren, Z., Liang, S., Li, P., Wang, S., de Rijke, M.: Social collaborative viewpoint regression with explainable recommendations. In: Proceedings of the Tenth ACM International Conference on Web Search and Data Mining, pp. 485–494. ACM (2017)
10. Huang, Z., Shijia, E., Zhang, J., Zhang, B., Ji, Z.: Pairwise learning to recommend with both users' and items' contextual information. IET Commun. **10**(16), 2084–2090 (2016)

Information Retrieval

Diversified Keyword Expansion
on Multi-labeled Graphs

Mohammad Hossein Namaki$^{(\boxtimes)}$, Yinghui Wu, and Xin Zhang

Washington State University, Pullman, USA
{mnamaki,yinghui,xzhang2}@eecs.wsu.edu

Abstract. Keyword search has been widely adopted to explore graph data. Due to the intrinsic ambiguity of terms, it is desirable to develop query expansion techniques to find useful and diversified information progressively in large graphs. To support exploration with keywords, we study the problem of *diversified keyword expansion* in graphs. Given a set of validated content nodes in a graph, it is to find a set of terms that maximizes the aggregated relevance of the validated nodes. Moreover, the terms should be diversified to cover different search interests. We develop a fast stream-based ($\frac{1}{2}$-ϵ)-approximation to suggest diversified terms, which guarantees a linear scan of the terms in the content nodes up to a bounded area with small update cost. Using real-world graphs, we experimentally verify the effectiveness and efficiency of our algorithms, and their applications in knowledge base exploration.

Keywords: Keyword query expansion · Submodular maximization

1 Introduction

Keyword search (KWS) has been used to explore and understand graph data [4,18,25,26]. A keyword query Q is a set of keywords $\{t_1, \ldots, t_l\}$. Given a keyword query Q and a graph G, an answer of Q is a subgraph of G that contains a set of *content nodes*. Each content node matches to a term in Q. The subgraph can be *e.g.,* a minimal rooted tree [29], a weighted Steiner tree [5], or an r-clique [14], when Q is distinct root-based (DR), Steiner tree-based (ST), or Steiner graph-based (SG), respectively [25]. In practice, answers can be ranked by established distance-based relevance, such as shortest paths in answers [5], semantic distances [33], and normalized Web distances [8].

To cope with the ambiguity of keywords, exploratory search [1,4,19,26] exploits interactive search sessions to find useful information progressively. Each session interleaves query suggestion and query evaluation [7]. Effective graph exploration with KWS is nevertheless challenging. (1) The interactive exploration requires fast query suggestion, which aims to discover both relevant and diversified terms [6,16]. While the nodes in real-world graphs are often multi-labeled, searching for good terms is usually expensive. (2) Real-world graphs are

© Springer International Publishing AG, part of Springer Nature 2018
Y. Cai et al. (Eds.): APWeb-WAIM 2018, LNCS 10987, pp. 219–235, 2018.
https://doi.org/10.1007/978-3-319-96890-2_19

also dynamic in that new terms are constantly inserted. Query suggestion needs to respond to such changes in a timely manner.

We consider an example from KWS-based knowledge exploration.

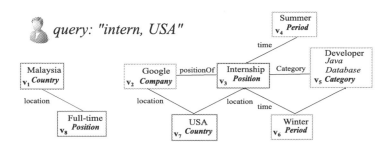

Fig. 1. Graph exploration with KWS (Color figure online)

Example 1. Figure 1 illustrates a fraction of a professional network G, such as LinkedIn knowledge graph. Each entity carries multiple terms, such as "Java, Database" that describe required skills, Developer as the title, and Category as its type. A user may search for internship opportunities with a query Q that contains terms "Intern" and "USA", and an answer $Q(G)$ that contains two entities with exactly matched labels marked in blue rectangles. While $Q(G)$ is not very informative, several terms can be suggested to help users to specify the search. (a) {Full-time, Malaysia} may be suggested due to co-occurrence of their types with the terms in Q following keyword suggestion for relational database [20], which finds job opportunities in Malaysia. This is not very relevant to the search intent as it ignores the semantic connection of entities in the information network. (b) The terms Summer and Winter are also relevant to $Q(G)$. But these are quite redundant due to that both suggest time, and lead to less informative queries.

A more desirable suggestion can be {Google, Developer}. Both terms are relevant to $Q(G)$ in distance-based semantic measures. Moreover, one covers a perspective of "USA companies" for the term "USA", and another specifies the titles for the term "internship". The user can issue a query expanded with either or both of the terms to perform diversified exploration.

The need of diversified query expansion for graphs is evident in knowledge base exploration [1], "why-not" queries [22], and may help to identify highly-correlated queries to reverse-engineering sensitive entity information [23]. However, discovering such terms can be expensive. For example, one may need to search for nearest neighbors of all validated nodes in $Q(G)$, and validate possible combinations of terms carried by these nodes to maximize diversity. Moreover, new terms can be inserted to G. For example, when a term "Tensorflow" is inserted to most relevant nodes with title "Developer" as a new required skill, a query suggestion scheme should respond fast and suggest the term.

Contribution. This paper studies diversified keyword expansion in graphs. We develop expansion algorithms with provable guarantees on term quality and time cost. We make the following contributions.

(1) Diversified KWS *Query Expansion.* Given a set V_C of validated content nodes, and a graph G, we identify a set of terms with maximized aggregated relevance to V_C, and suggest diversified search intent. We characterize the aggregated relevance of a term with minimum weighted paths from its content nodes to V_C. We also introduce a diversification measure by rewarding terms with new content nodes in V_C connected by such paths.

(2) Stream-Based Approximation. We show that the diversified query expansion problem is ($\frac{1}{2}$-ϵ)-approximable for constant $\epsilon > 0$, and develop a stream-style query expansion algorithm. The algorithm follows the Threshold Algorithm (TA) to generate a stream of terms with aggregated relevance via bounded traversal, and maintains lists to approximate top-k terms with early termination. The algorithm takes small update cost upon discovery of terms, and can naturally cope with newly inserted terms without more update cost.

(3) KWS-*Based Graph Exploration.* We specialize the algorithm for established KWS classes DR, ST, and SG. By simple adaption, we can devise diversified graph exploratory schemes that make use of these query classes, with matching performance guarantees. We evaluate the effectiveness and efficiency of our query expansion algorithms, using real world graphs (Sect. 4). We also show that these algorithms readily lead to flexible KWS for graph exploration, as verified by our case studies in knowledge bases and recommendation networks.

Related Work. We categorize the related work as follows.

Keyword Search in Graphs. Keyword search (KWS) has been studied extensively for unstructured [29] and structured [25] data. In graphs, KWS is to find subgraphs that contain content nodes that match given keyword queries. There are several established KWS query classes. (1) Distinct root-based KWS (DR) [11,13] defines $Q(G)$ as a minimal rooted tree that contains all content nodes at its leaves. (2) KWS with Steiner trees (ST) differs from DR queries in that it aims to find a weighted Steiner tree that minimizes the total edge weights. (3) Steiner graph-based KWS (SG) finds a set of content nodes such that the sum of pairwise distances is minimized (or bounded as r-cliques [14]).

Standard KWS follows "query-response" paradigms that assume accurate queries. We study diversified query expansion that can directly support graph exploration with established DR, ST, and SG queries, and provide parameters to enable tunable query expansion and graph exploration.

Query Suggestion. Keyword query suggestion and its variants (*e.g.*, query refinement [17] and reformulation [28]) have been studied to suggest new queries that better describe search intent. Prior work mostly adopts information retrieval methods that access query logs and user feedback [7] for Web search. Such log is scarce for emerging graph search applications. Co-occurring term retrieval (CoOcc) [20] discovers relevant keywords as those that appear most frequently in the results of original query Q. Tag cloud (TagCloud) [15] discovers terms that are highly relevant to the initial results with *e.g.*, tf-idf measures. These methods are designed for relational data rather than general graphs, where the relevance are often more involved with graph topology. Subgraphs are extracted to suggest structured queries by approximate KWS [24]. These methods cannot be directly applied for graph exploration with keywords.

Mismatch [30] aims to provide explanations for XML queries with empty answers. To this end, approximate answers of a query is computed by expanding its original answers with content nodes of the same type, and new terms are suggested to replace those without match. While Mismatch aims to find good replacement of terms over typed XML data, our work specifies diversified terms to expand original queries for general, multi-labeled graphs. Little has been done on KWS expansion to support diversified exploration in multi-labeled graphs.

Exploratory Search. Exploratory methods have been studied to query graphs [4, 18, 24, 26, 30]. It involves an interactive process of exploration and refinement as knowledge is progressively acquired. Visual interfaces are developed to support interactive KWS in XML [4], with a focus on query suggestion with mismatches [30]. TriniT [26] facilitates exploratory querying in incomplete knowledge graphs. It relaxes queries by replacing *e.g.*, its predicate to semantically similar ones using rules, and retrieves the relevant answers ranked with the weight of the rules. Query-by-example [12] refers to specified substructures of a graph as examples, and identifies relevant substructures using *e.g.*, (relaxed) subgraph isomorphism to approximate user's intent.

We develop efficient query expansion for established KWS queries and term relevance measures. Prior work does not provide efficiency guarantees for query expansion. Our diversified query expansion techniques can be readily adapted to enable data exploration with established KWS models, and can also be used in combination with existing data exploration schemes.

2   Keyword Query Expansion for Graphs

We consider a multi-labeled graph $G = (V, E)$ with a node set V and a set of weighted, undirected edges $E \subseteq V \times V$. Each node $v \in V$ is associated with a set of labels (terms) $L(v) \subseteq \Sigma$ that encodes its content, where Σ is a finite set of terms. Each edge $e = (u, v) \in E$ has a weight $w(e)$. In practice, a node v (resp. edge e) may represent a tuple [29] (resp. a dependency), an entity (resp. a fact) in a knowledge graph, or a location. The edge weight may characterize semantic closeness of entities [8], fact reliability, or spatial distance.

Keyword Search in Graphs. A keyword query Q is a set of terms $\{t_1, \ldots t_n\}$. Given graph $G=(V, E)$ and a term t_i, a match function determines a set of *content nodes* $V(t_i) \subseteq V$ that match t_i. For example, it can be a transformation [27] that finds the nodes with labels that are synonyms of t_i.

Existing matching semantics are based on either (1) lowest common ancestor (LCA), which finds a set of content nodes with common ancestor that minimizes a certain distance measure (such as root-leaf distances [11,13] or total edge cost [9]), or (2) dense subgraphs [14], which induce content nodes with small aggregated pairwise distances. We consider a general form of query answer $Q(G)$ of a keyword query Q, which is a subgraph of G that contains at least one node from $V(t_i)$ for each $t_i \in Q$. We denote the set of content nodes in $Q(G)$ as V_C.

Keyword Query Expansion. Query expansions are useful to interpret queries, understand results, and suggest new queries for graph exploration [24]. Given a keyword query Q, we define an *expansion* of Q as a set of terms Q_T not in Q.

Effective graph exploration requires to find meaningful new terms to trigger future search sessions toward desired information in graphs. The quality of such terms should be defined *relative to* query Q and validated content nodes V_C in $Q(G)$. We next introduce two relative measures that characterize term *relevance* and *diversity*, respectively. These measures have their conventional counterparts in IR-based query expansion for unstructured texts and documents.

Term Relevance. First, a suggested term should be relevant to validated content nodes V_C. A common practice in graph search quantifies the relevance between two nodes as a function dist of their (normalized) distances [8,28,33]. Distance-based measures are useful in its capacity to capture semantic distance between concepts, which is usually represented by the path connecting two entities in G. The more "closer" the two nodes are, the more relevant they are.

Given a term t and a content node $v_c \in V_C$ that matches a term $t' \in Q$, the *distance* of t to node v_c is defined as

$$\mathsf{dist}(t, v_c) = \min_{v_t \in V(t) \setminus V_C} \mathsf{dist}(v_t, v_c)$$

where $\mathsf{dist}(v_t, v_c)$ is a distance function from v_t to v_c in graph G. For example, $\mathsf{dist}(v_t, v_c)$ can be defined as $\sum_{e \in \rho} w(e)$, for a shortest path ρ between v_t and v_c in G. Here, we only consider those content nodes of term t that are not in V_C, which count for relevance from the "unseen" nodes to be explored.

Following this intuition, a term that is relevant to V_C should have content nodes with small aggregated distances. Given content nodes V_C and term t, the *term relevance* $\mathsf{rev}(t, V_C)$ of t *w.r.t.* V_C is further defined as

$$\mathsf{rev}(t, V_C) = \frac{1}{1 + \sum_{v_c \in V_C} \mathsf{dist}(t, v_c)}; \quad \mathsf{rev}(Q_T, V_C) = \sum_{t \in Q_T} \mathsf{rev}(t, V_C)$$

That is, the relevance of a term *w.r.t.* V_C is the aggregated relevance from its content nodes that are closest to the nodes in V_C. The relevance of a query expansion $Q_T = \{t_1, \ldots, t_m\}$ *w.r.t.* V_C is further aggregated from the relevance of its terms. Note that both are relative measures, which are determined by specific query, validated answers, and the observed terms. For example, among Google,

Developer, and Malaysia in Example 1, Google has the highest and Malaysia has the lowest relevance since as shown in Fig. 1, Google is closest to both validated content nodes and the Malaysia is far from and thus less relevant to them.

Remarks. We are aware of other Corpus-Based based relevance such as TF-IDF, node relevance [5], distinguishability [30], and surpriseness. Such terms can be readily integrated into $\mathsf{dist}(\cdot)$.

Term Diversity. The suggested terms should also be diversified to expand as many different aspects of answers $Q(G)$ as possible. Such diversity is preferred in *e.g.*, faceted search and result diversification [6,16,31].

We shall use the following notations. (1) We consider a reasonable *partition* $\mathcal{P}(\mathcal{V_C})$ of $V_C = \{V_{C_1}, \dots, V_{C_n}\}$, where each set $V_{C_i} \subseteq V_C$ represents a group of content nodes with similar contents that indicates an "aspect" of V_C. (2) Given a threshold r and a term t, the *cover set* of t w.r.t. r, denoted as $\mathsf{cov}(t, r)$, refers to the set $\{v | \mathsf{dist}(t, v) \leq r, v \in V_C\}$. Equally, $\mathsf{rev}(t, \{v\}) \geq \frac{1}{1+r}$.

We use the partition $\mathcal{P}(\mathcal{V_C})$ of V_C and cover set to characterize the diversity of a query expansion Q_T. Let $\mathcal{P}(\mathcal{V_C})$ contains n groups of content nodes. The diversity of Q_T w.r.t. $\mathcal{P}(\mathcal{V_C})$ and threshold r is defined as

$$\mathsf{div}(Q_T, \mathcal{P}(\mathcal{V_C}), r) = \frac{1}{n} \sum_{i=1}^{n} \sqrt{\sum_{t \in Q_T} \frac{|\mathsf{cov}(t, r) \cap V_{C_i}|}{|V_{C_i}|}}$$

Intuitively, the function $\mathsf{div}(\cdot)$ rewards the diversity in that there is more benefit to expand a query by a term that can cover content nodes from more "aspects", than its counterparts that cover fewer nodes and groups which are already covered by other terms. In practice, (1) $\mathcal{P}(\mathcal{V_C})$ can be constructed by grouping content nodes with the terms in Q they match, by node types, or by applying semantic clustering [32]; (2) the threshold r is tunable by users: larger r allows terms that are "further" from V_C to be considered for expansion.

Example 2. Consider a partition of content nodes V_C (Fig. 1) as two sets Country= $\{v_7\}$ and Position= $\{v_3\}$ based on their *node type*, and set $r = 1$. For term Winter (resp. Google), $\mathsf{cov}(\mathsf{Winter}, 1) = \{v_3\}$ (resp. $\mathsf{cov}(\mathsf{Google}, 1) = \{v_3, v_7\}$). Then, $\mathsf{div}(\{\mathsf{Summer}, \mathsf{Winter}\}) = 0.5(\sqrt{2})$; and $\mathsf{div}(\{\mathsf{Summer}, \mathsf{Google}\}) = 0.5(1+\sqrt{2})$, respectively. This indicates that $Q_T = \{\mathsf{Summer}, \mathsf{Google}\}$ is more diversified. Indeed, term Google has a larger reward if combined with Winter, since it increases the diversity of Q_T by covering the partition Country not being covered before.

Diversification. Given a query expansion Q_T, a partition $\mathcal{P}(\mathcal{V_C})$ of V_C and threshold r, we model the quality $F(Q_T)$ of Q_T as a bi-criterion objective function that combines relevance and diversity, which is defined as

$$F(Q_T, \mathcal{P}(\mathcal{V_C}), r) = \mathsf{rev}(Q_T, V_C) + \lambda * \mathsf{div}(Q_T, \mathcal{P}(\mathcal{V_C}), r)$$

When it is clear from the context, we refer to the quality function as $F(Q_T)$. We now formulate our diversified query expansion problem.

Problem Statement. Given graph G, keyword query Q, a distance threshold r and a set of validated content nodes V_C, the diversified query expansion problem is to find a set of query expansion Q_T^* with k terms, such that (1) for each node $v_c \in V_C$, there is a term $t \in Q_T^*$, such that $v_c \in \mathsf{cov}(t, r)$; and (2) Q_T^* satisfies

$$Q_T^* = \underset{Q_T \subseteq \Sigma, |Q_T| = k}{\arg\max} \ F(Q_T, \mathcal{P}(V_C), r)$$

For query expansion, we define the *marginal gain* $\mathsf{mg}(t, Q_T)$ of a term t to query expansion Q_T ($t \notin Q_T$) as $F(Q_T \cup \{t\}) - F(Q_T)$. The result below shows that function $F(\cdot)$ is well defined in terms of submodularity, which is widely used to characterize the quality of sets in Web search and database applications.

Lemma 1. *The function $F(\cdot)$ is a monotone submodular function for query expansion. That is, for any two query expansion Q_T and Q_T', if $Q_T \subseteq Q_T'$, then (1) $F(Q_T) \leq F(Q_T')$, and (2) for any term $l \notin Q_T'$, $\mathsf{mg}(t, Q_T') \leq \mathsf{mg}(t, Q_T)$.*

It is easy to verify Lemma 1 by definition of monotone submodularity and that the diversity reward function is defined in terms of square root function. Due to space limit, we omit the detailed proof.

The problem is NP-hard by reduction from set diversification. In practice, V_C is relatively small, while the size of terms and their content nodes are large and disk-based. This calls for fast algorithms to support query expansion for interactive data exploration, and time response for term insertions to G.

3   Diversified Query Expansion

Enumerating combinations of k terms to find Q_T^* is clearly not practical. We start with an alternative that uses an "explore-and-diversify" process. We then introduce a fast approximation algorithm that is based on stream diversification.

Explore-and-Diversify. The basic intuition is to first find all relevant terms T w.r.t. V_C, and diversify T to find Q_T. The algorithm capitalizes on data locality, and visits only the neighbors of the nodes up to a bounded distance.

The algorithm, denoted as Div, invokes an iterator $\mathsf{SSSP}(v)$ that performs a bounded shortest path traversal up to $N^r(v)$ (*i.e.*, the nodes with distance up to r) from v, for each content node $v \in V_C$. This conceptually constructs a node list ranked by their distance to v, following a sorted access to $N^r(v)$.

(1) For each content node $v \in V_C$, iterator $\mathsf{SSSP}(v).\mathsf{next}()$ outputs a pair $(v', \mathsf{dist}(v, v'))$, which finds the next node $v' \in N^r(v)$ following an increasing order of $\mathsf{dist}(v, v')$. For each term $t \in L(v')$, it computes term relevance $\mathsf{rev}(t, V_C)$, and $\mathsf{cov}(t, r) \cap V_C$. These are bookkept in a matrix dmap ($\mathsf{dmap}[t][v] = \mathsf{dist}(t, v)$).

(2) When no new term can be found, it collects the set of all the terms T, and invokes a standard greedy algorithm [10] to find top-k diversified terms in T. While $|Q_T| < k$, the process iteratively selects a term $t \in T \setminus Q_T$, by consulting dmap, such that t has a maximized marginal gain $\mathsf{mg}(t, Q_T)$. It then updates Q_T from iteration $i - 1$ (denoted as Q_T^{i-1}) as

$$Q_T^i = Q_T^{i-1} \cup \arg\max_{t \in T} \mathsf{mg}(t, Q_T^{i-1})$$

at iteration i. It is known that this strategy provides a $(1\text{-}\frac{1}{e})$-approximation.

The algorithm Div requires the discovery of all the terms. The greedy algorithm requires k passes over the term set T. Furthermore, it already takes $O(|V_C|(|N^r(V_C)| \log |N^r(V_C)| + |N^r(V_C)|))$ time to traverse the graph, thus it is expensive to recompute Q_T when new terms are inserted to G.

3.1   Stream-Based Diversification

We can do better by capitalizing on stream-based computation [3]. Our major idea is to treat terms as a weighted stream, and process each term at most once to find Q_T with optimality guarantee, and incur small update cost for each term.

Overview. Our stream-based algorithm, denoted as streamDiv, takes as input the graph G, query Q, partitioned content nodes $\mathcal{P}(\mathcal{V_C})$, integer k, threshold r. The algorithm interleaves two procedures: (1) *term generation*: a procedure GenTerm that follows a Threshold Algorithm (TA) style process to generate a stream $\mathcal{T}$ of terms with operator SSSP, and (2) *stream diversification*: a procedure DivTerm that dynamically constructs and maintains Q_T from $\mathcal{T}$. Each time GenTerm generates a term t, it adds t to stream $\mathcal{T}$. Upon receiving a new term t in $\mathcal{T}$, DivTerm updates Q_T, until no new terms can be added to $\mathcal{T}$.

We say query expansion Q_T is an α-approximation ($\alpha \in [0, 1]$) of the optimal Q_T^* if $F(Q_T, \mathcal{P}(\mathcal{V_C}), r) \geq \alpha * F(Q_T^*, \mathcal{P}(\mathcal{V_C}), r)$. We present our major result below.

Theorem 1. *Given a constant $\epsilon > 0$, algorithm* streamDiv

- *computes a $(\frac{1}{2} - \epsilon)$-approximation Q_T in $O(|V_C|(|N^r(V_C)| \log |N^r(V_C)| + |N^r(V_C)|)(\frac{\log k}{\epsilon}))$ time;*
- *performs a single pass of all terms in $N^r(V_C)$, and takes $O(k|V_C| + \frac{\log k}{\epsilon})$ time to process each term; and*
- *takes $O(k|V_C| + \frac{\log k}{\epsilon})$ time to update Q_T upon each term insertion to G.*

As a proof of Theorem 1, we introduce details of algorithm streamDiv. For the ease of presentation, we first present term stream diversification procedure DivTerm, with the performance guarantees in Theorem 1. We then show how procedure GenTerm optimizes the stream by only generating promising terms.

Procedure DivTerm($\mathcal{T}, \mathcal{P}(\mathcal{V}_C), r, Q_T, F_{max}$);

1.   **if** sieve set $O := \emptyset$ **then**
2. $O := \{(1 + \epsilon)^i | F_{max} \le (1 + \epsilon)^i \le k.F_{max}\}$;
3.   **for each** value $s \in O$ and each term $t \in \mathcal{T}$ **do**
4.   **if** $Q_T(s)$ is not defined **then** $Q_T(s) := \emptyset$;
5.   **if** $\text{mg}(t, Q_T(s)) \ge \frac{\frac{s}{2} - F(Q_T(s), \mathcal{P}(\mathcal{V}_C), r)}{k - |Q_T(s)|}$ **and** $|Q_T(s)| < k$ **then**
6. $Q_T(s) := Q_T(s) \cup \{t\}$; update minsieve;
7.   **if** $\forall s \in O$: $|Q_T(s)| = k$ **or Condition 2 then**
8. $Q_T := \arg\max_{Q_T(s)} F(Q_T(s), \mathcal{P}(\mathcal{V}_C), r)$;
9.   VoteforHalt(); **return** Q_T; /* *early termination* */

Fig. 2. Procedure DivTerm

3.2 Term Diversification

Upon receiving a stream $\mathcal{T}$ of elements (from procedure GenTerm) and their relevance, procedure DivTerm constructs Q_T by dynamically diversifying the elements from $\mathcal{T}$. We first review sieve-streaming for diversification.

Sieve-Streaming Revisited [3]. Given a monotone submodular function $F(\cdot)$, a constant $\epsilon > 0$, and the set of elements $\mathcal{D}$, sieve streaming finds top-k elements $\mathcal{S}$ that maximizes $F(\mathcal{S})$ as follows. If $F_{max} = \max_{e \in \mathcal{D}} F(\{e\})$ is known, it determines a set of sieve values $(1 + \epsilon)^i$ (i as integer) to discretizes the range $[F_{max}, k.F_{max}]$, and use the sieve values to approximate $F(\mathcal{S}^*)$. For each sieve value s, a corresponding sieve list S_v is dynamically maintained as the top-k terms with marginal gain at least $(\frac{s}{2} - F(S_v))/(k - |S_v|)$, falling into the specific range determined by s. It is shown that selecting $\mathcal{S}$ as the top-k elements from all the sieve sets generates a $(\frac{1}{2} - \epsilon)$-approximation [3].

Overview. The procedure DivTerm, illustrated in Fig. 2, solves a submodular optimization problem over the term stream $\mathcal{T}$, following sieve-streaming. It initializes a set of sieve values O determined by maximized value $F_{max} = \arg\max_{t \in \mathcal{T}} F(t)$ (line 2). Here, $F(t) = \text{rev}(t, V_C) + \lambda * \text{div}(t, \mathcal{P}(\mathcal{V}_C), r)$. DivTerm is activated upon seeing the maximum value F_{max} (see "Lazy sieving"). For each sieve value s, it maintains a list of top-k terms $Q_T(s)$ (lines 3–4). (1) For each term t arrived in $\mathcal{T}$, it dynamically inserts t into a sieve list $Q_T(s)$ if

$$\text{mg}(t, Q_T(s)) \ge (\frac{s}{2} - F(Q_T(s), \mathcal{P}(\mathcal{V}_C), r))/(k - |Q_T(s)|)$$

(2) Once there is no new term in stream $\mathcal{T}$, or all the sieve lists $Q_T(s)$ have size k, it terminates and constructs Q_T as the top-k terms from sieve lists (lines 7–9).

Analysis. Following [3], DivTerm is a $(\frac{1}{2} - \epsilon)$-approximation. (1) There exists a sieve value $s = (1 + \epsilon)^i \in [F_{max}, 2k * F_{max}]$ that is closest to $F(Q_T^*)$, say,

$(1 - 2\epsilon)F(Q_T^*) \leq s \leq F(Q_T^*)$; and (2) each set $Q_T(s)$ is a $(\frac{1}{2} - \epsilon)$ answer for an estimation of $F(Q_T^*)$ with sieve value s. Indeed, if $\mathsf{mg}(t, Q_T)$ satisfies the condition in DivTerm (line 5), then $F(Q_T(s), \mathcal{P}(\mathcal{V_C}), r) \geq \frac{s|Q_T(s)|}{2k} = \frac{s}{2}$ (when $|Q_T(s)|=k$). As there exists a value $s \in O$ that best estimates the optimal $F(\cdot)$, the top-k terms Q_T from sieve lists achieves approximation ratio $(\frac{1}{2} - \epsilon)$.

Optimization. The procedure DivTerm further optimizes sieve-streaming in [3] with two strategies, both interact with procedure GenTerm and makes use of an upper bound estimation of $F(t)$ for t in the "unseen" part of $\mathcal{T}$, denoted as $\overline{F}$.

Lazy Sieving. The first question is "how to find F_{max}?". When terms are inserted to $\mathcal{T}$ but F_{max} is unknown, sieve-streaming [3] eagerly maintains a dynamic number of lists for an increased value range $[F_{max}, 2k * F_{max}]$, determined by F_{max} in currently seen stream, to safely update results. Instead, we let procedure GenTerm "activates" the sieve streaming in DivTerm when it confirms the identification of F_{max}.

This is done by testing *"whether $\overline{F}$ is no larger than the smallest $F(t)$ for t in seen part of $\mathcal{T}$"* (**Condition 1**). When this happens, GenTerm sends F_{max} as the largest seen $F(t)$ to DivTerm, and activates sieve streaming. We found that such terms can be identified at quite early stage of streaming, thanks to the sorted access of $\mathsf{SSSP}(\cdot)$ and distance-based relevance measures.

Early Termination. Procedure DivTerm talks to GenTerm in turn to terminate the stream early. To this end, it keeps track of (1) the smallest seen threshold $\mathsf{minsieve} = \min_{s \in O}(\frac{s}{2} - F(Q_T(s), \mathcal{P}(\mathcal{V_C}), r))/(k - |Q_T(s)|)$; and (2) an estimated upper bound of all unseen terms t and sieve values s, denoted as $\overline{\mathsf{mg}}(\mathcal{T})$, and tests *"whether $\overline{\mathsf{mg}}(\mathcal{T}) < \mathsf{minsieve}$"* (**Condition 2**). If so, it sends a signal (VoteforHalt(), line 9) back to GenTerm to stop the term stream. Indeed, this indicates that no new term can be inserted to any sieve list and can make contribution to Q_T.

The following result shows that $\overline{\mathsf{mg}}$ can be set as $\overline{F}$ for early termination.

Lemma 2. $\overline{F}$ *is a valid upper bound of* $\mathsf{mg}(t)$ *for any unseen term t in $\mathcal{T}$.*

Proof sketch: For any term t in unseen part of $\mathcal{T}$ and any sieve value $s \in O$, we can verify that $\mathsf{mg}(t, Q_T(s)) \leq \mathsf{rev}(\{t\}, V_C) + \lambda \mathsf{div}(\{t\}, \mathcal{P}(\mathcal{V_C}), r) = F(\{t\}, \mathcal{P}(\mathcal{V_C}), r)$. As $\overline{F}$ is no less than the largest $F(\{t\}, \mathcal{P}(\mathcal{V_C}), r)$ for all unseen term t in $\mathcal{T}$, Lemma 2 follows.

Condition 1 and **Condition 2** verify that we can "activate" and "terminate" the stream diversification, by using only a simple upper bound estimation $\overline{F}$. The optimization is quite effective: they improve the efficiency of query expansion by 81%, as verified by our experimental study.

We next introduce procedure GenTerm, and show how to compute $\overline{F}$.

Procedure GenTerm

Input: Graph G, query Q, partition $\mathcal{P}(\mathcal{V}_C)$, integer k, threshold r;
Output: a stream of terms $\mathcal{T}$.

1.   stream $\mathcal{T} := \emptyset$, $F_{max} := 0$; $F_{min}:=0$; $\overline{F} := +\infty$;
       /* activating SSSP traversal */
2.   **for each** $v_i \in V_C$ **do** create iterator $\mathsf{SSSP}(v_i)$;
       /* TA-style term generation */
3.   **while** there exists $v_i \in V_C$: $\mathsf{SSSP}(v_i).\mathsf{peekDist}() \leq r$ **do**
4.   pair $<v', d> := \mathsf{SSSP}(v_i).\mathsf{next}()$;
5.   **for each** term $t \in L(v')$ **do**
6.   **if** $\mathcal{T}$ contains t **then continue** ; /*t is already seen*/
7. $\mathcal{T} := \mathcal{T} \cup \{t\}$; update $\mathsf{rev}(\cdot)$, $\mathsf{div}(\cdot)$ and $F(\cdot)$ for term t;
8. $F_{max} = \max_{t\in\mathcal{T}} F(\{t\}, \mathcal{P}(\mathcal{V}_C), r)$;
9. $F_{min} = \min_{t\in\mathcal{T}} F(\{t\}, \mathcal{P}(\mathcal{V}_C), r)$;
10.   estimate $\overline{F}$; /* estimate the best $F(\cdot)$ of unseen terms */
11.   **if** $\overline{F} \leq F_{min}$ **(Condition 1) then**
12. $\mathsf{DivTerm}(\mathcal{T}, \mathcal{P}(\mathcal{V}_C), r, F_{max})$; /* "activate" stream */
13.   **else if** $\mathsf{DivTerm}$ is active **then**
14. $\mathsf{DivTerm}(\{t\}, \mathcal{P}(\mathcal{V}_C), r, F_{max})$; /* process a new term */

Fig. 3. Procedure GenTerm

3.3   Term Stream Generation

Given graph G and partitioned content nodes $\mathcal{P}(\mathcal{V}_C)$, procedure GenTerm generates a stream of promising terms that are likely to contribute to Q_T. The procedure follows a Threshold-Algorithm style processing.

Overview. The procedure GenTerm, illustrated in Fig. 3, invokes iterator SSSP as in algorithm Div for each content node $v \in V_C$ (line 2). The bounded traversal is controlled by operator peekDist, which reports the distance of the next node in $N^r(v)$ by a sorted access to $N^r(v)$ following non-decreasing order of distances to v. This operator can be easily implemented using standard single source shortest path algorithm e.g., Dijkstra [5]. It then verifies new terms to be added to stream $\mathcal{T}$, and updates their relevance and diversity by leveraging an index such as pruned landmark labeling [2] (lines 5–7). Note that it never reprocesses a term t once it has been added to $\mathcal{T}$ earlier, as all the term information has been captured due to the visiting of its "closer" content node (line 6). It then estimates $F(\cdot)$ for unseen terms, and test whether F_{max} is identified to activate Div (line 12). While DivTerm is active, it sends a new term for processing (line 14).

Estimation of $\overline{F}$. To compute $\overline{F}$ (line 10), GenTerm estimates an upper bound of $\mathsf{rev}(\{t\}, V_C)$ and $\mathsf{div}(\{t\}, \mathcal{P}(\mathcal{V}_C), r)$, respectively. For $\mathsf{rev}(\{t\}, V_C)$ by consulting the current status of SSSP iterators, GenTerm estimates the potential closest distance for any unseen t, and updates $\mathsf{rev}(\{t\}, V_C) = (1 +$

$\sum_{v_i \in V_C}$ SSSP$_i$.peekDist$)^{-1}$. For div($\{t\}, \mathcal{P}(\mathcal{V}_C), r$), given each cluster $\mathcal{P}(\mathcal{V}_C)$, it estimates an upperbound for coverage of any unseen t as $|\{v_i \in \mathcal{P}(\mathcal{V}_C)|$SSSP$_i$.peekDist $\leq r\}|$.

Performance Analysis. Procedure GenTerm executes $|V_C|$ number of SSSP originated from $v \in V_C$ bounded by neighbors $N^r(v)$ up to distance r that each one takes $O(|N^r(v)| \log |N^r(v)| + |N^r(v)|)$. As each node carries a small constant number of terms, time cost of updating term information of a node is in $O(1)$. Each time a term is sent to DivTerm, it takes $O(\frac{\log k}{\epsilon})$ time to update the sets $Q_T(s)$'s. Thus, the total time is in $O(|V_C|(|N^r(V_C)| \log |N^r(V_C)| + |N^r(V_C)|)(\frac{\log k}{\epsilon}))$.

Coping with Term Insertion. Given a newly added term t to a node in G, streamDiv simply takes $O(k|V_C|)$ time to compute $F(t, \mathcal{P}(\mathcal{V}_C), r)$ by (1) update dist(t, v) for $v \in V_C$, using distance index [2]; and (2) update cov$(t, r) \cap V_C$. It then invokes DivTerm only if $F(t, \mathcal{P}(\mathcal{V}_C), r) \geq$ minsieve, and at most $O(\frac{\log k}{\epsilon})$ time to update sieve lists and update Q_T.

4   Experimental Evaluation

Using real-life graphs, we evaluate (1) the effectiveness and efficiency of our diversified query expansion algorithm, (2) the impact of factors, and (3) case studies to verify its application in knowledge base exploration.

Experimental Setting. We use the following setting.

Datasets. We use the following datasets. (1) DBpedia[1], a knowledge graph that contains 4.8M multi-labeled entities, in total 1.5M terms from both entity names and their type (*e.g.*, "Obama", "Place"), and 15M edges with 670 distinct relationships. Each node carries on average 11 terms. (2) IMDB[2] is an information network including 1.6M entities of movies, TV shows, and crews. It contains 5.1M edges and 1.4M tokens from *e.g.*, genre and titles of movies, and the name of crews. Each node has on average 3 terms. We assign edge weights following backward KWS in graph databases [13], and set dist($\cdot$) as the distance of shortest path between two nodes.

Partitioning. We created partitions of V_C grouped by the pair (t, type), where each group has the same type (*e.g.*, "Person", "Genre") and matched with the same term in query Q (See example 2).

KWS *queries.* The query generation is controlled by the size of Q ($|Q|$) and threshold r. We sample the queries using a random walk with restart [21]. To construct Q, we start from a random origin v in G and perform random walk to

---

[1] dbpedia.org.

[2] http://www.imdb.com.

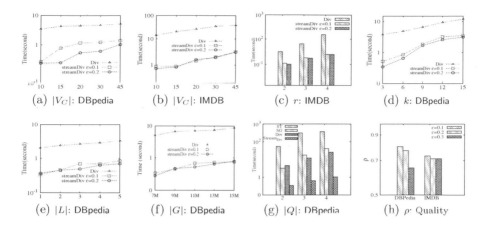

Fig. 4. Efficiency of query expansion

visit $N^r(v)$ multiple times. We construct Q with top keywords that have highest
TF-IDF score in its neighbors, to ensure the existence of reasonable answers.
We adopt established Bidirectional Search [13] for DR queries; GS 1-k [9] which
finds top answers for ST queries; and r-cliques [14] for SG queries.

Algorithms. We implement the "explore and diversify" algorithm Div and stream-
based algorithm streamDiv. We also implemented three term suggestion algorithms applicable for KWS in graphs. (1) CoOcc [20] suggests new terms to Q
that are most frequently co occurred in its answer $Q(G)$. (2) TagCloud [15] finds
"search entities" as star graphs induced by a center node and its neighbors.
It returns top keywords determined by TagCloud score, computed by TF − IDF
over search entities relevant to Q. (3) QBE [12] takes as input query tuples (keywords) and induces a query graph with the neighborhood of the answers to find
relevant triples with similar edge type. We take triples from $Q(G)$ as "examples" to QBE. The distance index of IMDB and DBpedia, constructed by pruned
landmark labeling [2], takes 1.16 GB and 4.34 GB, respectively.

All the tests are conducted on a machine with an Intel 2.3 GHz processor
with 64 GB memory. Each test is repeated 5 times and the average is reported.
We set $|Q| = 3$, $r = 3$, $|V_C| = 30$, $\epsilon = 0.1$, and $k = 5$, unless otherwise specified.

Exp-1: Efficiency of Query Expansion. We first evaluate the efficiency of
streamDiv, compared with two variants of Div with $\epsilon \in \{0.1, 0.2\}$. We use 50
queries, and evaluate the average response time per query. We evaluate the
impact of 5 factors: $|V_C|$, r, k, $|L|$ (the number of terms per node), and $|G|$.

Varying $|V_C|$. Fixing other parameters by default, we varied $|V_C|$ from 10 to 45.
Figure 4(a) (resp. Fig. 4(b)) shows the time required for query expansion over
DBpedia for DR (resp. IMDB for SG) queries. The results tell us the following. (1)
All the algorithms take more time as $|V_C|$ increases, due to that more terms are

processed. On the other hand, streamDiv outperforms Div by 17 and 3 times over IMDB and DBpedia, respectively. (2) streamDiv takes longer time for smaller ϵ, due to higher sieve lists maintenance cost. The impact of ϵ is higher over DBpedia than IMDB. Indeed, streamDiv pays more overhead from stream generation and sieve list maintenance in trade for quality over graphs with more labels per node.

Varying r. We evaluate the impact of threshold r over IMDB and DR queries. Figure 4(c) verifies that query expansion takes more time as the bound r increases, since the algorithms explore more parts of the graph.

Varying k. Figure 4(d) verifies the impact of number k. Varying k from 3 to 15 over DBpedia, streamDiv takes more time over larger k. The increased cost is due to that the stream takes longer to meet the early termination criteria when larger number of terms are required. On the other hand, streamDiv visits each term at most once, and is less sensitive to the change of k compared with Div.

Varying $|L|$. Varying the number of associated terms to each node in DBpedia, we investigate the impact of $|L|$ on performance of the query expansion. Figure 4(e) verifies that the more number of labels a node carries, the more time is needed for query expansion. While Div needs to collect all terms, streamDiv visits each term once, and never process a term that has been seen earlier in the stream.

Varying $|G|$. We sample 5 versions of DBpedia with edge size varied from $7M$ to $15M$. As shown in Fig. 4(f), the query expansion algorithms scale well with larger $|G|$ and streamDiv algorithms consistently outperform Div.

Session Response Time. We also report the response time of query expansion in practical graph exploration sessions. To this end, we implemented an interface that allows end users to explore real-life graphs with streamDiv. We let end users to start with queries with $|Q| = 2$, and execute multiple search sessions. Each session allows users to validate V_C, run our query expansion algorithm, and review results via query evaluation. We compare the query evaluation time (KWS) and expansion time per session, as two major bottlenecks for data exploration.

Figure 4(g) reports the results for ST queries over DBpedia. (1) Both KWS and streamDiv take more time as queries become larger, due to higher exploration cost. (2) The query expansion takes up to 4% of KWS evaluation for ST queries; similarly for other query classes we evaluated (not shown). These results indicate that streamDiv makes query expansion no longer a major bottleneck.

Exp-2: Quality of Query Expansion. We define a relative measure of quality for streamDiv *w.r.t.* Div, denoted as ρ. Given n queries in $\mathcal{Q}$, we define $\rho = \frac{\sum_{Q \in \mathcal{Q}} \frac{F(Q_T)}{F(Q'_T)}}{n}$ where Q_T and Q'_T are the set of expanded terms returned by streamDiv and Div, respectively. For Div, ρ is constantly 1.

Figure 4(h) shows that the quality of diversified terms is tunable by ϵ. When ϵ is changed from 0.1 to 0.3, the quality of terms from streamDiv is in general

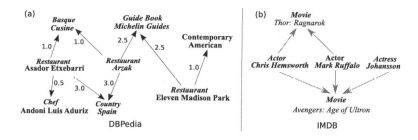

Fig. 5. Knowledge graph exploration with KWS (Color figure online)

not degrading too much *w.r.t.* Div (especially over IMDB). On the other hand, streamDiv becomes 15.48 and 2.83 times faster over IMDB and DBpedia, respectively. This verifies the tunable performance of streamDiv, which is desirable for data exploration under different time constraints.

Exp-3: Case Study. We next verify the applications of streamDiv.

Tunable Knowledge Base Exploration. We report a case in which an end user starts with a query $Q = \{$Asador, Arzak, Madison Park$\}$ to find famous restaurants. The answer (Fig. 5(a), in blue) is partitioned by their location (Spain and U.S).

(1) Setting $\lambda = 0.1$ and $k = 2$, streamDiv suggests top-2 terms $\{$Basque, Chef$\}$. While both are relevant, they state facts only about "Spain restaurants". The terms recommend a popular local Spanish dish and chefs who are known for it.
(2) The user enlarges λ to 0.6, and streamDiv identifies a set of more diversified terms $\{$Basque, American$\}$. Indeed, one suggests facts about US restaurants by recommending contemporary American food, and the other suggests local Spanish dish served by two Spanish restaurants.

Given the same query, QBE suggests other restaurants that are reviewed by Trade Magazine, CoOcc suggests a term $\{$Restaurant$\}$ which is not informative, and TagCloud suggests $\{$Bar$\}$ with content nodes less relevant to the restaurants.

"why-not" Queries. streamDiv can also be used to answer "why-not" questions. Given a query Q and answer $Q(G)$, "why-not" question aims to find a new query that contains the validated nodes in $Q(G)$, and has answers relevant to a set of "why-not" entities. We asked our end users to mark "why-not" entities in their exploration sessions, and show one such case in Fig. 5(b).

Our end user posed a query $\tilde{Q} = \{$Johansson, Ruffalo$\}$, searching for the actors played in the movie "*Thor: Ragnarok*" she can't recall. The result (marked in blue) nevertheless misses this movie. She wonders why the entity $\{$Thor: Ragnarok$\}$ is not in $Q(G)$, and adds it to V_C. The change of V_C now triggers streamDiv to explore another actor Hemsworth who played in "Thor: Ragnarok" with actor "Ruffalo", and suggests term "Hemsworth". Both the term and the entity can be suggested to user that "explains" the missing match.

5   Conclusion and Future Work

We have studied diversified query expansion for KWS in graphs. Our model maximizes the aggregate relevance to a set of validated content nodes with diversity guarantees over partitioned content nodes. We have developed a stream-style ($\frac{1}{2}$-ϵ) approximation that updates diversified terms with small update cost. We also showed that it readily copes with new insertion of terms. Our experimental results have verified the efficiency and effectiveness of diversified query expansion algorithm, and its applications in knowledge base exploration. Future topics include comparing with more graph exploration methods as well as investigating more relevancy measures and partitioning strategies.

Acknowledgment. Namaki and Wu are supported in part by NSF IIS-1633629 and Huawei Innovation Research Program (HIRP).

References

1. Achiezra, H., Golenberg, K., Kimelfeld, B., Sagiv, Y.: Exploratory keyword search on data graphs. In: SIGMOD, pp. 1163–1166 (2010)
2. Akiba, T., Iwata, Y., Yoshida, Y.: Fast exact shortest-path distance queries on large networks. In: SIGMOD, pp. 349–360 (2013)
3. Badanidiyuru, A., Mirzasoleiman, B., Karbasi, A., Krause, A.: Streaming submodular maximization: massive data summarization on the fly. In: SIGKDD, pp. 671–680 (2014)
4. Bao, Z., Zeng, Y., Jagadish, H., Ling, T.W.: Exploratory keyword search with interactive input. In: SIGMOD, pp. 871–876 (2015)
5. Bhalotia, G., Hulgeri, A., Nakhe, C., Chakrabarti, S., Sudarshan, S.: Keyword searching and browsing in databases using BANKS. In: ICDE, pp. 431–440 (2002)
6. Bouchoucha, A., He, J., Nie, J.-Y.: Diversified query expansion using conceptnet. In: CIKM, pp. 1861–1864 (2013)
7. Carpineto, C., Romano, G.: A survey of automatic query expansion in information retrieval. CSUR **44**, 1 (2012)
8. De Nies, T., Beecks, C., Godin, F., De Neve, W., Stepien, G., Arndt, D., De Vocht, L., Verborgh, R., Seidl, T., Mannens, E., et al.: A distance-based approach for semantic dissimilarity in knowledge graphs. In: ICSC, pp. 254–257 (2016)
9. Ding, B., Yu, J.X., Wang, S., Qin, L., Zhang, X., Lin, X.: Finding top-k min-cost connected trees in databases. In: ICDE, pp. 836–845 (2007)
10. Gollapudi, S., Sharma, A.: An axiomatic approach for result diversification. In: WWW, pp. 381–390 (2009)
11. He, H., Wang, H., Yang, J., Yu, P.S.: BLINKS: ranked keyword searches on graphs. In: SIGMOD, pp. 305–316 (2007)
12. Jayaram, N., Khan, A., Li, C., Yan, X., Elmasri, R.: Querying knowledge graphs by example entity tuples. TKDE **27**, 2797–2811 (2015)
13. Kacholia, V., Pandit, S., Chakrabarti, S., Sudarshan, S., Desai, R., Karambelkar, H.: Bidirectional expansion for keyword search on graph databases. In: VLDB (2005)
14. Kargar, M., An, A.: Keyword search in graphs: finding r-cliques. VLDB **4**, 681–692 (2011)

15. Koutrika, G., Zadeh, Z.M., Garcia-Molina, H.: Data clouds: summarizing keyword search results over structured data. In: EDBT, pp. 391–402 (2009)
16. Ma, H., Lyu, M.R., King, I.: Diversifying query suggestion results. In: AAAI (2010)
17. Mishra, C., Koudas, N.: Interactive query refinement. In: EDBT (2009)
18. Mottin, D., Müller, E.: Graph exploration: from users to large graphs. In: PODS, pp. 1737–1740 (2017)
19. Namaki, M.H., Wu, Y., Zhang, X.: GExp: cost-aware graph exploration with keywords. In: SIGMOD (2018)
20. Tao, Y., Yu, J.X.: Finding frequent co-occurring terms in relational keyword search. In: EDBT, pp. 839–850 (2009)
21. Tong, H., Faloutsos, C., Pan, J.-Y.: Fast random walk with restart and its applications (2006)
22. Tran, Q.T., Chan, C.-Y.: How to ConQueR why-not questions. In: SIGMOD, pp. 15–26 (2010)
23. Tran, Q.T., Chan, C.-Y., Parthasarathy, S.: Query reverse engineering. VLDB **23**, 721–746 (2014)
24. Tran, T., Wang, H., Rudolph, S., Cimiano, P.: Top-k exploration of query candidates for efficient keyword search on graph-shaped (RDF) data. In: ICDE, pp. 405–416 (2009)
25. Wang, H., Aggarwal, C.C.: A survey of algorithms for keyword search on graph data. In: Aggarwal, C., Wang, H. (eds.) Managing and Mining Graph Data. ADBS, vol. 40, pp. 249–273. Springer, Boston (2010). https://doi.org/10.1007/978-1-4419-6045-0_8
26. Yahya, M., Berberich, K., Ramanath, M., Weikum, G.: Exploratory querying of extended knowledge graphs. VLDB **9**, 1521–1524 (2016)
27. Yang, S., Wu, Y., Sun, H., Yan, X.: Schemaless and structureless graph querying. PVLDB **7**(7), 565–576 (2014)
28. Yao, J., Cui, B., Hua, L., Huang, Y.: Keyword query reformulation on structured data. In: ICDE, pp. 953–964 (2012)
29. Yu, J.X., Qin, L., Chang, L.: Keyword search in relational databases: a survey. IEEE Data Eng. Bull. **33**, 67–78 (2010)
30. Zeng, Y., Bao, Z., Ling, T.W., Jagadish, H., Li, G.: Breaking out of the mismatch trap. In: ICDE, pp. 940–951 (2014)
31. Zheng, B., Zhang, W., Feng, X.F.B.: A survey of faceted search. J. Web Eng. **12**, 041–064 (2013)
32. Zhou, Y., Cheng, H., Yu, J.X.: Graph clustering based on structural/attribute similarities. VLDB **2**, 718–729 (2009)
33. Zhu, G., Iglesias, C.A.: Computing semantic similarity of concepts in knowledge graphs. TKDE **29**, 72–85 (2017)

Distributed k-Nearest Neighbor Queries in Metric Spaces

Xin Ding[1], Yuanliang Zhang[1], Lu Chen[2], Yunjun Gao[1(✉)],
and Baihua Zheng[3]

[1] College of Computer Science, Zhejiang University, Hangzhou, China
{dingxin,yuanlz,gaoyj}@zju.edu.cn
[2] Department of Computer Science, Aalborg University, Aalborg, Denmark
luchen@cs.aau.dk
[3] School of Information Systems, Singapore Management University,
Singapore, Singapore
bhzheng@smu.edu.sg

Abstract. Metric k nearest neighbor (MkNN) queries have applications in many areas such as multimedia retrieval, computational biology, and location-based services. With the growing volumes of data, a distributed method is required. In this paper, we propose an <u>A</u>synchronous <u>M</u>etric <u>D</u>istributed <u>S</u>ystem (AMDS), which uniformly partitions the data with the pivot-mapping technique to ensure the load balancing, and employs publish/subscribe communication model to asynchronously process large scale of queries. The employment of asynchronous processing model also improves robustness and efficiency of AMDS. In addition, we develop an efficient estimation based MkNN method using AMDS to improve the query efficiency. Extensive experiments using real and synthetic data demonstrate the performance of MkNN using AMDS. Moreover, the AMDS scales sub-linearly with the growing data size.

Keywords: Metric space · k nearest neighbor query · Publish/subscribe
Query processing · Algorithm

1 Introduction

Metric k nearest neighbor (MkNN) queries find k objects most similar to a given query object under a certain criterion. Because metric spaces can support various data types (e.g., images, words, DNA sequences) and flexible distance metrics (e.g., L_p-norm distance, edit distance), this functionality has been widely used in real life applications. Here, we give two representative examples below.

Application 1 (Multimedia Retrieval). In an image retrieval system, the similarity between images can be measured using L_p-norm metric, earth mover's distance or other distance metrics between their corresponding feature vectors. Here, MkNN queries in metric space can help users to locate figures that are similar as a given one.

Y. Cai et al. (Eds.): APWeb-WAIM 2018, LNCS 10987, pp. 236–252, 2018.
https://doi.org/10.1007/978-3-319-96890-2_20

Application 2 (Nature Language Processing). In the WordNet, a knowledge graph for better nature language understanding, the similarity between two words could be measured by the shortest path, maximum flow or other distance metrics. Here, M*k*NN queries can help users to find the words that are closely related to a given one.

With the development of Internet, especially the widespread use of mobile devices, the volume, richness and diversity of data challenge the traditional M*k*NN query processing in both space and time. This calls for a scalable M*k*NN method to provide efficient query service. Hence, in this paper, we investigate the distributed M*k*NN queries.

Existing works on distributed processing in metric spaces [1–10] aim to accelerate M*k*NN queries in parallelism and try to build a suitable network topology to manage the large amount of data. However, the existing solutions are not sufficient because of following two main reasons. First, the ability to process a large quantity of M*k*NN queries simultaneously is in need nowadays. Second, the load balancing is also a basic need for distributed systems [11–13]. Motivated by these, we try to develop a distributed M*k*NN query processing system that takes the load balancing into consideration and aims at efficient query processing in large scale.

In order to design such a system, three challenges need to be addressed. The first challenge is how to ensure the load balancing of a distributed system? To ensure the load balancing, we uniformly divide the data into disjoint fragments using the pivot mapping technique, and then distribute each fragment to a computational node. The second one is how to efficiently process queries in large scale? To support synchronous process of large scale of queries, we utilize publish/subscribe communication model, and thus, massive queries can be executed with negligible time loss in message passing. The third challenge is how to reduce the cost of a single similarity query? We develop several pruning rules with the minimum bounding box (MBB) to save unnecessary verifications. In addition, an estimation based M*k*NN method is employed to further improve the query efficiency. Based on these, we develop the Asynchronous Metric Distributed System (AMDS) to support efficient M*k*NN queries in the distributed environment. To sum up, the key contributions in this paper are as follows:

- We present a pivot-mapping based data partition method, which first uses a set of effective pivots to map the data from a metric space to a vector space, and then uniformly divides the mapped objects into disjoint fragments.
- We utilize the publish/subscribe communication model to asynchronously exchange messages that saves time in network communication, and thus to support large scale of M*k*NN query processing simultaneously.
- We propose an estimation-based method to handle M*k*NN queries, where pruning rules with MBB are used to avoid redundant verifications.
- Extensive experiments using real and synthetic data evaluate the efficiency of AMDS and the performance of distributed M*k*NN queries using AMDS.

The rest of this paper is organized as follows. Section 3 reviews related works. Section 3 introduces the definitions of M*k*NN queries and the publish/subscribe communication model. Section 4 elaborates the system architecture. Section 5 presents an efficient algorithm for M*k*NN searches. Experimental results and findings are reported in Sect. 6. Finally, Sect. 7 concludes the paper with some directions for future work.

2 Related Work

We review briefly related work on distributed kNN queries in Euclidean and metric spaces.

2.1 Distributed Euclidean kNN Queries

Distributed kNN queries in Euclidean space have attracted a lot of attention since they are introduced. CAN [14] and Chord [9] build on top of DHT overlay network. LSH forest [15] uses a set of locality-sensitive hash functions to index data and perform (approximate) kNN queries on an overlay network. SWAM [16] consists of a family of distributed access methods for efficient kNN queries, which achieves the efficiency by bringing nodes with similar contents together. DESENT [17] is an unsupervised approach for decentralized and distributed generation of semantic overlay networks. VBI-Tree [18] is an abstract tree structure on top of an overlay network, which utilizes extensible centralized mapping methods. Mercury [19] is proposed to support multiple attributes as well as explicit load balancing. NR-Tree [20] is a P2P adaption of R*-Tree [12] to support kNN queries. FuzzyPeer [21] uses "frozen" technique to optimize query execution. A general and extensible framework in P2P network builds on the concept of hierarchical summary structure [12]. More recently, VITAL [22] employs a super-peer structure to exploit peer heterogeneity. However, all these above solutions focus on the vector space and they utilize the geometric properties (e.g., locality sensitive function [15], minimum bounding box [12]) that are unavailable in metric spaces, to distribute the data on the underlying overlay network and to accelerate the query processing. Hence, they are unsuitable for distributed MkNN queries.

2.2 Distributed MkNN Queries

Existing methods for distributed MkNN queries can be clustered into two categories. The first category utilizes basic metric partitioning principles to distribute the data over the underlying network. GHT* and VPT* [2] use ball and generalized hyperplane partitioning principles, respectively. Besides GHT* and VPT*, efficient peer splits based on ball and generalized hyperplane partitioning techniques are also investigated in [5]. The second category utilizes the pivot mapping technique to distribute the data. MCAN [23], relying on an underlying structured P2P network named CAN [14], maps data to vectors in a multi-dimensional space. M-Chord [24], relying on another underlying structured P2P network named Chord [9], uses iDistance [25] to map data into one-dimension values. M-Index [8] also generalizes iDistance technique to provide distributed metric data management. SIMPEER [6] works in autonomous manner, and uses the generated clusters obtained by the iDistance method to further summarize peer data at the supper peer level. In this paper, we adopt the pivot-mapping based method. This is because pivot-mapping based methods outperform metric partitioning based ones in terms of the number of distance computations [1, 26], one important criterion in metric spaces. As an example, MCAN and M-Chord utilizing the pivot mapping perform better than GHT* and VPT* using metric partitioning techniques [3, 4].

Apart from these, two general frameworks for distributed M*k*NN search are proposed. One, called MESSIF, is an implementation framework with code reusing of GHT*, VPT*, MCAN, M-Chord, Chord and Skip-Graphs [27]. The other utilizes a super-peer architecture, where super-peers are responsible for query routing [10].

However, all these above methods are not sufficient due to two reasons below. First, they cannot support synchronous processing of large scale of M*k*NN queries simultaneously, which is our main objective. To address it, we develop methods based on publish/subscribe communication model. Second, they do not take the load balancing into consideration, which is also important for distributed environment. To ensure the load-balancing, we develop a pivot-mapping based partition method to distribute the data uniformly among the computational nodes.

3   Preliminaries

In this section, we review the M*k*NN queries and publish-subscribe system. Table 1 summarizes the symbols frequently used throughout this paper.

Table 1. Symbols and description

Notation	Description
O or P	A set of objects or pivots
o or p	An object or a pivot
$\phi(o)$	A vector for object o after pivot-mapping
wp_i or mp_i	A worker peer or a master peer
MBB(wp_i) or MBB(mp_i)	The minimum bounding box for wp_i or mp_i
m	A mission used for communication among peers
M*k*NN(q, k)	A metric k nearest neighbor query w.r.t. q and k
$d()$	The distance function in a metric space
$RR(q, r)$	A metric range region centered at q with radius r
$q.d_k$	The distance between q to its k-th nearest neighbor
$q.d_k^i$	An estimation of $q.d_k$
deg_w	The number of worker peers that a master peer connects to
deg_m	The number of master peers that a root peer connects to
num_{wp}	The total number of worker peers

3.1   M*k*NN Queries

A metric space is denoted by a tuple (M, d), in which M is an object domain and d is a distance function to measure "similarity" between objects in M. In particular, the distance function d has four properties: (1) *symmetry*: $d(q, o) = d(o, q)$; (2) *non-negativity*: $d(q, o) \geq 0$; (3) *identity*: $d(q, o) = 0$ iff $q = o$; and (4) *triangle inequality*: $d(q, o) \leq d(q, p) + d(p, o)$. Based on these properties, we define M*k*NN queries.

DEFINITION 1 (MkNN QUERY). *Given an object set O, a query object q, and an integer k in M, a MkNN query finds k most similar objects from O for q, i.e., MkNN(q, k) = {R | R ⊆ O ∧ |R| = k ∧ ∀o_i ∈ R, ∀o_j ∈ O - R: d(q, o_j) ≥ d(q, o_i)}.*

An MkNN query can be regarded as a metric range (MR) query if the k-th nearest neighbor distance if known in advance (i.e., the search radius), as defined below.

DEFINITION 2 (MR QUERY). *Given an object set O, a query object q, and a search radius r in M, a metric range (MR) query finds objects from O with their distances to q are bounded by r, i.e., MR(q, r) = {o | o ∈ O ∧ d(q, o) ≤ r}.*

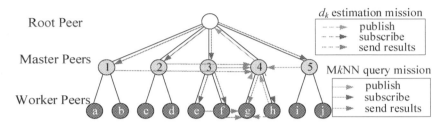

Fig. 1. AMDS structure and communications for MkNN processing

3.2 Publish/Subscribe

Publish/subscribe system, also termed as distributed event-bases system [28], is a system where publishers publish structured events to an event service and subscribers express interest in particular events through subscriptions [29]. Here, the interest can be arbitrary patterns over the structured events. Publish/subscribe systems are used in a wide variety of application domains, particularly in those related to the large-scale dissemination of events, such as financial information systems, monitoring systems and cooperative working systems where a number of participants need to be informed of events of shared interest. Hence, in this paper, we adopt the publish/subscribe communication model to support large scales of MkNN query processing simultaneously, which is required in real life applications.

Publish/subscribe systems have two main characteristics, heterogeneity and asynchronicity. Heterogeneity means that, components in a distributed system can work together as long as correct message is published and subscribed. Asynchronicity means that publishers and subscribers are time-decoupled, and message publishing and subscribing are performed independently. Hence, the asynchronicity, the heterogeneity, and the high degree of loose coupling suggest that publish/subscribe systems perform well in dealing with large scale of messages.

4   AMDS Architecture

In this section, we present system organization and data deployment of AMDS system.

4.1 System Organization

AMDS aims to answer a large scale of M*k*NN queries in a distributed environment simultaneously. In the following, we first introduce the system structure, and then describe the communications in the system.

System Structure. AMDS is a three-layer tree structure on top of the overlay network, consisting of three types of peers, termed as *root peers*, *master peers* and *worker peers*, as depicted in Fig. 1. Peers are physical entities with calculation and communication abilities. They are organized to index objects and to accomplish M*k*NN queries. Worker peers (e.g., wp_a, wp_b) directly index data objects and perform metric similarity queries locally; while root peers and master peers (e.g., mp_1, mp_2) manage children peers and distribute M*k*NN queries over the system.

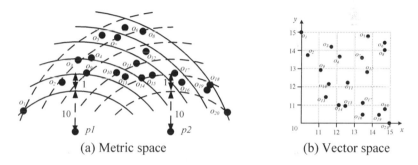

(a) Metric space   (b) Vector space

Fig. 2. Pivot mapping

In AMDS, deg_w/deg_m represent the number of worker peers/master peers that a master peer/a root peer connects to, respectively. The values of deg_w and deg_m depend on several factors, including the network environment and the storage ability. For simplify, in this paper, we assume there is one root peer, and each master peer maintains an equal number of worker peers. Hence, the value of deg_m equals to the number of master peers, and the value of $deg_m \times deg_w$ equals to the total number of worker peers, e.g., $deg_m = 5$ and $deg_w = 2$ for the example system depicted in Fig. 1. For clarity, we name a master peer with all the children worker peers as a *peer cluster*.

System Communication. To support communications between peers, we introduce the concept of *missions*. Missions are text messages exchanged among peers for communications. Data deployment, object updating operations, or M*k*NN queries can be packed into missions. The missions are published by worker peers in a bottom-up pattern, and subscribed by other worker peers in a top-down pattern, as illustrated in Fig. 1. More specifically, a worker peer, the owner of a mission, can publish a mission to its parent master peer and then to the root peer. Then, master peers can subscribe to the missions from the root peer and worker peers can subscribe to the missions from their master peers. Every master peer (or root peer) maintains a mission list to keep track of all the missions published by its children worker peers (or master peers).

4.2 Data Deployment

To achieve the load balancing, we divide the source data equally among worker peers, assuming that worker peers share the same calculation ability and storage capacity. Our framework of data deployment contains three phases, (i) the pivot-mapping of source data performed by root peers, (ii) the partitioning of mapped data performed by master peers, and (iii) the local index building performed by worker peers.

Pivot Mapping. In the first stage, we map the objects in a metric space to data points in a vector space, using well-chosen pivots. The vector space offers more freedom than the metric space when performing data partitioning and designing search approaches, since it is possible to utilize the geometric and coordinate information that is unavailable in the metric space. Given a pivot set $P = \{p_1, p_2, ..., p_n\}$, a general metric space (M, d) can be mapped to a vector space (R^n, L_∞). Specifically, an object o in a metric space is represented as a point $\phi(o) = \langle d(o, p_1), d(o, p_2), ..., d(o, p_n) \rangle$ in the vector space. For instance, consider the example in Fig. 2, where $O = \{o_i \mid 1 \leq i$ 20\}$ and L_2-norm is used. If $P = \{p_1, p_2\}$, O can be mapped to a two-dimensional vector space, in which the x-axis represents $d(o_i, p_1)$ and the y-axis represents $d(o_i, p_2)$ $(1 \leq i \leq 20)$. In particular, object o_1 is mapped to point $\langle 10, 15 \rangle$.

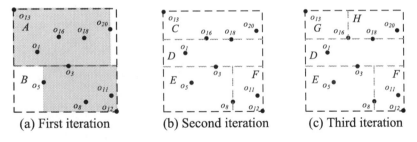

(a) First iteration   (b) Second iteration   (c) Third iteration

Fig. 3. Sample-based data partitioning

The quality of selected pivots has a marked impact on the search performance. It is shown that good pivots are far away from each other and from the rest of the objects in the database [30]. Based on this observation, we select pivots in a way such that (i) they are outliers, and (ii) the distances between each other are as large as possible. Moreover, theoretically, pivots do not need to be part of the object set. Consequently, the quality of pivots is highly related with the data distribution and we have the flexibility to insert/delete objects without changing the pivot set.

Data Partitioning. The root peer first samples the whole dataset and then maps the sampled data objects into a set of vectors using selected pivots as discussed above. After that, the root peer partitions the data objects into deg_m disjoint parts P_i ($1 \leq i$ deg_m) of equivalent size, with $BB(P_i)$ representing the bounding box corresponding to each part P_i. Here, $BB(P_i)$ is an axis aligned bounding box and it contains the part P_i such that $\forall 1 \leq i < j \leq deg_m$, $BB(P_i) \cap BB(P_j) = \emptyset$ and $\cup_{1 \leq i \leq degm} BB(P_i)$ covers the entire space. Then, each $BB(P_i)$ ($1 \leq i \leq deg_m$) is assigned to the corresponding master peer mp_i.

We give an example of sample-based data partition in Fig. 3. As depicted in Fig. 3 (a), we sort sampled objects o ($\in O$) in the mapped vector space according to their values on dimension y. In the first iteration, based on $deg_m = 5$ and $\lceil deg_m/2 \rceil / \lfloor deg_m/2 \rfloor = 3/2$, we partition the whole sampled dataset into two parts $A = \{o_1, o_3, o_{13}, o_{16}, o_{18}, o_{20}\}$ and $B = \{o_5, o_8, o_{11}, o_{12}\}$. The partition continues until five equal parts are obtained, i.e., $D = \{o_1, o_3\}$, $E = \{o_5, o_8\}$, $F = \{o_{11}, o_{12}\}$, $G = \{o_{13}, o_{16}\}$ and $H = \{o_{18}, o_{20}\}$, with corresponding bounding boxes (i.e., the dotted rectangle) depicted in Fig. 3(c). In the sequel, each data object o is associated to the corresponding master peer m_i with $\phi(o) \in BB(P_i)$. In addition, the minimum bounding box (MBB), i.e., the light gray rectangles depicted in Fig. 4(a), is built for each master peer m_i accordingly. Specifically, a MBB(mp_i) denotes the axis aligned *minimum bounding box* to contain all the mapped objects in mp_i. After that, each master peer further divides each P_i into deg_w disjoined equaled parts in a similar way and MBBs are also built for all the worker peers. The dark gray rectangles, depicted in Fig. 4(b), represent the MBBs for worker peers.

Local Index Construction. Finally, each worker peer builds a local metric index for all its objects. Here, we use M-tree to index the objects distributed to each worker peer in the mapped vector space.

Algorithm 1 *k*NN_*WP*

Input: *MkNN*(q, k) issued at wp_i
1: **if** $\phi(q) \subset MBB(wp_i)$ **then**
2:    $R := $ findkNN(wp_i, q, k)
3:    $q.d_k^i := d(q, R[k]_i)$
4:    $R := $ MR($q, q.d_k^i, wp_i$)
5: **else**
6:    $m :=$ newMission($ID, DEst, q, IP_i$)
7:    sendMessage($wp_i, m,$ IP($wp_i.parent$))
8:    $msg :=$ receiveMessage()
9:    **while** ($msg.ID \neq m.ID$ and
              $msg.type \neq$ DEst)
10:       $msg :=$ receiveMessage()
11:       $R := $ MR($q, msg.content, wp_i$)
12: **return** findkNN(R, q, k)

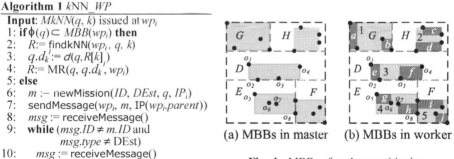

(a) MBBs in master   (b) MBBs in worker

Fig. 4. MBBs after data partitioning

---

5   Distributed Query Processing

In this section, we present how to support M*k*NN queries in AMDS. We first introduce the algorithms to support metric range query and metric *k*NN query, and then present the asynchronous execution of missions.

5.1 MkNN Query Processing

Two solutions exist to answer MkNN query. One possible solution is incrementally increasing the search radius until k nearest neighbor objects are retrieved [25, 31]. However, in a distributed environment, this method incurs very expensive communication cost due to too many message exchanges over the network. Alternatively, AMDS adopts a different approach. It performs a metric range query based on an estimated search radius with at most two round-trips message exchanges.

A metric range query retrieves the objects enclosed in the range region that is an area centered at q with a radius r. The range region of $MR(q, r)$ can also be mapped into the vector space [32]. Consider, for example, Fig. 5(a), where a *blue dotted circle* denotes a range region, and the *blue rectangle* in Fig. 5(b) represents the *mapped range region* using $P = \{p_1, p_2\}$. To obtain $MR(q, r)$, we only need to verify the objects o whose $\phi(o)$ are contained in the mapped range region, as stated below.

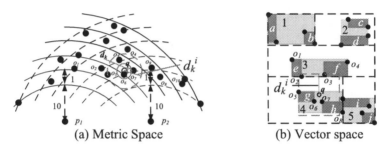

(a) Metric Space   (b) Vector space

Fig. 5. MkNN query (Color figure online)

Lemma 1. Given a pivot set P, if an object o is enclosed in $MR(q, r)$, then $\phi(o)$ is certainly contained in the mapped range region $RR(r)$, where $RR(r) = \{\langle s_1, s_2, ..., s_{|P|}\rangle \mid 1 \leq i \leq |P| \wedge s_i \geq 0 \wedge s_i \in [d(q, p_i) - r, d(q, p_i) + r]\}$.

Proof. Assume, to the contrary, that there exists an object $o \in MR(q, r)$ but $\phi(o) \notin RR$ (r), i.e., $\exists p_i \in P$, $d(o, p_i) > d(q, p_i) + r$ or $d(o, p_i) < d(q, p_i) - r$. According to the triangle inequality, $d(q, o) \geq |d(q, p_i) - d(o, p_i)|$. If $d(o, p_i) > d(q, p_i) + r$ or $d(o, p_i) < d(q, p_i) - r$, then $d(q, o) \geq |d(o, p_i) - d(q, p_i)| > r$, which contradicts with our assumption. Consequently, the proof completes. □

According to Lemma 1, if the MBB of a worker peer wp_i or a master peer mp_i does not intersect with $RR(r)$, we can avoid performing $MR(q, r)$ on wp_i or mp_i. For example, in Fig. 5, the master peer mp_5 does not need to perform $MR(q, r)$ as $M_5 \cap RR(r) = \varnothing$.

To obtain a good estimation of $q.d_k$ (i.e., the kth nearest neighbor distance), we perform a local $MkNN(q, k)$ on the worker peer wp_i with minimum $MIND(MBB(wp_i), \phi(q))$, and use $q.d_k^i$, the distance between q and the kth nearest neighbor returned by the local $MkNN(q, k)$ performed by worker peer wp_i as an estimation of $q.d_k$. We consider $q.d_k^i$ as a good overestimation of $q.d_k$. This is because as q is located nearest to worker peer wp_i, the value of $MIND(MBB(wp_i), \phi(q))$ reflects the likelihood that kNN objects of q are actually located within wp_i. Based on the definition of $q.d_k^i$, we can convert an MkNN search into a MR query, as stated in Lemma 2.

Lemma 2. Given a pivot set P, if an object o is an answer object for $MkNN(q, k)$, then $\phi(o)$ is certainly contained in the mapped range region $RR(q, q.d_k^i)$.

Proof. Assume, to the contrary, that there exists an object $o \in MkNN(q, k)$ but $\phi(o) \notin RR(q, q.d_k^i)$. According to the fact that $\phi(o) \notin RR(q, q.d_k^i)$, we have $d(q, o) > q.d_k^i$. Meanwhile, according to the definition of $q.d_k^i$, we have $q.d_k^i \geq q.d_k$, and hence $d(q, o) > q.d_k$, which contradicts with our assumption that $o \in MkNN(q, k)$. Consequently, the proof completes. $\square$

Based on Lemma 2, MkNN query processing in AMDS can also be partitioned into two phases, i.e., $q.d_k$ estimation phase and MkNN query phase. First, worker peer wp_i with the minimum $MIND(MBB(wp_i), \phi(q))$ is selected to perform a local MkNN query to obtain an estimation $q.d_k^i$ of $q.d_k$. Then, $MkNN(q, k)$ is transformed into a MR $(q, q.d_k^i)$. Note that, k is still needed because at most k objects will be sent back to the MkNN query poster to reduce the network communication volumes. For worker peers who receive such MkNN missions, they perform local $MR(q, q.d_k^i)$, but at most k nearest objects will be sent back to the mission poster. When all the contributors returned their query results, the *poster* will obtain the global kNN objects as the final result.

Algorithm 2 MkNNProcessing

Input: a peer p_i who is monitoring the mission publication
```
1: loop
2:    m := receiveMission()/receiveMessage()
3:    if m is a DEst mission then
4:      if p_i is a root peer then
5:        p_j := locateNearestMBB(p_i.MBBList, m.content)
6:        sendMessage(p_i, m, IP_j)
7:      else if p_i is a master peer then
8:        if φ(q) ⊂ MBB(p_i) or m.sender = root then
9:          wp_j := locateNearestMBB(p_i.MBBList, m.content)
10:         sendMessage(p_i, m, IP_j)
11:       else
12:         sendMessage(p_i, m, IP(p_i.parent))
13:     else if p_i is a work peer
14:       R = findKNN(p_i, q, k)
15:       q.d_k^i := d(q, R[k])
16:       sendMessage(p_i, (m.ID, DEst, q.d_k^i), m.IP_w)
```

We develop a MkNN_WP Algorithm to publish a mission when a MkNN(q, r) is issued at the worker peer wp_i, with the pseudo-code depicted in Algorithm 1. The algorithm takes MkNN(q, k) and the issuer wp_i as an input. If MBB of wp_i contains $\phi(q)$, work peer wp_i is confirmed to be the one with minimum $MIND(MBB(wp_i), \phi(q))$ value. A local MkNN search is performed to find the distance $q.d_k^i$ between q and its local k^{th} nearest object, and then we perform a MR query with radius set to $q.d_k^i$ (lines 1–4). Here, MR query searches on work peers whose MBBs are intersected with $RR(q.d_k^i)$ due to Lemma 1, which is simple, and thus, the codes are omitted. On the other hand, if the bounding box of the query issuer does not bound the query point, we need to find the worker peer wp_j with minimum $MIND(MBB(wp_j),\ \phi(q))$ value, via a mission with $type = DEst$ (lines 6–10). Once $q.d_k^i$ is located and returned, a metric range query based on q and $q.d_k^i$ is issued (line 11). Among the objects that are located within the search range, the top k objects with minimum distances to q are returned as the global kNN objects to complete the process.

We also develop an MkNNProcessing algorithm to explain how the estimation of the search radius can be performed by other peers, with its pseudo code listed in Algorithm 2. All the actions are triggered by new missions received, and the actions of different types of peers vary. For MkNN query processing, the objective of actions that are triggered by missions/messages with $type = DEst$ is to find a good estimation of q. d_k. As mentioned before, a $DEst$ mission is published only when the query object is not located within the MBB of the query issuer mission. The mission first reaches the parent master peer of the query issuer. If the parent master peer has its MBB bounding the query point (the first condition of the IF clause in line 8), it selects a child worker peer wp_j that has the minimum $MIND(MBB(wp_j),\ \phi(q))$ value via the function locateNearestMBB() and then informs wp_j to performs the estimation via a direct message (lines 9–10). Otherwise, the parent master peer is not able to confirm that it has a shorter $MIND$ to $\phi(q)$, as compared with other master peers. It has to ask the root peer for help (lines 11–12). The mission is then propagated to the root peer. The root peer locates the master peer mp_j with minimum $MIND$ distance $\phi(q)$ again via function locateNearestMBB() then informs mp_j to perform the estimation via a direct message (lines 4–6). The mission is then propagated to a master peer which might or might not be the parent master peer. Once a master peer receives the $DEst$ message from the root peer (the second condition of the IF clause in line 8), it is aware that itself is the nearest master peer to the query point, and it locates the nearest worker peer and informs the worker peer to continue the estimation task (lines 9–10). Now, the mission reaches the destination, the worker peer that is nearest to the query point. The worker peer performs a local kNN search, and the distance between q and its local k^{th} NN is returned to the query issuer as an estimation of $q.d_k$. The estimation is ended when a message containing the estimation is sent to the query issuer.

Example 1. We illustrate the MkNN query processing using the example shown in Fig. 5, with the corresponding communications depicted in Fig. 1. Suppose that worker peer wp_h raises a MkNN query MkNN$(q, 2)$ and it invokes kNN_WP algorithm. As the query object q locates outside its $MBB(wp_h)$, wp_h publishes a $q.d_k$ estimation mission m to its parent master peer mp_4. Once mp_4 receives m via MkNNProcessing Algorithm, it checks whether its MBB bounds q. As q falls inside $MBB(mp_4)$, it locates wp_g, the nearest worker peer among its children, and informs wp_g to continue the estimation via a direct message. Thereafter, worker peer wp_g performs a local MkNN query to obtain the result set S_R and $d(q, o_5)$ is returned to wp_h as an estimation of $q.d_k$. In the sequel, wp_h performs a range query with $r = q.d_k^i$. Once the result objects of the range query are received, worker peer can return the top-2 objects $\{o_6, o_7\}$ nearest to q as the result to complete the processing of MkNN query.

5.2 Asynchronous Execution of Missions

AMDS adopts the publish/subscribe communication model, which can support asynchronous execution of queries and thus can avoid waiting for communications with other peers. In AMDS, there are three types of characters during the query processing, i.e., the query initiator, the query broker and the query answerer. In particular, a query initiator is a peer which issues a query, a query answerer is a peer which performs the query and return the query answer to the query initiator, and a query broker is a peer that distributes the query to the correct answerers. A query can be divided into four main phases, query raising, query distributing, query processing and result collecting. Each of these four phases is processed by these characters independently, i.e., the query initiator, the query brokers, the query answerers and the query initiator, respectively. It is obvious that the four phases are loosely coupled, no strong relations between these phases exist, which is the premise of asynchronous execution.

Consider the example of asynchronous execution shown in Fig. 6. AMDS consists of two worker peers (i.e., wp_x and wp_y) and one master peer mp. Each of the worker peers issues a query (i.e., q_1 and q_2) that both wp_x and wp_y are related with the query. Although q_1 is finished earlier in Fig. 6(a) than that in Fig. 6(b), it is obvious that asynchronous fashion is more efficient overall. Note that, the performance of synchronous fashion will get worse as the number of queries increase.

Table 2. Parameter settings

Parameters	Value
Cardinality	250K, 500K, **1M**, 2M, 4M
The number of worker peers	1K, 2K, **4K**, 8K, 16K
k	1, 3, **9**, 27, 81

Table 3. Construction cost of AMDS

Dataset	Network
Title	397376 KB
CoPHIR	2157988 KB
VECTOR (250K)	200954 KB
VECTOR (500K)	398878 KB
VECTOR (1M)	792620 KB
VECTOR (2M)	1587004 KB
VECTOR (4M)	3174662 KB

6   Experimental Evaluation

In this section, we evaluate the effectiveness and efficiency of AMDS and M*k*NN queries via extensive experiments, using both real and synthetic datasets. AMDS and corresponding M*k*NN query algorithms are implemented in C++ with raw socket API. All the experiments are conducted on Intel E5 2620 processor and 64G RAM.

We employ two real datasets Title[1] and CoPHIR[2]. Title contains 800K PubMed paper titles, with strings whose length ranges from 8 to 666, resulting in an average length equaling to 71. The similarity between two strings is measured using edit-distance. CoPHIR consists of 1000 K standard MPEG-7 image features extracted

---

[1] Available at http://www.ncbi.nlm.nih.gov/pubmed.

[2] Available at http://cophir.isti.cnr.it/get.html.

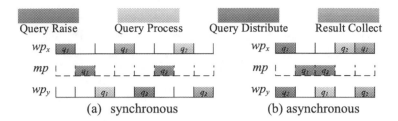

Fig. 6. Comparisons between execution modes

from Flickr[3], where the similarity between two features is measured as the L_2-norm. In addition, synthetic datasets VECTOR are generated with the cardinality varying from 250K to 4M, where L_∞-norm is the distance metric. Every dimension of VECTOR datasets is mapped to [0, 10000]. Each VECTOR dataset has 10 clusters and each cluster follows Gaussian distribution. In this paper, the number of pivots for each dataset is set to 5.

We investigate the performance of AMDS and MkNN algorithms under various parameters as summarized in Table 2. In each set of experiments, only one factor varies, whereas the others are fixed to their default values. As discussed in Sect. 4.2, if the number of worker peers is fixed, then the number of master peers will affect the efficiency of AMDS. Hence, in our experiments, the number of master peers is set as 32, 64 and 128 to evaluate the impact of the number of master peers. The main performance metrics include the CPU time and the network communication volume.

6.1 Construction Cost

The first set of experiments verifies the AMDS construction cost, i.e., the cost of data deployment of AMDS. Here, the network communication volume is used as the performance metric. We collected the construction cost on both real and synthetic datasets, with the results demonstrated in Table 3. The number of worker peers is set to 4 K as default, and the number of master peers is set to 64 as default. The first observation is that the data deployment in AMDS is efficient in terms of the network communication volume. This is because, the content of source dataset only copied twice in the data deployment process. It is first copied by root peer when passing data to master peers, and then copied by master peers when passing objects to worker peers. The second observation is that the larger dataset is, the higher construction cost is. This is because the network communication volume depends on the cardinality of dataset.

6.2 Evaluation of Metric Similarity Queries

The second set of experiments evaluates the performance of MkNN queries using real and synthetic datasets. We study the influence of several parameters, including (i) the value k, (ii) the number of worker peers num_{wp}, and (iii) the cardinality of dataset.

---

[3] Available at http://www.flicker.com.

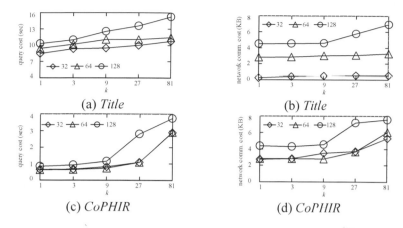

Fig. 7. Effect of *k*

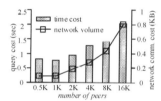

Fig. 8. Effect of *num_wp*

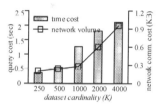

Fig. 9. Effect of cardinality

Effect of k. First, we investigate the performance of M*k*NN queries using real datasets. The CPU time and the network communication volume of M*k*NN queries are shown in Fig. 7 under various *k* values ranging from 1 to 81. The first observation is that the query cost increases with the growth of *k*. This is because, the search space grows as *k* increases, resulting in more related master peers and worker peers. Note that, on *CoPHIR*, the CPU time and the network communication volume grows rapidly when *k* exceeds 9 due to the corresponding distance distribution of the dataset.

Effect of Number of Worker Peers. Then, we evaluate the influence of number of worker peers. Figure 8 shows the results under various numbers of worker peers *num_wp* using synthetic datasets. Note that, the number of master peers is set to 64 as default. The first observation is that the query cost first decreases from 0.5K to 1K and then increases from 1K to 16K. This is because, with more worker peers, the objects managed by each worker peer become less, and thus the M*k*NN cost on each worker peer decreases. However, at the same time, more query cost is consumed on the managing of a larger number of peers and communications between peers. In this case, 1K worker peers performs the best for *VECOTOR* on AMDS.

Effect of Cardinality. After that, we study the impact of cardinality of using synthetic datasets, with the results depicted in Fig. 9. Here, we use 64 master peers as default. As expected, the query cost including CPU time and the network communication volume increases with the growth of cardinality.

7   Conclusions

In this paper, we present the Asynchronous Metric Distributed System (AMDS), which aims at dealing with a large scale of MkNN queries simultaneously. In the data deployment, AMDS uniformly partitions the data using the pivot-mapping technique to ensure the load balancing. During the MkNN query processing, AMDS utilizes the publish/subscribe communication model to support asynchronous processing and achieve the robustness at the same time. In addition, pruning rules are developed with the MBB technique to reduce the query cost. Furthermore, MkNN queries are solved using estimation to avoid high network communication cost. Finally, extensive experiments on real and synthetic datasets verify the efficiency of AMDS construction and MkNN search in both computational and communicational cost. In the future, we intend to use AMDS to support various metric queries, e.g., metric skyline queries.

Acknowledgements. This work was supported in part by the 973 Program No. 2015CB352502, the NSFC Grant No. 61522208, and the NSFC-Zhejiang Joint Fund under Grant No. U1609217. Yunjun Gao is the corresponding author of this work.

References

1. Batko, M., Gennaro, C., Zezula, P.: A scalable nearest neighbor search in P2P systems. In: Ng, W.S., Ooi, B.-C., Ouksel, Aris M., Sartori, C. (eds.) DBISP2P 2004. LNCS, vol. 3367, pp. 79–92. Springer, Heidelberg (2005). https://doi.org/10.1007/978-3-540-31838-5_6
2. Batko, M., Gennaro, C., Zezula, P.: Similarity grid for searching in metric spaces. In: Türker, C., Agosti, M., Schek, H.-J. (eds.) Peer-to-Peer, Grid, and Service-Orientation in Digital Library Architectures. LNCS, vol. 3664, pp. 25–44. Springer, Heidelberg (2005). https://doi.org/10.1007/11549819_3
3. Batko, M., Novak, D., Falchi, F., Zezula, P.: Scalability comparison of peer-to-peer similarity search structures. Future Gener. Comput. Syst. **24**(8), 834–848 (2008)
4. Batko, M., Novak, D., Falchi, F., Zezula, P.: On scalability of the similarity search in the world of peers. In: INFOSCALE, p. 20 (2006)
5. Dohnal, V., Sedmidubsky, J., Zezula, P., Novak, D.: Similarity searching: towards bulk-loading peer-to-peer networks. In: SISAP, pp. 87–94 (2008)
6. Doulkeridis, C., Vlachou, A., Kotidis, Y., Vazirgiannis, M.: Peer-to-peer similarity search in metric spaces. In: VLDB, pp. 986–997 (2007)
7. Traina Jr., C., Filho, R.F.S., Traina, A.J.M., Vieira, M.R., Faloutsos, C.: The Omni-family of all-purpose access methods: a simple and effective way to make similarity search more efficient. VLDB J. **16**(4), 483–505 (2007)
8. Novak, D., Batko, M., Zezula, P.: Large-scale similarity data management with distributed metric index. Inf. Process. Manag. **48**(5), 855–872 (2012)
9. Stoica, I., Morris, R.T., Karger, D.R., Kaashoek, M.F., Balakrishnan, H.: Chord: a scalable peer-to-peer lookup service for internet applications. In: SIGCOMM, pp. 149–160 (2001)

10. Vlachou, A., Doulkeridis, C., Kotidis, Y.: Metric-based similarity search in unstructured peer-to-peer systems. In: Hameurlain, A., Küng, J., Wagner, R. (eds.) Transactions on Large-Scale Data- and Knowledge-Centered Systems V. LNCS, vol. 7100, pp. 28–48. Springer, Heidelberg (2012). https://doi.org/10.1007/978-3-642-28148-8_2

11. Ares, L.G., Brisaboa, N.R., Esteller, M.F., Pedreira, O., Places, A.S.: Optimal pivots to minimize the index size for metric access methods. In: SISAP, pp. 74–80 (2009)

12. Beckmann, N., Kriegel, H., Schneider, R., Seeger, B.: The R*-tree: an efficient and robust access method for points and rectangles. In: SIGMOD, pp. 322–331 (1990)

13. Shen, H.T., Shu, Y., Yu, B.: Efficient semantic-based content search in P2P network. IEEE Trans. Knowl. Data Eng. **16**(7), 813–826 (2004)

14. Ratnasamy, S., Francis, P., Handley, M., Karp, R.M., Shenker, S.: A scalable content-addressable network. In: SIGCOMM, pp. 161–172 (2001)

15. Bawa, M., Condie, T., Ganesan, P.: LSH forest: self-tuning indexes for similarity search. In: WWW, pp. 651–660 (2005)

16. Banaei-Kashani, F., Shahabi, C.: SWAM: a family of access methods for similarity-search in peer-to-peer data networks. In: CIKM, pp. 304–313 (2004)

17. Doulkeridis, C., Nørvåg, K., Vazirgiannis, M.: DESENT: decentralized and distributed semantic overlay generation in P2P networks. IEEE J. Sel. Areas Commun. **25**(1), 25–34 (2007)

18. Jagadish, H.V., Ooi, B.C., Vu, Q.H., Zhang, R., Zhou, A.: VBI-tree: a peer-to-peer framework for supporting multi-dimensional indexing schemes. In: ICDE, p. 34 (2006)

19. Bharambe, A.R., Agrawal, M., Seshan, S.: Mercury: supporting scalable multi-attribute range queries. In: SIGCOMM, pp. 353–366 (2004)

20. Liu, B., Lee, W., Lee, D.L.: Supporting complex multi-dimensional queries in P2P systems. In: ICDCS, pp. 155–164 (2005)

21. Kalnis, P., Ng, W.S., Ooi, B.C., Tan, K.: Answering similarity queries in peer-to-peer networks. Inf. Syst. **31**(1), 57–72 (2006)

22. Ghanem, S.M., Ismail, M.A., Omar, S.G.: VITAL: structured and clustered super-peer network for similarity search. Peer-to-Peer Netw. Appl. **8**(6), 965–991 (2015)

23. Falchi, F., Gennaro, C., Zezula, P.: A content–addressable network for similarity search in metric spaces. In: Moro, G., Bergamaschi, S., Joseph, S., Morin, J.-H., Ouksel, Aris M. (eds.) DBISP2P 2005-2006. LNCS, vol. 4125, pp. 98–110. Springer, Heidelberg (2007). https://doi.org/10.1007/978-3-540-71661-7_9

24. Novak, D., Zezula, P.: M-Chord: a scalable distributed similarity search structure. In: INFOSCALE, p. 19 (2006)

25. Jagadish, H.V., Ooi, B.C., Tan, K., Yu, C., Zhang, R.: iDistance: an adaptive B+-tree based indexing method for nearest neighbor search. ACM Trans. Database Syst. **30**(2), 364–397 (2005)

26. Chávez, E., Navarro, G., Baeza-Yates, R.A., Marroquin, J.L.: Searching in metric spaces. ACM Comput. Surv. **33**(3), 273–321 (2001)

27. Batko, M., Novak, D., Zezula, P.: MESSIF: metric similarity search implementation framework. In: Thanos, C., Borri, F., Candela, L. (eds.) DELOS 2007. LNCS, vol. 4877, pp. 1–10. Springer, Heidelberg (2007). https://doi.org/10.1007/978-3-540-77088-6_1

28. Mühl, G., Fiege, L., Pietzuch, P.R.: Distributed Event-Based Systems. Springer, Heidelberg (2006). https://doi.org/10.1007/3-540-32653-7

29. Coulouris, G., Dollimore, J., Kindberg, T.: Distributed Systems - Concepts and Designs. International Computer Science Series, 3rd edn. Addison-Wesley-Longman, Boston (2002)

252   X. Ding et al.

30. Bustos, B., Navarro, G., Chávez, E.: Pivot selection techniques for proximity searching in metric spaces. Pattern Recognit. Lett. **24**(14), 2357–2366 (2003)
31. Yu, C., Ooi, B.C., Tan, K., Jagadish, H.V.: Indexing the distance: an efficient method to KNN processing. In: VLDB, pp. 421–430 (2001)
32. Chen, L., Gao, Y., Li, X., Jensen, C.S., Chen, G.: Efficient metric indexing for similarity search. In: ICDE, pp. 591–602 (2015)

Efficient Grammar Generation
for Inverted Indexes

Yan Fan, Xinyu Liu, Shuni Gao, Zhaohua Zhang, Xiaoguang Liu$^{(\boxtimes)}$,
and Gang Wang$^{(\boxtimes)}$

College of Computer and Control Engineering, Nankai University, Tianjin, China
{fanyan,liuxy,gaoshn,zhangzhaohua,liuxg,wgzwp}@nbjl.nankai.edu.cn

Abstract. Inverted indexes are commonly utilized in large-scale search engines to store lists of document identifies (docIDs) relevant to query terms, which are queried maybe thousands of times per second. Traditionally, optimized integer sequence encoding methods are applied to compress the inverted index while simultaneously maintaining reasonable query processing speeds. Recently, a context-free grammar-based method was introduced for inverted index compression, which is particularly useful for highly repetitive indexes.

Due to the high time and space cost of the traditional grammar generation (transform) algorithms designed for large inverted index collections with much redundancy, we propose a parallel generation algorithm for context-free grammar generation. We further propose a greedy dictionary pruning algorithm to reduce cache misses in query processing. We also implement encoding, list intersection, and WAND querying on the grammar index. Experimental results indicate that parallel grammar generation algorithm achieves a super-linear speedup with minor data overhead and nearly identical query efficiency compared to the single-threaded algorithm. For example, with 10 threads to process the data set, a speedup about 75 times faster is obtained with only 4.3% data overhead. Moreover, parallel grammar generation incurs negligible impact on query processing efficiency.

Keywords: Inverted index compression · Context-free grammar
Parallel grammar generation algorithm

1 Introduction

In the compression of inverted indexes, both the compression ratio and decompression speed are important for performances of search engines, because they measure the space cost and retrieval efficiency, respectively. Previous work on index compression focused on encoding methods to compress document identifiers (docIDs, which are integers). Typically, docIDs are stored in increasing

X. Liu and G. Wang—This work is partially supported by NSF of China (grant number: 61602266), Science and Technology Development Plan of Tianjin (grant numbers: 17JCYBJC15300, 16JCYBJC41900).

ⓒ Springer International Publishing AG, part of Springer Nature 2018
Y. Cai et al. (Eds.): APWeb-WAIM 2018, LNCS 10987, pp. 253–267, 2018.
https://doi.org/10.1007/978-3-319-96890-2_21

order, so the differences between consecutive docIDs are often smaller than the docIDs themselves. As such, it is natural to replace the inverted lists by lists of differences before encoding; we refer to such lists as *differential lists* (or *Delta lists* for short).

A different approach to compressing inverted indexes, called *grammar-based compression*, is to remove the duplicate data among lists in the index through the use of context-free grammar. Common subsequences, called *patterns*, of docIDs (or differences between successive docIDs) are replaced by a reference to a single record in the *dictionary*. In this paper, the grammar generated on *Delta lists* is called *Delta grammar*.

There are two drawbacks to grammar-based compression: (a) The grammar generation process is time consuming and requires a large amount of memory, especially for large-scale collections (for example, in modern search engines). (b) The performance, in terms of compression and query response time, is dependent on the grammar dictionary. A larger dictionary allows more patterns to be identified, which results in better compression but slower query processing. In this paper, our contributions are:

– We propose a parallel grammar generation algorithm called *PIPara* (Pattern Identification in Parallel) to speed up grammar generation using *PISequential* in [11].
– We design a greedy dictionary pruning strategy which benefits query processing.
– We implement Delta grammar encoding, TAAT AND query and DAAT WAND query methods.

2   Preliminaries and Related Work

2.1   Inverted-Index Compression

Here, we study some basic arithmetic coding schemes, one of which we use throughout this paper.

Binary interpolative coding [5] encodes an increasing sequence by recursively splitting the sequence into two parts and encoding the middle integer, which achieves a high compression ratio, but poor decompression speed. Simple (along with its variants) is also designed to use different bits to compress integer sequences, but achieves a better trade-off between compression and decompression speed, like the widely used compression method Simple16 [10].

Another popular encoder is PForDelta [12] where the least significant b bits are used to represent most integers (usually around 90%) and the integers requiring more bits (exceptions) are stored and compressed separately. OptP-ForDelta [8], a block-wise version of PForDelta, instead selects an appropriate number of exceptions for each block, with a block size usually 128 or 256 [5]. In this paper, OptPForDelta and Simple16 are used as the Delta grammar encoder for testing.

2.2 Grammar-Based Compression

A *context-free grammar* is a quadruple $G = (V, T, S, P)$ where T is a finite set with cardinality ≥ 2 called the *alphabet*, and elements in T are called *terminals*, V is another finite set (disjoint to T) whose elements are called *non-terminals* (NT), the unique non-terminal $S \in V$ is the *start symbol*. Any terminal or non-terminal is called a *symbol*. Set P is a finite set of *production rules* $A \to \alpha$, where $A \in V$ and α is a sequence of symbols called the *definition* or right-hand side of A. For any production like $A \to \alpha$, if we replace non-terminals in α recursively, we obtain a string only consisting of terminals, we call this operation *derivation*; We call the reverse *reduction*. For any NTs A and B, we say A equals B if the strings obtained by deriving A and B respectively are the same which we denote $A = B$. For the start symbol S, there is a production rule $S \to \beta$ in P, from which the original data can be recovered by replacing non-terminals recursively in β. In the context of grammar-based inverted index compression, however, the original data is a collection of posting lists rather than a single string, so we modify the notation for clarity. An index grammar is a quintuple $G = (V, T, P, R, S)$, where S becomes a set of non-terminals $\{l_1, l_2, \ldots, l_n\}$ corresponding to posting lists, so we call these non-terminals LNTs. Set R comprises of LNTs' production rules called *reduced lists*. We exclude S from V, so V contains only non-terminals that denote patterns called PNTs. Set P is the production set of PNTs, and we call it the *dictionary*. Because the maps between V and P, and between S and R are unique, in the following sections, we will denote G by P and R for simplicity. The size of an NT A is denoted $|A|$ which is the sum of the symbols in the definition of A and the size of the grammar G is denoted $|G|$ which is the total size of all unique NTs (including S) in G. The most basic unit of grammar reduction is a *digram*, which is defined as a pair of symbols next to each other appearing in an input string.

Grammar-based compression has been studied for several decades. Nevill-Manning and Witten [6] introduced a linear-time algorithm *Sequitur* that generates a context-free grammar to compress the input. Relevant to this paper is [9], where a greedy grammar generation algorithm *Sequential* is proposed which generates an irreducible grammar on any data sequence. In [9], an irreducible grammar G is defined which has the following three properties:

(1) Except for the start symbol S, all other non-terminals in grammar G appear at least twice in the range of G.
(2) There are no disjoint repeated substrings whose length greater than or equal to 2 in G.
(3) Different non-terminals are not equal.

For the properties (we call constraints later) of an irreducible grammar, we see that these constraints indicate the effectiveness of a grammar transform algorithm in the view of compression. In this paper, we relax the three constraints to design PIPara.

In the setting of inverted index compression, *Re-Pair* [4] is an offline algorithm that recursively finds the most frequent pair of symbols and replaces it with a new non-terminal, until no more replacements are valuable. Claude et al. [3] proposed *Repair-Skip*, which improves *Re-Pair* by compressing differential lists. They also add skip information for non-terminals to support fast skipping over compressed sequences during query processing.

Recently, Zhang et al. [11] proposed PISequential, a variant of Sequential to process posting lists of a general document collection directly, along with some optimizations. The present work takes PISequential as the baseline and use it as a calling process in PIPara.

3   Parallel Grammar Generation

The basic idea of the parallel grammar generation algorithm is to divide the original delta lists into c regions of almost the same size and assign each of them to a unique thread. Each thread will invoke PISequential [11] to generate a local irreducible grammar of the corresponding lists. After c local irreducible grammars have been generated, they are merged into a global grammar. This gives rise to an optimization problem: how to merge local irreducible grammars to obtain a small enough global grammar.

Irreducible grammar we describe above (in Sect. 2.2) gives the constraints of grammar reduction from the perspective of compression. However, considering the characteristic of parallel grammar, we weaken constraint 2 and get inferior properties a grammar have as follows:

Constraints 1 and 3 do not change because they are necessary for parallel grammar reduction.

We observed that if the input string only has duplicate digrams which appear only twice in the data without overlapping, then the generated irreducible grammar does not change the size of input string. Since two kinds of data are generated when the input string is transformed by PISequential, one is a grammatical form of the input string, which is the definition of S and the other is the dictionary containing all the definitions of NT except S. In practice, when the input string is long with many repetitive substrings, the size of S is much larger than the sum of the size of other NTs. Based on this we relax constraint 2 as follows:

Constraint 2: There are either no repeated substrings without overlapping or only substrings with length of 2 appearing twice in the definition of S (reduced lists).

The constraints above give a less strict form of an irreducible grammar which we call Sirreducible constraints.

3.1   Filtering and Rough Merging

After each thread generates a local irreducible grammar using PISequential, we use the main thread to merge all the local grammars generated to form a global grammar. Specifically, for each thread i ($i \in \{1, \ldots, c\}$), the initial local grammar is $G_i = (V_i, T_i, P_i, R_i, S_i)$ where V_i and P_i are both $\emptyset$, and $S_i = \{l_{i1}, l_{i2}, \cdots, l_{in_i}\}$

is a subset of LNTs and R_i contains their original posting lists. PISequential is used to identify patterns from R_i and transform it to an irreducible grammar G_i.

After this, each thread filters the irreducible grammar G_i by replacing all the PNTs occurring less than twice in R_i with their definitions and then removing these definitions. Then we relabel the remaining PNTs in V_i and update the relevant productions in D_i and R_i.

After this step, we perform a rough merging operation to form a global context-free grammar G. The local dictionaries P_i, for all $i \in \{1, \ldots, c\}$, are combined into P and likewise for V_i, R_i and S_i. We then reassign a global label to every PNT and update P and R accordingly. The following example illustrates this procedure.

Example 1. We invoke two threads T_1 and T_2 to process the inverted index $\{l_1, l_2, l_3, l_4, l_5\}$ shown below. We partition posting lists into two subsets $S_1 = \{l_1, l_2, l_3\}$ and $S_2 = \{l_4, l_5\}$ and assign them to T_1 and T_2 respectively.

$$l_1 = \langle 1, 2, 3, 14, 20, 21, 39, 40, 49, 57 \rangle \qquad l_2 = \langle 1, 2, 3, 9, 14, 21, 39, 40, 49 \rangle$$
$$l_3 = \langle 1, 14, 16, 21, 39 \rangle \qquad l_4 = \langle 1, 2, 3, 14, 20, 37, 42, 57, 58 \rangle$$
$$l_5 = \langle 1, 2, 3, 8, 15, 21, 39, 40, 49, 51 \rangle$$

After performing the grammar transform action on R_1 and R_2 using PISequential in T_1 and T_2, we obtain the irreducible grammars

$$G_1 = \{\{A_1 \rightarrow \langle 1, 2, 3 \rangle, B_1 \rightarrow \langle C_1, 40, 49 \rangle, C_1 \rightarrow \langle 21, 39 \rangle\},$$
$$\{l_1 \rightarrow \langle A_1, 14, 20, B_1, 57 \rangle, l_2 \rightarrow \langle A_1, 9, 14, B_1 \rangle, l_3 \rightarrow \langle 1, 14, 16, C_1 \rangle\}\}$$
$$G_2 = \{\{A_2 \rightarrow \langle 1, 2, 3 \rangle\},$$
$$\{l_4 \rightarrow \langle A_2, 14, 20, 37, 42, 57, 58 \rangle, l_5 \rightarrow \langle A_2, 8, 15, 21, 39, 40, 49, 51 \rangle\}\}$$

By filtering G_1 and G_2, we remove the PNT C_1 which appears once in R_1 and therefore the grammars become

$$G_1 = \{\{A_1 \rightarrow \langle 1, 2, 3 \rangle, B_1 \rightarrow \langle 21, 39, 40, 49 \rangle\},$$
$$\{l_1 \rightarrow \langle A_1, 14, 20, B_1, 57 \rangle, l_2 \rightarrow \langle A_1, 9, 14, B_1 \rangle, l_3 \rightarrow \langle 1, 14, 16, 21, 39 \rangle\}\}$$
$$G_2 = \{\{A_2 \rightarrow \langle 1, 2, 3 \rangle\},$$
$$\{l_4 \rightarrow \langle A_2, 14, 20, 37, 42, 57, 58 \rangle, l_5 \rightarrow \langle A_2, 8, 15, 21, 39, 40, 49, 51 \rangle\}\}$$

We obtain a global grammar G after a rough merging operation, where

$$G = \{\{A \rightarrow \langle 1, 2, 3 \rangle, B \rightarrow \langle 21, 39, 40, 49 \rangle, C \rightarrow \langle 1, 2, 3 \rangle\},$$
$$\{l_1 \rightarrow \langle A, 14, 20, B, 57 \rangle, l_2 \rightarrow \langle A, 9, 14, B \rangle, l_3 \rightarrow \langle 1, 14, 16, 21, 39 \rangle,$$
$$l_4 \rightarrow \langle C, 14, 20, 37, 42, 57, 58 \rangle, l_5 \rightarrow \langle C, 8, 15, 21, 39, 40, 49, 51 \rangle\}\}$$

In this example, we see that after this first step, no PNT appears less than twice in local grammars G_1 and G_2. However, the rough merged global grammar G has two equivalent PNTs, namely $A = C$, so it doesn't satisfy all three Sirreducible constraints.

3.2   Dictionary Merging

We remove duplicate PNTs in the rough global grammar $G = (V, T, P, R, S)$. Since this kind of duplication occurs only in the dictionary, in this section we tackle this problem by proposing a dictionary merging algorithm. We first derive all the PNTs, and then transform the resulting terminal strings to a grammar using PISequential, and finally eliminate duplication. More specifically, we perform the following five steps:

(1) We recursively derive all the PNTs according to their production rules in the dictionary.
(2) We use the resulting terminal strings as PISequential input, thereby generating a grammar $G_p = \{V_p, T, P_p, \emptyset, \emptyset\}$.
(3) We merge G_p into G by finding the correspondence between V_p and V.
(4) We find the PNTs of size 1 and replace any occurrence of them in R by their definition.
(5) We remove useless PNTs and their productions, and relabel the remaining PNTs and update R accordingly.

Example 2. For the rough global grammar G generated in Example 1, we obtain the grammar

$$G_p = \{\{A \to \langle D \rangle, B \to \langle 21, 39, 40, 49 \rangle, C \to \langle D \rangle, D \to \langle 1, 2, 3 \rangle\}, \{\}\}$$

We see that A and C have size 1, and therefore we replace all of their occurrences (in l_1, l_2, l_4, and l_5) by their definition D. The remaining PNTs B and D are then relabeled as A and B respectively and R is updated accordingly, we obtain

$$G = \{\{A \to \langle 21, 39, 40, 49 \rangle, B \to \langle 1, 2, 3 \rangle\},$$
$$\{l_1 \to \langle B, 14, 20, A, 57 \rangle, l_2 \to \langle B, 9, 14, A \rangle, l_3 \to \langle 1, 14, 16, 21, 39 \rangle,$$
$$l_4 \to \langle B, 14, 20, 37, 42, 57, 58 \rangle, l_5 \to \langle B, 8, 15, 21, 39, 40, 49, 51 \rangle\}\}$$

Since PISequential removes duplicate digrams in the input data, after the grammar transform on the dictionary, the definitions of equal PNTs are reduced to a single newly created PNT. Since any PNT appears at least twice in R, even after step 5, we still guarantee that the remaining PNTs appear at least twice in R. In Example 2, we replace A and C in l_1, l_2, l_3, l_4, and l_5, after which B and D appear at least twice in these reduced lists. Therefore, the first two Sirreducible constrains are satisfied.

3.3   Grammatical Iteration

From Example 2, we see that reduced lists generated by different threads may have common digrams after dictionary merging. In this section, we propose a grammatical iteration method to remove all such duplications. In brief, we perform a grammar transform on the reduced lists and the dictionary, which we see as a secondary reduction on the current grammar G. The main steps are as follows:

(1) Take P and R as the input and execute PISequential to generate a new version of G. Once the new version is generated, one of the following four cases holds for any PNT A, whose definition is $A \rightarrow \alpha$:

    (a) PNT A appears only once in the right-hand side of the productions in P and R.

    (b) Size of A equals 1.

    (c) Cases (a) and (b) hold simultaneously.

    (d) Both (a) and (b) do not hold.

(2) For any PNT A in V, if (d) doesn't hold, we replace every occurrence of A with its definition.

(3) Once all the PNTs have been processed, we remove useless PNTs and their productions. Finally, we relabel the remaining PNTs and update P and R accordingly.

Example 3. For the resulting grammar $G = (V, T, P, R, S)$ in Example 2, we take P and R as input and perform the secondary reduction. The new version is

$$G = \{\{A \rightarrow \langle D \rangle, B \rightarrow \langle 1, 2, 3 \rangle, C \rightarrow \langle B, 14 \rangle, D \rightarrow \langle 21, 39, 40, 49 \rangle\},$$
$$\{l_1 \rightarrow \langle C, 20, A, 57 \rangle, l_2 \rightarrow \langle B, 9, 14, A \rangle, l_3 \rightarrow \langle 1, 14, 16, 21, 39 \rangle,$$
$$l_4 \rightarrow \langle C, 20, 37, 42, 57, 58 \rangle, l_5 \rightarrow \langle B, 8, 15, D, 51 \rangle\}\}$$

We see that $|A| = 1$, so its occurrences are replaced by D and it is removed. The final version of G is thus

$$G = \{\{A \rightarrow \langle 1, 2, 3 \rangle, B \rightarrow \langle A, 14 \rangle, C \rightarrow \langle 21, 39, 40, 49 \rangle\},$$
$$\{l_1 \rightarrow \langle B, 20, C, 57 \rangle, l_2 \rightarrow \langle A, 9, 14, C \rangle, l_3 \rightarrow \langle 1, 14, 16, 21, 39 \rangle,$$
$$l_4 \rightarrow \langle B, 20, 37, 42, 57, 58 \rangle, l_5 \rightarrow \langle A, 8, 15, C, 51 \rangle\}\}$$

After grammatical iteration, we have three conclusions below.

1: G satisfies all three Sirreducible constraints.

    After the first step of grammatical iteration, the repeated digrams in the reduced lists have been removed so as to satisfy Sirreducible constraint 2. However, constraints 1 and 3 do not hold when (a), (b), or (c) hold. In the second step we identify PNTs that do not satisfy (d) and replace all the occurrences of these PNTs with their definitions. So after the end of grammatical iteration, G satisfies all the three Sirreducible constraints.

2: The size of the grammar G obtained by the parallel grammar generation algorithm is $\mathcal{O}\left(\text{size}/\log(\text{size}/c)\right)$, where size is the size of the original data. Yang and Kieffer [9] prove that the grammar size obtained by Sequential is $\mathcal{O}(\text{size}/\log \text{size})$ when size $\rightarrow \infty$. When we process c local grammars with c threads, the total size of grammars obtained by these c threads is $\mathcal{O}(c \times ((\text{size}/c)/ (\log(\text{size}/c)))) = \mathcal{O}\left(\text{size}/\log(\text{size}/c)\right)$. Because after grammatical iteration, G satisfies the Sirreducible constraints and Sirreducible constraints force the size of the global grammar to be smaller than the total size of c local grammars, so conclusion 2 holds.

3: The time complexities of PIPara are $\mathcal{O}(\tau)$ and $\omega\left(\tau/c^2\right)$ where c is the number of threads and τ is the time complexity of PISequential.

In practice, the execution time of PISequential is mainly consumed in digram searching and production matching. Both operations are based on the hashing operations in hash tables. The hash buckets are organized as linked lists to resolve collisions. In PISequential, assuming that a hash bucket B has length k, the average time for accessing it is $\mathcal{O}(k/2)$. In the parallel grammar generation algorithm PIPara, the number of access operations on B issued by each thread becomes $1/c$ times that of PISequential. Moreover, the number of values inserted into B becomes $1/c$ times that of PISequential on average, so the number of comparison operations on B of each thread in PIPara is $\Omega\left(\frac{k}{2c^2}\right)$ and time consumed in hashing is $\Omega\left(\tau/(2c^2)\right)$. As for dictionary merging and grammatical iteration, for the same hash bucket B, the average length of a linked list is $\sum_{1 \leq i \leq c} k_i$, where k_i is the length of the linked list in the corresponding bucket in thread i and $\sum_{1 \leq i \leq c} k_i$ must be less than k because the input comprises of reduced lists and/or the dictionary. Similarly, there are fewer access operations on B than in PISequential. So the total time consumed by comparison operations in dictionary merging and grammatical iteration is $\mathcal{O}(\tau)$ and conclusion 3 holds.

4   Greedy Dictionary Pruning

In the process of list intersection and WAND query processing, the dictionary is frequently accessed to recover associated lists. To reduce cache misses in the derivation of PNT, we consider how to expand dictionary properly by fully deriving every PNT while keeping the grammar as small as possible.

The main idea is a greedy strategy; we optimize the Delta grammar by greedy iteration pruning, namely each round we prune some PNTs in the current grammar to obtain a smaller grammar under the premise that all PNTs in V have been derived in the current round.

We define the *top PNT* as follows: for a Delta grammar $G = (V, T, P, R, S)$ and a non-empty subset V_s of V, we call symbol $A \in V_s$ the top PNT in set V_s if A satisfies $C \to \alpha \in P$ for all $C \in V_s$, A does not appear in α. The proposed greedy pruning algorithm proceeds as follows:

1. We recursively remove all top PNTs that appear fewer than 2 times in R and update G accordingly.
2. For any top PNT A in V, if A satisfies

$$freq(A) + \|A\| + \textit{offset} \geq freq(A) \times |A|$$

where $freq(A)$ is the frequency of which A appears in D, and $|A|$ and $\|A\|$ are the length of the right-hand side of the production of A before and after deriving A, respectively, and *offset* is an external parameter, then we remove A by replacing the occurrences of A with its definition.

3. We update grammar G by relabeling the remaining PNTs and derive all PNTs for the complete dictionary.

Since the final to-be-compressed index is a Delta grammar G with a fully expanded dictionary, if a PNT A appears at least twice in G while it appears less than twice in R, we consider A and its production to be redundant. In this case, the position of A appearing in R is replaced by the right-hand side of its production, which saves time when accessing the dictionary and also saves one element of space. The reason for choosing the top PNTs to take into account is that it does not introduce repeated PNTs. To illustrate, assuming there are two productions $A \rightarrow \langle 1, 2, 3 \rangle$ and $B \rightarrow \langle A, 7, 8 \rangle$ which both appear once in R, PNT A is not a top PNT according to our previous definition. If we first replace A and then replace B with $\langle A, 7, 8 \rangle$, then A is reintroduced into R. Instead, if we replace top PNT B first, then A will appear twice in R and is not removed, which is beneficial to storage cost. From this example we see that the pruning order has an influence on the size of the grammar after dictionary expansion.

For step 2, the left- and right-hand sides of the inequality represent the cost of storage related to A before and after A is replaced, respectively. If there is a decrease in size after replacing A, then we remove A and replace it by its definition.

5   Data Layout and Query Implementations on Delta Grammar Index

5.1   Data Structure of Delta Grammar Index

In the dictionary, for each PNT, instead of storing its corresponding d-gap list, we store the prefix sum results. Since the dictionary is frequently accessed during query process, such a format reduces the time spent on recovering lists, and moreover, we encode the dictionary via the interpolative scheme at a small storage cost.

To support fast query operations, we make use of a skip-table data structure and a block-based encoding method. For each reduced list, which are compressed using context-free grammar on delta lists, we split it into 128-element blocks, except possibly for the last block which may contain anywhere between 1 and 127 elements. The sample data layout for a reduced list is depicted in Fig. 1, and it is similar to that in [11].

Each reduced list is made up of two parts, namely a *list head* and the *block info*. For a *list head*, if the number of elements in a list exceeds 128, we store len, the length of the reduced list, and store an array of pairs, each pair of which corresponds to one block in the reduced list. The first value in each pair is the offset to the next block in bytes, and the other value gives the maximum docID in the block. For each block in *block info*, the first value nT, is the number of terminals in the block, and the "i-th pos" is the offset of the i-th terminal in the block. The last part of each block, *compressed data*, contains the encoded symbols of the reduced block. For the second encoding, we choose OptPForDelta and Simple16 to encode *compressed data*.

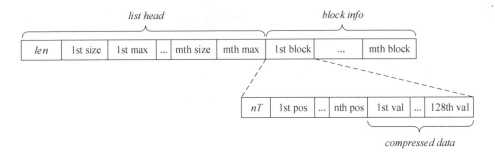

Fig. 1. Data layout for a reduced index

5.2   List Intersection and WAND Query Processing

For list intersection, motivated by the layout of reduced indexes, we design a TAAT intersection method which mainly involves searching docID in a block B_i consisting of PNTs and terminals. For any docID e, after assuring that it is smaller than the max docID in B_i, it is compared with the symbols in B_i sequentially. If the current symbol is a terminal, say t, and a prefix sum ss is computed from the first element in B_i to the element right before t, and $ss + t$ is compared with e and e is kept in the temporary result of this intersection operation if $e = ss+t$. Otherwise if the symbol is a PNT (whose maximum value is smaller than e), say A, we fetch the definition of A and compare e with the sums of values between each value in A and ss sequentially, and e is kept if it equals one of these sums.

The WAND [2] algorithm is an early stop technique using DAAT query processing. For WAND query, we only need to re-implement the routine Next_GEQ (d) in the WAND procedure, which returns the next docID greater than or equal to d in the current reduced list. We achieve this using a method similar to that described in the preceding paragraph.

6   Experiments

We design experiments to focus on the impact of parallel grammars with varying number of threads on the compression ratio and query response time in addition to the effectiveness of greedy dictionary pruning on the grammar size. The following experiments test the efficiency of the parallel algorithm by comparing the single-threaded and multi-threaded grammar transform in terms of grammar generation time, grammar size, Delta grammar compression ratio, and response time for both AND and WAND queries and the grammar size with/without greedy dictionary pruning on the dictionary. The empirical parameter *offset* (described in Sect. 4) is set to 10 on all three indexes.

6.1 Experimental Setup

All experiments are carried out on a PC server; its statistics are listed in Table 1. Experiments are performed on three indexes of the GOV2 collection reordered by URL, IBDA [1] and TRM [7] with corresponding term frequency information, denoted GOV2URL, GOV2IBDA and GOV2TRM, respectively.

Table 1. The experimental platform details

OS version	Red Hat 4.4.7-4
GCC version	gcc-4.8.4
CPU clock	2.20 GHz
Logical CPU cores	48
Cache size	30720 KB
Memory size	504 GB

6.2 Effectiveness and Efficiency of Grammar Generation and Dictionary Pruning

The first step of grammar compression, namely grammar generation, is a time consuming process which, along with the size of the generated grammar, varies according to the data size and the redundancy in the collection. In this paper, to relieve hash collisions among string matching in PISequential, the first three symbols of a string to be matched are combined into one lookup key to compute a hash value, instead of one symbol used in the original implementation and the updated version of PISequential is used as the call function in PIPara. In the C++ implementation of PIPara, CPU threads are employed in parallel and there is not any communication between threads. To generate the Delta grammar, first we compute d-gaps of the three indexes of GOV2 data sets. Then we compare two kinds of PISequential algorithms (the single-threaded original, and the updated one) with the proposed PIPara. Figures 2(a) and (b) respectively plot experimental results for time consumption and space usage as the number of threads varies.

In Fig. 2(a), the highest speedup ratio 75.4 of PIPara over PISequential (original) is achieved on GOV2URL with 10 threads and we see that the proposed PIPara method significantly outperforms PISequential (both the original and updated versions) in terms of time consumption and achieves super-linear speedup as the number of threads increases, which we prove in Sect. 3.3.

For each data set, as the number of threads increases, the generated grammar size slightly increases and the trend is relatively flat, which we see in Fig. 2(b). The maximum observed storage overhead is 9% for GOV2URL with 10 threads while the minimum overhead is 1.4% for GOV2IBDA with 2 threads. This indicates that the impact of PIPara on grammar size is modest when we require satisfying Sirreducible constraints in Sect. 3.

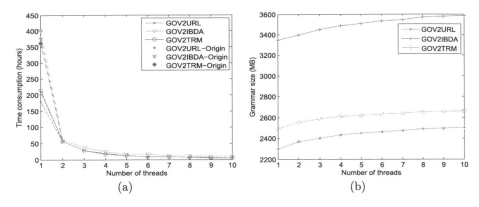

Fig. 2. Time consumption (left) and grammar size space (right) for single-threaded PISequential (original and updated versions) and PIPara as the number of threads varies from 2 to 10. In (a), the time consumptions of the original PISequential on GOV2URL, GOV2IBDA, and GOV2TRM are denoted by GOV2URL-Origin, GOV2IBDA-Origin, and GOV2TRM-Origin, respectively. For thread 1 of GOV2URL, GOV2IBDA, and GOV2TRM, drawn using solid lines, the updated PISequential is used. For (b), the space usage of grammar index includes the dictionary and reduced lists without any pattern pruning strategies or arithmetic encoding

Table 2 tabulates the observed grammar sizes of the Delta grammar on three datasets obtained from PISequential before and after the greedy dictionary pruning, in addition to directly deriving $PNTs$, and are refered as Origin, Greedy, and Direct. For GOV2URL, GOV2IBDA, and GOV2TRM, the greedy pruing strategy reduce the space usage of the grammars by 31%, 19%, and 17%, respectively, compared to directly deriving all $PNTs$ assuming the definition of every PNT contains only terminals, which indicates the effectiveness of the proposed greedy dictionary pruing strategy.

Table 2. Grammar size (MB) after directly deriving and applying greedy dictionary pruning.

Type	GOV2URL	GOV2IBDA	GOV2TRM
Origin	2297	3346	2492
Direct	3493	4357	3079
Greedy	2408	3529	2564

6.3   Compression and Query Processing

In Table 3 we list the final compression ratio of the compressed index. Here OptPForDelta is used for the second encoding.

Table 3. Space usage (bits/docID) of compressed indexes as the number of threads varies.

Threads	1	2	3	4	5	6	7	8	9	10
GOV2URL	4.509	4.592	4.620	4.632	4.656	4.665	4.681	4.677	4.687	**4.702**
GOV2IBDA	6.968	**7.021**	7.094	7.137	7.156	7.191	7.204	7.224	7.238	7.236
GOV2TRM	4.932	5.013	5.037	5.047	5.056	5.069	5.077	5.080	5.088	5.087

From Table 3, we make the following observations. The biggest difference between PISequential and PIPara is 4.3% (on GOV2URL with 10 threads) and the smallest difference is 0.7% (on GOV2IBDA with 2 threads). Compared with the range 9% to 1.4% of grammar storage overhead we see in Fig. 2, we observe that second encoding further reduces the overhead. For each data set, the compression ratio difference between PISequential and PIPara varies slowly as the number of threads increases, which indicates the effectiveness of the proposed parallel grammar generation method.

To verify the query processing performance of indexes generated using PIPara and PISequential, we test two different query operations that are widely used, TAAT AND and DAAT WAND.

The experimental results are listed in Tables 4 and 5, where we compare the average response time (ms) for query processing on the indexes compressed by PISequential and PIPara.

Table 4. Average response time (ms) for TAAT AND query as the number of threads varies.

Threads	1	2	3	4	5	6	7	8	9	10
GOV2URL	3.941	3.924	3.957	3.894	3.944	3.931	3.872	**3.749**	3.981	3.927
GOV2IBDA	7.272	7.242	7.103	6.981	7.061	7.106	7.125	6.991	6.983	7.128
GOV2TRM	4.598	4.632	4.609	4.625	4.621	**1.636**	4.502	4.461	4.493	4.467

In Tables 3 and 4, we find that a higher compression ratio leads to higher efficiency of list intersection. The reason for this is that a smaller grammar implies fewer blocks in reduced lists, which accelerates the docID locate speed. For each data set, the average response time varies only slightly as the number of threads varies. The maximum difference between PIPara and PISequential is from −4.90% on GOV2URL with 8 threads and the smallest is 0.83% on GOV2TRM with 6 threads, which shows that PIPara has little additional impact on TAAT AND query performance (compared with PISequential). For DAAT AND query, the results are similar and are not discussed.

For WAND query, in Table 5, we arrive at the same conclusions as that for TAAT AND query, there is little difference in terms of time efficiency as the number of threads increases. On all these data sets, the difference between PIPara

Table 5. Average response time (ms) for DAAT WAND query with the number of threads varies

Threads	1	2	3	4	5	6	7	8	9	10
GOV2URL	10.431	10.279	9.938	9.782	9.901	9.873	9.919	9.782	**9.761**	9.777
GOV2IBDA	16.046	**16.273**	16.229	16.183	16.183	16.171	16.153	16.135	16.067	16.078
GOV2TRM	7.951	7.904	7.954	7.982	7.843	7.782	7.797	7.631	7.607	7.704

and PISequential varies from -6.42% to 1.41%. We see that the impact of the parallel grammar generation method on WAND query processing is minor, which also indicates the high efficiency of the Next_GEQ(d) function we rewrite.

When Simple16 is used as the second encoder of Delta grammar, similar results are obtained and not discussed.

7   Conclusions and Future Work

The proposed parallel grammar generation algorithm PIPara, along with the proposed greedy pruning strategy, reinforces and improves upon previous related work [9,11]. Compared with single-threaded Sequential [9] and PISequential [11], we obtain 75.4 speedup in the grammar generation process with minor overhead (almost 9% and 4.3% before and after the secondary encoding). For list intersection and WAND query, PIPara increases the response time by up to 0.83% and 1.41% respectively compared to PISequential.

Since Sirreducible constrains are weaker than irreducible constraints, in the future we intend to consider other constraints which are more rigorous than Sirreducible constrains to obtain better parallel algorithms. In PIPara, each thread performs linear scanning on lists, so the lists' order will affect the grammar transform's performance. Therefore, it would be worthwhile studying how best to rearrange the original index lists.

In our experiments, we design a parallel algorithm by multi-threading. However, in order to make efficient use of memory, it is natural to extend PIPara for multiprocessing in distributed systems. In this case, PIPara could extend to the large data to generate a context-free grammar. In addition, it maybe beneficial to rewrite our algorithm for GPU to further speed up the grammar generation process.

Context-free grammar is widely used in data compression, natural language processing, biological DNA matching and pattern recognition, so studying how to use PIPara in these fields would open this research to a wide range of applications.

Acknowledgement. Thank Rebecca J. Stones for her guidance of writing in this paper.

References

1. Arroyuelo, D., González, S., Oyarzún, M., Sepulveda, V.: Document identifier reassignment and run-length-compressed inverted indexes for improved search performance. In: Proceedings of the SIGIR, pp. 173–182 (2013)
2. Broder, A.Z., Carmel, D., Herscovici, M., Soffer, A., Zien, J.Y.: Efficient query evaluation using a two-level retrieval process. In: Proceedings of the CIKM, pp. 426–434 (2003)
3. Claude, F., Fariña, A., Martínez-Prieto, M.A., Navarro, G.: Indexes for highly repetitive document collections. In: Proceedings of the CIKM, pp. 463–468 (2011)
4. Larsson, N.J., Moffat, A.: Off-line dictionary-based compression. In: Proceedings of the DCC, pp. 296–306 (1999)
5. Moffat, A., Stuiver, L.: Binary interpolative coding for effective index compression. Inf. Retr. 3(1), 25–47 (2000)
6. Nevill-Manning, C.G., Witten, I.H.: Compression and explanation using hierarchical grammar. Comput. J. 40, 103–116 (1997)
7. Shi, L., Wang, B.: Yet another sorting-based solution to the reassignment of document identifiers. In: Hou, Y., Nie, J.-Y., Sun, L., Wang, B., Zhang, P. (eds.) AIRS 2012. LNCS, vol. 7675, pp. 238–249. Springer, Heidelberg (2012). https://doi.org/10.1007/978-3-642-35341-3_20
8. Yan, H., Ding, S., Suel, T.: Inverted index compression and query processing with optimized document ordering. In: Proceedings of the WWW, pp. 401–410 (2009)
9. Yang, E.-H., Kieffer, J.C.: Efficient universal lossless data compression algorithms based on a greedy sequential grammar transform-part one: without context models. IEEE Trans. Inf. Theory 46, 755–777 (2000)
10. Zhang, J., Long, X., Suel, T.: Performance of compressed inverted list caching in search engines. In: Proceedings of the WWW, pp. 387–396 (2008)
11. Zhang, Z., Tong, J., Huang, H., Liang, J., Li, T., Stones, R.J., Wang, G., Liu, X.: Leveraging context-free grammar for efficient inverted index compression. In: Proceedings of the SIGIR, pp. 275–284 (2016)
12. Zukowski, M., Heman, S., Nes, N., Boncz, P.: Super-scalar RAM-CPU cache compression. In: Proceedings of the ICDE, pp. 59–70 (2006)

LIDH: An Efficient Filtering Method for Approximate k Nearest Neighbor Queries Based on Local Intrinsic Dimension

Yang Song$^{(\boxtimes)}$, Yu Gu, and Ge Yu

School of Computer Science and Engineering,
Northeastern University, Shenyang 110819, Liaoning, China
songyang1610558@stumail.neu.edu.cn, {guyu,yuge}@mail.neu.edu.cn

Abstract. Due to the so-called "curse of dimensionality" causing poor performance when querying in the high-dimensional space, the high-dimensional approximate kNN (AkNN) query has been extensively explored to trade accuracy for efficiency. In this paper, we propose a Local Intrinsic Dimension-based Hashing (LIDH) method for the high-dimensional AkNN query which locates a definite searching range by Local Intrinsic Dimensionality for filtering data points. Specifically, we propose a filter-refinement model for the AkNN query to avoid the virtual rehashing with fewer index space. Experimental evaluations demonstrate that our method can provide higher I/O and CPU efficiency while retaining satisfactory query accuracies.

Keywords: High dimension · Approximate kNN query
Local Intrinsic Dimension · Hashing

1 Introduction

With the advent of the big data age, the amount of information is constantly increasing also with the growing dimensionality of data. The substantially increasing dimensionality leads to the phenomenon called "curse of dimensionality" which makes the accurate kNN query solutions inefficient. Therefore, the Approximate k Nearest Neighbor (AkNN) query in the high dimensional space has become a hot research issue.

A representative AkNN solution is Locality Sensitive Hashing (LSH). In recent years, many methods as variants of LSH such as C2LSH [1], QALSH [2] are devised to solve the AkNN query problem in the high-dimensional space efficiently. Nevertheless, these methods require building more indexes and we need to conduct "virtual rehashing" to tune the widths of hash buckets. In terms of similarity search, ID is usually referred to the expansion dimension of Karger and Ruhl [3] and its variant named Generalized Expansion Dimension (GED) [4] which is considered as a representation of the extreme-value-theoretic Local Intrinsic Dimension (LID) [5] of distance distribution [6,7]. There are several

© Springer International Publishing AG, part of Springer Nature 2018
Y. Cai et al. (Eds.): APWeb-WAIM 2018, LNCS 10987, pp. 268–276, 2018.
https://doi.org/10.1007/978-3-319-96890-2_22

multi-step methods for similarity search [8–10] which rely on the termination of an expanding ring search based on distance functions for queries with existing index structures.

In this paper, we introduce a method named Local Intrinsic Dimension-based Hashing (LIDH) to solve the problem of AkNN. We map data points into hash values. In the filter process, we locate the search range in the one-dimensional mapping space by GED and mine the mapping relation between the mapping space and the original one. Benefiting from this, we can obtain a definite searching range to avoid the virtual rehashing. In the refinement process, we return the candidate points obtained by filtering to the original space and evaluate them for the AkNN results. The experimental evaluations indicate that our method can provide competitive query accuracies. Meanwhile, it can significantly reduce the I/O cost and the CPU time with fewer indexes. The major contributions of this paper include:

- We exploit the relation between LID and the searching range satisfying the query requirement to avoid the virtual rehashing in traditional LSH methods.
- We propose a novel filter-refinement model for the AkNN query based on GED as a representation of LID. The definite searching range in the mapping space served as the filter condition is presented and its correctness and feasibility are proved.
- Experimental evaluations on four data sets confirm that our method can provide competitive query accuracies and efficiency with fewer indexes.

2   Related Work

LSH is a traditional solution for the similarity search in the high-dimensional space. Many disk-based methods for the AkNN query are investigated. C2LSH [1] and its variant QALSH [2] provide both satisfactory results in accuracy and efficiency with probability guarantees. Especially, QALSH defines new hash functions and proposes the concept of the "anchor" to improve the efficiency. But the process of "virtual rehashing" incurs much time. Meanwhile, building too many hash tables results in the efficiency reduction.

LID and its representation GED have recently been adopted in similarity search. [8] proposes a heuristic multi-step method for the kNN search without probability guarantees. Likewise, [9,10] introduce multi-step methods for the aggregate similarity search and the subspace similarity search with guarantees, respectively. However, they focus on the termination condition of search without proposing specific index structure to accelerate the query process.

3   Preliminaries

3.1   Mapping by Hashing

In this paper, we denote R^d as the Euclidean space of representational dimension $d \in N$ ($d > 0$) and identify data points with their feature vectors in R^d. Let

$d(x, y)$ refer to the Euclidean distance between two data points x and y in R^d. Given a query point $q \in R^d$ and a query range r $(r > 0)$, we denote S $(S \subseteq R^d)$ as the range centered at q with radius r. Let $B_S(q, r) = \{x \in S : d(x, y) \leq r\}$ refer to the subset of points from S.

Definition 1. *In R^d, given a data point o whose associated feature vector is $\boldsymbol{o}$, and a random vector whose entries are drawn independently from the standard normal distribution is $\boldsymbol{a}$, the hash function is defined as $h_{\boldsymbol{a}}(o) = \boldsymbol{a} \cdot \boldsymbol{o}$.*

Given a query point q, we can obtain its hash value and the value is regarded as the "anchor". The searching range is specified as w. If a data point's hash value satisfies $|h_{\boldsymbol{a}}(q) - h_{\boldsymbol{a}}(o)| \leq w$, it is a target point.

3.2   Intrinsic Dimension

A measure of ID for similarity search is proposed in [3]. Given two spheres $B_S(q, r_1)$ and $B_S(q, r_2)$ $(0 < r_1 < r_2)$ centered at the query point q, we can obtain the volumes of these balls $V_i = \int_{B_S(q, r_i)} dx = \frac{\pi^{d/2}}{\Gamma(\frac{d}{2}+1)} r_i^d$, where $i = \{1, 2\}$. According to the equation above, $d = \frac{\log(V_2/V_1)}{\log(r_2/r_1)}$.

In fact, the generation of a data set can be regarded as a statistical process. LID can be interpreted in terms of the extreme value theory. The volume V is replaced by $F(r)$ which represents the probability measure of the ball centered at q with radius r. In this way, LID can be represented as below:

$$LID = \lim_{r \to 0^+} \lim_{\epsilon \to 0^+} \frac{\log(F((1 + \epsilon)r))/F(r)}{\log(1 + \epsilon)}$$

In order to represent the LID, Houle et al. propose the definition of GED [4,7]. For extremely small values of r_1 and r_2, we represent the LID by substituting V_1 and V_2 by the numbers of points k_1 and k_2 contained in their respective balls. The form of GED is defined as:

$$GED(B_S(q, r_1), B_S(q, r_2)) = \frac{\log(k_2/k_1)}{\log(r_2/r_1)}$$

4   LIDH Index Construction and Query Process

4.1   Computing Searching Range Based on LID

We consistently compute an effective query threshold to settle the AkNN query problem. When a query point is given, we utilize the aforementioned LID to mine the relation between the number of points and the radius in a certain space and compute the searching range satisfying the query requirement.

Theorem 1. *Given a data set S and a query point q, the user-specified number of the nearest neighbor points is k, and the MaxGED is M. Given the furthest neighbor point $o_b \in S$ as the benchmark point, the distance between o_b and q is*

r'. There are k' points within the range centered at q with the radius r' and we assume $k' \geq k$. It can be obtained that the kNN points of q are within the range centered at q with the radiusr $= r'(\frac{k'}{k})^{-\frac{1}{M}}$.

Proof. In the following, we extend LID in terms of extreme value theory to a certain space of neighborhood ball within the framework of the GED. For M is the upper bound of GED, we obtain:

$$M \geq GED(B_S(q, r'), B_S(q, r)) = \frac{\log(k'/k)}{\log(r'/r)}$$

Rearranging the inequality above, we obtain the searching range $r \leq r'(\frac{k'}{k})^{-\frac{1}{M}}$

4.2   Searching in Mapping Space

In Theorem 1, we consider the furthest neighbor point $o_b \in S$ as the benchmark point. In practice, we map the high-dimensional data points into the one-dimensional hash values and these hash values are indexed by B^+ trees. Based on this, we can locate the furthest neighbor point in the mapping space easily and compute the searching range in the mapping space to handle the query problem in the low-dimensional space according to Theorem 2 below. Figure 1 intuitively indicates the concept of LIDH.

Theorem 2. *Given a data set S and a query point $q \in S$, let α denote the GED in the original space with the maximum value α_{max} and β denote the GED in the mapping space with the minimum value β_{min}. Any kNN point of q in the original space is contained in the k' NN ($k' \geq k$) set of q in the mapping space where $k' = (\alpha_{max} - \beta_{min})k$.*

Proof. In the original space, assuming the distance between q and the kth NN point is r, and the distance between q and the nth NN point is R, we can obtain:

$$GED(B_S(q, R), B_S(q, r)) = \frac{\log(n/k)}{\log(R/r)} = \alpha$$

According to the equation above, we have $\frac{n}{k} = (\frac{R}{r})^{\alpha}$. Let H denote the mapping space, w denote the distance between q and the k'^{th} NN point, and W denote the distance between q and the nth NN point. Similarly, we obtain:

$$GED(B_H(q, W), B_H(q, w)) = \frac{\log(n/k')}{\log(W/w)} = \beta$$

and $\frac{n}{k'} = (\frac{W}{w})^{\beta}$. According to above, we have $\frac{R}{W} = \frac{r}{w}$. Therefore, it implies $\frac{k'}{k} = \alpha - \beta$. Because $\alpha - \beta \leq \alpha_{max} - \beta_{min}$, the maximum value of k' is $(\alpha_{max} - \beta_{min})k$. Therefore, we have to search k' NN points in the mapping space at most for kNN results in the original space.

We denote the maximum value of GED in the mapping space as β_{max}, and choose the furthest neighbor point of q as the benchmark in the mapping space whose distance to q is W. On the basis of Theorem 1, the searching range in the mapping space satisfies:

$$w \leq W(\frac{n}{k'})^{-\frac{1}{\beta_{max}}} = W(\frac{n}{(\alpha_{max} - \beta_{min})k})^{-\frac{1}{\beta_{max}}}$$

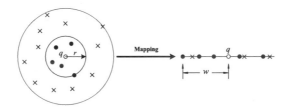

Fig. 1. Relation between original space and mapping space

4.3   LID and Parameter

LID Estimation. As mentioned above, we consider GED as the estimation of LID and utilize the Maximum Likelihood Estimation (MLE) [7] for estimating LID. According to the MLE estimator described in [7].

Parameter Setting. The accuracy of the kNN result is affected by several factors containing the given confidence when estimating the LID $1 - \theta$, the given probability guarantee p, the number of LIDH index m and the mark threshold t. θ affects the query range. The probability that a data point is marked larger than t is represented:

$$Pr(Mark \geq t) = \sum_{i=t}^{m} C_m^i p^i (1-p)^{m-i}$$

We expect that $Pr(Mark \geq t) \geq 1 - \theta$. As mentioned in [1,2], based on Hoeffding's Inequality [11], we obtain:

$$m \geq \lceil \frac{1}{2(p - t/m)^2} \ln \frac{1}{1-p} \rceil$$

We can figure out the relation between m and t:

$$t \geq \lfloor pm + \frac{\sqrt{2 \ln \frac{1}{1-p}}}{4} \rfloor$$

In addition, it's necessary to satisfy $m \geq t$ and it implies that:

$$m \geq \frac{\sqrt{2 \ln \frac{1}{1-p}}}{4(1-p)}$$

4.4 A*k*NN Query

The query process is based on a filter-refinement model. Given a query point q, we can compute the searching range w in the mapping space served as the filter condition by the LIDH index. We search the range centered at q with width w and each point appearing in the searching range will be marked once. We collect the mark number denoted as *Mark*. When a points *Mark* is larger than the pre-specified mark threshold t, the point is regarded as a candidate point and inserted into the candidate set. When all the candidate points are located, we evaluate the first k points according to their distances to q in the original space and obtain the kNN results.

5 Experiments

5.1 Experiment Setup

We choose Collision Counting LSH (C2LSH) [1] and Query-Aware LSII (QALSH) [2] as the benchmark methods and select four real-life data sets called Audio[1], Mnist[2], Forest[3] and Sift[4] for the experiment. For all four data sets, we randomly select 50 objects to form the query sets.

5.2 Parameter Setting

In our method, we randomly select 50 samples for estimating the LID. The required probability guarantee is 0.9 leading to $\theta - 0.1$. The default value of probability guarantee is 0.9. Based on the value of p, we can obtain the minimum value of the indexes $m = 6$ and infer the mark threshold $t = 5$. For two small data sets Audio and Mnist, we set $m = 6$ and $t = 5$. For the medium data set Forest and the large data set Sift, we set $m = 11$, $t = 10$ and $m = 16$, $t = 14$, respectively.

5.3 Index Size and Indexing Time

We list the index sizes of all the three testing method over four data sets in Table 1, where m represents the number of LIDH indexes.

The indexing construction time of our method is the smallest among three methods. For the data set Sift as an example, C2LSH takes about 5 min, and QALSH takes about 3 min, while our method takes about 30 s.

---

[1] http://www.cs.princeton.edu/cass/audio.tar.gz.

[2] http://yann.lecun.com/exdb/mnist/.

[3] http://archive.ics.uci.edu/ml/databases/Covertype.

[4] http://corpus-texmex.irisa.fr/.

Table 1. Index size

	Audio	Mnist	Forest	Sift
m	114	117	133	153
C2LSH	25.1 MB	28.4 MB	312.6 MB	619.7 MB
m	64	66	75	86
QALSH	14.1 MB	16.1 MB	176.3 MB	348.3 MB
m	6	6	11	16
LIDH	1.32 MB	1.46 MB	25.9 MB	64.8 MB

5.4   Overall Ratio

Overall ratio $[1,2]$ is defined as $\frac{1}{k}\sum_{i=1}^{k}\frac{d(o_i,q)}{d(o_i^*,q)}$, where o_i is the i-th point returned and o_i^* is the accurate i-th NN point, $i = 1, 2, \ldots k$. The results are shown in Fig. 2. All the three methods demonstrate comparable performances. QALSH and C2LSH perform slightly better than our method. This is because the other two methods build more hash tables to check more data points.

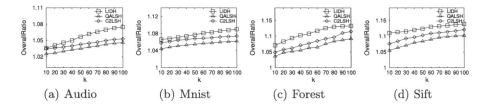

(a) Audio   (b) Mnist   (c) Forest   (d) Sift

Fig. 2. Overall ratio

5.5   I/O Cost

I/O cost is defined as the number of accessed disk pages whose page size is set to 4 KB. The results are plotted in Fig. 3. Compared to QALSH and C2LSH, our method requires smaller I/O costs. This is because the definite searching range can filter out more useless data points which contributes to reading fewer data points from the disk during the refinement process.

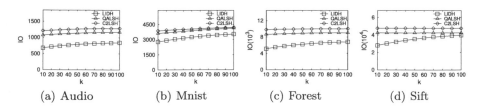

(a) Audio   (b) Mnist   (c) Forest   (d) Sift

Fig. 3. I/O cost

5.6   Running Time

In Fig. 4, we further report the test results of running time evaluation. In all testing cases, our method spends less time than QALSH and C2LSH because of building fewer indexes which give rise to less time during searching and marking the data points. Besides, the definite searching range in our method avoids the virtual rehashing which incurs much running time when constantly expanding the searching range.

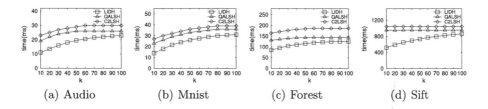

(a) Audio   (b) Mnist   (c) Forest   (d) Sift

Fig. 4. Running time

6   Conclusions

In this paper, we propose a LID-based Hashing method for AkNN query in the high-dimensional Euclidean space. We map high-dimensional data points into one-dimensional hash values and estimate the LID by Maximum Likelihood Estimation to compute the searching range in the mapping space as the filter condition avoiding the virtual rehashing. Our method alleviates the phenomenon of "curse of dimensionality" while providing competitive query quality and high query efficiency.

Acknowledgements. This work is supported by the National Natural Science Foundation of China (61472071 and 61433008), the Fundamental Research Funds for the Central Universities (N171605001) and the Natural Science Foundation of Liaoning Province (2015020018).

References

1. Gan, J., Feng, J., Fang, Q., Ng, W.: Locality-sensitive hashing scheme based on dynamic collision counting. In: Proceedings of the ACM SIGMOD International Conference on Management of Data, SIGMOD 2012, Scottsdale, AZ, USA, 20–24 May 2012, pp. 541–552 (2012)
2. Huang, Q., Feng, J., Zhang, Y., Fang, Q., Ng, W.: Query-aware locality-sensitive hashing for approximate nearest neighbor search. PVLDB **9**(1), 1–12 (2015)
3. Karger, D.R., Ruhl, M.: Finding nearest neighbors in growth-restricted metrics. In: Proceedings on 34th Annual ACM Symposium on Theory of Computing, 19–21 May 2002, Montréal, Québec, Canada, pp. 741–750 (2002)

4. Houle, M.E., Kashima, H., Nett, M.: Generalized expansion dimension. In: 12th IEEE International Conference on Data Mining Workshops, ICDM Workshops, Brussels, Belgium, 10 December 2012, pp. 587–594 (2012)
5. Casanova, G., Englmeier, E., Houle, M.E., Kröger, P., Nett, M., Schubert, E., Zimek, A.: Dimensional testing for reverse k-nearest neighbor search. PVLDB 10(7), 769–780 (2017)
6. Houle, M.E.: Dimensionality, discriminability, density and distance distributions. In: 13th IEEE International Conference on Data Mining Workshops, ICDM Workshops, TX, USA, 7–10 December 2013, pp. 468–473 (2013)
7. Amsaleg, L., Chelly, O., Furon, T., Girard, S., Houle, M.E., Kawarabayashi, K., Nett, M.: Estimating local intrinsic dimensionality. In: Proceedings of the 21th ACM SIGKDD International Conference on Knowledge Discovery and Data Mining, Sydney, NSW, Australia, 10–13 August 2015, pp. 29–38 (2015)
8. Houle, M.E., Ma, X., Nett, M., Oria, V.: Dimensional testing for multi-step similarity search. In: 12th IEEE International Conference on Data Mining, ICDM 2012, Brussels, Belgium, 10–13 December 2012, pp. 299–308 (2012)
9. Houle, M.E., Ma, X., Oria, V.: Effective and efficient algorithms for flexible aggregate similarity search in high dimensional spaces. IEEE Trans. Knowl. Data Eng. 27(12), 3258–3273 (2015)
10. Houle, M.E., Ma, X., Oria, V., Sun, J.: Efficient similarity search within user-specified projective subspaces. Inf. Syst. 59, 2–14 (2016)
11. Hoeffding, W.: Probability inequalities for sums of bounded random variables. Publ. Am. Stat. Assoc. 58(301), 13–30 (1963)

Query Performance Prediction and Classification for Information Search Systems

Zhongmin Zhang, Jiawei Chen, and Shengli Wu[✉]

Jiangsu University, Zhenjiang 212013, China
swu@ujs.edu.cn

Abstract. Automatic performance prediction and classification for information search results is useful in different scenarios. In this paper, we propose two score-based post-retrieval performance prediction methods. Both of them take magnitude and variance of resultant document scores into consideration at the same time. We also try to classify queries into three different classes: easy, medium, and hard by using a support vector machine-based approach. The experimental results show that the proposed predictors in this paper are very competitive compared with other predictors in the same category, and the support vector machine-based approach is effective for query classification.

Keywords: Information retrieval · Query performance prediction
Support vector machine · Query classification

1 Introduction

Query Performance Prediction (QPP), also known as query difficulty prediction, has been identified as an important issue for information search [1–3, 14, 19]. For example, performance of any Web search engine varies significantly from one query to next due to the ad hoc nature of the queries submitted. It would be very good if the search engine can detect those poor query results automatically at low cost before presenting them to the user. Then some other techniques such as query reformation can be used to try to improve the results. From the search system administrator's point of view, it might be worthy of adding some more relevant documents into the collection for those poorly performing queries. In a federated search environment, if the search agent can detect which results are good and which results are poor, then such information is very useful for the agent to carry out some tasks including resource selection and results merging more effectively.

A lot of methods have been proposed for query performance prediction [4–10, 20]. Query performance predictors can be categorized into two categories: pre-retrieval predictors [3, 4] and post-retrieval predictors [5, 6]. As their name indicates, pre-retrieval predictors can be done before the retrieval stage, because this type of predictors relies on some query and collection-related factors but not results-related features. Post-retrieval predictors are carried out after the search results are generated. In this type of predictors all features including results-related ones can be used. In the following we only review some related work on post-retrieval predictors, because the methods presented in this paper fall into this category.

© Springer International Publishing AG, part of Springer Nature 2018
Y. Cai et al. (Eds.): APWeb-WAIM 2018, LNCS 10987, pp. 277–285, 2018.
https://doi.org/10.1007/978-3-319-96890-2_23

Cronen-Townsend et al. [5] proposed a method that computes the relative entropy between the models of the query and the collection. A few more clarity-based predictors were proposed by other researchers [15, 16].

Zhou et al. [6] proposed a ranking robustness framework to predict query performance. Robustness based approaches evaluate how robust the results are to perturbations in query, result list and retrieval model. Related research by others can be found in [1, 17].

One branch of the post-retrieval prediction methods is score distribution. Zhou et al. [8] proposed a predictor, Weighted Information Gain (WIG), which measures the divergence between the mean retrieval score of some top-ranked documents and that of a typical document in the entire collection. Shtok et al. [9] proposed another predictor, Normalized Query Commitment (NQC), to estimate query drift in the list of top-ranked and/or bottom-ranked documents. Both [7, 18] focus on standard deviation of scores and some of its variants. Tao and Wu [10] proposed a method that considers both score magnitude and variance for performance prediction.

In this paper, we focus on score distribution for performance prediction and query classification. However, if required, other features can also be used in our approach.

2 Methodology

In this section we discuss related methodology. First we review some score distribution-based methods and some extensions to them, and then discuss how to use support vector machine to classify queries based on their difficulty level.

2.1 Score Distribution-Based Methods

Let q, C, and s denote a query, a collection of documents, and an information search system, respectively. We use $L(q, s)$ and D_q^k to denote the result list returned in response to query q by s over C and the top-k documents in the resultant list $L(q, s)$, respectively. k is a free parameter, set to an arbitrary natural number prior to the search. The goal of performance prediction is to establish a predictor for evaluating the quality of the ranking list returned by s over C for a given query q without relevance judgment information.

SD2 [7] uses Eq. 1

$$\text{SD2}(q, s) = \frac{1}{\sqrt{|q|}} \sqrt{\frac{1}{k-1} \sum_{d \in D_q^k} (score(d) - \hat{\mu})^2} \tag{1}$$

to calculate scores for a given query q. Here $score(d)$ is the score that document d is awarded by s. $\hat{\mu} = \frac{1}{k} \sum_{d \in D_q^k} score(d)$, which is the average score of all k results in D_q^k. $|q|$ is the number of query terms and $\sqrt{|q|}$ serves as a scale factor to make SD2 values of different queries comparable.

WIG [8] uses Eq. 2

$$\text{WIG}(q,s) = \frac{1}{k}\sum_{d\in D_q^k}\frac{1}{\sqrt{|q|}}(\text{score}(d) - \text{score}(C)) \tag{2}$$

to calculate scores for a given query q. Here score(C) is the score that an average document in C would be given by s. $\sqrt{|q|}$ has the same effect as in SD2.

NQC [9] uses Eq. 3

$$\text{NQC}(q,s) = \frac{1}{\text{score}(C)}\sqrt{\frac{1}{k}\sum_{d\in D_q^k}(\text{score}(d) - \hat{\mu})^2} \tag{3}$$

to calculate scores for a given query q. NQC can be regarded as a variation of standard deviation, which is investigated by Perez-Iglesias and Araujo [18] and Cummins et al. [7].

SMV [10] uses Eq. 4

$$\text{SMV}(q,s) = \frac{\frac{1}{k}\sum_{d\in D_q^k}\left(\text{score}(d) \times \left|\ln\frac{\text{score}(d)}{\hat{\mu}}\right|\right)}{\text{score}(C)} \tag{4}$$

to calculate scores for a given query q. SMV considers both score magnitude and variance at the same time. There are two components inside the summation of Eq. 4: score(d) and $\left|\ln\frac{\text{score}(d)}{\hat{\mu}}\right|$. They are used to represent score magnitude and variance, respectively. The two components are combined by multiplication.

Apart from the above-mentioned methods, we present two other methods that also consider both score magnitude and variance as SMV does. They are referred to as C_1 and C_2.

C_1 uses Eq. 5

$$C_1(q,s) = (1 - \lambda) \times \text{SD2}(q,s) + \lambda \times \text{WIG}(q,s) \tag{5}$$

to calculate scores. It is a linear combination of SD2 and WIG scores. Note that SD2 concerns the variance of scores, but WIG concerns the magnitude of scores.

C_2 uses Eq. 6

$$C_2(q,s) = \frac{\frac{1}{k}\sum_{d\in D_q^k}(\text{score}(d) \times (\text{score}(d) - \hat{\mu}))^2}{\text{score}(C)} \tag{6}$$

to calculate scores. C_2 is quite similar to SMV. The difference is: instead of using $\left|\ln\frac{\text{score}(d)}{\hat{\mu}}\right|$, we use the Brier score[1] $(\text{score}(d) - \hat{\mu})^2$ to represent score variance.

---

[1] https://en.wikipedia.org/wiki/Brier_score.

2.2   Support Vector Machine for Classifying Queries

Support Vector Machine (SVM) was proposed by Cortes and Vapnik in 1995 [12]. It is a type of supervised learning for classification of two categories in machine learning. One major property of SVM is it rather than considering all the points in each category, it considers only those points that are very close to the border referred to as the support vectors.

When used for query classification, we may take some features such as calculated scores from WIG and NQC. Different categories can be defined such as difficult & not-difficult queries, or easy & not-easy queries. The intersection of not-easy and not-difficult queries comprises medium queries.

3   Experiment Settings and Results Analysis

Experiments are carried out to evaluate the effectiveness of the proposed methods. We choose a group of runs submitted to the TREC 2004 Robust Track [11]. The major reason of doing so is that this task includes 250 queries, which are the most in all historical tasks of TREC so far. Using a data set with a large set of queries will make the results of the experiments more reliable. Because one query (number 672) does not have any relevant documents, it is removed in the experiment.

14 research groups participated in the Robust Track in TREC 2004. We choose the best performing one from each research group apart from University of Illinois at Chicago. The document scores provided by that group are not reasonable. Information of the selected runs is shown in Table 1.

Table 1. Runs selected from 13 research groups

Run selected	Submitted by	MAP
input.apl04rsTDNfw	Johns Hopkins University	0.3172
input.fub04TDNge	Fondazione Ugo Bordoni	0.3405
input.humR04t5e1	Hummingbird	0.2768
input.icl04pos2f	Peking University	0.2160
input.JuruTitDes	IBM Research, Haifa	0.2803
input.mpi04r07	Max-Planck-Institute for Computer Science	0.1755
input.NLPR04NcA	Chinese Academy of Sciences(CAS-NLPR)	0.2833
input.pircRB04td2	Queens College, CUNY	0.3586
input.polyudp5	Hong Kong Polytechnic University	0.2455
input.SABIR04BA	Sabir Research, Inc.	0.2944
input.uogRobLWR10	University of Glasgow	0.3201
input.vtumlong436	Virginia Tech	0.3280
input.wdoqla1	Indiana University	0.2914

As these runs submitted are different in many aspects such as retrieval models, representation of documents and queries, and so on, the scores of the retrieved documents from these runs are various on magnitude and distribution. Some runs include positive scores only, but some others include negative scores. However, predictors SD2, C_1, C_2, and NQC require that all the scores are positive. Therefore, score normalization is required to let these predictors work properly. Two measures are taken. First, if the minimum document score (score_min) in a run is less than 0, then all the scores in the run are converted to zero and positive scores (by new_score = score-score_min). Second, if the maximum document score (score_max) in the run is greater than 1, then all the scores in the run are updated to (score = score/score_max). Thus the scores in all the runs are comparable because they are inside the interval [0, 1].

3.1 Performance Prediction

In this experiment we are going to evaluate the effectiveness of the proposed predictors C_1 and C_2, along with other predictors SD2 [7], WIG [8], NQC [9], and SMV [10]. All these predictors need to set some parameters. $\sqrt{|q|}$ is used in SD2, WIG, and C_1. Because it is not clear how the query contents are for generating these runs, $\sqrt{|q|}$ is set to 1 in all the cases. k is set to 5 for WIG, which is the same as in [14]; k is set to 100 for SD2, SMV, NQC, and C_2. λ is set to 0.5 for C_1. $score(C)$ is defined as the average score appears in the resultant list of 1000 documents after normalization treatment.

For any run, the performances of all the queries are estimated by a performance predictor, and their performances are also evaluated by relevance judgment that is provided by TREC, thus we obtain two ranked lists of queries that are performance-based. We compare these two lists to see how similar they are by calculating their correlation coefficients. Table 2 shows the Spearman correlation coefficients of them.

Table 2. Spearman coefficients for several predictors on 13 runs (the highest value for each run is shown in boldface)

RunId	Baseline QPPs				Proposed QPPs	
	SD2	NQC	WIG	SMV	C_1	C_2
apl04rsTDNfw	0.408	0.403	0.406	0.405	0.425	**0.434**
fub04TDNge	0.523	0.545	0.544	0.556	0.543	**0.558**
humR04t5e1	0.426	0.512	0.564	0.553	**0.567**	0.542
icl04pos2f	0.598	0.628	0.579	0.613	0.589	**0.643**
JuruTitDes	0.541	0.463	0.544	0.461	0.550	**0.560**
mpi04r07	0.361	**0.425**	0.420	0.411	**0.425**	0.411
NLPR04NcA	0.513	0.630	**0.639**	0.603	0.626	0.630
pircRB04td2	0.573	0.493	0.596	0.486	0.599	**0.608**
polyudp5	0.254	0.465	0.257	0.473	0.312	**0.478**
SABIR04BA	0.511	0.572	0.538	0.552	0.543	**0.610**
uogRobLWR10	0.471	0.592	0.450	**0.594**	0.456	0.521
vtumlong436	0.445	0.612	0.566	**0.613**	0.554	0.591
wdoqla1	0.544	**0.621**	0.540	0.609	0.544	0.604

Table 3. Recall on the hard category (the highest value for each run is in bold)

RunId	SVM	SD2	NQC	WIG	SMV	C_1	C_2
apl04rsTDNfw	**0.682**	0.145	0.000	0.158	0.000	0.227	0.000
fub04TDNge	**0.672**	0.058	0.000	0.128	0.000	0.128	0.000
humR04t5e1	**0.632**	0.342	0.059	0.450	0.110	0.460	0.059
icl04pos2f	**0.848**	0.722	0.745	0.718	0.754	0.718	0.801
JuruTitDes	**0.709**	0.450	0.000	0.505	0.000	0.506	0.113
mpi04r07	**0.976**	0.873	0.894	0.812	0.889	0.798	0.917
NLPR04NcA	**0.770**	0.317	0.508	0.556	0.480	0.546	0.508
pircRB04td2	**0.729**	0.106	0.039	0.252	0.013	0.252	0.013
polyudp5	**0.858**	0.045	0.524	0.202	0.501	0.274	0.544
SABIR04BA	**0.633**	0.368	0.365	0.396	0.316	0.418	0.466
uogRobLWR10	**0.768**	0.000	0.058	0.000	0.048	0.000	0.000
vtumlong436	**0.729**	0.000	0.027	0.208	0.029	0.135	0.000
wdoqla1	**0.788**	0.295	0.331	0.312	0.274	0.310	0.022
Average	**0.753**	0.286	0.273	0.361	0.263	0.367	0.265

Table 4. Precision on the hard category (the highest value for each run is in bold)

RunId	SVM	SD2	NQC	WIG	SMV	C_1	C_2
apl04rsTDNfw	0.433	0.576	0.000	0.503	0.000	**0.591**	0.000
fub04TDNge	0.509	0.533	0.000	**0.687**	0.000	0.620	0.000
humR04t5e1	0.657	0.714	0.600	0.779	**0.800**	0.769	**0.800**
icl04pos2f	0.701	0.787	0.761	0.779	0.751	0.789	**0.792**
JuruTitDes	0.648	0.781	0.000	0.799	0.000	0.778	**0.800**
mpi04r07	0.700	0.739	0.748	0.748	0.736	**0.749**	0.727
NLPR04NcA	0.655	0.753	0.769	0.769	0.763	**0.783**	0.769
pircRB04td2	0.557	0.427	0.233	**0.833**	0.200	**0.833**	0.200
polyudp5	0.612	0.125	**0.701**	0.550	0.695	0.634	0.665
SABIR04BA	0.589	0.641	**0.743**	0.661	0.735	0.688	0.729
uogRobLWR10	**0.549**	0.000	0.367	0.000	0.167	0.000	0.000
vtumlong436	0.513	0.000	0.167	0.648	0.133	**0.670**	0.000
wdoqla1	0.608	0.761	0.716	0.746	**0.858**	0.167	0.000
Average	0.595	0.526	0.447	**0.654**	0.449	0.621	0.422

Table 5. Average F-score of all seven methods (the highest value for each category is in bold)

Category	SVM	SD2	NQC	WIG	SMV	C_1	C_2
Hard	**0.665**	0.371	0.339	0.465	0.332	0.461	0.326
Easy	**0.476**	0.326	0.375	0.369	0.368	0.373	0.370

In Table 2, we can see that all the pairs are positively correlated and the strength of the correlation is moderate. If we compare those different predictors, then C_2 is the best performer on 7 out of 13 runs. SMV, NQC, and C_1 are the best on 2 out of 13 runs. WIG is the best on 1 run while SD2 does not achieve the best on any of the runs.

3.2   Classification

In some applications such as Web search, it is useful if the search engine can estimate automatically how difficult a query is. In this experiment we address this issue. We divide all the queries into three categories based on its difficulty level: Easy, Medium, and Hard. And 0.2 and 0.4 are set as thresholds, which mean that a query is classified as medium if its performance in average precision is between 0.2 and 0.4, a query is hard if its average precision is below 0.2, and a query is easy if its average precision is above 0.4. When using such thresholds in this data set, there is roughly equal number of queries in each category.

The SVM program for prediction is downloaded from the Libsvm-3.21 [13] software package[2]. Two features are used for SVM to work: the scores obtained from WIG and the scores obtained from NQC. They are complementary because WIG mainly considers magnitude of scores while NQC considers variance of scores of resultant documents. All other 6 predictors SD2, WIG, SMV, NQC, C_1, and C_2 are also involved. For a given method such as SD2, we estimate its average precision by using the linear regression with its SD2 score (y = a + b * SD2-score), and the query is classified to one of the categories based on the estimated average precision values. The five-fold cross validation is used. As for evaluation, we use three measures including precision, recall, and F-score. The results are shown in Tables 3 and 4 for category Hard, which is the most important among three categories.

From Table 3 we can see that, SVM is the best when recall is considered. It is much better than the others and all the others are quite close. From Table 4 we can see that WIG performs better than the others when precision is considered. However, the difference between it and the others is not big and all of them are quite close.

To have a balanced view on these methods, we use F_1 to measure all these methods involved. The results are shown in Table 5. From Table 5, we can see that SVM is the best in both categories Hard and Easy. Another noticeable thing is that SVM does better in category Hard than in Easy; but the performances of all the others are close in both categories.

4   Conclusion

In this paper, we have investigated the problem of performance prediction for information search systems. Two proposed score-based methods, especially C_2, are very competitive compared with other state-of-the-art methods in the same category. We have also tried to classify queries into different categories based on their performance.

---

[2] http://www.csic.ntu.edu.tw/~cjlin/libsvm/.

Experiments show that the support vector machine-based method works better than any individual performance predictors significantly when both precision and recall are considered at the same time.

In terms of future work, we shall focus on two aspects: firstly, for the SVM-based method, there are a lot of options for features to be taken. Apart from those different performance predictors, we may choose other types of features such as collection-related and query-related features; secondly, other types of classification methods such as naive Bayes, random forests may also be investigated for the query classification task.

References

1. Yom-Tov, E., Fine, S., Carmel, D., Darlow, A.: Learning to estimate query difficulty: including applications to missing content detection and distributed information retrieval. In: 28th SIGIR, pp. 512–519 (2005)
2. Lang, H., Wang, B., Li, J., et al.: Predicting query performance for text retrieval. J. Softw. **19** (2), 291–300 (2008)
3. Hauff, C., Hiemstra, D., de Jong, F.: A survey of pre-retrieval query performance predictors. In: 17th CIKM, pp. 1419–1420 (2008)
4. Katz, G., Shtok, A., Kurland, O., Shapira, B., Rokach, L.: Wikipedia-based query performance prediction. In: 37th SIGIR, pp. 1235–1238 (2014)
5. Cronen-Townsend, S., Zhou, Y., Croft, W.B.: Predicting query performance. In: 25th SIGIR, pp. 299–306 (2002)
6. Zhou, Y., Croft, W.B.: Ranking robustness: a novel framework to predict query performance. In: 15th CIKM, pp. 567–574 (2006)
7. Cummins, R., Jose, J.M., O'Riordan, C.: Improved query performance prediction using standard deviation. In: 34th SIGIR, pp. 24–28 (2011)
8. Zhou, Y., Croft, W.B.: Query performance prediction in web search environments. In: 30th SIGIR, pp. 543–550 (2007)
9. Shtok, A., Kurland, O., Carmel, D., Raiber, F., Markovits, G.: Predicting query performance by query-drift estimation. ACM Trans. Inf. Syst. **30**(2), 305–312 (2009)
10. Tao, Y., Wu, S.: Query performance prediction by considering score magnitude and variance together. In: 23th CIKM, pp. 1891–1894 (2014)
11. Voorhees, E.M.: Overview of the TREC 2004 robust track. In: 23rd TREC, 13 (2004)
12. Cortes, C., Vapnik, V.: Support-vector networks. Mach. Learn. **20**(3), 273–297 (1995)
13. Chang, C.-C., Lin, C.-J.: LIBSVM: a library for support vector machines. ACM Trans. Intell. Syst. Technol. **2**(27), 1–27 (2011)
14. Zhou, Y.: Retrieval performance prediction and document quality. University of Massachusetts, Massachusetts (2007)
15. Diaz, F., Jones, R.: Using temporal profiles of queries for precision prediction. In: 27th SIGIR, pp. 18–24 (2004)
16. He, B., Ounis, I.,: Inferring query performance using pre-retrieval predictors. In: 11th SPIRE, pp. 43–54 (2004)
17. Vinay, V., Cox, I.J., Milic-Frayling, N., Wood, K.R.: On ranking the effectiveness of searches. In: 29th SIGIR, pp. 398–404 (2006)

18. Perez-Iglesias, J., Araujo, L.: Standard deviation as a query hardness estimator. In: 17th SPIRE, pp. 207–212 (2010)
19. Mizzaro, S., Mothe, J.: Why do you think this query is difficult? A user study on human query prediction. In: 39th SIGIR, pp. 1073–1076 (2016)
20. Shtok, A., Kurland, O., Carmel, D.: Query performance prediction using reference lists. ACM Trans. Inf. Syst. 34(4), 19–52 (2016)

Aggregate Query Processing
on Incomplete Data

Anzhen Zhang$^{(\boxtimes)}$, Jinbao Wang, Jianzhong Li, and Hong Gao

Department of Computer Science and Technology,
Harbin Institute of Technology, Harbin, China
{azzhang,wangjinbao,lijzh,honggao}@hit.edu.cn

Abstract. Incomplete data has been a longstanding issue in database community, and yet the subject is poorly handled by both theory and practice. In this paper, we propose to directly estimate the aggregate query result on incomplete data, rather than imputing the missing values. An interval estimation, composed of the upper and lower bound of aggregate query results among all possible interpretation of missing values, are presented to the end-users. The ground-truth aggregate result is guaranteed to be among the interval. Experimental results are consistent with the theoretical results, and suggest that the estimation is invaluable to better assess the results of aggregate queries on incomplete data.

Keywords: Aggregate query · Incomplete data · Estimation

1 Introduction

Incomplete data (missing value) has been a longstanding issue in database community. The scale of the problem is such that it is common to find critical information missing from databases. Missing values make it difficult for data scientists, both in commercial enterprises and in academia, to perform reliable data analysis. One common way to cope with missing values in practice, is to complete their imputation (filling in) as a preprocessing step before analysis. A number of different imputation methods have been reported in the literature [1–7]. The development of the new methods was mainly driven by a need to improve accuracy of the imputation. However, not a single imputation method could impute all missing values correctly in all cases. Besides, the query result on such 'complete' data without any confidence guarantee could hardly be trusted by data analysts.

As a motivating example, consider an incomplete movie reviews in Fig. 1(a), where values in parentheses are considered missing. A natural approach for inferring missing values in relational databases is to capture attribute dependencies with graphical models, e.g., Baysesian network. In a Bayesian network, each random variable is associated with a probability distribution conditioned on the other variables. In this example, there are two *conditional probability*

id	user	movie	rating	time
t_1	Mike	Room	(5)	2015
t_2	John	(Your Name)	3	2016
t_3	Mary	Your Name	3	2016
t_4	Emily	Room	1	2015
t_5	Lisa	(Trolls)	3	2016
t_6	James	Your Name	3	2016
t_7	Mark	Your Name	(5)	2016
t_8	Kate	Trolls	(1)	2016

id	user	movie	rating	time
t_1	Mike	Room	**1**	2015
t_2	John	**Your Name**	3	2016
t_3	Mary	Your Name	3	2016
t_4	Emily	Room	1	2015
t_5	Lisa	**Your Name**	3	2016
t_6	James	Your Name	3	2016
t_7	Mark	Your Name	**3**	2016
t_8	Kate	Trolls	**3**	2016

(a) Incomplete *reviews*   (b) Bayesian imputation

Fig. 1. An example of movie reviews

distribution (CPD), $P(movie|time)$ and $P(rating|movie)$. After imputation, a 'complete' reviews table is shown in Fig. 1(b). Note that $t_5[movie]$ is filled with *YourName*, since it appears most frequently when $time = 2016$. Similarly, the missing *rating* values of t_1, t_7 and t_8 are filled with the corresponding most frequent values given their *movie* value, i.e., $t_1[rating] = 1$ and $t_7[rating] = 3$, $t_8[rating] = 3$. Comparing with Fig. 1(a), we find out that 4 out of 5 values in Bayesian imputation are totally wrong.

Now we move on to show how such imputation errors significantly influence aggregate queries. Consider an aggregate query AQ_1: SELECT COUNT(*rating*) FROM *reviews* WHERE *movie = Trolls*, which aims to find the number of ratings for *Trolls*. The query result from Fig. 1(b) is 1, while the ground-truth is 2 in Fig. 1(a). Let's consider another aggregate query AQ_2: SELECT AVG(*rating*) FROM *reviews* WHERE *movie = YourName*, which computes the averating rating for *YourName*. The average rating of *YourName* in Fig. 1(b) is 3, while the ground-truth is 3.4 in Fig. 1(a). Note that, in addition to such tremendous query errors, we have no idea of how far such query results are from the ground-truth.

Motivated by this, we study the problem of estimating aggregate query results on incomplete data. This work is a first step towards developing techniques to estimate aggregate query results on incomplete data under *close-world* assumption (CWA). In particular, we provide an interval estimation of the query result, denoted by $[lb, ub]$, where lb and ub are the lower bound and upper bound of aggregate result over all possible realizations of the missing values, respectively. Therefore, the ground-truth result is definitely among the interval. The main challenge is that, we cannot explicitly enumerate all possible realizations of missing values since it could be exponentially large in the size of the missing values. We will show that, in a careful manner, aggregate query answers can be estimated efficiently during the process of aggregate query.

The remainder of this paper is organized as follows. We first present the problem definition in Sect. 2 and then introduce aggregate query evaluation in

Sect. 3. Next, we present techniques to estimate SUM queries in Sect. 4. In Sect. 5 we evaluate our techniques, followed by conclusion.

2 Problem Definition

Our primary focus is answering aggregate numerical queries of the form: SELECT AGG(attr) FROM table WHERE predicate. *Predicate attributes* are attributes involved in the predicate condition of the aggregate query, and *aggregate attributes* are attributes involved in the aggregate computation (AGG). AGG could be SUM, COUNT, AVG, etc. Note that group-by queries can be implemented by adding group-by keys into predicates. Given an incomplete table $R = \{D_1, D_2, \ldots, D_n\}$, and the aggregate query AQ. We are interest in the estimating the range of $AQ(R)$ over all possible interpretations of nulls under the Close World Assumption (CWA).

The *semantics* of an incomplete database R, denoted by $\|R\|$, which is the set of all complete database $R^{'}$ that R can represent [8]. Under *close-world* assumption [9], we believe that an incomplete database represents information fully, except some missing values. The key notion of CWA semantics is a *valuation* of missing values (denote $\perp$). In the remainder of this paper, we use null and $\perp$ for missing value interchangeably. A *valuation* of $\perp$ in D_i is a mapping $v : \perp \rightarrow D_i$. That is, each null is associated with a constant value from the same domain. It naturally extends to databases: $v(R)$ is simply the result of replacing each null $\perp$ in R by $v(\perp)$. Therefore, the CWA semantics is defined as follows:

$$\|R\|_{CWA} = \{R^{'} | R^{'} = v(R), v \text{ is a valuation}\} \tag{1}$$

Definition 1. *(Interval estimation) is an estimation of the aggregate query result, denoted by $[lb, ub]$. lb and lb are the lower and upper bound of the aggregate query result among all possible valuations of R under CWA.*

$$lb = min\{AQ(R^{'})|R^{'} \in \|R\|_{CWA}\}$$
$$ub = max\{AQ(R^{'})|R^{'} \in \|R\|_{CWA}\} \tag{2}$$

Note that $[lb, ub]$ covers all possible aggregate results of possible $R^{'}$ in $\|R\|_{CWA}$. Clearly, the ground-truth aggregate result is definitely among the interval, that is, $AQ(R^*) \in [lb, ub]$. Instead of enumerating all possible $R^{'}$, we develop techniques that can directly obtain lb and ub during the process of aggregate query evaluation.

3 Aggregate Query Evaluation

Traditional aggregate query evaluation on complete data works as follows. We first filter those tuples which satisfy the predicate and then aggregate on those tuples. When it comes to incomplete data, the evaluation of predicate and aggregation need some modifications, as shown in the following sections.

3.1 Predicate Evaluation

The predicate in AQ is denoted by P. The predicate corresponds to predicate calculus formulas with free variables, i.e. $P = F(p_1, p_2, \ldots, p_n)$. The atomic formulas or primitive terms in a query have the form: $p_i = (D \; op \; V)$, where D is the predicate attribute, op is one of the predicate symbols $=, \neq, <, \leq, >, \geq$ and V is a value.

Next, we introduce how to evaluate each primitive term in P. For a tuple $t \in R$ and $p_i = (D \; op \; V)$, if $t[D] \neq \perp$, then the result of $p_i(t)$ is identical to the result of the conventional query. However, if $t[D] = \perp$, things are different. In particular, $p_i(t)$ is evaluated to be $true$ if for all possible constants in D, $p_i(t) = true$, and it is evaluated to be $false$ if for all constants in D, $p_i(t) = false$. Otherwise, $p_i(t) = unknown$. Based on the result of each primitive term, we can evaluate the predicate P according to Codd's three-valued truth table [10]. Clearly, if $P(t) = false$, then t should be excluded from next-stage aggregate computation. Otherwise, t is included in either $true \; result$ or $maybe \; result$ of the $AQ(R)$.

Definition 2. *(true result) Given an aggregate query AQ and an incomplete table R, a tuple will be in the true result of AQ(R), if and only if the predicate of AQ is evaluated to be true for the tuple.*

Definition 3. *(maybe result) Given an aggregate query AQ and incomplete table R, a tuple is in the maybe result of AQ(R), if and only if the predicate is evaluated to be unknown for the tuple.*

3.2 Aggregate Evaluation

Now we evaluate the aggregate function (denote AGG) on the aggregate attribute for tuples in $true \; result$ and $maybe \; result$. We first introduce some notations that will be used in the following sections.

Definition 4. *(True aggregate values set) contains aggregate attribute values in true result, denoted by TA.*

Definition 5. *(Maybe aggregate values set) contains aggregate attribute values in maybe result, denoted by MA.*

Definition 6. *(Whole aggregate values set) includes all aggregate attribute values in in true result and maybe result, denoted by A, $A = TA \cup MA$.*

The missing part of TA is denoted by $TA_\perp$ and complete part is TR_c. That is, $TA_\perp$ contains all nulls in TA, and TA_c contains all constant values in TA. Similarly, we have MA_c, $MA_\perp$, A_c and $A_\perp$.

Based on these notations, we can transform the aggregate query $AQ(R)$ to the aggregate computation on the whole aggregate values set A; that is, $AQ(R) = AGG(A)$. The ground-truth $AQ(R^*) = AGG(A^*)$, where A^* is the truth aggregate value set, $A^* \subseteq A$. The interval estimation of $AQ(R)$, $[lb, ub]$, can be reformed as:

$$[lb, ub] = [min\{AGG(A)\}, max\{AGG(A)\}] \tag{3}$$

4   SUM Query

In this section, we provide estimation techniques for SUM query, and the estimations for COUNT and AVG query are omitted due to the page limits. The SUM of aggregate values in A is:

$$SUM(A) = SUM(TA_c) + SUM(TA_\perp) + SUM(MA) \tag{4}$$

In order to obtain interval estimation for SUM, we need to resolve the minimum and maximum value of $SUM(\perp)$ for $\perp$ in $A_\perp$. Throughout this paper, we assume the aggregate attribute of AQ is D_1 without loss of generality. Suppose the values range of D_1 is $[a, b]$, then we have the following theorem.

Theorem 1. *In SUM query, if $A_\perp \neq \emptyset$ and $D_1 \in [a, b]$, $lb = SUM(TA_c) + a \cdot |TA_\perp|$, and $ub = SUM(A_c) + b \cdot |A_\perp|$.*

Proof. We start by the proof of the upper bound. Note that $ub = SUM(TA_c) + max\{SUM(TA_\perp)\} + max\{SUM(MA)\}$. In order to obtain $max\{SUM(TA_\perp)\}$, $SUM(\perp)$ should take the maximum value of D_1, i.e., $SUM(\perp) = b$. Thus we have $max\{SUM(TA_\perp)\} = b \cdot |TA_\perp|$. Next, we need to resolve $max\{SUM(MA)\}$. Obviously, $SUM(MA)$ is maximum when all values in MA are involved in SUM computation (i.e., all of the tuples in *maybe result* satisfies the predicate), and all null values in $MA_\perp$ take the maximum value in D_1. That is, $max\{SUM(MA)\} = b \cdot |MA_\perp| + SUM(MA_c)$. Therefore, we can reduce ub to $SUM(A_c) + b \cdot |A_\perp|$. The lower bound can be obtained in a similar way.

Before we close this section, we give a brief discussion on the effectiveness of the interval estimation for SUM and COUNT. The width of the interval estimation of $AQ(R)$ is denoted by w, $w(AQ, R) = ub - lb$. We will show that, for SUM and COUNT queries, w is effected by the missing rate of predicate attributes, which quantifies the level of missingness in predicate attributes.

Definition 7. *(Missing Ratio of Predicates) ϵ is a ratio of the number of tuples whose predicate attributes are missing to that of all tuples, $\epsilon \in [0, 1]$.*

Theorem 2. *Given the missing ratio of predicates ϵ, and $D_1 \in [a, b]$, $w(COUNT, R) \leq \epsilon \cdot m$, and $w(SUM, R) \leq \epsilon \cdot b \cdot m$.*

Proof. There are totally $\epsilon \cdot m$ tuples whose predicate attributes are missing. For COUNT query, $ub - lb = |MA|$. In extreme condition, the predicate evaluation result for $\epsilon \cdot m$ tuples with predicate attributes missing all equals *unknown*, that is, these tuples are all involved in *maybe result*, $|MA| = \epsilon \cdot m$. For SUM query, $ub - lb = SUM(MA_c) + b \cdot |A_\perp| - a \cdot |TA_\perp|$. Since $SUM(MA_c) \leq b \cdot |MA_c|$, we can infer that $ub - lb \leq b \cdot |MA| - (b - a) \cdot |TA_\perp|$. Thus, we have $ub - lb \leq \epsilon \cdot b \cdot m$.

5   Experiments

We conducted a set of experiments on real and synthetic data to evaluate the following characteristics: (1) Comparing our framework with an representative missing data imputation approach, Bayesian approach [11]. (2) Evaluating the effect of missing ratio of predicates (i.e., ϵ) on the width of the estimation interval (i.e., $w(AQ, R)$) for SUM and COUNT queries (3) Evaluating the effect of θ on the width of the estimation interval for AVG query.

5.1   Experimental Setup

Datasets: We use both real-world and synthetic workloads to evaluate our method.

Reviews[1] is a real-world movie reviews from Amazon, with 7911684 reviews from 889176 users. We construct a table *reviews* with four attributes: *user, movie, rating* and *time*. Note that *rating* is from 1 to 5.

TPC-H[2] We generate a 1GB TPC-H benchmark dataset (6,001,199 records in *lineitem* table). The *lineitem* table schema simulates industrial purchase order records with 12 attributes, *quantity, returnflag, linestatus*, etc.

Let r denote the *whole missing ratio* of the dataset. We simulate two types of missingness, missing at random (MAR) and not missing at random (NMAR) [12]. For MAR, we randomly remove tuples in both *reviews* and *lineitem* tables. For NMAR, we randomly censor *rating* values larger than 4 for *movie*, and *quantity* values larger than 20 for *lineitem*. We conduct the following aggregate queries on *reviews* and *lineitem*, where f represents any of aggregations, e.g., SUM, COUNT and AVG.

   Q1: SELECT $f(rating)$ FROM reviews GROUP BY *movie*;
   Q2: SELECT $f(quantity)$ FROM *lineitem* WHERE *returnflag*="A".

5.2   Accuracy

To begin with, we show how the *whole missing ratio* r affects the aggregate query results. Let f equal AVG, and we ran Q1 on *reviews* with r varies from 0.1 to 0.5 in both MAR mode and NMAR mode. The aggregate results for one of the movies, m_1 are shown in Fig. 2(a) and (b). Next, let f equal COUNT, and we ran Q2 on *lineitem* with r varies from 0.1 to 0.5 in both MAR mode and NMAR mode. The results are shown in Fig. 2(a) and (b).

From the results, we can see that the interval width w increases with the growth of r for both MAR and NMAR modes. In addtion, the query result of the Bayesian approach deviates from the ground-truth result, and such deviation increases gradually with the increase of the whole missing ratio r. In particular, the query results of the Bayesian approach are larger than the ground-truth value in MAR mode, while they are smaller than the ground-truth in NMAR mode.

[1] Reviews: https://snap.stanford.edu/data/web-Amazon.html.
[2] TPC-H: http://www.tpc.org/tpch.

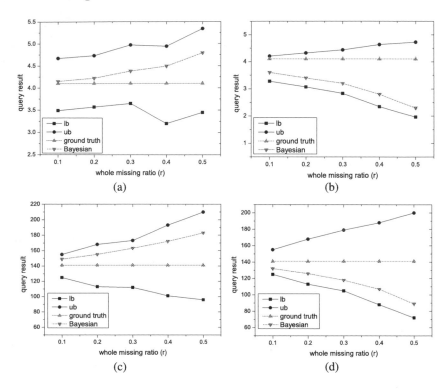

Fig. 2. Accuracy comparison. (a) Query result of Q1 in MAR mode. (b) Query result of Q1 in NMAR mode. (c) Query result of Q2 in MAR mode (d) Query result of Q2 in NMAR mode

We give detailed explanation for *reviews* table and the situation for *lineitem* table is similar. When we dive into the distribution of the ratings of m_1, we find that most ratings are larger than 4. It means that, the imputation of the missing values in Bayesian approach in MAR mode tends to fill in larger rating values, e.g., 4 or 5. However, in NMAR mode, we randomly censored values larger than 4, and thus the distribution of ratings changed. In this case, the rating value that occurs most frequently is 2, and thus the Bayesian approach tends to fill in the missing values with 2, which leads to the underestimation of the average rating of m_1. Note that our interval estimation approach, on the contrary, performs better than Bayesian approach in both MAR and NMAR mode since the interval always involves the ground-truth.

5.3   Effect of Parameters

We conduct experiments to evaluate the effect of *missing ratio of predicates* ϵ on $w\,(COUNT, R)$, and the effect of *missing ratio of aggregate value set* θ on $w(AVG, R)$. Let f equal COUNT, and we ran Q1 on *reviews* with ϵ varies from 0.1 to 0.5 in NMAR mode; the results are shown in Fig. 3(a). Let f equal AVG

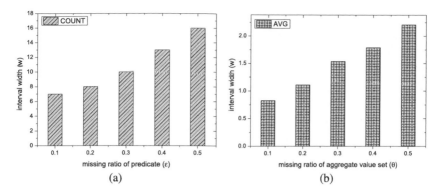

Fig. 3. The effect of parameters on the interval width. (a) The effect of ε (b) The effect of θ.

and we ran Q1 on *reviews* with θ varies from 0.1 to 0.5 in NMAR mode; the results are shown in Fig. 3(b). We can see that, for both AVG and COUNT query, the interval width w increases linearly with the parameter, which agrees with our theoretical analysis.

6   Conclusion

In this paper, we studied the problem of aggregate query on incomplete data. We show that missing data imputation approaches have a lot of limitations, such as imputation accuracy and efficiency. We develop techniques to efficiently estimate the aggregate query result in a form of interval estimation, which covers the ground-truth at all cases. Through theoretical analysis and experimental verification, our approach are more reliable than missing data imputation methods, and we believe that our approach is invaluable to better assess the results of aggregate queries on incomplete data.

References

1. Osborne, J.W.: Best Practices in Data Cleaning: A Complete Guide to Everything You Need to Do Before and After Collecting Your Data. Sage, Thousand Oaks (2012)
2. Rahm, E., Do, H.H.: Data cleaning: problems and current approaches. IEEE Data Eng. Bull. **23**(4), 3–13 (2000)
3. Ebaid, A., Elmagarmid, A.K., Ilyas, I.F., Ouzzani, M., Quiané-Ruiz, J.-A., Tang, N., Yin, S.: NADEEF: a generalized data cleaning system. PVLDB **6**(12), 1218–1221 (2013)
4. Deng, T., Fan, W., Geerts, F.: Capturing missing tuples and missing values. ACM Trans. Database Syst. **41**(2), 10:1–10:47 (2016)
5. Guagliardo, P., Libkin, L.: Correctness of SQL queries on databases with nulls. SIGMOD Rec. **46**(3), 5–16 (2017)

6. Fahandar, M.A., Hüllermeier, E., Couso, I.: Statistical inference for incomplete ranking data: the case of rank-dependent coarsening. In: Proceedings of the 34th International Conference on Machine Learning, ICML 2017, Sydney, NSW, Australia, 6–11 August 2017, pp. 1078–1087 (2017)
7. Sarabia, J.M., Shahtahmassebi, G.: Bayesian estimation of incomplete data using conditionally specified priors. Commun. Stat. Simul. Comput. **46**(5), 3419–3435 (2017)
8. Lipski Jr., W.: On semantic issues connected with incomplete information databases. ACM Trans. Database Syst. **4**(3), 262–296 (1979)
9. Reiter, R.: On closed world data bases. In: Logic and Data Bases, Symposium on Logic and Data Bases, Centre d'études et de recherches de Toulouse, pp. 55–76 (1977)
10. Codd, E.F.: Extending the database relational model to capture more meaning. ACM Trans. Database Syst. **4**(4), 397–434 (1979)
11. Mayfield, C., Neville, J., Prabhakar, S.: ERACER: a database approach for statistical inference and data cleaning. In: Proceedings of the ACM SIGMOD International Conference on Management of Data, SIGMOD 2010, Indianapolis, Indiana, USA, June 6–10, 2010, pp. 75–86 (2010)
12. Rubin, D.B., Little, R.J.A.: Statistical Analysis with Missing Data. Wiley, Hoboken (2014)

Machine Learning

Travel Time Forecasting with Combination of Spatial-Temporal and Time Shifting Correlation in CNN-LSTM Neural Network

Wenjing Wei[1], Xiaoyi Jia[1], Yang Liu[1], and Xiaohui Yu[1,2(✉)]

[1] School of Computer Science and Technology, Shandong University,
Jinan 250101, Shandong, China
migowei0621@163.com, jiaxiaoyishd@163.com, {yliu,xyu}@sdu.edu.cn
[2] School of Information Technology, York University, Toronto, ON M3J 1P3, Canada

Abstract. The problem of short-term travel time estimation has been intensively investigated recently. However, accurate travel time predicting is still a challenge due to dynamic changes of the traffic and the difficulty of extracting urban traffic data features. In this paper, we mainly focus on time shifting feature of urban roads, which represents the impact of the upstream sections that will be conveyed to the downstream sections after a certain period of time Δt. Firstly, we obtain the spatial relationships of the traffic time with Kullback-Leibler divergence (KL-divergence) and urban road networks. Then a Convolutional Neural Network (CNN) module is adopted to extract the spatial-temporal and time shifting information of the target road. Finally, a novel deep architecture combined CNN and Long-short Term Memory Recurrent Neural Network (LSTM) is utilized to predict the short-term travel time. The experimental result on the real data set shows that the proposed model is more effective than other existing approaches.

Keywords: Travel time prediction · Time shifting feature
CNN-LSTM neural network

1 Introduction

Advanced traffic information systems (ATIS) and Intelligent Transportation Systems (ITS) have become the significant standard in urban areas recently. Since the rapid growth of urban vehicles, there are still many obstacles to overcome to build a Wisdom City. The accurate and reliable forecast of travel time is the premise and basement of traffic monitoring [3], driving directions finding [24] and taxi dispatching [25]. Besides, people also show special solicitude for travel time when they want to choose a better travel route. Therefore, the travel time information is a valuable feature for everyday road users. The investigation of the travel time forecasting can be divided into the following two categories: (1)

Long-term travel time forecasting, which refers to the forecast of travel time in the coming months or years, is usually applied for long-term transport planning. (2) Short-term travel time forecasting is often used to predict the next few minutes or hours. In this paper, we mainly focus on estimation of the short-term travel time, and the experimental dataset is collected from traffic surveillance system of a major metropolitan area.

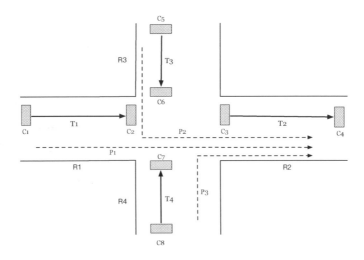

Fig. 1. The road structure and travel time distribution

Figure 1 shows an example of urban road structure and travel time distribution [4–6]. The rectangles in the figure represent a group of cameras at each intersection, and these cameras capture the traffic data of passing vehicles. The road is defined as the segment between two intersections, as shown, there are four roads in figure, and the road R_2 is the target road that we aim at forecasting the travel time, R_1, R_3, R_4 are the upstream roads of R_2. Besides, P_1, P_2 and P_3 are the vehicle trajectories. In the diagram, only one direction (right direction) has been considered. We record the average passing time of all the vehicles which have passed the certain road during a pre-defined period and defined it as travel time.

Current majority of researches on travel time prediction are based on time series theory, e.g., moving average methods (MA), Auto-Regressive Integrated Moving Average (ARIMA) [2] or Space-Time Auto-Regressive Integrated Moving Average (STARIMA) model [15], k-Nearest Neighbor (kNN) technique [16], [11], Support Vector Machines (SVMs) [13], and neural networks [9]. On the other hand, deep learning has drawn a lot of academic and industrial attention recently [1], and it has been applied with success in the tasks of classification, dimensionality reduction, and traffic flow estimation, etc. There are also some attempts to apply deep learning, especially CNN and LSTM, to travel time forecasting [14,22]. Unfortunately, these approaches commonly rely on the historical, spatial, or spatial-temporal features of travel time series. But indeed, the travel time of an urban road not only relies on the aforementioned features, but also depends on the *time shifting* information of its relevant roads.

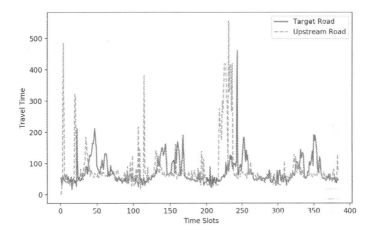

Fig. 2. Travel time series of relevant roads

Through extensive observations and experimental analysis on real data sets, we first propose the time shifting information of an urban region, in which the roads conform to the hierarchy of urban road networks and have the maximum similarities. For example, if we define the n slots travel time series as $T_R = \{t_1, t_2, ..., t_n\}$, where r represents the road in urban road networks, then as shown in Fig. 2, there are two travel time series of two roads, the solid line represents the target (downstream) road's travel time series T_t, and the dashed line represents the upstream road's travel time series T_u. Obviously, it can be observed that the two series would be extremely similar if we perform a right-shift action to the dashed one by a certain period of time Δt, and we mark the dashed line after time shifting as $\tilde{T}_u$. To verify this observation, we calculate KL-divergence values between them, and the experimental result indicates that the KL-divergence between T_t and $\tilde{T}_u$ is much smaller than the others. Note that, the smaller KL-divergence value is, the more similar the two travel time distributions are.

Based on the above analysis, we propose a hybrid deep learning framework focused on time shifting feature. In order to figure out the most top-k relevant roads of the target section, we perform a method combining KL-divergence method and urban road networks. Then a spatial-temporal and time shifting matrix can be built up using the k roads aforementioned. Finally, we extract the time shifting feature vector through CNN module and forecast the travel time of target road with LSTM neural network.

The main contributions of this paper can be summarized as follows.

- We propose a novel deep learning architecture for short-term travel time forecasting, which combines CNN and LSTM neural networks. To the best of our knowledge, it is the first time that the CNN and LSTM are concentrated to predict the short-term travel time.

- We use KL-divergence and urban road networks to calculate the similarity between roads and then obtain the most top-k relevant roads of target section. Based on the results obtained above, we can greatly reduce the complexity of the input data.
- We identify a time shifting relationship between different sections. We first extract the information of time shifting feature, and integrate this feature into the existing spatial-temporal methods.

The remainder of this paper is organized as follows. Section 2 reviews previous research associated with our work. Section 3 provides a brief description of preliminary preparations. Section 4 gives a detailed description of our method. In Sect. 5, the experiment details are discussed. Finally, Sect. 6 concludes the paper and suggests future directions.

2 Related Work

For the past two decades, a considerable effort has made to develop efficient travel time prediction methods, which is backed by a huge number of literatures in this fields. In summary, the various techniques can be classified into the following four major categories [17]: naïve, parametric, non-parametric and hybrid.

Naïve methods, such as historical average (HA), usually don't include any advanced mathematical model for the exploitation of traffic data. Such methods are commonly characterized by low computational complexity but poor prediction accuracy.

Parametric techniques are based on specific models whose general structure has been defined in advance, the only thing has to be determined are the exacting values of a given set of parameters. Common examples of parametric models include Markov Chain [19], MA, ARIMA [2], and STARIMA [15], all of them proposed a model that dealing with the supposed stationarity of the system and the constant relationship among the neighboring links.

Non-parametric methods mean that both the exact model structure and its parameters are need to be specified during the training phase of the model, e.g., kNN technique [16,18], Principal Component Analysis (PCA), Artificial Neural Networks (ANNs) [12]. Deng and Qu [8] fuse the spatial relationship into the kNN method. Besides, Salamanis et al. [17] are more concerned on the spatial graph dependencies in large volumes data. On the other hand, a great deal of researches have been done on deep learning neural networks, Cosme and Vega [7] propose State Space Neural Network (SSNN) to identify non-linear model, which can also be utilized for travel time forecasting [21]. Since LSTM has been widely recognized in time series prediction, there are also several researches based on LSTM in traffic forecasting [20]. Duan et al. [10] make an attempt to predict short-term travel time with LSTM neural network using the data of road time series. In addition, CNN can be utilized for mining features of high-dimensional data, and it has been proven to be successful in the field of traffic prediction recently, e.g., Yu et al. introduce a Data Grouping Convolutional Neural Network (DGCNN) algorithm, which aiming at analyzing the influence between locations [23].

Hybrid techniques has gained wide attention over the past decade, which exploit the advantages of both parametric and non-parametric ones. In this respect, Zhang [26] propose a methodology combined ARIMA and ANN processes. Wu and Tan [22] first extract the spatial feature of traffic data by CNN module, and the temporal feature of traffic data by LSTM module, finally, the features extracted above are concentrated to forecasting traffic flow. The same approach is also adopted by Liu et al. [14], who propose Convolution-LSTM module and Bi-LSTM module which can extract the spatial-temporal feature and periodicity feature respectively. The short-term travel time prediction approach presented in this paper also falls into this category.

3   Preliminaries

In this section, we define the key terms that are required for the succeeding discussion more formally.

Definition 1 (Trajectory). *A trajectory P_k is defined as a series of roads that a vehicle k has passed, and these roads are arranged in chronological order, like $P_k = \{(R_1, t_1), (R_2, t_2), ..., (R_{s-1}, t_{s-1}), (R_s, t_s)\}$, where R represents a road and s is the length of the trajectory.*

Definition 2 (Travel Time). *For each vehicle, we regard the time difference between the vehicle's 'out-road' time and 'in-road' time as the vehicle's travel time, where in-road time indicates when the vehicle is entering the road, out-road time indicates when the vehicle is leaving the road as well. Then we calculate the average vehicle travel time as the travel time of this road for every 15 min. In the rest of this paper, travel time refers to the road travel time, and can be expressed as, $T = \{x_1, x_2, ..., x_{t-1}, x_t\}$.*

Definition 3 (Time Shifting). *Time shifting refers that the impact of the upstream roads that will be conveyed to the downstream roads after a certain period of time Δt (see the example shown in Fig. 2).*

Definition 4 (Relevant Roads). *Given a road r, if the roads in urban road networks satisfy all the following conditions, we call these roads are the relevant roads U_r of r. (1) The roads are the upstream roads of r. (2) After time shifting, the roads will be greatly similar to r, where the similarity are measured by KL-divergence.*

4   Travel Time Modeling

To solve the problem of short-term travel time forecasting, we propose a novel deep learning framework combined CNN and LSTM, which is suitable to extract the spatial-temporal and time shifting information of the urban traffic data. Via a lot of analysis on the large traffic datum, we notice that in a dense traffic network, the travel time of different roads may have strong relationship between each other, e.g., shown in Fig. 1. Sometimes, the travel time of one road is directly affected by the immediate upstream section, especially in the rush hour of a day; sometimes we cannot intuitively observe the correlations between roads in an urban network, because this influence may be caused by all the upstream roads together. The relationship mentioned above mainly includes spatial-temporal relation and time shifting relation. Therefore, we need to extract the relevant roads and construct the relationship matrix in our model.

Multi-layers Convolutional Neural Network (CNN) has been proven to have excellent precision and performance in image processing and traffic forecasting. With the assist of convolution layers and pooling layers, CNN can accurately extract the feature from the input matrix. Moreover, LSTM module shows astonishing results in the problem of time series prediction, using the historical time series data. This is mainly because LSTM has the ability to extract and retain useful information from the historical input sequences. In conclusion, both CNN and LSTM can be powerful tools for forecasting the short-term travel time.

Our method consists basically of three parts. In Subsect. 4.1, we obtain the relevant roads of the target road by KL-divergence, and the results are corrected with the urban network information. Then we construct the spatial-temporal and time shifting matrix using the output of the first step in Subsect. 4.2. The third part will be described in the Subsect. 4.3, in which we combine CNN and LSTM module to predict travel time, and the input of whole network is the feature matrix we constructed in the second step.

4.1   Extracting Relevant Roads

For the purpose of constructing an efficient feature matrix, the relevant roads of the target need to be extracted from the urban traffic network. Therefore it's necessary for us to gain a comprehensive understanding of the roads. We regard the urban network as a multi-tree with the target as the root node. As shown in Fig. 3, R_0 is the target road, the 'children' are the upstream roads of it. Meanwhile, solid circles represent the relevant roads we extracted from the urban networks, dashed circles in level -1 represent the downstream roads of the target, which we throw away straightly because these roads have no influence on the target road R_0, and dashed circles in level $0-n$ is the irrelevant roads we filtered by KL-divergence and the urban network.

Kullback-Leibler divergence (KL-divergence) is known as the relative entropy, and usually utilized for the measurement of similarity between different distributions. Given the target road's travel time distribution $p(x)$ and another road's travel time distribution $q(x)$, we perform the right-shift action of the distribution $q(x)$ by $q(x_{i+m\Delta t})$, where Δt is the time shifting unit, m is the number of time shifting units. Then we extend the length of two distributions to $N + m\Delta t$, padding the new distribution $q(\tilde{x})$ with 0 in the head of the sequence and the

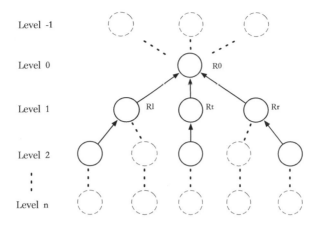

Fig. 3. The relevant roads

new distribution $p(\tilde{x})$ with 0 in the tail of the sequence. The KL-divergence between distributions p and q is expressed as:

$$D(p||q) = \sum_{i=0}^{N+m\Delta t} p(\tilde{x}_i)\ln\{q(\tilde{x}_i)/p(\tilde{x}_i)\} \tag{1}$$

There are still a problem of the results obtained from above step: some roads may have the similar distributions to the target, but indeed, they are extremely far from the target in the urban network. Therefore, we adopt the traffic trajectory data to modify the result. The trajectory data are expressed as follows:

$$P = \begin{bmatrix} p_0\ p_1\ ...\ p_i \end{bmatrix}^{\mathrm{T}}, \quad p_i = \{(r_0, t_0), (r_1, t_1), ..., (r_s, t_s)\} \tag{2}$$

Now, we obtain the relevant roads U_r, which are the directly upstream sections of the target road. Note that, the KL-divergence value within 0.5 is considered to be similar in this paper.

4.2 Time Shifting Feature Matrix

To integrate the spatial-temporal and time shifting information into the input of our model, we construct a feature matrix. Since the N-slots travel time series of each road can be represented as $T = \{x_1, x_2, ..., x_N\}$, the $N + m\Delta t$ slots travel time series after time shifting can be the form as follows:

$$\tilde{T}_u = \{\underbrace{x_p, ..., x_p}_{m\Delta t}, x_1, x_2, ..., x_N\} \tag{3}$$

$$\tilde{T}_t = \{x_1, x_2, ..., x_N, \underbrace{x_p, ..., x_p}_{m\Delta t}\} \tag{4}$$

where $\tilde{T}_u$, $\tilde{T}_t$ represent the travel time series of relevant roads and the target road respectively, and x_p is the padding value.

We put travel time series of the prediction road into the bottom of the matrix, and fill the matrix by travel time series of relevant roads according to the distance from the prediction road. Supposed that the number of relevant roads is 4, the following figure shows the configuration of the feature matrix.

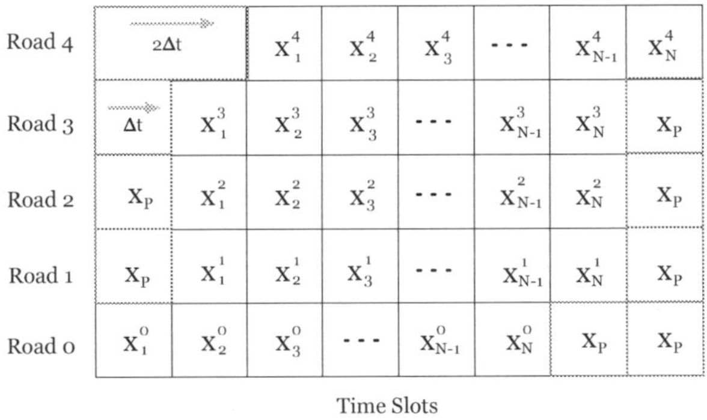

Road 4	$2\Delta t$	X^4_1	X^4_2	X^4_3	- - -	X^4_{N-1}	X^4_N	
Road 3	Δt	X^3_1	X^3_2	X^3_3	- - -	X^3_{N-1}	X^3_N	X_p
Road 2	X_p	X^2_1	X^2_2	X^2_3	- - -	X^2_{N-1}	X^2_N	X_p
Road 1	X_p	X^1_1	X^1_2	X^1_3	- - -	X^1_{N-1}	X^1_N	X_p
Road 0	X^0_1	X^0_2	X^0_3	- - -	X^0_{N-1}	X^0_N	X_p	X_p

Time Slots

Fig. 4. Feature matrix (Color figure online)

In Fig. 4, Road 0 is the target road. Road 1–3 are the upstream sections of road 0 of one hop in the urban network, moreover, the travel time series of road 1–3 shift by one time unit Δt. Besides, road 4 is the upstream section of two hops, and its travel time series shift by two Δt, because there are also time shifting relationships between road 4 and road 1–3. The whole matrix have $N + 2\Delta t$ columns, and the red value x_p is the padding value 0 to fill the vacancy. Note that, we should consider about the spatial feature, the padding value can not be omitted.

4.3   Modeling and Training

In this subsection, we adopt the Convolutional Neural Network (CNN) to extract the 2-dimensional spatial-temporal and time shifting features, and Long-short Term Memory Recurrent Neural Network (LSTM) for travel time forecasting. We can illustrate them graphically as Fig. 5.

Here, CNN in our method is comprised of two convolutional layers and one pooling layer. Supposed that the shape of feature matrix we built in Subsect. 4.2 is $M \times N$, where M is the number of relevant roads plus 1, and N is the number of time slots. Hence the input matrix to a convolution layer is $M \times N \times c$, c represents the channel of the input, here c is 1. The convolution layer have the filter of size $n \times n \times q$, where n is smaller than the dimension of the input matrix and q can either be the same as the number of channels c or smaller and may vary for each kernel, here, filter size is $2 \times 2 \times 1$ because it's limited by the size of feature matrix. The stride of filter is 1 and the type of pooling layer is average pooling in our work. The input and output of convolution layer are both feature maps, and if we denote the k-th feature map at a given layer as h^k, whose filters are determined by the weights W^k and bias b_k, then the feature map h^k is obtained as follows:

$$h^k = f(W^k \cdot x + b_k) \tag{5}$$

Here, $f(\cdot)$ is ReLU function, it solves the gradient vanishing problem and has efficient calculation and convergence efficiency. The CNN level in Fig. 5, shows the structure of CNN in our model.

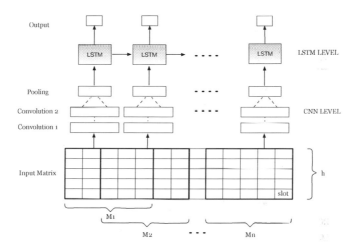

Fig. 5. Model structure

As shown in Fig. 5, the feature maps extracted from CNN level is the input of LSTM level. The traditional forecasting models mainly suffer from a drawback in handing time mode information of travel time [22], the drawback is that traditional methods especially traditional Recurrent Neural Networks (RNNs) are difficult to train if the travel time series has long time slots. To tackle this issues, LSTM has been proposed in recent years.

Similar to traditional RNNs, a LSTM structure is composed of one input layer, one or more hidden layers and one output layer. The key to LSTMs is the cell state, and the LSTM does have the ability to remove or add information to the cell state, carefully regulated by structures called gates. In Fig. 5, LSTM level have several LSTM cells, and the Fig. 6 illustrates the structure of a LSTM cell. At time t, the input is x_t, the hidden layer output is h_t and its former output is h_{t-1}, the cell input state is $\tilde{C}_t$, the cell output state is C_t and its former state is C_{t-1}, the three gates' states are i_t, f_t and o_t. Finally, the equations of the three gates' states and the cell state are as follows, input gate:

$$i_t = \sigma(W_1^i \cdot x_t + W_h^i \cdot h_{t-1} + b_i), \tag{6}$$

forget gate:

$$f_t = \sigma(W_1^f \cdot x_t + W_h^f \cdot h_{t-1} + b_f), \tag{7}$$

output gate:

$$o_t = \sigma(W_1^o \cdot x_t + W_h^o \cdot h_{t-1} + b_o), \tag{8}$$

cell input:

$$\tilde{C}_t = \tanh(W_1^C \cdot x_t + W_h^C \cdot h_{t-1} + b_C). \tag{9}$$

Here, b_i, b_f, b_o and b_C are the bias terms. W_1^i, W_1^f, W_1^o and W_1^C are the weight matrices among the three gates and cell input, and W_h^i, W_h^f, W_h^o and W_h^C are the matrices connecting h_{t-1} to the three gates and cell input. In our work, the batch size of LSTM is 50, the layer of LSTM is 2, and we use Adam optimization algorithm in the training phase.

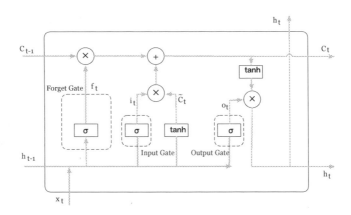

Fig. 6. The structure of a LSTM cell

Now, let's go back to the whole model shown in Fig. 5. The input matrix is what we constructed in Sect. 4.2, we split it into several sub-matrices with the size of $M_i \times h$ as the input of each step. And we utilize the Mean Square Error (MSE) as the loss function of our model, and the AdamOptimizer to optimize the iterative process. MSE is calculated by:

$$MSE = \frac{1}{N} \sum_{i=1}^{N} (y_i - \hat{y}_i)^2$$

where y_i is the observed value and $\hat{y}_i$ is the forecast value of observation i.

5   Experiments

In this section, we first introduce the dataset of our experiments, followed by the evaluation matrices we apply to evaluate the performance. Finally, we show the performances of our method and baselines.

5.1   Dataset Description

In the experiments, we use a real dataset of vehicle passage records from December 1, 2015 to February 29, 2016, which were collected from the traffic surveillance systems in a city of China. The dataset consists of 479,451,848 passage records, and involves 1585 mainly roads in urban networks. The time interval for counting travel time is 15 min, that is also the minimum unit of time shifting distance Δt. Therefore, there are a total number of 8736 time slots in each original travel time series. We split the dataset into two parts, training set and testing set. The training set covers the data from December 1, 2015 to February 9, 2016, which is about 80% of the original dataset. And the traffic records of testing set is from February 10 to February 29, 2016, 20% of the original dataset.

5.2   Performance Metrics

In order to evaluate the performance of the proposed methods, we adopt three performance metrics to measure the prediction results. Here, y_i and $\hat{y}_i$ are the observed and the forecast values of the i-th travel time data respectively.

(1) The Mean Absolute Error (MAE) is the measurement of difference between two continuous variables and is calculated by

$$MAE = \frac{1}{N} \sum_{i=1}^{N} (|y_i - \hat{y}_i|)$$

(2) The Root Mean Squared Error (RMSE) represents the sample standard deviation of the differences between predicted values and observed values. The equation of RMSE is

$$RMSE = \sqrt{\frac{1}{N} \sum_{n=1}^{N} (y_i - \hat{y}_i)^2}$$

(3) The Mean Absolute Percentage Error (MAPE) is a measure of prediction accuracy of forecasting algorithms in statistics and is calculated by

$$MAPE = \frac{1}{N} \sum_{n=1}^{N} \frac{|y_i - \hat{y}_i|}{y_i} \times 100\%$$

To avoid the denominator is zero, we use the average observed value to replace the denominator when $y_i = 0$.

5.3   Performance of Methods

In order to measure the performance of our method, we compare it with the other existing methods for short-term travel time forecasting.

(1) Auto Regressive Integrated Moving Average Model (ARIMA) Model: We adopt a general ARIMA model of order (p, d, q), where the orders p and q are the AR and MA operators respectively, and d is the order of differencing.

(2) LSTM Model: Traditional LSTM model has the ability to fit the non-linear relationship between input and output, and it is often used for time series processing.

(3) CNN Model: CNN is often utilized for extracting features of high-dimensional data, such as image data and voice. Lately, many researchers also adopt the CNN model to process traffic data.

(4) CNN-LSTM Model: The hybrid architecture combining CNN and LSTM module has also been considered in the evaluation.

(5) TimeShifting-CNN Model (T-CNN): We apply the time shifting matrix into the traditional CNN model, which is comprised of one input layer, two convolution layers, one pooling layer, and one full connection layer.

We choose the parameters for each model by observing the best fitting for forecasting and compare their predictions on the same testing set. In ARIMA

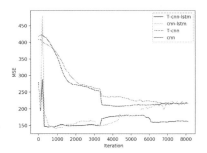

Fig. 7. MSEs of different methods

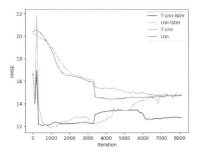

Fig. 8. RMSEs of different methods

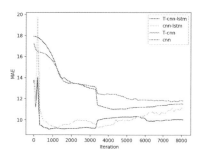

Fig. 9. MAEs of different methods

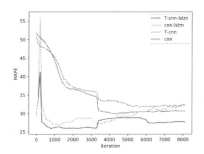

Fig. 10. MAPEs of different methods

model, we set parameters as (2, 0, 3), which is the most optimal parameter. The batch size of the neural networks above is 50, and the number of training iterations is 8000. The architecture of the LSTM model is set to be a three-layer neural network, which contains 24 input neurons, 10 hidden neurons and 1 output neuron. In addition, the filter size of CNN is $2 \times 2 \times 1$. Figures 7, 8, 9 and 10 shows the MSEs, RMSEs, MAEs, MAPEs of different methods.

Table 1 shows the performances of different forecasting models on the testing set. As we can see, ARIMA is the worst, and the LSTM and CNN are also inaccurate. It' s clear that time shifting CNN (T-CNN) outperforms the traditional CNN, as it considers about the time shifting factor. And T-CNN-LSTM performs better than CNN-LSTM, since it not only considers about the spatial-temporal feature, but also uses the time shifting feature of the travel time.

Table 1. Performance comparison

Method	MAE	RMSE	MAPE %
ARIMA	17.29	20.2531	37.1985
LSTM	13.2838	22.9698	29.0727
CNN	12.7581	16.2162	27.7015
CNN-LSTM	13.5931	18.3231	19.8431
T-CNN	12.1572	14.8857	27.4971
T-CNN-LSTM	**11.4421**	**14.7152**	**17.2432**

Another important parameter in our model is the number of time shifting units. Here, we set shifting units as $0 \cdot \Delta t$, which means the time series does not shift (tradition CNN or tradition CNN-LSTM), $1 \cdot \Delta t$, $2 \cdot \Delta t$ and $3 \cdot \Delta t$, where Δt is 15 min. The difference is obviously shown in Table 2, and it's clear that when shifting units is $1 \cdot \Delta t$, we obtain better results in MAE and RMSE.

Table 2. Performance of different time shifting units

Shifting units (Δt)	MAE	RMSE	MAPE %
0	13.5931	18.3231	19.8431
1	**11.4421**	**14.7152**	17.2432
2	11.8274	14.7722	**16.6970**
3	11.6942	14.8697	19.4282

6   Conclusions

In this paper, we studied the problem of forecasting short-term travel time in urban. We propose a novel hybrid deep learning architecture, which has two advantages: (1) The similarity of roads and urban networks are used to extract the relevant roads of the target road. (2) The spatial-temporal and time shifting information are considered in our method. (3) CNN module and LSTM module are combined for handling the problem of short-term travel time forecasting. To evaluate the performance of our proposed method, we have compared it with several state-of-art methods on a real dataset. For future work, we will consider the periodicity feature of the travel time data, which will help to make more accurate prediction.

Acknowledgment. This work was supported in part by the National Basic Research 973 Program of China under Grant No. 2015CB352502, the National Natural Science Foundation of China under Grant Nos. 61272092 and 61572289, the Natural Science Foundation of Shandong Province of China under Grant Nos. ZR2012FZ004 and ZR2015FM002, the Science and Technology Development Program of Shandong Province of China under Grant No. 2014GGE27178, and the NSERC Discovery Grants.

References

1. Bengio, Y.: Learning deep architectures for AI. Found. Trends Mach. Learn. **2**(1), 1–127 (2009)
2. Billings, D., Yang, J.: Application of the ARIMA models to urban roadway travel time prediction - a case study. In: Proceedings of the IEEE International Conference on Systems, Man and Cybernetics, Taipei, Taiwan, 8–11 October 2006, pp. 2529–2534 (2006)
3. Chawla, S., Zheng, Y., Hu, J.: Inferring the root cause in road traffic anomalies. In: 12th IEEE International Conference on Data Mining, ICDM 2012, Brussels, Belgium, 10–13 December 2012, pp. 141–150 (2012)
4. Chen, M., Liu, Y., Yu, X.: NLPMM: a next location predictor with markov modeling. In: Tseng, V.S., Ho, T.B., Zhou, Z.-H., Chen, A.L.P., Kao, H.-Y. (eds.) PAKDD 2014. LNCS (LNAI), vol. 8444, pp. 186–197. Springer, Cham (2014). https://doi.org/10.1007/978-3-319-06605-9_16
5. Chen, M., Yu, X., Liu, Y.: Mining moving patterns for predicting next location. Inf. Syst. **54**, 156–168 (2015)
6. Chen, M., Yu, X., Liu, Y.: Mining object similarity for predicting next locations. J. Comput. Sci. Technol. **31**(4), 649–660 (2016)
7. Cosme, J.M.Z., Vega, P.: State space neural network. properties and application. Neural Netw. **11**(6), 1099–1112 (1998)
8. Deng, M., Qu, S.: Road short-term travel time prediction method based on flow spatial distribution and the relations. Math. Prob. Eng. **2016**(4), 1–14 (2016)
9. Dia, H.: An object-oriented neural network approach to short-term traffic forecasting. Eur. J. Oper. Res. **131**(2), 253–261 (2001)
10. Duan, Y., Lv, Y., Wang, F.: Travel time prediction with LSTM neural network. In: 19th IEEE International Conference on Intelligent Transportation Systems, ITSC 2016, Rio de Janeiro, Brazil, 1–4 November 2016, pp. 1053–1058 (2016)
11. Jiwon, M., Kim, D.K., Kho, S.Y., Park, C.H.: Travel time prediction using k nearest neighbor method with combined data from vehicle detector system and automatic toll collection system. Transp. Res. Rec. J. Transp. Res. Board **20**(2256), 51–59 (2011)

12. Kamarianakis, Y., Vouton, V.: Forecasting traffic flow conditions in an urban network: comparison of multivariate and univariate approaches. Transp. Res. Rec. **1857**(1), 74–84 (2003)
13. Liu, X., Fang, X., Qin, Z., Ye, C., Xie, M.: A short-term forecasting algorithm for network traffic based on chaos theory and SVM. J. Netw. Syst. Manag. **19**(4), 427–447 (2011)
14. Liu, Y., Zheng, H., Feng, X., Chen, Z.: Short-term traffic flow prediction with Conv-LSTM. In: 9th International Conference on Wireless Communications and Signal Processing, WCSP 2017, Nanjing, China, 11–13 October 2017, pp. 1–6 (2017)
15. Min, W., Wynter, L.: Real-time road traffic prediction with spatio-temporal correlations. Transp. Res. Part C Emerg. Technol. **19**(4), 606–616 (2011)
16. Rice, J., van Zwet, E.: A simple and effective method for predicting travel times on freeways. IEEE Trans. Intell. Transp. Syst. **5**(3), 200–207 (2004)
17. Salamanis, A., Kehagias, D.D., Filelis-Papadopoulos, C.K., Tzovaras, D., Gravvanis, G.A.: Managing spatial graph dependencies in large volumes of traffic data for travel-time prediction. IEEE Trans. Intell. Transp. Syst. **17**(6), 1678–1687 (2016)
18. Smith, B.L., Demetsky, M.J.: Multiple-interval freeway traffic flow forecasting. Transp. Res. Rec. J. Transp. Res. Board **1554**(1), 136–141 (1996)
19. Tian, D., Yuan, Y., Xia, H., Cai, F., Wang, Y., Wang, J.: A route travel time distribution prediction method based on Markov chain. In: IEEE First International Smart Cities Conference, ISC2 2015, Guadalajara, Mexico, 25–28 October 2015, pp. 1–5 (2015)
20. Tian, Y., Pan, L.: Predicting short-term traffic flow by long short-term memory recurrent neural network. In: 2015 IEEE International Conference on Smart City/SocialCom/SustainCom 2015, Chengdu, China, 19–21 December 2015, pp. 153–158 (2015)
21. Wei, S., Wei, S., Wynter, L.: Rejoinder: Real-time Road Traffic Forecasting Using Regime-Switching Spacetime Models and Adaptive Lasso. Wiley, Hoboken (2012)
22. Wu, Y., Tan, H.: Short-term traffic flow forecasting with spatial-temporal correlation in a hybrid deep learning framework. CoRR abs/1612.01022 (2016)
23. Yu, D., Liu, Y., Yu, X.: A Data Grouping CNN Algorithm for Short-Term Traffic Flow Forecasting. Springer International Publishing, Cham (2016)
24. Yuan, J., Zheng, Y., Xie, X., Sun, G.: T-drive: enhancing driving directions with taxi drivers' intelligence. IEEE Trans. Knowl. Data Eng. **25**(1), 220–232 (2013)
25. Yuan, N.J., Zheng, Y., Zhang, L., Xie, X.: T-finder: a recommender system for finding passengers and vacant taxis. IEEE Trans. Knowl. Data Eng. **25**(10), 2390–2403 (2013)
26. Zhang, G.P.: Time series forecasting using a hybrid ARIMA and neural network model. Neurocomputing **50**(1), 159–175 (2003)

$DMDP^2$: A Dynamic Multi-source Based Default Probability Prediction Framework

Yi Zhao[1], Yong Huang[2], and Yanyan Shen[1(✉)]

[1] Shanghai Jiao Tong University, Shanghai, China
{zhaoyizhaoyi,shenyy}@sjtu.edu.cn
[2] Hong Kong University of Science and Technology, Hong Kong, China
yonghuang826@yahoo.com

Abstract. In this paper, we propose a dynamic forecasting framework, named $DMDP^2$ (Dynamic Multi-source based Default Probability Prediction), to predict the default probability of a company. The default probability is a very important factor to assess the credit risk of listed companies on a stock market. Aiming at aiding financial institutions in decision making, our $DMDP^2$ framework not only analyses financial data to well capture the historical performance of a company, but also utilizes Long Short-Term Memory model (LSTM) to dynamically incorporate daily news from social media to take the perceptions of market participants and public opinions into consideration. The study of this paper makes two key contributions. First, we make use of unstructured news crawled from social media to alleviate the impact of financial fraud issue made on default probability prediction. Second, we propose a neural network method to integrate both structured financial factors and unstructured social media data with appropriate time alignment for default probability prediction. Extensive experimental results demonstrate the effectiveness of $DMDP^2$ in predicting default probability for the listed companies in mainland China, compared with various baselines.

1 Introduction

Granting loans to potential borrowers is considered as one of the core business activities for financial institutions. Though loans can help these institutions gain profits, they may also cause huge loss, which is often known as financial risks. For instance, the 2008 financial crises resulted in huge losses globally. Hence, nowadays financial institutions devote more and more attention to evaluating risks before granting loans. In particular, most financial institutions are now cognizant of the need to adopt rigorous credit risk assessment models when determining whether or not to grant loans to specific borrowers.

In the early stage, various classical statistical approaches [1,2] such as logistic regression [3], multivariate adaptive regression splines [4,5] and linear discriminate analysis [6,7] were proposed for credit risk prediction. However, statistical

© Springer International Publishing AG, part of Springer Nature 2018
Y. Cai et al. (Eds.): APWeb-WAIM 2018, LNCS 10987, pp. 312–326, 2018.
https://doi.org/10.1007/978-3-319-96890-2_26

approaches are typically based on certain assumptions, e.g., multivariate nor-
mality for independent variables and non-multicollinearity of data, which make
the proposed solutions theoretically invalid for finite samples [8]. Fortunately,
with the advent of machine learning algorithms, many studies demonstrated that
neural networks (NN) [5,9,10], support vector machines (SVM) [11–15], decision
trees (DT) [16,17], random forests (RF) [18,19] and Naive Bayes (NB) [1,20,21]
can be used to build credit scoring models for measuring default risks with high
accuracy. Some practical works [22–26] have focused on classifier ensembles and
demonstrated that ensemble classifiers constantly outperform single classifier in
terms of prediction accuracy.

Although machine learning methods can automatically learn hidden and critical factors based on the past observations and do not require specific prior
assumption, the performance of these supervised methods greatly rely on the
quality of training data. To be more specific, the accuracy of the risk evaluation
results is typically affected by the trustworthiness and the comprehensiveness of
the available historical data. At corporation level, the data for default probability
prediction involve basic financial indicators such as industry section, geographical area, and financial statements. One of the major problems encountered in
adopting financial indicators for credit risk assessment is that the companies
might commit accounting fraud in order to artificially improve the appearance
of the financial reports, which impedes the effectiveness of the learned prediction models. Furthermore, almost all the prior works focus on static models that
leverage the most recent indicator values for prediction, but do not consider the
temporal trend of indicators that is valuable to reflect the long-term financial
status of a company.

In practice, for credit risk assessment, most lenders take advantage of the
information from social media networks, such as Twitter and Facebook, to decide
whether their potential borrowers are credit-worthy. However, probably due to
the difficulty in grabbing daily financial news, we do not find any study that
leverages social media analysis to improve the prediction performance in terms
of the default probabilities for companies. It is important to notice that besides
financial statements, social media data contain subjective appraisals of the firm's
prospects which are discriminate indicators to assess the default probability of
a publicly traded company.

In this paper, we propose a dynamic multi-source based default probability
predication framework named $DMDP^2$, to predict the default probability of the
listed companies. In order to relieve the exertion of potentially flawed financial
data and enhance the accuracy of machine learning based default probability
prediction method, we make use of social media data to trace the latest development of a company. Through mining the public opinions (randomly updated) as
well as the financial indicators (quarterly updated), we can make a comprehensive evaluation of the observed companies in terms of their default probabilities.
More importantly, aiming at prior evaluation for group loans, our framework is
designed to handle the evolving data and continuously produce default probability prediction results based on the up-to-date company statuses, thus helping

the financial institutions to make quick response when the borrowers experience a drastic market decline.

Compared with the previous works [27–30], our study makes the following key contributions.

- To the best of our knowledge, this work is the first to adopt public perception factors from social media in default probability evaluation. These factors are supplementary to the basic financial indicators and help boost the performance of the prediction model.
- We propose an end-to-end default probability prediction framework named $DMDP^2$ to incorporate both structured financial factors and unstructured social media data in a unified manner. $DMDP^2$ dynamically aligns the timelines of multiple data sources, and adopts an LSTM-based neural network model to continuously learn discriminative features as well as long-term trend of company status from multi-source data for prediction.
- We have conducted extensive experiments to verify the effectiveness of our default probability prediction framework $DMDP^2$. The results demonstrate that $DMDP^2$ consistently outperforms traditional machine learning models in terms of prediction accuracy.

The remainder of this paper is organized as follows. Section 2 presents preliminaries. Section 3 introduces our proposed framework with two major components: dynamic multi-source data alignment and neural network based prediction model. Section 4 provides the experimental results. We review the related works in Sect. 5 and conclude this paper in Sect. 6.

2   Preliminaries

In general, the objective of this study is to effectively distinguish "bad" corporations from "good" ones, which can be considered as a classification problem where a company is categorized to class "1" if it is predicted to be default, i.e., receive delisting risk warning (*ST). Otherwise, it will be labeled by "0". In this work, we use sequences of historical financial indicators and unstructured news data in the previous time periods to predict whether the target company will be default or not in the future.

2.1   Definitions

We first introduce the definitions used in this paper. The information of each company during time period T can be recorded in a tuple of $(FIN_T, TEXT_T, y_T)$, where FIN_T contains the values of a set of financial indicators collected at the end of period T. $TEXT_T$ is the representation of unstructured news released during period T. y_T is the binary response variable to indicate whether the company is labeled (*ST) in the period.

Financial Indicators (FIN). Financial indicators are commonly included in the financial statements, and they review important characteristics of an company. For instance, the indicator "cash flow to liabilities ratio", directly reflects a company's ability to cover its liability within a time window and hence is critical when predicting the default probability of the company in near future. By similar logic, we extract several key financial indicators from financial statements of a company at the end of a financial period. Suppose we extract P indicators in total. We denote them by $FIN = \{X_1, X_2, \ldots, X_P\}$.

News Data Representation ($TEXT$). $TEXT$ stands for the embedding-based representation of news data that are relevant to the company during a period. The news are crawled from social media. We map each news to the corresponding company based on the stock symbol contained in the news content.

Class Label (y). y stands for the class label of the target company in a financial period. That is, $y = 1$ if the company receives delisting risk warning (*ST) during the period. Otherwise, $y = 0$. In this work, we do not take the sequence of the previously observed label values as the input for our prediction model as the discriminative power of the previous *ST values is quite limited.

2.2   Problem Statement

We now formally define the problem studied in this paper as follows.

Definition 1 (Problem Statement). *Given a target company, an observation financial period T, sequences of financial indicators FIN and news $TEXT$ before time period T (T is included) in chronological order, we want to develop a framework to predict the default probability of the company during time period $T + 1$. The predicted value could be 0 or 1, representing relatively low default probability or high default probability respectively.*

3   Methodology

Our $DMDP^2$ framework consists of two major components: (1) dynamic multi-source data alignment; (2) neural network based prediction model. Figure 1 provides the illustration of our $DMDP^2$ framework.

3.1   Dynamic Multi-source Data Alignment

To predict whether a company c will be labeled (*ST) or not at time period $T + 1$, we extract the sequences of financial indicators $\{FIN_1^c, \ldots, FIN_T^c\}$ and unstructured news representations $\{TEXT_1^c, \ldots, TEXT_T^c\}$, where FIN_1^c and $TEXT_1^c$ are the inputs in the first financial period after company c gets listed. Because a listed company is required to publish its financial statements for each financial period according to regulation, the sequence of the historical financial indicators is complete. We denote the sequence of financial indicators for a company c before time period T by $\{FIN_t^c\} = \{X_{it}^c\}$, where $i \in \{1, \ldots, P\}$ and

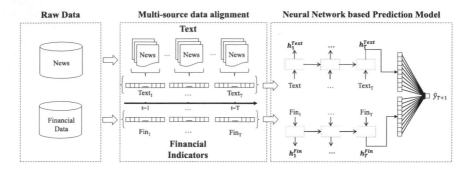

Fig. 1. $DMDP^2$ framework

$t \in \{1, \ldots, T\}$. For the sequence of news representations, we perform preprocessing and alignment over the raw news data to deal with the problems of data
missing and irregularity. On the one hand, the number of news within a specific
financial period varies from one company to another. On the other hand, even
for the same company, the number of relevant news released in different financial periods varies greatly, from zero to several dozens. For instance, a company
may receive multiple news within one day while it might take months or even
quarters to get one relevant news for certain time period.

To handle the missing and irregularity problems of the news sequence, we
apply the following method to preprocess the raw news data and align them
properly. First, we concatenate all the news contents for a company within a
period into a single text. Second, we extract the top K keywords using TF-IDF
values. Third, we retrieve the embeddings of these keywords from a pre-trained
Word2Vec model and calculate the average embedding to represent the concatenated news in the period. Therefore, for company c at the end of time period T,
the sequence of news data could be represented by $\{TEXT_t^c\} = \{\frac{1}{K}\sum_{j=1}^{K} \mathbf{v}_{jt}^c\}$,
where $t \in \{1, \ldots, T\}$, $\mathbf{v}_j^c = \langle v_1, \ldots, v_d \rangle$. Here, K is the number of the extracted
keywords, and d is the number of dimensions of the word embeddings. Finally,
we align the sequence of financial indicators and that of the news representations by imputation. For instance, given a company c, we want to predict its
default probability in time period T_5. Suppose we have collected the financial
indicators sequence in T_1, T_2, T_3, T_4. But for the news representation sequence,
we only observe the news data in T_2 and T_4. In this situation, we could impute
the news representations for T_1 and T_3 by vectors of all zeros.

We summarize the pseudocode for the multi-source data alignment method
in Algorithm 1. After aligning multi-source data, we obtain the final input to
our prediction model $\mathbf{X}_T^c = (\{FIN_{t_{fin}}^c\}, \{TEXT_{t_{text}}^c\})$, where $\{t_{fin}\}$ is not necessarily the same with $\{t_{text}\}$ based on whether we use a multi-source LSTM or
a single LSTM to handle the input sequences.

Algorithm 1. AlignMultiSourceData

Input : Sequence of financial indicators $\{FIN_1^c, \ldots, FIN_T^c\}$; Sequence of news for a company.

Output: Aligned results for the two sequences,
$\mathbf{X}_T^c = (\{FIN_{t_{fin}}^c\}, \{TEXT_{t_{text}}^c\})$.

1 Concatenate all the news for company c during a period into a single text.
2 Extract top K keywords using TF-IDF values.
3 Get the embeddings of the keywords from a pre-trained Word2Vec model and calculate the average embedding. We denote the sequence of text representation by $\{TEXT_t^c\} = \{\frac{1}{K}\sum_{j=1}^{K}\mathbf{v}_{jt}^c\}$.
4 Align the sequence of financial indicators and that of the news representation by imputation. We denote it by $\mathbf{X}_T^c = (\{FIN_{t_{fin}}^c\}, \{TEXT_{t_{text}}^c\})$.
5 **return** $\mathbf{X}_T^c = (\{FIN_{t_{fin}}^c\}, \{TEXT_{t_{text}}^c\})$.

3.2 Neural Network Based Default Probability Prediction Model

The architecture of our neural network based model is illustrated in Fig. 2. In this section, we describe each layer of the model in details.

Input Layer. The first layer is the input layer, which contains the aligned sequences of financial indicators and news representations, during time periods $1, \cdots, T$. Formally, the input layer is defined as $\mathbf{X}_T^c = (\{FIN_{t_{fin}}^c\}, \{TEXT_{t_{text}}^c\})$.

LSTM Layer. Long short-term memory (LSTM) was proposed in [31]. It is a variant of vanilla RNN, which includes cell states with the gating mechanism. With different types of gates, LSTM is capable of controlling how much information flow through the cell state and thus is capable of learning long-term dependencies. In this work, we follow the implementation of LSTM used in [32]:

$$\mathbf{f}_t = \sigma(\mathbf{W}_f[\mathbf{h}_{t-1}; x_t] + \mathbf{b}_f) \tag{1}$$

$$\mathbf{i}_t = \sigma(\mathbf{W}_i[\mathbf{h}_{t-1}; x_t] + \mathbf{b}_i) \tag{2}$$

$$\mathbf{o}_t = \sigma(\mathbf{W}_o[\mathbf{h}_{t-1}; x_t] + \mathbf{b}_o) \tag{3}$$

$$\mathbf{s}_t = \mathbf{f}_t \odot \mathbf{s}_{t-1} + \mathbf{i}_t \odot \tanh(\mathbf{W}_s[\mathbf{h}_{t-1}; x_t] + \mathbf{b}_s) \tag{4}$$

$$\mathbf{h}_t = \mathbf{o}_t \odot \tanh(\mathbf{s}_t) \tag{5}$$

where $\mathbf{f}_t, \mathbf{i}_t, \mathbf{o}_t, \mathbf{s}_t, \mathbf{h}_t$ refer to the forget gate, input gate, output gate, memory cell state and hidden state, respectively; x_t refers to the input to the LSTM unit, δ represents the standard sigmoid function; $\odot$ denotes the Hadamard product; $\mathbf{W}$ and $\mathbf{b}$ terms are the weight matrices and bias terms to be learned, respectively.

In our context, we use LSTM to capture temporal information and latent feature representations from the input sequences for default probability prediction. Specifically, we propose the following two different ways to implement this layer given the aligned sequences from the input layer.

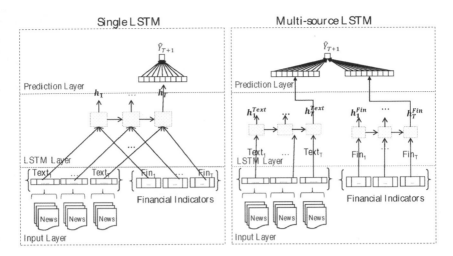

Fig. 2. The architecture of the neural network based prediction model.

- **Single LSTM:** In this implementation, we concatenate $\{FIN_t^c\}$, $\{TEXT_t^c\}$ from the aligned sequences for each time period t. By doing this, we obtain a single sequence represented by $\{[FIN_t^c, TEXT_t^c]\}$. We feed this concatenated sequence into a single LSTM. The hidden state of the LSTM in the last time period T, denoted by $\mathbf{h}_T^c$ will be passed to the next layer to predict the default probability in the next financial period.
- **Multi-source LSTM:** In this implementation, we feed two sequences from different sources to separate LSTMs. We extract the hidden states from different LSTMs in the last time period T and concatenate them into a single vector, denoted by $\mathbf{h}_T^c = [\mathbf{h}_T^c(FIN), \mathbf{h}_T^c(TEXT)]$. To accelerate the learning process of the Multi-source LSTM, an alternative is to allow using two sequences with different lengths. That is, we may skip all the zero paddings for missing news when feeding the news representation sequence to the LSTM.

Since data from multiple sources are generally heterogeneous, concatenating them into a single model may bring noise and we expect the model with the Multi-source LSTM layer to perform better. In the experiments, we evaluate the performance of the two implementations on a real dataset.

Prediction Layer. At last, we feed the output from the LSTM layer to a fully connected layer. The output of the fully connected layer is the predicted response value indicating whether the company c will be default in time period $T + 1$. Formally, we have:

$$\hat{y}_{T+1}^c = \mathbf{w}\mathbf{h}_T^c + b \tag{6}$$

where $\mathbf{w}$ is the weight vector and b is the bias to be learned.

3.3   Learning and Optimization

Our prediction problem is essentially a binary classification problem. Hence, we choose to use cross entropy as the loss function, which is defined as follows.

$$\mathbf{L} = \sum_{\substack{c\in\{1,...,C\}, \\ T\in\mathbf{T}_c}} -y_{T+1}^c \log(\sigma(\hat{y}_{T+1}^c)) - (1 - y_{T+1}^c)(1 - \log(\sigma(\hat{y}_{T+1}^c))) \quad (7)$$

where y_{T+1}^c is the actual class label of company c at time period $T + 1$ and $\mathbf{T}_c$ is the collection of financial periods for company c after it gets listed. We use stochastic gradient descent and Adam optimizer [33] in our training process. The Adam optimizer dynamically adapts the learning rate and makes the convergence faster via adaptive estimates of the lower-order moments. It is computationally efficient with little memory requirements. Besides, it is invariant to diagonal rescaling of the gradients and is well suited for problems that involve data or parameters in large size.

4   Experiments

4.1   Experimental Settings

Datasets. We obtained the financial indicators for the listed companies using **tushare**[1] package. The dataset covers 30 financial indicators for $2,877$ listed companies in mainland China, from January 2014 to September 2017. It contains $29,303$ records of financial indicators updated in each quarter. The news data are crawled from Sina Finance[2] and contains more than $600,000$ pieces of news, describing the historical financial performance and the public opinions of the observed companies. Among the $2,877$ companies, the vast majority are in the "normal" class and only 209 companies bearing 238 delisting risk warnings are categorized as "relative high default risk". To even-up the extremely imbalanced classes in our dataset, we adopt random over-sampling before model training procedure. The financial variables are updated every 3 months according to the quarterly financial statements of the companies. We chose 30 critical financial indicators reflecting the statuses of the listed companies. The indicators are summarized in Table 1.

Preprocessing News Text. Data preprocessing procedures mainly require four steps. To begin with, for each company, we bin the news in one season into a quarterly concatenated news records, which reduces over $600,000$ pieces of news to $27,411$ records for companies in each quarter. Then we adopt Jieba[3] for Chinese word segmentation. Next we remove the stop words and extract the top 50 keywords from the quarterly concatenated text according to their TF-IDF

---

[1] http://tushare.org.
[2] http://finance.sina.com.cn.
[3] https://github.com/fxsjy/jieba.

Table 1. The numerical financial indicators used in our experiments.

Category	Financial indicators
Cash flow ability	Cash flow to sales ratio
	Cash flow to net profit margin ratio
	Cash flow to liabilities ratio
	Cash flow ratio
Operation ability	Account receivable turnover
	Account receivable turnover days
	Inventory turnover
	Inventory turnover days
	Current asset turnover
	Current asset turnover days
Profitability	Return of equity(roe), EPS
	Net profit ratio, Net profits
	Gross profit rate, Business income
	Business income per share
Solvency	Current ratio, Cash ratio
	Quick ratio
	Interest coverage ratio
	Shareholders equity ratio
Growth	Main business rate of growth
	Net profit rate of growth
	Net asset
	Total asset rate of growth
	EPS rate of growth
	Shareholders equity rate of growth

values. Embeddings are looked up from the pretrained Word2Vec model provided by Facebook's FastText module [34]. We represent the concatenated text with a 300-dimensional embedding vector, which is the average of the embedding vectors of the top 50 keywords. After that, we remove the Null value from the financial dataset and standardize the numerical data into range $(0, 1)$ with Max-Min Scalar. We align the two sequences of financial indicators and news representation using Algorithm 1.

Implementation Details. In the model training stage, we randomly partition the data sets into the training, validation and test sets. The three disjoint sets include 7%, 15% and 15% of the data, respectively. To minimize the influence of the variability of the training set, model training process is repeated 10 times for each setting. We use Tensorflow to implement and train our prediction model.

Generally, we test different learning rates ranging from 0.1 to 0.001. We also test different batch sizes. Besides, we evaluate the effects of different numbers of LSTM hidden units, i.e., {16, 32, 64, 128, 256, 512}. In our experiments, we try to lessen the influence of class imbalance using random over-sampling. We control the parameter "resampling ratio" in our experimental results, which stands for the percentage of positive samples in a training batch.

Compared Methods. In this work, we compare our $DMDP^2$ framework with the baseline GAM model. Also we run our prediction model on FIN sequence or $TEXT$ sequence only. The compared methods are described as follows:

(1) *GAM* (generalized additive model) [27] is treated as the baseline in our study. In general, GAM is used to deal with time series data with a fixed window size. Enabling the discovery of a nonlinear fit between a variable and the response, the model makes use of the idea that time series could be decomposed as a plenty of individual trends, denoted by a sum of smooth functions [35].
(2) *LSTM* on FIN data only. The LSTM here is an ordinary LSTM with variable-length inputs. The inputs to this model is the 30-dimensional financial indicators sequence.
(3) *LSTM* on $TEXT$ data only. The LSTM here is also an ordinary LSTM with variable-length inputs. The inputs to this model is the 300-dimensional embedding representation of text sequence.

Metrics. We adopt area under receiver operating characteristic curve (AUC) as our evaluation criterion. Commonly used in selecting the optimal classifier that predicts the classes best, AUC weights errors on the two classes separately and tells a more truthful story when working with the imbalanced dataset. The random predictor will produce the AUC value with 0.5, the more powerful the classifier is, the larger AUC value will be. We also report the prediction accuracy of different methods on the test set.

4.2   Comparison Results

To verify the efficiency of our $DMDP^2$ framework, we compare the results of $DMDP^2$ with several baseline methods. The average accuracy and AUC are summarized in Table 2.

From the table, we find that $DMDP^2$ with Multi-source LSTM achieves the highest AUC of 0.85, higher than all the baselines. Also it is notable that given only $TEXT$ information, we can achieve AUC of 0.78, which is also higher than the baseline GAM. It demonstrates the effectiveness of information contained in the news. However, it seems that the LSTM model with FIN variables as its inputs only achieves comparable AUC performance with GAM. Moreover, the results show that $DMDP^2$ framework with Single LSTM has inferior performance compared to other methods. It somehow suggests that direct concatenation of the multiple sequences is not an effective data alignment method to

Table 2. Comparison of different methods. Note that each method was trained 10 times and we report their average performance and standard deviations (in brackets) for comparison.

Framework	Model	Data	Accuracy	AUC
-	GAM	FIN	0.68(0.01)	0.75(0.01)
-	LSTM	FIN	0.75(0.06)	0.75(0.03)
-	LSTM	TEXT	0.68(0.10)	0.78(0.05)
$DMDP^2$	Single LSTM	FIN + TEXT	0.79(0.03)	0.66(0.06)
$DMDP^2$	Multi-source LSTM	FIN + TEXT	0.94(0.03)	**0.85**(0.03)

Table 3. The results of different learning rates in $DMDP^2$ with Multi-source LSTM. Here the number of hidden units is 16 and the resampling ratio is 0.2.

Learning rate	Accuracy	AUC
0.1	0.99(0.00)	0.60(0.05)
0.01	0.94(0.03)	**0.85**(0.03)
0.001	0.99(0.00)	0.76(0.03)

combine the two heterogeneous data. However, by feeding input sequences from different sources into Multiple LSTMs, our $DMDP^2$ model is able to fully utilize the FIN and $TEXT$ information to predict the default probability with high accuracy. Therefore, our $DMDP^2$ framework with proper data alignment and Multi-source LSTM is superior to the baseline GAM. In addition, the results also illustrate the importance to align data properly and feed heterogeneous data into different LSTMs to effectively make use of multi-source data.

4.3   Parameter Tuning

Learning Rate. We test different values of learning rate from $\{0.1, 0.01, 0.001\}$ in our models. The results are presented in Table 3. We can see that our model configured with learning rate as 0.01 outperforms the other settings.

Hidden Units. We also evaluate the effects of different numbers of hidden units on model performance. The results are presented in Table 4. We can see that the models with 16 hidden units gives the best performance. However, the performance variance with different hidden units of LSTM in $DMDP^2$ is relatively small, indicating that the model is robust.

Resampling Ratio. The data, which has no more than 1% of records as *ST samples, is highly imbalanced. To alleviate this problem, We randomly over-sample the positive cases during the training process. We manually set the percentage of positive samples in the mini-batch as a parameter and tune this parameter to achieve the best performance. We call this parameter "the resampling ratio". In our experiments, we tested the resampling ratio ranging from

Table 4. The results of different numbers of hidden units in $DMDP^2$ with Multi-source LSTM. Here the learning rate is set to be 0.01 and the resampling ratio is 0.2.

#Hidden units	Accuracy	AUC
16	0.94(0.03)	**0.85**(0.03)
32	0.95(0.03)	0.81(0.03)
64	0.94(0.02)	0.83(0.04)
128	0.94(0.02)	0.81(0.04)
256	0.94(0.01)	0.83(0.05)
512	0.94(0.02)	0.82(0.03)

Table 5. The results of different resampling ratios in $DMDP^2$ with Multi-source LSTM. Here the learning rate is set to be 0.01 and the number of hidden units is 16.

Resampling ratio	Accuracy	AUC
0.1	0.97(0.01)	0.80(0.04)
0.2	0.94(0.03)	**0.85**(0.03)
0.3	0.89(0.03)	0.81(0.04)
0.4	0.86(0.04)	0.81(0.04)
0.5	0.82(0.05)	0.81(0.05)
0.6	0.82(0.04)	0.83(0.02)
0.7	0.74(0.14)	0.84(0.02)
0.8	0.63(0.20)	0.81(0.05)
0.9	0.61(0.21)	0.76(0.04)

0.1 to 0.9. The results could be seen in Table 5 The results show that we achieve the best result when the resampling ratio is equal to 0.2. From the results we can find that as the increase of resampling ratio, AUC does not change much. But accuracy decreases from 0.97 to 0.61. The reason behind this is to sacrifice the accuracy of one class to increase the accuracy of the other class. And AUC is a trade-off between the two classes.

5   Related Work

Recently, RNN variants such as LSTM [31] have been very successful in modeling the long-range sequential dependencies. And they have been applied to many time series forecasting or classification tasks. In [36], the authors used LSTM to predict whether a stock (6 typical stocks) would increase 0%–1% (class 1), above 1% (class 2) or not increasing (class 3) within next three hours with the highest accuracy at 59.5%. In [37], the authors proposed a novel SFM model, which incorporates Discrete Fourier Transform (DFT) into LSTM, to predict values in the future series. They argued that by decomposing the hidden states of LSTM into multi-frequency components, they could capture different latent patterns behind the original time series. In [38], the authors tried to combine LSTM and CNN into a single framework called TreNet as the author argued that CNNs extract salient features from local raw data while LSTM captures long-term dependency. The results demonstrated that the combined network outperforms both CNN and LSTM as well as various kernel based models in predicting the trend in time series. However, to the best of our knowledge, no prior work has studied the problem of default probability prediction which is a critical task to perform risk assessment for listed companies. Moreover, none of the existing time series prediction methods leverages the informative social media data to enhance prediction accuracy.

6   Conclusion

This paper has developed a default probability prediction framework $DMDP^2$, which leverages both structured financial factors and unstructured news from social media, to capture default risk states of the observed corporations. $DMDP^2$ involves a data alignment component to absorb multi-source data with different timestamps. We further adopt the LSTM to learn discriminative latent features as well as long-term financial status trend from the aligned sequences for final default probability prediction. In the experiments, we considered over 30 financial indicators including the profitability, solvency, operation ability, cash flow ability and potential growth ability of records for over 2000 listed corporations in mainland China. The results show that compared to the existing risk assessment approach that only considers financial factors, our neural method with additional indicators from social media news improves the accuracy of the default probability prediction results. As future work, we will investigate the following research directions: (1) the effects of public opinions among affiliated companies on a company's default value; (2) the importance of different features on default probability prediction performance; (3) incorporating text information in a more effective way other than extracting keywords only.

Acknowledgments. The work is partially supported by the Hong Kong RGC GRF Project 16214716, NSFC with grant No. 61602297 and 61729201.

References

1. Baesens, B., Van Gestel, T., Viaene, S., Stepanova, M., Suykens, J., Vanthienen, J.: Benchmarking state-of-the-art classification algorithms for credit scoring. J. Oper. Res. Soc. **54**(6), 627–635 (2003)
2. Desai, V.S., Crook, J.N., Overstreet, G.A.: A comparison of neural networks and linear scoring models in the credit union environment. Eur. J. Oper. Res. **95**(1), 24–37 (1996)
3. Hand, D.J., Henley, W.E.: Statistical classification methods in consumer credit scoring: a review. J. Roy. Stat. Soc. Ser. A (Stat. Soc.) **160**(3), 523–541 (1997)

4. Lee, T.S., Chiu, C.C., Chou, Y.C., Lu, C.J.: Mining the customer credit using classification and regression tree and multivariate adaptive regression splines. Comput. Stat. Data Anal. **50**(4), 1113–1130 (2006)
5. Lee, T.S., Chen, I.F.: A two-stage hybrid credit scoring model using artificial neural networks and multivariate adaptive regression splines. Expert Syst. Appl. **28**(4), 743–752 (2005)
6. Altman, E.I.: Financial ratios, discriminant analysis and the prediction of corporate bankruptcy. J Finance **23**(4), 589–609 (1968)
7. Altman, E.I., Saunders, A.: Credit risk measurement: developments over the last 20 years. J. Bank. Finance **21**(11–12), 1721–1742 (1997)
8. Huang, Z., Chen, H., Hsu, C.J., Chen, W.H., Wu, S.: Credit rating analysis with support vector machines and neural networks: a market comparative study. Decis. Support Syst. **37**(4), 543–558 (2004)
9. West, D.: Neural network credit scoring models. Comput. Oper. Res. **27**(11), 1131–1152 (2000)
10. Zhao, Z., Xu, S., Kang, B.H., Kabir, M.M.J., Liu, Y., Wasinger, R.: Investigation and improvement of multi-layer perceptron neural networks for credit scoring. Expert Syst. Appl. **42**(7), 3508–3516 (2015)
11. Chen, W., Ma, C., Ma, L.: Mining the customer credit using hybrid support vector machine technique. Expert Syst. Appl. **36**(4), 7611–7616 (2009)
12. Harris, T.: Credit scoring using the clustered support vector machine. Expert Syst. Appl. **42**(2), 741–750 (2015)
13. Hens, A.B., Tiwari, M.K.: Computational time reduction for credit scoring: an integrated approach based on support vector machine and stratified sampling method. Expert Syst. Appl. **39**(8), 6774–6781 (2012)
14. Huang, C.L., Chen, M.C., Wang, C.J.: Credit scoring with a data mining approach based on support vector machines. Expert Syst. Appl. **33**(4), 847–856 (2007)
15. Schebesch, K.B., Stecking, R.: Support vector machines for classifying and describing credit applicants: detecting typical and critical regions. J. Oper. Res. Soc. **56**(9), 1082–1088 (2005)
16. Bijak, K., Thomas, L.C.: Does segmentation always improve model performance in credit scoring? Expert Syst. Appl. **39**(3), 2433–2442 (2012)
17. Yap, B.W., Ong, S.H., Husain, N.H.M.: Using data mining to improve assessment of credit worthiness via credit scoring models. Expert Syst. Appl. **38**(10), 13274–13283 (2011)
18. Verikas, A., Gelzinis, A., Bacauskiene, M.: Mining data with random forests: a survey and results of new tests. Pattern Recogn. **44**(2), 330–349 (2011)
19. Brown, I., Mues, C.: An experimental comparison of classification algorithms for imbalanced credit scoring data sets. Expert Syst. Appl. **39**(3), 3446–3453 (2012)
20. Tsai, C.F., Chen, M.L.: Credit rating by hybrid machine learning techniques. Appl. Soft Comput. **10**(2), 374–380 (2010)
21. Yeh, I.C., Lien, C.: The comparisons of data mining techniques for the predictive accuracy of probability of default of credit card clients. Expert Syst. Appl. **36**(2), 2473–2480 (2009)
22. Doumpos, M., Zopounidis, C.: Model combination for credit risk assessment: a stacked generalization approach. Ann. Oper. Res. **151**(1), 289–306 (2007)
23. Tsai, C.F., Wu, J.W.: Using neural network ensembles for bankruptcy prediction and credit scoring. Expert Syst. Appl. **34**(4), 2639–2649 (2008)
24. Twala, B.: Multiple classifier application to credit risk assessment. Expert Syst. Appl. **37**(4), 3326–3336 (2010)

326   Y. Zhao et al.

25. Wang, G., Hao, J., Ma, J., Jiang, H.: A comparative assessment of ensemble learning for credit scoring. Expert Syst. Appl. **38**(1), 223–230 (2011)
26. Yu, L., Wang, S., Lai, K.K.: Credit risk assessment with a multistage neural network ensemble learning approach. Expert Syst. Appl. **34**(2), 1434–1444 (2008)
27. Sousa, M.R., Gama, J., Brandão, E.: A new dynamic modeling framework for credit risk assessment. Expert Syst. Appl. **45**, 341–351 (2016)
28. Wang, W.Y., Hua, Z.: A semiparametric gaussian copula regression model for predicting financial risks from earnings calls. In: ACL, vol. 1, pp. 1155–1165 (2014)
29. Klinkenberg, R.: Learning drifting concepts: example selection vs. example weighting. Intell. Data Anal, **8**(3), 281–300 (2004)
30. Sousa, M.R., Gama, J., Gonçalves, M.J.S.: A two-stage model for dealing with temporal degradation of credit scoring. arXiv preprint arXiv:1406.7775 (2014)
31. Hochreiter, S., Schmidhuber, J.: Long short-term memory. Neural Comput. **9**(8), 1735–1780 (1997)
32. Zaremba, W., Sutskever, I., Vinyals, O.: Recurrent neural network regularization. arXiv preprint arXiv:1409.2329 (2014)
33. Kingma, D.P., Ba, J.: Adam: A method for stochastic optimization. arXiv preprint arXiv:1412.6980 (2014)
34. Joulin, A., Grave, E., Bojanowski, P., Douze, M., Jégou, H., Mikolov, T.: Fasttext.zip: compressing text classification models. arXiv preprint arXiv:1612.03651 (2016)
35. Hastie, T., Tibshirani, R.: Generalized Additive Models. Wiley, Hoboken (1990)
36. Gao, Q.: Stock market forecasting using recurrent neural network. Ph.D thesis. University of Missouri-Columbia (2016)
37. Zhang, L., Aggarwal, C., Qi, G.J.: Stock price prediction via discovering multi-frequency trading patterns. In: Proceedings of the 23rd ACM SIGKDD International Conference on Knowledge Discovery and Data Mining, pp. 2141–2149. ACM (2017)
38. Lin, T., Guo, T., Aberer, K.: Hybrid neural networks for learning the trend in time series. In: Proceedings of the Twenty-Sixth International Joint Conference on Artificial Intelligence, IJCAI-17, pp. 2273–2279 (2017)

Brain Disease Diagnosis Using Deep Learning Features from Longitudinal MR Images

Linlin Gao[1], Haiwei Pan[1(✉)], Fujun Liu[2], Xiaoqin Xie[1], Zhiqiang Zhang[1], Jinming Han[3], and the Alzheimer's Disease Neuroimaging Initiative

[1] Department of Computer Science and Technology, Harbin Engineering University, Harbin, People's Republic of China
panhaiwei@hrbeu.edu.cn
[2] Department of Electrical and Computer Engineering, University of Florida, Gainesville, USA
[3] Department of Neurology and Neuroscience Center, 1st Hospital of Jilin University, Changchun, China

Abstract. Deep learning-based brain disease diagnoses utilizing magnetic resonance (MR) images has attracted increasing attention in the field of computer-aided diagnosis. However, most existing methods require computationally expensive preprocessing before feature extraction, such as 3D MR image registration and landmark detection. Additionally, these methods only employ cross-sectional MR images. Recent studies have demonstrated that longitudinal images acquired at different time points can comprehensively reflect the pathological changes of diseases. To date, effectively capturing information from variable numbers of longitudinal MR images has not been adequately investigated. In this study, we propose a deep learning method taking advantage of longitudinal MR images for disease diagnoses. In particular, we first extract features from slice images employing a Deep Convolutional Neural Network (DCNN) in an end-to-end manner. This avoids 3D image registration and landmark detection. We then generate longitudinal-level features by using the Bag-of-Words (BoW) model. Lastly, we devise a Recurrent Neural Network (RNN) to capture the pathological changes for facilitating disease diagnoses. We evaluate the proposed method on the public Alzheimer's Disease National Initiative (ADNI) dataset. Extensive experiments show that the proposed method is superior to baseline methods and is robust to both the Alzheimer's disease (AD) and mild cognitive impairment (MCI) diagnoses. Moreover, the proposed method can effectively learn pathological changes from the longitudinal MR images for disease diagnosis.

Keywords: Disease diagnoses · Slice images · DCNN
Longitudinal MR images · RNN

1   Introduction

An increasing number of deep learning-based studies have been developed for brain disease diagnoses [1–4]. However, most of them utilize multimodal information [1,2,5,6] that is typically pricey to obtain in practice. Recent studies reveal that MR images has been an effective tool for brain disease diagnoses because of its capability to visualize anatomical structures [3]. *Suk et al.* proposed a deep learning method for Alzheimer Disease (AD) and Mild Cognitive Impairment (MCI) diagnoses using one-time-point brain MRI scans [3]. They devised a convolutional neural network (CNN) to fuse results from different sparse regression models, but not to extract features from images. Moreover, the method required 3D image registration and hand-crafted landmark extraction that are computationally expensive to obtain. *Korolev et al.* also devised CNNs for the AD and MCI diagnoses using cross-sectional MR images [4]. Different from [3], they utilized DCNNs to extract features without any hand-crafted landmark. However, it still required time-consuming 3D image registration. The method is the most similar to the proposed method since both of them extract features using deep learning methods directly for binary classification. The method achieves the accuracy of 0.80 for the AD diagnosis and 0.59 for the MCI diagnosis [4] which is very close to random classification. It is worth noting that both of the above methods only use cross-sectional MR images for disease diagnoses.

A growing number of studies have shown that longitudinal data taken at different time points can significantly contribute to disease diagnoses [7–9]. This is because longitudinal data can demonstrate the pathological changes of diseases. However, most longitudinal studies just focus on analyzing longitudinal data with fixed length [7,8], which is uncommon in practice partly due to data missing at some time points during data collection. To handle the problem, Zhang. *et al.* proposed a landmark-based method to fuse variable numbers of longitudinal MR images for disease diagnoses [9]. The method extracted landmarks at the first place and then utilized the Jacobian map to capture pathological changes of these landmarks. However, the method required costly preprocessing such as 3D image registration, patch selection, and landmark extraction *etc.*.

To relieve computationally expensive preprocessing and to effectively capture the pathological changes of diseases, we propose a deep learning method employing longitudinal MR images to improve diagnostic results. We are motivated by the following evidence: (1) Clinic doctors first extract diagnostic information from slice images and then integrate the slice-level diagnostic information to make diagnoses; (2) Transfer learning has been a powerful tool to enable training a large network without overfitting because of the general features in the lower layers [10]; (3) The success of RNNs in many tasks such as video classification and caption generation has proven the impressive ability of RNNs to learn from arbitrarily long sequences. The framework of the proposed method is schematically illustrated in Fig. 1. Specifically, we extract slice-level features by fine-tuning the state-of-the-art ResNet [11] at the first place. This avoids computationally expensive 3D registration or landmark detection. We then integrate the slice-level features of each scan into a longitudinal-level feature using

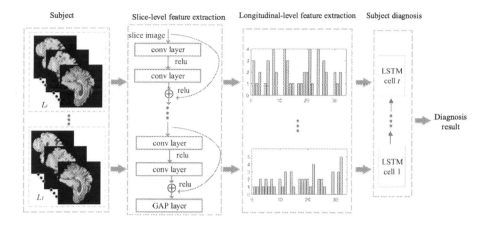

Fig. 1. Framework of the proposed method.

the BoW model [12]. At last, we devise a RNN model to learn the pathological changes from the longitudinal-level features for disease diagnoses. Experimental results on the public ADNI dataset show the effectiveness of the proposed method in both the AD and MCI diagnoses. To the best of our knowledge, this is the first study that employs deep learning methods to extract features from slice images and to capture pathological changes of diseases from longitudinal medical images for diagnoses.

2   Materials and Methods

2.1   Dataset and Preprocessing

The ADNI is launched in 2003 to test whether serial MRI, positron emission tomography, other biological markers, and clinical and neuropsychological assessment can be combined to measure the progression of MCI and early AD[1]. In this study, we downloaded the MR images of 415 subjects from the ADNI. All the MR images were acquired using 1.5T scanners with the protocol of sagittal T1-weighted MPRAGE. The 415 subjects consist of 111 AD, 150 MCI, and 154 nomal control (NC). Each subject has up to 6 MR scans obtained at different time points. The general time interval among longitudinal MI images is 3 months, but it is not fixed because of data missing and absence of participants during data collection. Each MR scan has 176 ± 13 sagittal slices and each original sagittal slice has the width of 243 ± 10 and height of 223 ± 33. We conduct skull stripping, dura and neck removal on each scan using the brain extraction tool (BET)[2]. After that, each slice is made up of two portions: background and the intracranial portion. Since there is no slice-wise label, we assign each slice the

---

[1] http://adni.loni.usc.edu/.
[2] https://fsl.fmrib.ox.ac.uk/fsl/fslwiki/BET.

label of its subject. However, this leads to numerous noisy labels. To handle the problem, we extract the intracranial portion along its external vertical rectangle, and select the top-50 large intracranial portions from each scan as our experimental slice images. After this, most noisy and useless slice images are removed. The value of 50 is determined based on the guidance of neurologists, because the slices of patients including the top-50 large intracranial portions contain most diagnostic information and simultaneously the labels of these slices are almost correct.

2.2   Slice-Level Feature Extraction by Fine-Tuning a DCNN

Recent research shows that DCNNs have gained great success in learning rich hierarchical features [13]. However, DCNNs require a large image set to train from scratch, such as the ImageNet [14]. It is relatively rare to have such a sufficient dataset in the field of medical image analysis. In practice, it is common to use a pre-trained DCNN as initialization for new tasks [15]. This is because the lower layers of DCNNs contain more generic features that are useful to many tasks and can be transferred from one application domain to the others [10,16]. Therefore, we fine-tune the notable ResNet [11] to generate a slice-level feature extractor.

ResNet was pre-trained on the ImageNet and won first place on the ILSVRC 2015 classification task. It has been proven easy to optimize and can gain accuracy from considerably increased depth. The architecture of the ResNet that we adopt is shown in Table 1, which is 18 layers and is deep enough for our tasks. Max-pooling is performed after conv1. Downsampling is performed by conv3_1, conv4_1, and conv5_1 with a stride of 2. Two output layers are depicted in Table 1. The above one comprises a global average pooling (GAP) layer, a fully connected (FC) layer, and a Softmax layer. The FC layer has N units representing the number of categories in the classification task. For instance, $N = 2$ in binary classification. This output layer with all the preceding layers is used to fine-tune the ResNet-18. The bottom output layer only contains a GAP layer. The GAP is useful for localizing discriminative image regions for classification [17]. It with all the preceding layers is thereby used as our feature extractor to generate slice-level features. From Table 1, we can see that each slice image can be represented by a 512-dimensional slice-level feature.

Training Details: We resize each slice image into the size of 224×224 as the input of the ResNet-18. Stochastic gradient descent (SGD) is employed for optimization with a mini-batch size of 128. Cross-entropy is used as the cost function. The initial learning rate is set to 0.001 and divided by 10 every 10 epochs with 0.9 momentum. The model converges after 60 epochs. The detailed implementation of the ResNet-18 can be found in the open source code[3].

---

[3] http://pytorch.org/docs/0.3.0/modules/torchvision/models/resnet.htmlresnet18.

Table 1. The architecture of the ResNet used in our study.

Layer name	Layers	Output size
conv1	7×7, 64, stride 2	112×112
conv2_x	$\begin{bmatrix} 3 \times 3, 64 \\ 3 \times 3, 64 \end{bmatrix} \times 2$	56×56
conv3_x	$\begin{bmatrix} 3 \times 3, 128 \\ 3 \times 3, 128 \end{bmatrix} \times 2$	28×28
conv4_x	$\begin{bmatrix} 3 \times 3, 256 \\ 3 \times 3, 256 \end{bmatrix} \times 2$	14×14
conv5_x	$\begin{bmatrix} 3 \times 3, 512 \\ 3 \times 3, 512 \end{bmatrix} \times 2$	7×7
Output	GAP, FC, Softmax	$N \times 1$
	GAP	512×1

2.3 Longitudinal-Level Feature Integration Using the BoW Model

Since each slice image is represented by a slice-level feature, each scan is made up of 50 slice-level features. In this section, we employ the simple yet effective BoW model [12] to integrate the slice-level features of each scan into a longitudinal-level feature.

Let $\mathcal{F}_{ij} \in \mathbb{R}^{P \times Q}$ denote the slice-level feature set of the jth scan l_{ij} of the ith subject L_i, where $P = 50$ is the number of slices of l_{ij} and $Q = 512$ is the dimension of a slice-level feature. All the slice-level features of the training data constitute $F \in \mathbb{R}^{M \times Q}$, where M is the total number of the slices in the training data. That is, $F = \{\mathcal{F}_{ij}\}$. We first cluster the slice-level features in F using K-means [18] and then build a code book $\mathcal{B} = \{b_1, ..., b_k, ..., b_K\}$, where $\mathcal{B} \in \mathbb{R}^{K \times Q}$, and $b_k \in \mathbb{R}^{1 \times Q}$ representing the kth clustering center. The K-dimentional longitudinal feature $X_{ij} = \{x_{ij1}, ..., x_{ijk}, ..., x_{ijK}\}$ of l_{ij} is calculated based on Eq. 1.

$$x_{ijk} = \sum_{f \in \mathcal{F}_{ij}} ((\underset{m \in \{1,2,...,K\}}{argmin} ||f - b_m||_2^2) == k), \tag{1}$$

where $f \in \mathbb{R}^{1 \times Q}$ is a slice-level feature. Thus, the subject L_i can be represented by the longitudinal-level features: $X_{i1}, ..., X_{ik}, ..., X_{it}$, where t is the total number of scans of L_i and t varies among subjects. It is noted that $X_{i1}, ..., X_{ik}, ..., X_{it}$ are ordered by the time that the corresponding scans are acquired, i.e., X_{i1} is the longitudinal-level feature of the first-acquried scan and is followed by X_{i2}, and X_{it} is that of the last-acquired scan.

2.4 Disease Diagnosis Using the Devised RNN Model

RNNs are extremely powerful to address tasks using arbitrary sequences. In particular, a RNN produces a new hidden state at each time-step based on

Table 2. The architecture of the devised RNN.

Layer name	Layers	Output size
LSTM	128×1	128×1
Output	Softmax	$N \times 1$

both the current input and previous hidden state in a feed-forward fashion. The process is formalized by:

$$h_k = f(W_{xh}X_k + W_{hh}h_{k-1}), \tag{2}$$

where X_k is the input at the kth time-step, h_k and h_{k-1} are the kth and $(k-1)$th hidden states, respectively, and W_{xh} and W_{hh} are the weights to be learned. The feed-forward propagation favors the "remembering" ability of RNNs. RNNs then learn the dependencies of a sequence by utilizing backpropagation. However, vallina RNNs generally suffer from the vanishing or exploding gradient during backpropagation. Therefore, the long short-term memory (LSTM) model is generated by replacing each node in the hidden layer with a memory cell to handle long-sequence dependency [19]. In this study, we employ the LSTM to explore the pathological changes among longitudinal MR images for disease diagnoses. Pathological changes captured from early-acquired MR images to later-acquired MR images provide parallel information compared to that captured from later-acquired MR images to early-acquired MR images. Therefore, the bidirectional LSTM is not necessary in our study.

The architecture of the devised RNN model is shown in Table 2. It consists of a LSTM layer and an output layer. The LSTM layer has 128 units. Because of the up to 6 longitudinal-level features of subjects, the LSTM layer has 6 hidden states. The longitudinal-level features with size less than 6 are padded with 0. The output layer utilizes the Softmax function to achieve classification with N categories.

Training Details: The input of the RNN model is the longitudinal-level features of subjects, *i.e.*, $X_{i1}, ..., X_{ik}, ..., X_{it}$, where X_{ik} is the kth longitudinal-level feature of the ith subjects and it corresponds to X_k in Eq. 2. RMSProp [20] is employed for optimization with a mini-batch size of 32. Cross-entropy is used as the cost function. The learning rate is set to 0.01 with no decay nor momentum. The model is stopped after 200 epochs.

3   Experimental Results and Analysis

We evaluate the proposed method with two classification tasks: NC vs. AD (*i.e.*, the AD diagnosis) and NC vs. MCI (*i.e.*, the MCI diagnosis), where AD and MCI subjects are regarded as the positive category and NC subjects as the negative one. For each task, 40%, 20%, and 40% of subjects are selected

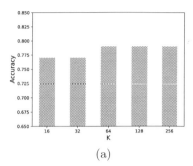 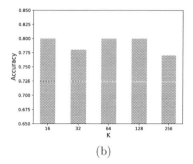

(a)   (b)

Fig. 2. Accuracy of the proposed method with different K. (a) NC vs. AD. (b) NC vs. MCI

to form training, validation, and test data, respectively. Specifically, for the AD diagnosis, 106, 53, and 106 subjects constitute training, validation, and test sets, respectively. For the MCI diagnosis, 123, 60, and 121 subjects are used as training, validation, and test sets, respectively. Four common metrics are adopted to evaluate the performance of the proposed method. They are (1) Accuracy (ACC), (2) Balanced accuracy (BAC) [9], (3) Mathews Correlation Coefficient (MCC) [21], and (4) area under receiver operating characteristic (ROC) curve.

3.1 Evaluation of the Parameter K

In this section, we investigate the parameter K of K-means in Sect. 2.3, which is used to build the code book in the BoW model. The K is selected based on the accuracy of the proposed method performed on the validation set, as shown in Fig. 2. From Fig. 2(a), we can see that the accuracy of the AD diagnosis achieves the highest point when the K is equal to 64, 128, and 256. From Fig. 2(b), we can see that the accuracy of the MCI diagnosis is the best when the K equals 16, 64, and 128. Here, we set the K values of both the AD and MCI diagnoses to 64 in order to keep the consistency of the K in the two tasks and to keep the length of the longitudinal-level as short as possible for reducing the computational complexity.

3.2 Comparison Results with Baseline Strategies

To evaluate the effectiveness of the proposed method in capturing the pathological changes, we compare the proposed method with three baseline strategies on the test sets of the two tasks. All of the three strategies utilize longitudinal MR images for diagnoses. However, they take different ways to organize the features of longitudinal MR images. The details of the three strategies are as below:

1. MIL: The method utilizes multiple instance learning [22] for disease diagnoses. Let $\mathcal{D}_\mathcal{T} = \{(S_1, y_1), ..., (S_i, y_i), ..., (S_\mathcal{T}, y_\mathcal{T}) | S_i = \{X_{i1}, ..., X_{ik}, ..., X_{it}\}\}$ represent the training data, where S_i is a bag representing all the longitudinal-level features of the subject L_i and $\mathcal{T}$ is the total number of subjects in $\mathcal{D}_\mathcal{T}$; y_i is

Table 3. ACC, MCC, and BAC of the four methods.

	NC vs. AD			NC vs. MCI		
	ACC	BAC	MCC	ACC	BAC	MCC
MIL	84.8%	83.4%	68.6%	63.3%	63.3%	26.6%
WMV	85.7%	84.2%	70.7%	50.0%	50.6%	1.7%
BowSVM	87.6%	87.4%	74.7%	65.0%	65.1%	30.6
Proposed	**89.5%**	**88.1%**	**78.8%**	**81.7%**	**81.4%**	**66.2%**

the label of the subject S_i; X_{ik} is an instance of S_i representing the longitudinal-level feature of the scan l_{it} and t is the number of scans of the subject L_i; A MIL classifier is trained based on $\mathcal{D}_T$ for disease diagnoses.

2. WMV: The method utilizes weighted majority voting for diagnoses. Let w_{ij} denote the output of the ith slice of a subject via the softmax layer of the fine-tuned ResNet, i.e., the score probability. Since the two tasks are binary classification, w_{ij} is a 2-dimensional vector. The diagnostic result of a subject is calculated by:

$$result = \arg \max_{j \in \{0,1\}} \sum_{i=1}^{P \times t} w_{ij}, \qquad (3)$$

where t is the number of the scans of the subject and $P = 50$ is the number of the slices in each scan. Here $result = 0$ indicates the subject is positive, $result = 1$ indicates the subject is negative.

3. BowSVM: The method utilizes the BoW and SVM [23] models for diagnoses. We first calculate the longitudinal-level features of subjects as the way in Sect. 2.3, i.e., using the BoW model to integrate the slice-level features into the longitudinal-level features. An SVM classifier is trained based on the longitudinal-level features for disease diagnoses.

The performance of the three strategies and the proposed method is summarized in Table 3 and Fig. 3. Among the methods, the MIL, BowSVM, and the proposed method utilize K-means to build a code book. To keep the consistency of experimetal setting, we fix the K values of the three strategies to 64. As can be seen from the Table 3, the proposed method achieves the best ACC, BAC, and MCC values in both tasks. Moreover, for the MCI diagnosis, the results of the proposed method outweigh that of the three strategies significantly. From Fig. 3, we can see that (1) for the AD diagnosis (i.e., Fig. 3(a)), the ROC curve of the proposed method (the red curve) is a little higher than that of the MIL-subject-scan and intersects that of the other two strategies. These results indicate the proposed method for the AD diagnosis is superior to the MIL-subject-scan and is comparable with the other two strategies. (2) For the MCI diagnosis (i.e., Fig. 3(b)), the proposed method (the red line) outperforms the three strategies notably. The result reveal the effectiveness of the proposed method in capturing the pathological changes from longitudinal MR images for

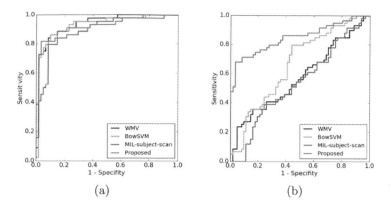

Fig. 3. ROC curves of the four methods. (a) NC vs. AD. (b) NC vs. MCI (Color figure online)

the MCI diagnosis. Moreover, even though the pathological changes among MCI subjects are more challenging to detect compared with that among AD subjects, the proposed method can achieve inspiring results. From Table 3 and Fig. 3, we can conclude that (1) the proposed method is robust to both the AD and MCI diagnoses because of the more similar results for both the AD and MCI diagnoses of the proposed method than the results for both the AD and MCI diagnoses of the other three strategies; (2) the proposed method benefits the MCI diagnosis a lot.

3.3 Evaluation of the Features Extracted Using the Proposed Method

The performance of the BowSVM strategy is the most similar to that of the proposed method. Here, we utilize the t-SNE visualization [24] to evaluate the features generated from the LSTM layer of the proposed method and the features input to the SVM classifier in the BowSVM strategy, as shown in Fig. 4. Figure 4(a) and (c) are the t-SNE visualization of features generated using the BowSVM strategy for the AD and MCI diagnoses, respectively, and (b) and (d) are that using the proposed method for the AD and MCI diagnoses. In these four figures, the blue stars represent the features from NC subjects and the red circles represent the features from AD subjects (in Fig. 4(a) and (b)) or features from MCI subjects (in Fig. 4(c) and (d)). Comparing Fig. 4(a) and (b), we can see that the number of blended blue stars and red circles in Fig. 4(a) are more than that in Fig. 4(b). This indicates that features used for the AD diagnosis generated using the proposed method are more discriminative than that using the BowSVM strategy. Comparing Fig. 4(c) and (d), we can see that the number of blended blue stars and red circles in Fig. 4(c) are more than that in Fig. 4(d). This indicates that features used for the MCI diagnosis using the proposed method are more discriminative than that using the BowSVM strategy.

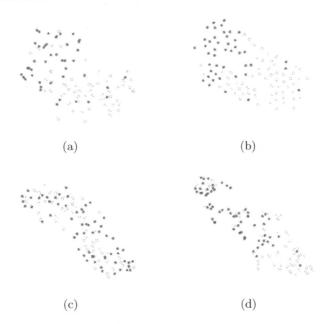

Fig. 4. t-SNE visualization of features generated using the proposed method and the BowSVM strategy for the AD and MCI diagnoses. (a) For NC vs. AD using the BowSVM strategy. (b) For NC vs. AD using the proposed method. (c) For NC vs. MCI using the BowSVM strategy. (b) For NC vs. MCI using the proposed method. (Color figure online)

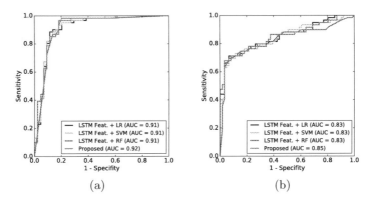

Fig. 5. ROC curves of the six methods. (a) NC vs. AD. (b) NC vs. MCI.

Therefore, the proposed method can more effectively capture the pathological changes of diseases from the longitudinal MR images for both AD and MCI diagnoses compared with the BowSVM strategy.

To explore the stability of the pathological changes captured from longitudinal MR images using the proposed method, we further demonstrate the ROC curves of different classifiers using the features from the LSTM layer. These classifiers are the SVM, logistic regression (LR), random forest (RF), and the proposed method which utilizes the Softmax function for classification. Their ROC curves are displayed in Fig. 5. It can be seen that these ROC curves, for both the AD and MCI diagnoses, are very close. This indicates that the features learned from the devised RNN model have strong stability.

4 Conclusion

In this study, we proposed a deep learning-based method to capture pathological changes from longitudinal MR images for disease diagnoses. Unlike most of existing methods, the proposed method does not require computationally expensive preprocessing steps. In particular, we first extract slice-level features by fine-tuning the state-of-the-art ResNet-18 in an end-to-end manner. We then integrate the slice-level features of scans into the longitudinal-level features by using the BoW model. At last, we devise a RNN model to capture the pathological changes of diseases for more accurate disease diagnoses. Experimental results on the ADNI dataset show that (1) the proposed method outweighs the state-of-the-art classifiers; (2) the proposed method is robust to both the AD and MCI diagnoses. These results indicate can effectively take advantage of longitudinal MR images to facilitate both the AD and MCI diagnoses.

Since the proposed method requires the guidance of doctors to set the N in the process of automatic slice image selection, our future work will focus on selecting slice images without doctors' guidance, i.e., reducing noisy data without any priori knowledge. Moreover, since ordinal information is a nature of the development of the NC, MCI, and AD diseases, we will focus on multi-classification task combining ordinal information.

Acknowledgement. The paper is partly supported by the National Natural Science Foundation of China under Grant No. 61672181, 51679058, Natural Science Foundation of Heilongjiang Province under Grant No. F2016005, the Numerical Tank Innovative Project (Phase I) and the Council Scholarship of China. Data collection and sharing was funded by ADNI (National Institutes of Health Grant U01 AG024904) and DOD ADNI (Department of Defense award number W81XWH-12-2-0012). ADNI is funded by the National Institute on Aging, the National Institute of Biomedical Imaging and Bioengineering, and through generous contributions. We are also thankful for Mason Mcgough (his introduction is here https://www.bme.ufl.edu/labs/yang/group.html) for his serious presentation modification.

References

1. Liu, S., Liu, S., Cai, W., Pujol, S., Kikinis, R., Feng, D.: Early diagnosis of Alzheimer's disease with deep learning. In: ISBI, pp. 1015–1018. IEEE (2014)
2. Shi, J., Zheng, X., Li, Y., Zhang, Q., Ying, S.: Multimodal neuroimaging feature learning with multimodal stacked deep polynomial networks for diagnosis of Alzheimer's disease. IEEE J. Biomed. Health Inf. **22**(1), 173–183 (2018)

3. Suk, H.-I., Lee, S.-W., Shen, D., et al.: Deep ensemble learning of sparse regression models for brain disease diagnosis. Med. Image Anal. **37**, 101–113 (2017)
4. Korolev, S., Safiullin, A., Belyaev, M., Dodonova, Y.: Residual and plain convolutional neural networks for 3D brain MRI classification. arXiv preprint arXiv:1701.06643 (2017)
5. Suk, H.-I., Shen, D.: Deep learning-based feature representation for AD/MCI classification. In: Mori, K., Sakuma, I., Sato, Y., Barillot, C., Navab, N. (eds.) MICCAI 2013. LNCS, vol. 8150, pp. 583–590. Springer, Heidelberg (2013). https://doi.org/10.1007/978-3-642-40763-5_72
6. Suk, H.-I., Lee, S.-W., Shen, D., et al.: Hierarchical feature representation and multimodal fusion with deep learning for AD/MCI diagnosis. NeuroImage **101**, 569–582 (2014)
7. Chincarini, A., Sensi, F., Rei, L., Gemme, G., Squarcia, S., Longo, R., Brun, F., Tangaro, S., Bellotti, R., Amoroso, N., et al.: Integrating longitudinal information in hippocampal volume measurements for the early detection of Alzheimer's disease. Neuroimage **125**, 834–847 (2016)
8. Farzan, A., Mashohor, S., Ramli, A.R., Mahmud, R.: Boosting diagnosis accuracy of Alzheimer's disease using high dimensional recognition of longitudinal brain atrophy patterns. Behav. Brain Res. **290**, 124–130 (2015)
9. Zhang, J., Liu, M., An, L., Gao, Y., Shen, D.: Alzheimer's disease diagnosis using landmark-based features from longitudinal structural mr images. IEEE J. Biomed. Health Inf. **21**, 1067–1616 (2017)
10. Yosinski, J., Clune, J., Bengio, Y., Lipson, H.: How transferable are features in deep neural networks? In: Advances in Neural Information Processing Systems, pp. 3320–3328 (2014)
11. He, K., Zhang, X., Ren, S., Sun, J.: Deep residual learning for image recognition. In: Proceedings of the IEEE Conference on CVPR, pp. 770–778 (2016)
12. Sivic, J., Zisserman, A.: Video Google: a text retrieval approach to object matching in videos. In: Null, p. 1470. IEEE (2003)
13. Yang, J., Price, B., Cohen, S., Lee, H., Yang, M.-H.: Object contour detection with a fully convolutional encoder-decoder network. In: Proceedings of the IEEE Conference on Computer Vision and Pattern Recognition, pp. 193–202 (2016)
14. Deng, J., Dong, W., Socher, R., Li, L.-J., Li, K., Fei-Fei, L.: ImageNet: a large-scale hierarchical image database. In: IEEE Conference on Computer Vision and Pattern Recognition, CVPR 2009, pp. 248–255. IEEE (2009)
15. Noh, H., Hongsuck Seo, P., Han, B.: Image question answering using convolutional neural network with dynamic parameter prediction. In: Proceedings of the IEEE Conference on CVPR, pp. 30–38 (2016)
16. Cai, J., Lu, L., Xie, Y., Xing, F., Yang, L.: Pancreas segmentation in MRI using graph-based decision fusion on convolutional neural networks. In: Descoteaux, M., Maier-Hein, L., Franz, A., Jannin, P., Collins, D.L., Duchesne, S. (eds.) MICCAI 2017. LNCS, vol. 10435, pp. 674–682. Springer, Cham (2017). https://doi.org/10.1007/978-3-319-66179-7_77
17. Zhou, B., Khosla, A., Lapedriza, A., Oliva, A., Torralba, A.: Learning deep features for discriminative localization. In: Proceedings of the IEEE Conference on Computer Vision and Pattern Recognition, pp. 2921–2929 (2016)
18. Hartigan, J.A., Wong, M.A.: Algorithm as 136: a k-means clustering algorithm. J. Roy. Stat. Soc. Ser. C Appl. Stat. **28**(1), 100–108 (1979)
19. Hochreiter, S., Schmidhuber, J.: Long short-term memory. Neural comput. **9**(8), 1735–1780 (1997)

20. Hinton, G., Srivastava, N., Swersky, K.: RmsProp: divide the gradient by a running average of its recent magnitude. Neural Networks For Machine Learning, Coursera Lecture (2012)

21. Matthews, B.W.: Comparison of the predicted and observed secondary structure of t4 phage lysozyme. Biochim. et Biophys. Acta (BBA) Protein Struct. **405**(2), 442–451 (1975)

22. Doran, G., Ray, S.: A theoretical and empirical analysis of support vector machine methods for multiple-instance classification. Mach. Learn. **97**(1–2), 79–102 (2014)

23. Joachims, T.: Making large-scale SVM learning practical. SFB 475: Komplexitätsreduktion in Multivariaten Datenstrukturen, Universität Dortmund, Technical Report (1998)

24. Maaten, L., Hinton, G.: Visualizing data using t-SNE. J. Mach. Learn. Res. **9**(Nov), 2579–2605 (2008)

Attention-Based Recurrent Neural Network for Sequence Labeling

Bofang Li[1,2], Tao Liu[1,2(✉)], Zhe Zhao[1,2], and Xiaoyong Du[1,2]

[1] School of Information, Renmin University of China, Beijing, China
{libofang,tliu,helloworld,duyong}@ruc.edu.cn
[2] Key Laboratory of Data Engineering and Knowledge Engineering,
MOE, Beijing, China

Abstract. Sequence labeling is one of the key problems in natural language processing. Recently, Recurrent Neural Network (RNN) and its variations have been widely used for this task. Despite their abilities of encoding information from long distance, in practice, one single hidden layer is still not sufficient for prediction. In this paper, we propose an attention architecture for sequence labeling, which allows RNNs to selectively focus on every useful hidden layers instead of irrelative ones. We conduct experiments on four typical sequence labeling tasks, including Part-Of-Speech Tagging (POS), Chunking, Named Entity Recognition (NER), and Slot Filling for Spoken Language Understanding (SF-SLU). Comprehensive experiments show that our attention architecture provides consistent improvements over different RNN variations.

1 Introduction

Nowadays, analyzing and extracting useful information from plain text (especially web content) is one of the most important research areas. For many applications, sequence labeling is a fundamental pre-processing step. It is also one of the most well-studied tasks in natural language processing. As shown in Table 1, sequence labeling tasks aim at automatically assigning words in texts with labels.

Traditionally, Hidden Markov Models (HMM), Conditional Random Fields (CRFs), and Support Vector Machine (SVM) has been widely used for sequence labeling tasks [9,10,14,15]. Compared with these models, Recurrent Neural Networks (RNNs) are able to capture information from a fairly long distance. Recently, with the help of extra resources and feature engineering, the combination of RNN and other models achieves state-of-the-art results [3,8,12,13].

For sequence labeling, each target word and its corresponding label are explicitly aligned. Previous RNNs predict label solely based on each hidden layer of the corresponding target word. However, in practice, using one single hidden layer is not sufficient for prediction, even with sophisticated variations like Bidirectional Recurrent Neural Network (Bi-RNN) [16], Long Short-Term Memory (LSTM) [7], and Gated Recurrent Unit (GRU) [4].

© Springer International Publishing AG, part of Springer Nature 2018
Y. Cai et al. (Eds.): APWeb-WAIM 2018, LNCS 10987, pp. 340–348, 2018.
https://doi.org/10.1007/978-3-319-96890-2_28

Table 1. An example of sequence labeling tasks.

Words	Flight from Boston to New York					
POS	NN	IN	NNP	TO	NNP	NNP
Chunking	B-NP	B-PP	B-NP	B-PP	B-NP	I-NP
NER	O	O	B-loc	O	B-loc	I-loc
SF-SLU	O	O	B-dept	O	B-arr	I-arr

This paper proposes an Attention-based Recurrent Neural Network for Sequence Labeling (ARNN-SL), which allows RNNs to "focus" not only on the aligned hidden layer, but other informative hidden layers as well.

Different from other tasks such as machine translation [1], image caption [19], and speech recognition [5], where attention mechanism has been successfully applied, sequence labeling has its own characteristics for deciding which hidden layer is informative or not. Intuitively, the closer a hidden layer is to target word, the more information it contains. Moreover, the aligned hidden layer is always most important to this end. A windowing technique is introduced by limiting our model to selectively focus on hidden layers in a small window size, instead of irrelative hidden layers far way. ARNN-SL explicitly leverages the information from the aligned and attention-focused hidden layers for prediction.

2   Model

2.1   Simple RNN for Sequence Labeling

Formally, sequence labeling aims at finding the most probable label sequence $\mathbf{y} = \{y_1, ..., y_T\}$ for a given input word sequence $\mathbf{w} = \{w_1, ..., w_T\}$, where T is the sequence length.

The overall architecture of simple RNN (sRNN) for sequence labeling is depicted in Fig. 1. In this figure, w_t represents word at time step t and $x_t \in \mathbb{R}^n$ is w_t's word embedding. $y_t \in \mathbb{R}^L$ is the probability over L labels of word at position t and is defined as[1]:

$$y_t = softmax(Vh_t) \tag{1}$$

where $h_t \in \mathbb{R}^m$ is the hidden state at time step t. h_t encodes the information in previous time steps and is computed as:

$$h_t = \sigma(Wx_t + Uh_{t-1}) \tag{2}$$

where V, W and U are weight matrices. σ is active function and is often set to *sigmoid*.

---

[1] For simplicity, we omit bias terms in all the equations.

2.2 RNN Variations

Most of the variations of RNN focus on modifying the way of calculating hidden layer h_t in Eq. 2. For example, Long Short-Term Memory (LSTM) [7] and Gated Recurrent Unit (GRU) [4] introduces additional memory cells and gates to depict the long-range dependency information much better. For local context window technique [13], instead of using x_t for computing h_t, it uses a weighted sum of all x_j in a context window of wn as input: $\sum_{i=-wn}^{wn} U_i x_{t+i}$, which allows RNN to consider more local context dependency information.

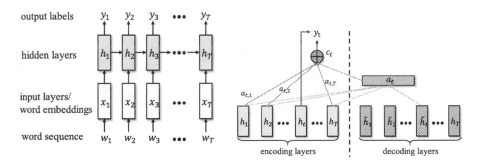

Fig. 1. Illustration of sRNN and ARNN-SL for sequence labeling.

Bi-directional Recurrent Neural Network [16] is also commonly used for improving the performance of sRNN. It computes the forward hidden layer h_t^f and backward hidden layer h_t^b using h_{t-1}^f and h_{t+1}^b respectively, and concatenates these two types of layers to form the final hidden layer h_t. In this way, both past and future information is preserved.

2.3 Proposed Attention Architecture

In practice, h_t alone is still not sufficient for encoding all the information needed for predicting y_t, even with sophisticated variations like Bi-RNN, LSTM, and GRU. In this paper, we propose Attention-based Recurrent Neural Network for Sequence Labeling (ARNN-SL), which allows RNN to selectively use multiple hidden vectors' information instead of using h_t alone.

As shown in Fig. 1, ARNN-SL has two types of hidden layers: encoding layer h_t and decoding layer $\tilde{h}_t$. The same as most encoder-decoder frameworks, the last encoding layer is used as the input of the first decoding layer, two sets of parameters are used for encoder and decoder respectively.

If we ignore the attention component c_t and all decoding layers, the architecture is exactly the same as sRNN and its variations. Attention component c_t is used to selectively gather information from encoding layers for prediction, which is computed as:

$$c_t = \sum_{j=1}^{T} a_{t,j} h_j \tag{3}$$

where $a_{t,j}$ is the weight for h_j at time step t:

$$a_{t,j} = \frac{\exp(e_{t,j})}{\sum_{k=1}^{T} \exp(e_{t,k})} \tag{4}$$

where $e_{t,j}$ is attention score. The larger $e_{t,j}$ is, the larger $a_{t,j}$ becomes and the more h_j contributes to c_t and y_t. For sequence labeling task, y_t is mainly decided by encoding layer h_t. In order to make full use of its information, h_t is directly used for prediction and $e_{t,t}$ should be set to 0. Since encoding layers which are far from current time step may be noisy, we pre-define a window size wn and directly set $e_{t,j}$ to 0 when j is outside the window. To summarize, $e_{t,j}$ is calculated as:

$$e_{t,j} = \begin{cases} score(\widetilde{h}_t, h_j) & t - wn < j < t + wn \text{ and } j \neq t \\ 0 & \text{else} \end{cases} \tag{5}$$

the input of the *score* function is the current time step's decoding layer $\widetilde{h}_t$ and h_j. As proposed in [11], there are mainly two types of *score* functions which can be used:

$$score(\widetilde{h}_t, h_j) = \begin{cases} \widetilde{h}_t^T W_e h_j & general \\ V_e \tanh(\widetilde{W}_e \widetilde{h}_t + W_e h_j) & concat \end{cases} \tag{6}$$

Finally, the calculation of the output layer y_t in Eq. 1 is defined as:

$$y_t = softmax(V_h h_t + V_c c_t) \tag{7}$$

encoding hidden layer h_t is most informative for predicting y_t and is explicitly leveraged with attention component c_t.

Since our attention architecture does not change the way of computing hidden layers, it can be directly built upon sRNN and LSTM. When ARNN-SL is built upon bi-directional RNNs, two sets of weights and variables are used for forward and backward directions respectively. For example, two attention components c_t^f and c_t^b selectively focus on h_j^f and h_j^b. They are then concatenated to form a new attention component c_t for the final prediction.

3 Related Work

Recently, attention mechanism leads to state-of-the-art results on many complex tasks such as machine translation [1,11], image caption task [19], and speech recognition [2,5]. However, due to the characteristics such as align strategy, directly applying the same mechanism to sequence labeling is not feasible.

The main difference of our specially designed ARNN-SL with attention architectures in other tasks is the way of calculating the attention component c_t, as shown in Fig. 2. In machine translation [1], a translated word could be aligned with a word at any position of the sentence. Attention architecture should selectively focus on hidden layers at every position. In image caption task [19], in order to generate current caption word, hidden layers corresponding to all image segments should be focused on for the same reason.

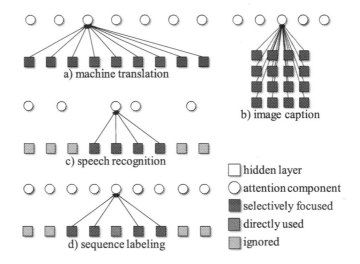

Fig. 2. Attention architectures used in different tasks.

The most similar attention architecture to ARNN-SL is used in speech recognition [2,5]. Instead of using all hidden layers, only the layers corresponding to the most probable consecutive k acoustic frames are focused on. The same idea is also implemented in machine translation [11], but it performs worse than attention for all layers. Compared with these architectures, ARNN-SL does not dynamically decide which consecutive hidden layers should be used for attention (Eq. 5). The useful hidden layers are always close to the current position for sequence labeling.

For sequence labeling, the label is mostly affected by the word and hidden layer in current time step. To the best of our knowledge, ARNN-SL is the first attention architecture that explicitly leverages the contribution from the current hidden layer and attention component (Eq. 7).

4   Experiments

4.1   Datasets and Experimental Setup

ARNN-SL is evaluated on four commonly used tasks for sequence labeling: Part-Of-Speech Tagging (POS), Chunking[2], Named Entity Recognition (NER)[3], and Slot Filling for Spoken Language Understanding (SF-SLU) [6,17,18]. Since there is no pre-defined development data for the datasets of Chunking and SF-SLU tasks, we randomly choose 20% of training data for validation. AdaDelta is used to control learning rate [21]. We use the same dropout strategy as that in [20] and the dropout rate is set to 0.5. Word embedding size and hidden layer size

---

[2] CoNLL 2000 shared task: http://www.cnts.ua.ac.be/conll2000/chunking.

[3] CoNLL 2003 shared task: http://www.cnts.ua.ac.be/conll2003/ner.

are set to 500. Word embeddings are either randomly initialized or pretrained using Word2Vec toolkit[4] on English Wikipedia (August 2013 dump). Models are trained for 25 epochs, and we report the results on epoch which achieves the highest performance on development data. We do not use features which are derived from lexical resources or other NLP systems. The only pre-processing we use is lowercasing.

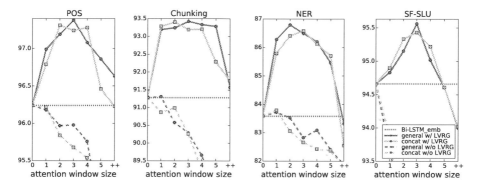

Fig. 3. Illustration of the impact of attention window sizes and score functions. ARNN-SL is built upon Bi-LSTM. Pretrained word embeddings are used. **Attention window size**: "0" indicates that attention architecture is not used, "++" indicates that windowing technique is not used and attention selectively focuses on all encoding layers. **w/o LVRG**: (dotted line) traditional attention mechanism. **w/LVRG**: (solid line) ARNN-SL which explicitly leverage the contribution from the current hidden layer and attention component (Eq. 7).

4.2 Main Results

As shown in Fig. 3, compared to traditional attention mechanism (w/o LVRG), models that explicitly leverage the information from the aligned and attention-focused hidden layers (w/LVRG) perform consistently better. In most cases, models without LVRG actually perform worse than Bi-LSTM baselines, especially on SF-SLU task and when the window size is large. The weighted sum of all encoding layers (Eq. 3) is likely to bring noises. The current encoding layer should always be directly used for prediction.

Our proposed attention architecture consistently improves the performance of Bi-LSTM on all tasks. The best performance is usually achieved at window size 2 or 3. The performance drops when the window size is bigger than 3 and reaches the minimum when attention focuses on all encoding layers. Windowing technique is indispensable for the good performance on sequence labeling.

The trend of the curve for the *concat* score function is similar to that of *general*. However, *general* score function often performs better than *concat*.

---

[4] http://code.google.com/p/word2vec/.

We highly recommend using *general* score function in practice, since it's also easier to implement and is 2–3 times faster.

Overall, ARNN-SL obtains 1.14%, 2.14%, 3.21% and 0.90% improvement on POS, Chunking, NER and SF-SLU respectively compared to sophisticated bi-direction Bi-LSTM baselines.

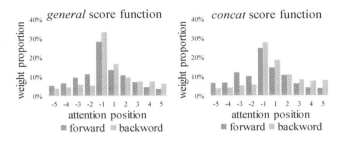

Fig. 4. Illustration of which position of encoding layers ARNN-SL focus on. ARNN-SL is built upon Bi-LSTM and is tested on SF-SLU task. For each window position j, we sum up variable $a_{t,j}$ in all training examples at all time steps. Each bar represents the percentage of this summed value over corresponding positions.

4.3   Attention Visualization

Since attention window size is crucial for ARNN-SL's performance, it's worth to explore which position of encoding layers the attention really focused on. As shown in Fig. 4, *general* and *concat* score functions both tend to focus on positions near the center. This explains the performance's decline when attention window size is bigger than 3 in the above section. This also further strengthens our claim: encoding layers which are far from the current time steps are noisy.

Another interesting phenomenon from this figure is that the forward direction of ARNN-SL mainly focuses on both position -1 and position 1. While the backward direction mainly focuses only on position -1. This may caused by that backward RNN's hidden layers contains only future information, so hidden layer at position 1 has no information about word at position -1. While word at this position may contains most useful information.

5   Conclusion and Future Work

This paper presents a novel attention architecture called ARNN-SL, designed for sequence labeling. We demonstrate its effectiveness on POS, Chunking, NER, and SF-SLU tasks. More precisely, we conclude that for sequence labeling tasks: (1) it's crucial to explicitly leverage the contribution from the current hidden layer and attention component, (2) *general* score function is a better choice than *concat*, (3) using windowing technique to restrict the attention is indispensable and the window size should be small.

The aim of this paper is investigating the impact of the attention architecture. We keep our model as simple and reproducible as possible. Note that the state-of-the-art results on POS, Chunking and NER tasks are all obtained by models combination, extra resources and feature engineering [3,8,12]. In the future, it's promising to implement ARNN-SL under the same sophisticated configurations for further improvements on these tasks (e.g. combining ARNN-SL and CRF/CNN, make using of different features and DBpedia knowledge).

Acknowledgments. This work is supported by the Fundamental Research Funds for the Central Universities, the Research Funds of Renmin University of China, National Natural Science Foundation of China with grant No. 61472428.

References

1. Bahdanau, D., Cho, K., Bengio, Y.: Neural machine translation by jointly learning to align and translate. CoRR abs/1409.0473 (2014)
2. Bahdanau, D., Chorowski, J., Serdyuk, D., Brakel, P., Bengio, Y.: End-to-end attention-based large vocabulary speech recognition. In: 2016 IEEE International Conference on Acoustics, Speech and Signal Processing, ICASSP, pp. 4945–4949. IEEE (2016)
3. Chiu, J.P.C., Nichols, E.: Named entity recognition with bidirectional LSTM-CNNs. TACL **4**, 357–370 (2016)
4. Cho, K., van Merrienboer, B., Gulcehre, C., Bahdanau, D., Bougares, F., Schwenk, H., Bengio, Y.: Learning phrase representations using RNN encoder-decoder for statistical machine translation. In: EMNLP (2014)
5. Chorowski, J., Bahdanau, D., Serdyuk, D., Cho, K., Bengio, Y.: Attention-based models for speech recognition. CoRR abs/1506.07503 (2015)
6. Hemphill, C.T., Godfrey, J.J., Doddington, G.R.: The ATIS spoken language systems pilot corpus. In: DARPA Speech and Natural Language Workshop, pp. 96–101 (1990)
7. Hochreiter, S., Schmidhuber, J.: Long short-term memory. Neural Comput. **9**(8), 1735–1780 (1997)
8. Huang, Z., Xu, W., Yu, K.: Bidirectional LSTM-CRF models for sequence tagging. arXiv preprint arXiv:1508.01991 (2015)
9. Kudo, T., Matsumoto, Y.: Chunking with support vector machines. In: NAACL, pp. 1–8. ACL (2001)
10. Lafferty, J.D., McCallum, A., Pereira, F.C.N.: Conditional random fields: probabilistic models for segmenting and labeling sequence data. In: ICML, pp. 282–289 (2001)
11. Luong, T., Pham, H., Manning, C.D.: Effective approaches to attention-based neural machine translation. In: EMNLP (2015)
12. Ma, X., Hovy, E.H.: End-to-end sequence labeling via bi-directional LSTM-CNNs-CRF. In: ACL, pp. 147–155. ACL (2016)
13. Mesnil, G., Dauphin, Y., Yao, K., Bengio, Y., Deng, L., Hakkani-Tur, D., He, X., Heck, L., Tur, G., Yu, D., et al.: Using recurrent neural networks for slot filling in spoken language understanding. IEEE/ACM Trans. Audio Speech Lang. Process. **23**(3), 530–539 (2015)
14. Ratinov, L., Roth, D.: Design challenges and misconceptions in named entity recognition. In: ACL, pp. 147–155. ACL (2009)

15. Raymond, C., Riccardi, G.: Generative and discriminative algorithms for spoken language understanding. In: INTERSPEECH, pp. 1605–1608 (2007)
16. Schuster, M., Paliwal, K.K.: Bidirectional recurrent neural networks. IEEE Trans. Signal Process. **45**(11), 2673–2681 (1997)
17. Tur, G., Hakkani-Tur, D., Heck, L.: What is left to be understood in ATIS? In: Spoken Language Technology Workshop, pp. 19–24. IEEE (2010)
18. Wang, Y.Y., Acero, A., Mahajan, M., Lee, J.: Combining statistical and knowledge-based spoken language understanding in conditional models. In: COLING/ACL, pp. 882–889. ACL (2006)
19. Xu, K., Ba, J., Kiros, R., Cho, K., Courville, A.C., Salakhutdinov, R., Zemel, R.S., Bengio, Y.: Show, attend and tell: neural image caption generation with visual attention. In: ICML (2015)
20. Zaremba, W., Sutskever, I., Vinyals, O.: Recurrent neural network regularization. CoRR abs/1409.2329 (2014)
21. Zeiler, M.D.: Adadelta: an adaptive learning rate method. CoRR abs/1212.5701 (2012)

Haze Forecasting via Deep LSTM

Fan Feng[1], Jikai Wu[1], Wei Sun[1], Yushuang Wu[1], HuaKang Li[1],
and Xingguo Chen[1,2(✉)]

[1] Jiangsu Key Laboratory of Big Data Security and Intelligent Processing,
Nanjing University of Posts and Telecommunications, Nanjing,
People's Republic of China
ffeng1017@gmail.com, wujikai983@gmail.com, vicsun0330@gmail.com,
yswunjupt@gmail.com, {huakanglee,chenxg}@njupt.edu.cn
[2] National Key Laboratory for Novel Software Technology,
Nanjing University, Nanjing, China

Abstract. $PM_{2.5}$ is a crucial indicator of haze pollution, which can cause problems in respiratory systems. Accurate $PM_{2.5}$ concentration forecasting systems are essential for human beings to take precautions. State-of-the-art methods including support vector regression (SVR), artificial neural network (ANN) and Bayesian, try to forecast $PM_{2.5}$ concentrations of the following 3 days via building an approximation from weather features to $PM_{2.5}$ concentration. However, the performances of these methods are poor because they ignore the essence of the problem: $PM_{2.5}$ concentration is the product of a time series.

This paper aims to propose more accurate forecasting algorithms to forecast $PM_{2.5}$ concentration. First, we employ the recurrent neural network with Long Short Term Memory kernel to handle the time series forecasting. Secondly, in order to further improve the performance, a convolutional neural network (CNN) is utilized as feature extractor to generate input for LSTM. Two models are proposed to handle the forecast for the following 3 and 7 days: (i) based on 2 days' weather features and $PM_{2.5}$ concentrations; (ii) based on 4 days' (including 2 days of this year, the day of last year, and the day two years ago) weather features and $PM_{2.5}$ concentrations. Finally, all experiments are compared with the root of mean squared errors (RMSE) for each city and averaged root of mean squared errors (ARMSE) of all cities. Experiments are tested on two datasets: one with hourly meteorological data and daily air-pollution data of 104 cities in east China from 2013 to 2017, the other with both hourly meteorological and air-pollution data in 5 cities from 2010 to 2015. Experimental results show that the proposed methods significantly outperform the state-of-the-art.

Keywords: Haze forecasting · Convolutional neural network · LSTM

1 Introduction

Haze is an atmospheric phenomenon where smoke, dust, moisture, and vapor suspended. It is called "the pollution people see" that reduces visibility [1].

© Springer International Publishing AG, part of Springer Nature 2018
Y. Cai et al. (Eds.): APWeb-WAIM 2018, LNCS 10987, pp. 349–356, 2018.
https://doi.org/10.1007/978-3-319-96890-2_29

The appearance of haze is an indicator of the high concentration of particulate matters in the air in addition with multiple chemicals. $PM_{2.5}$ (fine particles) working as one of leading indexes of the measurement of particles, refers to fine particles with a diameter of $2.5\,\mu m$ or less. In China, intensive emissions of air pollution contribute to the pollution condition in the city-clusters such as the Yangtze River Delta, Beijing-Tianjin Area and Pearl River Delta Area [2].

Fine particles have an adverse impact on human health. Short-term particulate exposures contributed to acute coronary events, especially among patients with underlying coronary artery diseases [3]. In the long term, with a fact that mortalities of cancers are influenced by the concentration of fine particles, each $10\,\mu g/m^3$ elevation in fine particulate air pollution was associated with approximately a 4%, 6%, and 8% increased risk of all-cause, cardiopulmonary, and lung cancer mortality, respectively [4].

It is almost impossible for human to be isolated from air pollution. However, forecasting haze condition indicates a warning of pollution level to the public. The public informed is able to take necessary initiatives guided by advice to inhibit parts of health hazards from air pollution. For individuals, wearing a gauze mask or avoiding going out when receiving the forecast reduces harm caused by pollution. And for authorities, proper emergency policies issued in events of pollution prevent the damage caused by haze to a maximum extent. Therefore, haze forecasting is significant to human beings and society.

Recently, haze forecasting methods based on machine learning have been extensively studied. The Support vector regression (SVR) method is proposed to figure out the relationship between the $PM_{2.5}$ concentrations and other weather features [5]. An application of SVR was proposed for the real-time prediction of $PM_{2.5}$ concentrations in Beijing. Given dataset $\{(x_i, y_i), i, \cdots, m\}$, where x denotes weather features, y denotes the concentration of $PM_{2.5}$, the functional relationship between x and y is given by $y = f(x) = w^T \phi(x) + b$, where w is the weight vector, $\phi(x)$ is the kernel function, b is a constant term. SVR minimizes the loss function, which is given by $L(f) = \frac{1}{2}\|w\|^2 + C\sum_{i=1}^{m}\frac{1}{2}(y_i - f(x_i))^2$. The Result of SVR method shows that the model based on SVR can forecast daily $PM_{2.5}$ concentration within 6 days and hourly $PM_{2.5}$ concentrations within 72 h. A Bayesian model method was proposed to establish a space-time predictive model of $PM_{2.5}$ concentrations in Jing-Jin-Ji area, which does not only take account into the temporal and spatial variability of $PM_{2.5}$ concentration, but also utilizes the meteorological data as covariates to make forecast [6]. A novel hybrid model combining air mass trajectory analysis and wavelet transform to improve the artificial neural network (ANN) forecast accuracy of daily average concentrations of $PM_{2.5}$ 2 days in advance were presented [7].

However, these algorithms show imprecise experimental results of $PM_{2.5}$ concentrations and their algorithms can only forecast the haze in a short time and in specific areas, relatively. The reason based on our analysis is that they considers only the relationship between weather features and $PM_{2.5}$ concentrations.

Actually, forecasting the magnitude of $PM_{2.5}$ concentrations can be seen as a time series forecasting model. Therefore, Long Short Memory Network (LSTM) is utilized for time series prediction. Besides, deep learning has been successfully applied in many fields, such as computer vision [8], digital images processing [9], natural language processing [10], computer games [11], etc. To further improve the performance, we choose deep neural network structures.

The rest of this paper is organized as follows. Preliminaries of data and LSTM are given in Sect. 2. Section 3 proposed the time series forecasting model, deep LSTM, and its optimization method. The experimental settings, results and some analysis are given in Sect. 4. Finally, the conclusion and future work are provided in Sect. 5.

2   Preliminaries

2.1   Data Overview

In this paper, two datasets are utilized to test algorithms.

We accessed to hourly meteorological and daily air-pollution data of 104 cities in east China from 2013 to 2017. We collected air-pollution data originally issued by Ministry of Environmental Protection of the People's Republic of China and meteorological data obtained from SYNOP and BUFR messages issued by official weather stations. The air-pollution data is an average of everyday air pollutants, and all data is categorized by location, containing air quality index (AQI), PM_{10} and $PM_{2.5}$ concentration and other chemical pollution levels. Meteorological data consists of the date, temperature, wind, cloud condition of each location.

We also obtained a dataset from UCI dataset with hourly data of meteorological and air-pollution data in Beijing, Shanghai, Shenyang, Chengdu and Guangzhou ranging from 2010 to 2015. Dimensions of the data are identical with the first dataset.

2.2   Long Short Term Memory

Many variations have been developed in RNN. Ordinarily, the recurrent network has an input layer x, a hidden layer or content layer s and an output layer y. A function f is used on each input x and the output f in the previous time step, see Fig. 1.

However, learning to store information over extended time intervals by recurrent propagation takes too much time. A novel method named LSTM was proposed to address the problem. LSTM can learn to bridge minimal time lags in excess of 1000 discrete-time steps by enforcing constant error flow through constant error carousels within special units [12]. LSTM creates a new concept of a memory cell which essentially acts as an accumulator of the state information. If the input gets i_t is activated, information of new inputs will be accumulated. One of the advantages of LSTM is that its convergence performance is better compared to the basic RNN. A basic LSTM unit with a forget gate can be described in following equations: $f_t = g(W_f \cdot x_t + U_f \cdot h_{t-1} + b_f)$, $i_t = g(W_i \cdot x_t + U_i \cdot h_{t-1} + b_i)$, $k_t = \tanh(W_k \cdot x_t + Uk \cdot h_{t-1} + b_k)$, $c_t = f_t \times c_{t-1} + i_t \times k_t$,

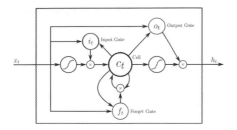

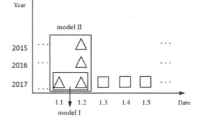

Fig. 1. LSTM block at time t **Fig. 2.** Forecasting models

$o_t = g(W_o \cdot x_t + U_o \cdot h_{t-1} + b_o)$, and $h_t = o_t \times \tanh(c_t)$, where x_t is the input vector at time t and g is an activation function. W, U are matrices of weight and b is the bias vector. h_t and c_t are the output and the cell state vector at time t. f_t is a forget gate and i_t is an input gate.

3   Deep LSTM

In this section, Deep LSTM and two time models for forecasting $PM_{2.5}$ concentrations are presented.

3.1   Deep LSTM

Although the LSTM layer has proven powerful in handling temporal correlation, it show indisposed performance for feature learning. To address this problem, we propose an extension of LSTM which has convolutional neural networks (CNN) structures before LSTM structure. We call this structure as "Deep LSTM".

Our CNN structure consists of 2 convolutional, 2 max-pooling layers followed by two fully connected layers with both 128 neurons. At the input level, our training data of size $24 \times 24 \times 2$ are fed to the network as input. At the first convolutional layer, 20 kernels of size $5 \times 5 \times 3$ with stride of 1 pixel applied in the input data. After local response normalization and max pooling operations, the feature maps of size $12 \times 12 \times 32$ are obtained. Then we add a flatten layer to make multidimensional input into one-dimensional input, which is commonly used in transition from convolution layer to fully connection layer. The 3×3 convolutional kernel is $\begin{bmatrix} 1 & 0 & 1 \\ 0 & 1 & 0 \\ 1 & 0 & 1 \end{bmatrix}$.

Hence, the whole structure of Deep LSTM is shown in Fig. 3. Firstly, the training sequences data X was input into the CNN structure to extract the features of weather data. Then, the output of CNN structure were entered to LSTM structure to do the sequence learning and get the prediction vector H. By forming and using this structure, a network model was built not only for the precipitation forecasting problem but also for more general spatiotemporal sequence forecasting problems.

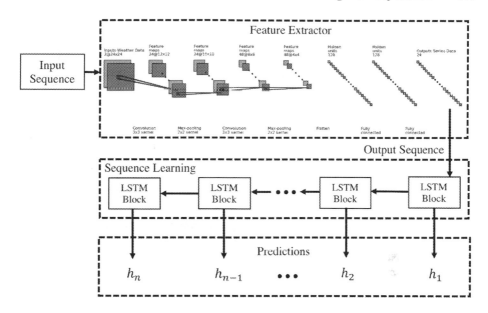

Fig. 3. The Deep-LSTM structure.

3.2 Loss Function and Stochastic Optimization

In our structure, Root Mean Square Error (RMSE) was chosen as our loss function, which is defined as follows: $RMSE = \sqrt{\frac{1}{N} \sum_{i=1}^{N}(y_i - h_i)^2}$, where y_i represents the actual observation value at time t, h_i represents the forecast value for the same period, and N is the number of forecast.

We choose Adam as our optimization method for proposed network. Adam is an algorithm for first-order gradient-based optimization of stochastic objective functions, based on adaptive estimation of lower-order moments. Adam performs well when dealing with big-scale data [13]. The parameter updating process is $\theta_{t+1} = \theta_t - \frac{\alpha}{\sqrt{\hat{v}_t}+\varepsilon}\hat{m}_t$. $\hat{m}_t$ is mean gradient at time t and $\hat{v}_t$ is the non-center variance value of gradient. We choose that $\alpha = 0.01$, $\varepsilon = 10^{-8}$.

3.3 Forecasting Model

We proposed two time series forecasting models, namely model I and model II. Model I only uses data in the same year by date order to forecast haze in the subsequent days. Model II implements a method that uses data both from the same year and identical dates from previous years. Details of model I and model II using 2-day sequence to forecast $PM_{2.5}$ concentrations of 3 days are shown in Fig. 2. For 2-day sequence forecasting $PM_{2.5}$ concentrations of 7 days, we extend the forecasting period from 3 days to 7 days and other conditions remain unchanged.

4   Experiments

We first compare LSTM method with SVR and Bayesian Methods in to figure out the performances among these three models. Then, we conduct experiments on two Time Prediction Models: Model I and Model II, respectively. At last, we train the dataset with Deep-LSTM model on two Forecasting Models to show the performance of Deep-LSTM. The data we use is both 5 cities' dataset on UCI and our data of 104 cities.

4.1   Experimental Settings

The overall evaluation is the averaged root of mean squared errors (ARMSE) for all cities: $ARMSE = \frac{1}{N} \sum_{i=1}^{N} RMSE(city_i)$.

To compare give comparisons, we reimplement the state of the art, i.e., the SVR method and the Bayesian methods (The ANN method was unable to reimplement due to the lack of its details). We use 80% data as training set and 20% data as test set. In SVR method, we choose sigmoid kernel $k(x, y) = \tanh(gx^\top y + r)$, where the parameter $g = 0.125$, and $r = 0$. We use cross validation to divide training and test sets. In the Bayesian method, we choose Naive Bayes with a priori Gaussian distribution. We choose RMSE as loss function and Adam as optimizer.

We divided all weather data and $PM_{2.5}$ concentrations data into training set and test set. The training set is 80% and test set is 20%.

In both CNN and LSTM structure, the batch-size is 32 and learning rate is 0.01. We choose RMSE as loss function and Adam as Optimizer. In CNN structure, we choose a 3×3 convolutional kernel, 2×2 max-pooling kernel and 128 units for 2 hidden layers. In LSTM structure, dropout is 0.2 and time-step is 24. The input data is 80% weather data and $PM_{2.5}$ concentrations data and the output data is to forecast 3 or 7 days' $PM_{2.5}$ concentrations data in test set.

4.2   Experimental Results and Analysis

In two experiments, we use SVR method, Bayesian method, LSTM and Deep-LSTM model with two forecasting models.

In experiment 1, we use data of 2 days as input series x to forecast $PM_{2.5}$ concentrations of 3 days' output sequence y. In experiment 2, the length of output sequence is 7 days. The ARMSE of 104 cities and 5 cities in both experiment 1 and experiment 2 are given in Table 1. From the results, we can conclude that in most cities, the performance of Bayesian method is better than the SVR method and LSTM methods show better performances than both SVR and Bayesian methods. In LSTM structure, model II is superior to model I by a limited extent. In deep-LSTM, model I is practically better than all previous methods, and model II has a better effect than model I.

In summary, model II of the deep LSTM has a significant improvement among all methods except several locations that get lower RMSE in model I of the deep LSTM. We can conclude that Deep LSTM has the best performance among these algorithms and model II is better than model I in accuracy.

Table 1. Forecasting ARMSE.

Dataset	Forecasting period	SVR	BAYES	LSTM		Deep LSTM	
				Model I	Model II	Model I	Model II
104 cities	3 days	41.526	34.168	25.881	25.241	21.430	**20.860**
	7 days	52.232	43.845	37.609	37.007	32.498	**31.901**
5 cities	3 days	29.881	29.105	27.066	26.051	24.100	**23.797**
	7 days	41.023	40.155	36.051	34.689	32.272	**31.686**

5   Conclusions and Future Work

A $PM_{2.5}$ concentration forecast algorithm based on Deep LSTM is proposed. Deep LSTM structure includes two parts: CNN to extract features and LSTM to realize sequence learning. We do experiments on our forecast model to forecast $PM_{2.5}$ concentration in 3 days and 7 days with 2 days weather features and $PM_{2.5}$ concentration input series set. The results illustrate that LSTM has a good performance in time series prediction and deep learning can improve accuracy.

Future work includes: (1) Our data can only access to daily mean $PM_{2.5}$ concentrations so there may exist some defect on our data. We can improve our algorithm when we obtain more detailed data. (2) We can choose some other deep learning structures to extract the features and do the sequences learning. Like using residual network to extract features and using seq2seq to complete sequences learning [14]. (3) The haze data is collected hourly. It is exhausted and inconvenient to train the model hourly, daily, or monthly in off-line mode. Thus, on-line or incremental haze forecasting algorithms are necessary.

Acknowledgements. The work was supported by National Natural Science Foundation of China (Grant No. 61403208, No. 61502247), China Postdoctoral Science Foundation (Grant No. 2016M600434), Natural Science Foundation of Jiangsu Province (BK20161516), China Postdoctoral Science Foundation (Grant No. 2016M600434), Scientific and Technological Support Project (Society) of Jiangsu Province (BE2016776), and Science Foundation of Nanjing University of Posts and Telecommunications (NY214014).

References

1. Hyslop, N.P.: Impaired visibility: the air pollution people see. Atmos. Environ. **43**(1), 182–195 (2009)
2. Cheng, Z., Wang, S., Jiang, J., Fu, Q., Chen, C., Xu, B., Yu, J., Fu, X., Hao, J.: Long-term trend of haze pollution and impact of particulate matter in the Yangtze River Delta, China. Environ. Pollut. **182**, 101–110 (2013)
3. Pope, C.A., Muhlestein, J.B., May, H.T., Renlund, D.G., Anderson, J.L., Horne, B.D.: Ischemic heart disease events triggered by short-term exposure to fine particulate air pollution. Circulation **114**(23), 2443–2448 (2006)

4. Pope III, C.A., Burnett, R.T., Thun, M.J., Calle, E.E., Krewski, D., Ito, K., Thurston, G.D.: Lung cancer, cardiopulmonary mortality, and long-term exposure to fine particulate air pollution. JAMA **287**(9), 1132–1141 (2002)
5. Zhu, Y., Li, Q., Hou, J., Feng, X., Fan, J.: Real time prediction of $PM_{2.5}$ concentration based on support vector regression algorithms. Sci. Surv. Mapp. **41**(1), 12–17 (2016)
6. Zhu, Y., Li, Q., Hou, J., Fan, J., Feng, X.: Spatio-temporal prediction of $PM_{2.5}$ concentration based on bayesian models. Sci. Surv. Mapp. **41**(2), 44–48 (2016)
7. Feng, X., Li, Q., Zhu, Y., Hou, J., Jin, L., Wang, J.: Artificial neural networks forecasting of $PM_{2.5}$ pollution using air mass trajectory based geographic model and wavelet transformation. Atmos. Environ. **107**, 118–128 (2015)
8. Nguyen, A., Yosinski, J., Clune, J.: Deep neural networks are easily fooled: high confidence predictions for unrecognizable images. In: Proceedings of the IEEE Conference on Computer Vision and Pattern Recognition, pp. 427–436 (2015)
9. Ciregan, D., Meier, U., Schmidhuber, J.: Multi-column deep neural networks for image classification. In: 2012 IEEE Conference on Computer Vision and Pattern Recognition, CVPR, pp. 3642–3649. IEEE (2012)
10. Collobert, R., Weston, J.: A unified architecture for natural language processing: deep neural networks with multitask learning. In: Proceedings of the 25th International Conference on Machine Learning, pp. 160–167. ACM (2008)
11. Silver, D., Schrittwieser, J., Simonyan, K., Antonoglou, I., Huang, A., Guez, A., Hubert, T., Baker, L., Lai, M., Bolton, A., et al.: Mastering the game of go without human knowledge. Nature **550**(7676), 354–359 (2017)
12. Hochreiter, S., Schmidhuber, J.: Long short-term memory. Neural Comput. **9**(8), 1735–1780 (1997)
13. Kingma, D.P., Ba, J.: Adam: a method for stochastic optimization. arXiv preprint arXiv:1412.6980 (2014)
14. Sutskever, I., Vinyals, O., Le, Q.V.: Sequence to sequence learning with neural networks. In: Advances in Neural Information Processing Systems, pp. 3104–3112 (2014)

Importance-Weighted Distance Aware Stocks Trend Prediction

Zherong Zhang[1], Wenge Rong[2(✉)], Yuanxin Ouyang[2], and Zhang Xiong[2]

[1] Sino-French Engineer School, Beihang University, Beijing, China
zherongzhang@buaa.edu.cn
[2] School of Computer Science and Engineering, Beihang University, Beijing, China
{w.rong,oyyx,xiongz}@buaa.edu.cn

Abstract. A great deal of work has been proposed on the application of neural networks (NN) to stock price prediction. Since correlation between stocks has been proven to be crucial in stock price prediction, researchers presented plenty of clustering algorithms to retrieve similar stocks. However, most of the existing clustering approaches have the issue of information-loss and may sometimes offer an oversized cluster, both of which will cause a drop of prediction performance. In this paper, we propose a multidimensional similarity-calculating function to seek out relevant stocks, the technical indicators of which are then fed into NN to predict the stocks' price trend. Evaluation is carried out on two stock indexes and 18 individual stocks, demonstrating that the proposed model provides a promising alternative to stock price prediction.

Keywords: Stocks similarity · Trend prediction · Weighted distance

1 Introduction

Through decades, both statistical approaches and soft computing approaches have been widely employed to predict stock prices. The former contains several basic mathematical models such as the moving average model [1] and autoregressive integrated moving average model [2]. However, these models consider neither nonlinear relations nor the high volatility of stocks, thereby making the prediction usually less precise [3]. The latter approach is increasingly popular for its ability to simulate complex price patterns, thus diverse neural network models have been presented recently [3,4].

The correlation between stocks, meaning that past stock price could also probably affect the price of other stocks [5], plays an important role in stock price prediction. To this end, clustering has been employed a lot to retrieve correlated stocks. Since stock string is multidimensional while clustering models only inclined to focus on a certain dimension of the data [6], proven effective in real applications, clustering still has the information-loss problem. Furthermore, the stock sample number is quite limited, so as an unsupervised algorithm, clustering method always gives a large cluster, therefore more correlated stocks have

Y. Cai et al. (Eds.): APWeb-WAIM 2018, LNCS 10987, pp. 357–365, 2018.
https://doi.org/10.1007/978-3-319-96890-2_30

to be considered. However, an over-sized input may add more noise to the NN model, making it more difficult to "learn" the proper information.

To address these challenges, we follow the concept of retrieving similar stocks to predict stock price and propose a multidimensional similarity-based prediction method. We define a multidimensional similarity-computing function to make full use of the information contained in the stock data. Recognizing the importance-inhomogeneity of the stock string [7], we propose a Importance-Weighted Distance (IW-D) metric to capture the corresponding feature.

2   Related Works

In stock price prediction research, similarity-based models can be divided into two categories in terms of the way the features are selected. The first school of thought concentrates on the selection of features. Similar stocks are firstly retrieved based on some simple metrics. By calculating the features of the target stock and also of its similar stocks, different algorithms are proposed to select "useful" features to predict the future price of the target stock. For instance, Kwon et al. [5] found the other companies which were highly correlated with the target company and then collected all the input variables generated from not only the target company but also the selected companies. Afterwards, a genetic algorithm was used to select a subset of features among a number of input variables. Meanwhile, other researchers believed that the selection of similar stocks outweighs that of features. They proposed several methods to retrieve similar stocks, features of which were directly fed to NN. To find similar stocks, clustering becomes a common scheme. For example, Hsu et al. [8] used the self-organizing map (SOM) to decompose the whole input space into regions where data points with similar statistical distributions are grouped together, so as to contain and capture the non-stationary property of financial series.

3   Methodology

The overview of our proposed framework is shown in Fig. 1. First, ten technical indicators of the target stock are computed and fed into an LSTM to extract temporal features. Subsequently, we proposed an Importance-Weighted Distance (IW-D) by which we can compute the similarity between our prediction target and every other stock in the market. Afterwards, we choose the top K stocks in the similarity rankings as similar stocks (K is determined later by experiments). Next, technical indicators of the K similar stocks are computed and concatenated with extracted features in the first step. Lastly, a MLP is employed to fuse two different types of features together and to give the final prediction result.

(1) **Trend Deterministic Indicators and Feature Extraction.** Technical indicators are adopted extensively in stock market analysis and are usually used to predict the market trend [9]. In this study, we use 10 technical indicators proposed in [10]. Afterwards, we applied the LSTM to extract features

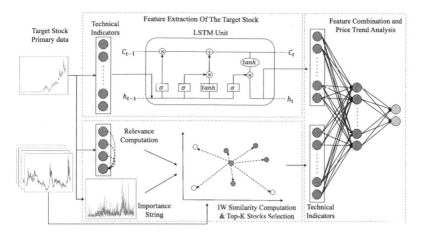

Fig. 1. Similarity based-stock market prediction framework (components related to the target stock/similar stocks are colored in red/blue) (Color figure online)

from the technical indicators of the target stock. Since technical indicators already include trade information in the past, to avoid the over-influence of the past information of similar stocks, LSTM is not applied to similar stocks. The technical indicators of similar stocks are directly combined with extracted features of the target stock in the last step and then together sent to a MLP.

(2) **Stock Similarity Computation.** In this study, each stock S is defined as a vector containing four components, $S = (S_{open}, S_{close}, S_{highest}, S_{lowest})$. Each component in S is composed of n scalar entries (corresponding to n days' trading data). To calculate the distance between stocks, S_i is firstly encoded into trend-strings T_i with the following rule: for each element $S_i[j]$ in S_i, if $S_i[j] > S_i[j-1]$, this position is encoded as 1 and -1 otherwise. We compute the T_i for each S_i respectively. Afterwards, the relevancy between different elements in S is computed using the cos formula: $R_{S_i,S_j} = \frac{T_i * T_j}{|T_i||T_j|}$. These relevancies are then used as weights for the computation of the similarity between the target stock and other stocks in the market. We employ *Cosine* function as the similarity weight for its outstanding performance in high-dimensional spaces compared to other metrics [11]. Besides, *cosine* has a unique property that its value is inherently normalized and it equals 1 when computing the relevance of itself. The similarity is given by the following formula:

$$Similarity(target, stock) = \frac{1}{n} \sum_{i \in I} R_{S_{close}, S_i} * D_{IW-D}(S_i^{target}, S_i^{stock}) \quad (1)$$

where n is the length of S_i^{target} and $D_{IW-D}(.,.)$ is the proposed IW-D metric.

On the basis of the characteristics of stocks market, the importance of each point in the time series of a stock is not always equal [7]. A great increase will weaken the upward momentum during the following days while a fews days of vibration will lead to a great possibility to go up or down. Hence, some key points in time series are much more important than the others. For each string S, we define an attached importance string:

Definition 1. $\forall$ *string* $S = \{S[1], S[2], ..., S[n]\}$, *we can define an attached importance string* Im, *where* $Im[k] = |S[k-1] - S[k]| + |S[k+1] - S[k]|$.

Based on this definition, points with a large change rate compared to its adjacent points will be assigned a greater importance. And finally, we define the Importance-Weighted Distance:

$$D_{IW-D}(S_i^{target}, S_i^{stock}) = \sqrt{\sum_{j=1}^{n} \rho_j (T_i^{target}[j] - T_i^{stock}[j])^2} \tag{2}$$

where we set

$$\rho_j = \begin{cases} v_1 & if\ Im_{target}[i] \in [min(Im_{target}), min(Im_{target}) + \alpha|M|] \\ v_2 & if\ Im_{target}[i] \in [min(Im_{target}) + \alpha|M|, max(Im_{target}) - \alpha|M|] \\ v_3 & if\ Im_{target}[i] \in [max(Im_{target}) - \alpha|M|, max(Im_{target})] \end{cases}$$

In the above definition, $M = max(Im_{target}) - min(Im_{target})$. v_1, v_2, v_3, α are constants and $v_1 < v_2 < v_3$. The values of all the constants will be determined in following parts. Compared to the naive Euclidean Distance, this metric considers the string's morphology, and gives a larger penalization to the difference at "key points" of the string. Hence, stocks which have similar trends with the target stock at points with a high importance score will be given a smaller distance.

(3) **Feature Combination and Price Trend Analysis.** The technical indicators of the K similar stocks from the previous step are end-to-end combined and then connected with the output of the last time step of LSTM. Subsequently, this fused vector is fed into a MLP. The output layer of the MLP above has 2 units y_l ($l = 0, 1$), corresponding to 2 price trends (up/down). The class probability is computed by taking the softmax: $y_l = \frac{\exp(y_l)}{\sum_{j \in L} \exp(y_j)}$, where L denotes up/down trends and the label with the maximum probability will be the expected label. We chose the Cross Entropy to be the loss function.

4   Experimental Study

(1) **Data Sets and Evaluation Metrics.** To guarantee the generality, we randomly selected 9 stocks from Shanghai 50 Stock Index Futures, 9 stocks

Table 1. Start date, end date and percentage of increase of each selected stock

Stock	NoS	PoI	Stock	NoS	PoI
SH600000	4013	49.32%	SZ000333	664	44.88%
SH600048	2399	53.77%	SZ000063	4495	48.52%
SH600887	4912	46.15%	SZ000858	4393	48.01%
SH600893	4893	40.02%	SZ000800	4613	48.91%
SH601601	2056	54.81%	SZ000876	4840	46.03%
SH601668	1669	38.98%	SZ000001	6200	50.11%
SH601766	1899	40.23%	SZ002570	1259	46.15%
SH601857	2093	41.43%	SZ000060	4700	47.53%
SH601985	246	41.87%	SZ002230	1969	45.51%
SH000016	3021	49.52%	SZ399330	2522	45.64%

SH000016 is the Shanghai 50 Stock Index Futures while SZ399330 is the Shenzhen 100 Stock Index Futures. We collected the trading information of each stock from its listing date to a fixed date 6/8/2016. NoS and PoI are respectively abbreviations for Number of samples and Percentage of increase.

from Shenzhen 100 Stock Index Futures and these two indexes themselves to be our dataset (see Table 1). We randomly choose 80% of the dataset as the training set, 10% as the validation set and on which we find the best value of K. The remaining 10% is the test set. Being similar to [10], we use accuracy and $F1$ Score to evaluate our models.

(2) **Benchmark Models.** The following baseline scenarios were considered, i.e., (a) Since we used technical indicators proposed by Patel et al. [10], we treated models proposed in [10] as the baseline, including Support Vector Machine (SVM) and Random Forest (RF). (b) To verify the performance of our proposed multidimensional similarity-computing algorithm, we also compare our results against a stock price prediction model sharing the viewpoint of using similar stocks [12], by applying a single unidimensional SOM-based clustering method. (c) To evaluate the effect of the LSTM-based feature extraction, a model considering similar stocks but using naive MLP is considered. Afterwards, the effect of the proposed Importance-Weighted Distance (IW-D) is compared with two well-known distance metrics: Euclidean Distance (ED) and Correlation Coefficient (CC). Finally, we analyze the impact of different values of K on the model's performance.

(3) **Experiment Settings.** In Sect. 3, when computing IW-D, the values of all the constants (v_1, v_2, v_3 and α) were not assigned. These constants may vary from stock to stock and we determine them by experiments. We test the different numbers of LSTM hidden units and find that 16 hidden units can be selected to achieve the best results. The batch size, the learning rate and the dropout factor value are respectively set to 20, $1e - 3$ and 0.5.

(4) **Results and Discussion.** In Fig. 2, we compare our model with all the baseline models. Our similarity-based model improves notably the stock prediction performance compared two models proposed in [10], reproving stock correlations' existence. In most cases, our model has a better performance compared to the unidimensional clustering model, indicating that the proposed model solve the issue of information-loss to some extent. However, outlier still exists (eg. SH600893). For this stock, SOM Clustering model [12] has a better performance, maybe it is because that some non-selected stocks of our model or certain dimensions of them are valuable for our prediction. The result also demonstrates the superiority of the extraction action in comparison to naive MLP. Besides, since stock index is the weighted average of individual stocks and the list of selected individual stocks is renewed every three or six months, we can not find an "exact" correlated stock of the stock index when constructing the model (because the list of correlated stocks we find is static while the real list is not). Therefore, our proposed model performs bad when applied to stock indexes.

Figure 3 shows the trading curves of the target stock SH600887 and of the similar stocks we find using different distance functions. For this stock, we set $v_1 = 0.7$, $v_2 = 1.0$, $v_3 = 1.3$ and $\alpha = 0.2$, these are the best values we found for this stock. In Fig. 3, we see that IW-D has a better performance than the other

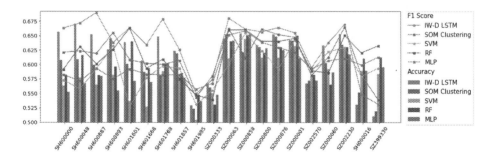

Fig. 2. Accuracy and F1 Score comparison of different predictors (IW-D LSTM represents our proposed model)

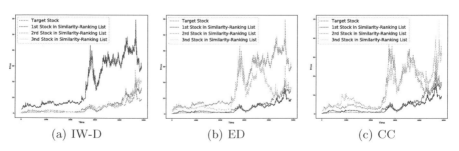

(a) IW-D   (b) ED   (c) CC

Fig. 3. Top-3 stocks in the similarity ranking list of different distance metrics for stock SH600887

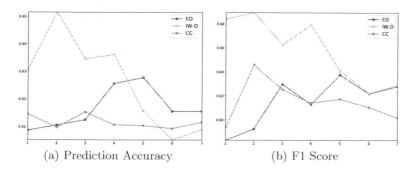

(a) Prediction Accuracy   (b) F1 Score

Fig. 4. Prediction performance of models with different K value for stock SH600887

Table 2. Accuracy/F1 Score comparison experiments result

Dist func	SH600000 Acc/F1	SH600048 Acc/F1	SH600887 Acc/F1	SH600893 Acc/F1	SH601601 Acc/F1
ED	60.94/63.51	61.72/64.2	62.78/63.75	57.4/58.57	59.51/57.57
IW-D	**65.63/66.31**	**67.03/67.15**	**65.16/68.97**	**62.58/63.74**	**63.82/66.39**
CC	61.91/62.59	66.97/63.48	61.53/64.59	61.78/63.98	61.38/62.42
Dist func	SH601668 Acc/F1	SH601766 Acc/F1	SH601857 Acc/F1	SH601985 Acc/F1	SZ000333 Acc/F1
ED	56.59/61.45	61.47/63.22	58.46/61.28	53.19/53.02	54.51/56.65
IW-D	60.63/63.37	**64.84/67.85**	62.37/62.58	**52.88/54.97**	**56.69/57.39**
CC	**62.8/64.29**	62.37/63.56	**64.79/66.83**	51.12/55.85	55.73/58.59
Dist func	SZ000063 Acc/F1	SZ000858 Acc/F1	SZ000800 Acc/F1	SZ000876 Acc/F1	SZ000001 Acc/F1
ED	60.73/62.93	58.94/59.02	59.24/61.25	63.73/63.83	62.51/61.04
IW-D	**64.47/68.01**	**65.32/66.12**	**63.34/65.63**	**65.21/66.37**	64.22/65.47
CC	58.36/60.71	63.73/64.91	62.01/63.62	62.52/65.73	**66.06/68.87**
Dist func	SZ002570 Acc/F1	SZ000060 Acc/F1	SZ002230 Acc/F1	SH000016 Acc/F1	SZ399330 Acc/F1
ED	56.78/58.48	**63.36/64.66**	62.64/64.1	52.25/55.64	53.32/55.98
IW-D	**56.71/59.41**	63.29/63.74	**65.24/66.93**	**53.03/54.78**	51.06/53.31
CC	55.25/56.05	6.47/63.22	63.86/66.09	52.85/52.43	**53.63/55.89**

two, which also lead to an forecasting outperformance later in the prediction part. This verifies the inhomogeneity of the importance of each point in stock time series and demonstrates that the morphology characteristic influences the overall prediction performance. All of the three algorithms select similar stocks whose curves are "far" from that of the target stock, that is because our distance function focuses on the price trend but not on the price level. Figure 4 shows the enumeration method which aims to find value K of SH600887. It shows the prediction accuracy and F1 Score of different $K \in [1, 7]$. With the increase of K, the accuracy and the F1 Score rise until a certain K value and then remain unchanged or slightly fall down. This is because at the beginning, we add some neighbor stocks to the model and new correlation information is received, but when K is too large, some less-related stocks are added and the model gets more noise and hence the forecast performance becomes worse. For each stock and each distance function, we only need to find a K with the highest accuracy value and F1 Score. From this figure, we see that the IW-D has the highest accuracy and the highest F1 Score, for the reason that it selected the best similar stocks in previous step. To illustrate the generality of our method, Table 2 shows the prediction results of models based on different distance functions on all the 10 selected stocks.

5   Conclusion

In this paper, we proposed a similarity-based model to predict the next day's stock price trend. Considering the morphology characteristic of stock strings, an Importance-Weighted Distance metric is proposed when computing the similarity. Based on the experimental results obtained, it should be emphasized that the correlation exists and truly influences the stock price prediction. Experiments show that our model significantly improves the prediction performance.

Acknowledgements. This study was partially supported by the National Natural Science Foundation of China (No. 61332018).

References

1. Wei, L., Cheng, C., Wu, H.: A hybrid ANFIS based on n-period moving average model to forecast TAIEX stock. Appl. Soft Comput. **19**, 86–92 (2014)
2. Ariyo, A.A., Adewumi, A.O., Ayo, C.K.: Stock price prediction using the ARIMA model. In: Proceedings of the 16th International Conference on Computer Modelling and Simulation, UKSim 2014, pp. 106–112 (2014)
3. Zhou, T., Gao, S., Wang, J., Chu, C., Todo, Y., Tang, Z.: Financial time series prediction using a dendritic neuron model. Knowl. Based Syst. **105**, 214–224 (2016)
4. Zhang, L., Aggarwal, C., Qi, G.: Stock price prediction via discovering multi-frequency trading patterns. In: Proceedings of the 23rd ACM SIGKDD International Conference on Knowledge Discovery and Data Mining, pp. 2141–2149 (2017)
5. Kwon, Y., Choi, S., Moon, B.R.: Stock prediction based on financial correlation. In: Proceedigns of 2005 Genetic and Evolutionary Computation Conference, pp. 2061–2066 (2005)
6. Aghabozorgi, S.R., Teh, Y.W.: Stock market co-movement assessment using a three-phase clustering method. Expert Syst. Appl. **41**(4), 1301–1314 (2014)
7. Liu, Q., Li, S., Fang, Y., Long, T., Cao, J., Liu, H.: An effective similarity measure algorithm for time series based on key points. In: Proceedings of 8th International Conference on Intelligent Human-Machine Systems and Cybernetics, pp. 17–20 (2016)

8. Hsu, S., Hsieh, J.J.P., Chih, T., Hsu, K.: A two-stage architecture for stock price forecasting by integrating self-organizing map and support vector regression. Expert Syst. Appl. **36**(4), 7947–7951 (2009)

9. Kara, Y., Boyacioglu, M.A., Baykan, Ö.K.: Predicting direction of stock price index movement using artificial neural networks and support vector machines: the sample of the Istanbul stock exchange. Expert Syst. Appl. **38**(5), 5311–5319 (2011)

10. Patel, J., Shah, S., Thakkar, P., Kotecha, K.: Predicting stock and stock price index movement using trend deterministic data preparation and machine learning techniques. Expert Syst. Appl. **42**(1), 259–268 (2015)

11. Qian, G., Sural, S., Gu, Y., Pramanik, S.: Similarity between Euclidean and cosine angle distance for nearest neighbor queries. In: Proceedings of 2004 ACM Symposium on Applied Computing, pp. 1232–1237 (2004)

12. Zhang, J., Rong, W., Liang, Q., Sun, H., Xiong, Z.: Data augmentation based stock trend prediction using self-organising map. In: Liu, D., Xie, S., Li, Y., Zhao, D., El-Alfy, E.S. (eds.) Proceedings of 24th International Conference on Neural Information Processing Part II, pp. 903–912. Springer, Cham (2017). https://doi. org/10.1007/978-3-319-70096-0_92

Knowledge Graph

Jointly Modeling Structural and Textual Representation for Knowledge Graph Completion in Zero-Shot Scenario

Jianhui Ding[1], Shiheng Ma[1], Weijia Jia[1,2(✉)], and Minyi Guo[1]

[1] Shanghai Jiao Tong University, Shanghai, China
{ding-jh,ma-shh}@sjtu.edu.cn, {jia-wj,guo-my}@cs.sjtu.edu.cn
[2] University of Macau, Macau, China
jiawj@umac.mo

Abstract. Knowledge graph completion (KGC) aims at predicting missing information for knowledge graphs. Most methods rely on the structural information of entities in knowledge graphs (In-KG), thus they cannot handle KGC in zero-shot scenario that involves Out-of-KG entities, which are novel to existing knowledge graphs with only textual information. Though some methods represent KG with textual information, the correlations built between In-KG entities and Out-of-KG entities are still weak. In this paper, we propose a joint model that integrates structural information and textual information to characterize effective correlations between In-KG entities and Out-of-KG entities. Specifically, we construct a new structural feature space and build combination structural representations for entities through their most similar base entities. Meanwhile, we utilize bidirectional gated recurrent unit network to build textual representations for entities from their descriptions. Extensive experiments show that our models have good expansibility and outperform state-of-the-art methods on entity prediction and relation prediction.

Keywords: Knowledge representation
Knowledge graph completion · Zero-shot learning

1 Introduction

Knowledge graphs (KGs) including Freebase [1], NELL [5] and WordNet [15] provide effective structured information and have been widely used in many applications, such as question answering [7], web search and information extraction. A typical knowledge graph is represented as a mass of triplets (*head entity, relation, tail entity*), (h, r, t) for short, indicating the relation between two entities. However, most existing knowledge graphs are built through semi-automatic or manual methods, which result in two issues: (1) incompleteness, plenty of potential relations between In-KG entities have not been fully

© Springer International Publishing AG, part of Springer Nature 2018
Y. Cai et al. (Eds.): APWeb-WAIM 2018, LNCS 10987, pp. 369–384, 2018.
https://doi.org/10.1007/978-3-319-96890-2_31

explored; (2) poor scalability, KGs cannot be automatically extended with Out-of-KG entities.

Recently, a variety of methods have been proposed to deal with KGC, which aims to predict the missing elements in the triplets, such as $(?, r, t), (h, ?, t), (h, r, ?)$. To conduct normal KGC that only involves In-KG entities, many successful translation-based methods, including TransE [3], TransH [21], TransR [12], and PTransE [11], follow the principle $h + r \approx t$, and then embed entities and relations into continuous low-dimensional vector space. However, these methods rely on the topology structure of KGs, hence they cannot learn representations for Out-of-KG entities.

To extend KGs with Out-of-KG entities automatically, Xie et al. [23] propose a non-trivial zero-shot scenario for KGC: when predicting missing element in (h, r, t), there exists at least one Out-of-KG entity (h or t) and all entities have additional textual information, such as descriptions and types, which are contained in most KGs. This scenario plays an important role in practical large-scale KGC since Out-of-KG entities are encountered when adding new entities to KGs. For example, Fig. 1 shows three entities with their descriptions and types. At test time, we receive an Out-of-KG entity called "Nightfall" and hope to infer possible relations between In-KG entity "Asimov" and Out-of-KG entity "Nightfall". The essence of this scenario is a zero-shot learning problem [17], which focuses on recognizing objects of new labels that only appear at test time. In-KG/Out-of-KG entities are the source/target domain in zero-shot learning, respectively. It is noteworthy that the correlations built between source domain and target domain determine the effectiveness of zero-shot learning.

There are several methods [20,23,26] representing KGs with textual information. However, these methods have two disadvantages in zero-shot scenario: (1) They can only utilize textual information, such as descriptions and names, to build the correlations between In-KG entities and Out-of-KG entities; (2) There are many words that only appear in the textual information of Out-of-KG entities, meanwhile, these methods build textual representations on word-level rather

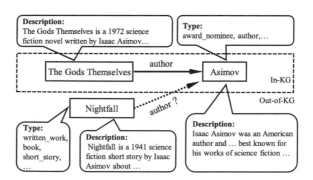

Fig. 1. Example of zero-shot scenario for KGC. Type sequence can be regarded as a kind of labels. For example, as for entity "Asimov", its type sequence contains "author", "award nominee".

than character-level, which could result in "Out-of-Vocabulary" problem. Since these new words cannot obtain proper representations without training and the proportion of new words is not small, it is difficult to build effective correlations between In-KG entities and Out-of-KG entities when only employing textual information.

To address above challenges for zero-shot scenario, we propose a joint model to learn representations for knowledge graphs with textual and structural information. First, we consider that similar entities should have similar representations, both in structural and textual feature space. According to the performance of translation-based methods, structural representations are beneficial for KGC. Thus, we propose a two-layer structural representation for entity: (1) **base layer**, we treat all In-KG entities as base-entity set and the structural representations of base entities are regarded as base structural representations; (2) **combination layer**, we design an entity similarity algorithm with two kinds of textual information: descriptions and types. With entity similarity algorithm, we can find the most similar base entities for all entities, and then build combination structural representations for entities through their most similar base entities. Compared with the convolutional neural network (CNN), bidirectional gated recurrent unit network (Bi-GRU) is better at language sequential modeling. Hence, we utilize Bi-GRU to encode valuable information from entity descriptions. Finally, we combine structural and textual representations to build joint representations for entities. Extensive experiments on two public datasets show that our models achieve promising results, indicating that the correlations between In-KG entities and Out-of-KG entities are enhanced.

Our contributions in this paper are summarized as follows:

- We propose a joint model to make full use of structural and textual information, which could build effective correlations between In-KG entities and Out-of-KG entities.
  We are the first to build combination structural representations for entity in zero-shot scenario. Meanwhile, this combination structural representation can be easily extended to other knowledge representation learning models.
- We achieve state-of-the-art results on entity and relation prediction.

2   Related Work

Most knowledge graphs have rich textual information, such as types, descriptions, and names, which can be regarded as an important supplement for representation learning. Hence, some methods utilize these textual information to help knowledge graph completion.

Wang et al. [20] make use of entity names and Wikipedia anchors to align knowledge and textual information in the same space, bringing improvements to the accuracy of predicting facts. However, this method relies on Wikipedia anchors, which limits the practical application scope. Thus, Zhong et al. [26] propose a novel alignment model that only aligns entities with the corresponding text descriptions. Zhang et al. [25] represent an entity with its name or the average of word embeddings in its descriptions. However, their use of descriptions neglects word order, and the use of name suffer from ambiguity problem. SSP [22] extend TransH [21] with textual descriptions. The key component of SSP is constructing semantic vector of entity by using topic distribution of its description. Cao et al. [4] consider each mention may contain one or multiple senses, and propose a joint model that learns the representations of mentions, entities and words in same space.

Notably, some methods take advantage of neural network to build textual representations for entities. Socher et al. [18] use neural tensor networks to represent an entity as the average of word vectors in its name, which allows the sharing of textual information between similar entity names. DKRL [23] adopt convolutional neural network to build description-based representations for entities. Xu et al. [24] propose a joint model that employs long short-term memory network (LSTM) with an attention mechanism [13] to adaptively build textual representations for entities. Hamaguchi et al. [8] employ graph neural network to build representations for entities, but this method needs auxiliary triplets to obtain the representations for new entities.

There also have been plenty of works [2,16,19] focused on relation semantic representation. They embed knowledge graph relations and textual relations into same space and achieve significant improvements.

In zero-shot scenario, all of the above works cannot learn structural representations for Out-of-KG entities. However, our models can build combination structural representations for all entities, which allow different entities share a same structural feature space and enhance the correlations between them.

3   Methodology

3.1   Problem Formulation

We first show the notations and definitions used in this paper. E is the set of In-KG entities and E' is the set of Out-of-KG entities. R stands for the set of relations. The description of entity $e \in E \cup E'$ is denoted as a word sequence $D_e = (w_1, w_2, \cdots, w_n)$, where $n = |D_e|$. In addition, the type of specific entity e is represented as a type sequence $T_e = (t_1, t_2, \cdots, t_l)$, where $l = |T_e|$. A triplet (h, r, t), where $h, t \in E$ and $r \in R$, is regarded as a **golden triplet** if it reflects true fact. Given two elements of a triplet, the goal of KGC in zero-shot scenario is to find new **golden triplet** (h_*, r_*, t_*) based on D_{h_*}, T_{h_*}, D_{t_*}, T_{t_*}, where $h_*, t_* \in E \cup E'$, $r_* \in R$ and at least one entity (h_* or t_*) $\in E'$.

3.2   Overall Architecture

To deal with KGC in zero-shot scenario, our models build combination structural representations and textual representations for In-KG/Out-of-KG entities. Our models learn the base structural representation $e_s^{(b)}$ for base entity at training time and build combination structural representation e_s for all entities. The textual representation of entity e is denoted as $\boldsymbol{e}_d$, which is built from entity description by Bi-GRU network. Each of the two representations has its advantages: the former can capture the information hidden in triplet structures, while the latter can obtain valuable semantic information from description.

To make full use of these two kinds of representations, we integrate them into a joint representation. We first suppose that each dimension of structural representation and textual representation represents same semantics. Then, we make linear combination on every dimension to form joint representation:

$$\boldsymbol{e}_{joint} = \alpha * \boldsymbol{e}_s + (1 - \alpha) * \boldsymbol{e}_d,$$

where $*$ is a multiplication operator, $\boldsymbol{e}_{joint} \in \mathbb{R}^{k \times 1}$ represents joint representation, $\boldsymbol{e}_s \in \mathbb{R}^{k \times 1}$, $\boldsymbol{e}_d \in \mathbb{R}^{k \times 1}$ where k is the dimension of representation, α is the weight of combination structural representation and its value domain is $[0, 1]$.

Given a triplet (h, r, t), we denote $\boldsymbol{e}_{joint}^{(h)}/\boldsymbol{e}_{joint}^{(t)}$ as the joint represetation of its head/tail entity and $\boldsymbol{r}$ as its relation representation. The score function of our models are based on TransE and TransH:

$$\begin{cases} E(h,r,t)_{(transe)} = \left\| \boldsymbol{e}_{joint}^{(h)} + \boldsymbol{r} - \boldsymbol{e}_{joint}^{(t)} \right\|_{L_1/L_2}, \\ E(h,r,t)_{(transh)} = \left\| (\boldsymbol{e}_{joint}^{(h)} - \boldsymbol{w}_r^T \boldsymbol{e}_{joint}^{(h)} \boldsymbol{w}_r) + \boldsymbol{r} - (\boldsymbol{e}_{joint}^{(t)} - \boldsymbol{w}_r^T \boldsymbol{e}_{joint}^{(t)} \boldsymbol{w}_r) \right\|_{L_1/L_2}, \end{cases}$$

$$(1)$$

where $\|\cdot\|_{L_1/L_2}$ is L1-norm or L2-norm function, $\boldsymbol{w}_r \in \mathbb{R}^{k \times 1}$ denotes the relation-specific hyperplane for relation r. The score of **golden triplet** is expected to lower than incorrect triplet.

Our models can be divided into two parts: one builds combination structural representation $\boldsymbol{e}_s$ through base entities, the other builds textual representation $\boldsymbol{e}_d$ based on entity description.

3.3   Structural Encoder

Since Out-of-KG entities only contain textual information, we need to construct a new structural feature space, which is shared by In-KG entities and Out-of-KG entities. We first present the strategy of constructing base-entity set. Then, we introduce an entity similarity algorithm. Finally, we propose an inference function to build combination structural representations for entities (Fig. 2).

Base-Entity Set. It is obvious that entity description is made up of several words in "Word Set" and the correlations between different entity descriptions depend on the same or similar words they share. Inspired by the mechanism of entity description, we first construct the basic elements of structural feature space, which also can be regarded as "Base-Entity Set". Then, based on this set, we can build "Base-Entity Sequence" for each entity. In order to enable every element in the "Base-Entity Set" to be fully trained, we treat all In-KG entities(E) as "Base-Entity Set"($E^{(b)}$).

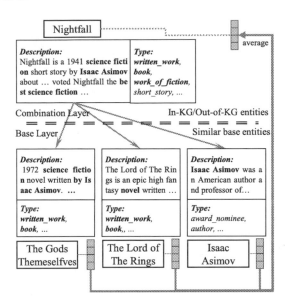

Fig. 2. Combination structural representations for entities.

Entity Similarity. In practical zero-shot scenario, there may exist data missing problem. To improve the robustness of entity similarity algorithm, we utilize two kinds of textual information: descriptions and types. The former can represent entity from various aspects, while the latter can accurately depict the entity. As for the description, there are many new words only appearing at test time. Thus, we only preserve the words that are shared by In-KG entities and Out-of-KG entities. We utilize Jaccard [9] to compute the similarity of entities e_i and e_j:

$$S_d(e_i, e_j) = \frac{|D_{e_i} \cap D_{e_j}|}{|D_{e_i} \cup D_{e_j}|}, S_t(e_i, e_j) = \frac{|T_{e_i} \cap T_{e_j}|}{|T_{e_i} \cup T_{e_j}|}, \qquad (2)$$

$$S(e_i, e_j) = \beta * S_d(e_i, e_j) + (1 - \beta) * S_t(e_i, e_j), \qquad (3)$$

where S_d represents similarity based on description, S_t represents similarity based on type, β is a hyperparameter to balance two similarities and its value domain is $[0, 1]$, S is the final similarity for entity e_i and e_j.

Inference Function. We assume that similar entities should have similar base entities and the characteristic of specific entity can be expressed by its similar base entities. Thus, we utilize entity similarity algorithm to find K most similar base entities for every In-KG/Out-of-KG entity e to form its "Similar Base-Entity Sequence"- $P(e)$. Then, we utilize inference function I to build combination structural representations for entity e with $P(e)$. The "average" inference function is defined as:

$$e_s = I_{average}(e) = \frac{1}{K} \sum_{e^{(b)} \in P(e)} e_s^{(b)}. \qquad (4)$$

Since we take all In-KG entities as "Base-Entity Set", as for specific In-KG entity e_i, it is obvious that the most similar base entity of e_i is $e_i^{(b)}$, which makes the "Similar Base-Entity Sequence" of e_i must contain $e_i^{(b)}$. Thus, every element in "Base-Entity Set" can be fully trained through corrsponding "Similar Base-Entity Sequence".

3.4 Textual Encoder

There are rich textual descriptions for entities in or out of KGs. We utilize Bi-GRU network with pooling to encode descriptions, and build the correlations between In-KG entities and Out-of-KG entities in the textual feature space.

A gated recurrent unit (GRU) [6] is used to capture features of sequential data. There are two gates in a GRU, a reset gate p, and an update gate q, to adaptively capture dependencies of different time scales. The reset gate defines the weights of the current state and the previous memory. The update gate defines the weight of the previous memory when calculating the current state.

Given an entity description $D_e = (w_1, w_2, \cdots, w_n)$, we first embed these words into a vector space. $\boldsymbol{w}_i \in \mathbb{R}^{n_w \times 1}$ denotes the vector of word w_i where n_w is the dimension of word vector. Then, we define the GRU that processes D_e from left to right as follows:

$$
\begin{aligned}
\boldsymbol{q}_t &= \sigma_g \left(\boldsymbol{W}_q \boldsymbol{w}_t + \boldsymbol{U}_q \boldsymbol{h}_{l-1} + \boldsymbol{b}_q \right) \\
\boldsymbol{p}_t &= \sigma_g \left(\boldsymbol{W}_p \boldsymbol{w}_t + \boldsymbol{U}_p \boldsymbol{h}_{t-1} + \boldsymbol{b}_p \right) \\
\boldsymbol{h}_t &= \boldsymbol{q}_t \circ \boldsymbol{h}_{t-1} + (1 - \boldsymbol{q}_t) \circ \sigma_h \left(\boldsymbol{W}_h \boldsymbol{w}_t + \boldsymbol{U}_h \left(\boldsymbol{p}_t \circ \boldsymbol{h}_{t-1} \right) + \boldsymbol{b}_h \right),
\end{aligned}
\tag{5}
$$

where t represents the t-th time step in GRU, here it also denotes the ordinal number of words in entity description, thus the value domain of t is $[1, n]$. $\boldsymbol{q}_t \in \mathbb{R}^{d \times 1}$, is the update gate vector of GRU at t-th time step. $\boldsymbol{p}_t \in \mathbb{R}^{d \times 1}$, is the reset gate vector of GRU at t-th time step. $\boldsymbol{h}_t \in \mathbb{R}^{d \times 1}$, is the output of GRU at t-th time step. $\boldsymbol{W} \in \mathbb{R}^{d \times n}$, $\boldsymbol{U} \in \mathbb{R}^{d \times n}$, and $\boldsymbol{b} \in \mathbb{R}^{d \times 1}$ denote parameter matrices and vectors. σ_g is a sigmoid activation function, σ_h is a tanh activation function and $\circ$ denotes the Hadamard product.

To capture the features from the opposite direction synchronously, we employ Bi-GRU which combines the outputs of the two directions. We denote $\overrightarrow{\boldsymbol{h}_t}$ as the output of the GRU shown in (5). The GRU with the opposite direction is denoted by exchanging t and $t-1$ in (5), and $\overleftarrow{\boldsymbol{h}_t}$ denotes its output. The output of Bi-GRU at t-th time step is the concatenation of the two vectors:

$$
\overleftrightarrow{\boldsymbol{h}_t} = \left[\overrightarrow{\boldsymbol{h}_t} : \overleftarrow{\boldsymbol{h}_t} \right].
\tag{6}
$$

We denote this Bi-GRU network as

$$
\overleftrightarrow{\boldsymbol{h}} = \left(\overleftrightarrow{\boldsymbol{h}_1}, \cdots, \overleftrightarrow{\boldsymbol{h}_n} \right) = \text{BIGRU} \left(w_1, \cdots, w_n \right),
\tag{7}
$$

where $\overleftrightarrow{\boldsymbol{h}} \in \mathbb{R}^{2d \times n}$ is a combination of Bi-GRU outputs for all the n time steps, the i-th column of $\overleftrightarrow{\boldsymbol{h}}$ is $\overleftrightarrow{\boldsymbol{h}_i}$.

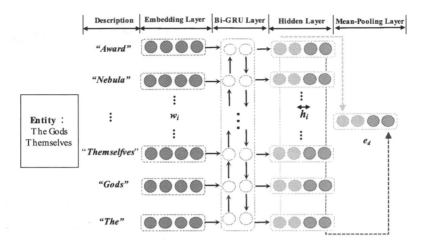

Fig. 3. Bi-GRU encoder with pooling.

In order to preserve more information of the earlier unit, we add a pooling layer after hidden layer to take advantages of all output vectors. In this paper, we utilize mean-pooling to obtain final textual representation:

$$\boldsymbol{e}_d = \text{enc}(e) = \sum_{t=1}^{n} \overleftrightarrow{\boldsymbol{h}_t}/n. \tag{8}$$

The encoding process is shown in Fig. 3.

3.5   Training and Implementation Details

Given a training set M, we denote the i-th triplet as (h_i, r_i, t_i), ($i = 1, 2, \cdots, |M|$). Each triplet has a label y_i to indicate whether the triplet is positive ($y_i = 1$) or negative ($y_i = 0$). The positive and negative sets are denoted as $L = \{(h_i, r_i, t_i)|y_i = 1\}$ and $L' = \{(h_i, r_i, t_i)|y_i = 0\}$, respectively. We assume that each triplet in M is a positive triplet and construct negative triplets by randomly replacing elements in positive triplets.

To keep consistent with DKRL [23], we utilize replacements of entities and relations in negative sampling. Specifically, "bern" strategy [21] is applied to replace the head or tail entity. We construct two negative triplets (h', r, t') and $(h, r', t) \in L'$ for each positive triplet $(h, r, t) \in L$. With the negative sampling, the objective function is defined as a margin-based ranking criterion [3]:

$$\sum_{(h,r,t)\in L} \sum_{(h',r',t')\in L'} \max(0, E(h,r,t) + \gamma - E(h',r',t')), \tag{9}$$

where $\gamma > 0$ is a margin between positive triplets and negative triplets. $E(h,r,t)$ is the score funtion defined as (1). In addition, we utilize following constraints when minimizing the objective function:

$$||e_s||_2 \leq 1, ||e_d||_2 \leq 1, ||\boldsymbol{r}||_2 \leq 1, ||\boldsymbol{w}_r||_2 \leq 1. \qquad (10)$$

4 Experiment

4.1 Experimental Setup

Dataset. Two public benchmark datasets, FB15K and FB20K [23], are adopted to evaluate the performance of our models on two tasks, entity prediction and relation prediction. We take entities in FB15K as In-KG entities and the extra entities in FB20K as Out-of-KG entities. FB20K contains four sets: both the head and the tail are In-KG (e-e), the head is Out-of-KG but the tail is In-KG (d-c), the tail is In-KG but the head is Out-of-KG (e-d), both the head and the tail are Out-of-KG (d-d). The statistics of datasets are listed in Table 1.

Table 1. Statistics of datasets

Dataset	#Ent	#Rel	#Train	#Valid	#Test
FB15K	14,904	1,341	472,860	48,991	57,803
Dataset	#Ent	#e-e	#d-e	#e-d	#d-d
FB20K	19,923	57,803	18,753	11,586	151

Preprocessing. We directly utilize the descriptions and types information provided in DKRL [23]. After removing specific symbols, the average length of entity description is 147. Then, each word is represented by a word embedding as the input of Bi-GRU. In our experiments, we train the word embeddings on the descriptions of In-KG entities through word2vec [14] and utilize the results to initialize word vectors.

Baselines. The joint model proposed in [20] utilizes entity name to obtain the representations of Out-of-KG entities. However, in FB20K, some Out-of-KG entity names do not appear in the training corpus, thus this joint model cannot learn representations for these Out-of-KG entities. The joint model proposed in [26] aligns entity and the words in its description in the same space. However, this model cannot learn the representations for Out-of-KG entities directly. Hence, we choose four state-of-the-art methods proposed in [23] as our baselines: Partial-CBOW, CBOW, Partial-CNN, and CNN.

Evaluation Protocol. We adopt the same protocol used in previous work [3]. Two standard measures are considered as evaluation metrics: One is HITS@N, the proportion of valid entities (or relations) ranked in top N. In this task, we utilize HITS@10/HITS@1 for entity/relation prediction. The other is mean reciprocal rank (MRR), the average reciprocal rank of valid entities (or relations).

We follow DKRL [23] and present the results on "Filter" setting. It is noteworthy that the negative triplets are randomly generated during training process. Thus, we report the average results of five times to make the results more reliable. A higher HITS@10 or MRR means better performance.

Experiment Settings. As the datasets are the same, we directly reuse the results of baselines from literature [23]. We employ adam optimizer [10] to minimize the objective function and the mini-batch size is 1024. We choose top 10 similar base entities to build combination structural representations for entities. Then, we select margin γ among $\{1, 2\}$, word vector dimension n_w among $\{50, 100\}$, learning rate λ among $\{0.0005, 0.001\}$, entity/relation dimension k among $\{50, 100\}$, weight β for description information among $\{0, 0.3, 0.5, 0.7, 1\}$, weight α for structural representation among $\{0, 0.5, 1\}$. We choose the configurations that achieve best HITS@10 of entity prediction in valid set. Other parameters are randomly initialized on a uniform distribution between $[-\frac{6}{\sqrt{k}}, \frac{6}{\sqrt{k}}]$. The optimal configurations are: $\gamma = 1, n_w = 100, \lambda = 0.0005, k = 100, \beta = 0.5$ and $\alpha = 0.5$. We train our models for 1000 epochs.

4.2 Entity Prediction

The goal of entity prediction is to infer the missing entity when given an entity and a relation, i.e. we predict t given $(h, r, *)$, or predict h given $(*, r, t)$.

Table 2. HITS@10 results of entity prediction on FB20K

Test set	e-e	d-e	e-d
Metric	Filter	Filter	Filter
DKRL(Partial-CBOW)	-	26.5	20.9
DKRL(CBOW)	51.8	27.1	21.7
DKRL(Partial-CNN)	-	26.8	20.8
DKRL(CNN)	57.6	31.2	26.1
TransE-Single(CNN, des-only)	51.4	28.1	22.5
TransE-Single(Bi-GRU, des-only)	58.8	29.2	24.1
TransE-Single(AVG, struct-only)	45.4	27.6	21.6
TransH-Single(AVG, struct-only)	45.0	26.5	20.6
TransE-Jointly(Bi-GRU+AVG)	**60.5**	**34.3**	**29.6**
TransH-Jointly(Bi-GRU+AVG)	58.6	33.0	28.6

Results. Evaluation results on FB20K are shown in Tables 2 and 3, "Bi-GRU" represents our Bi-GRU textual encoder. "AVG" denotes the way of building combination structural representations for entities. Single(Bi-GRU) only uses

Table 3. MRR results of entity prediction on FB20K

Test set	e-e		d-e		e-d	
Metric	Filter		Filter		Filter	
	Head	Tail	Head	Tail	Head	Tail
DKRL(Partial-CNN)	-	-	3.2	36.0	23.1	3.8
DKRL(CNN)	30.7	37.3	3.4	37.1	24.3	3.9
TransE-Single(CNN, des-only)	27.0	33.6	2.9	36.3	23.4	3.4
TransE-Single(Bi-GRU, des-only)	32.0	38.6	2.8	35.0	24.0	3.3
TransE-Single(AVG, struct-only)	23.7	30.3	2.9	34.4	23.4	2.9
TransH-Single(AVG, struct-only)	23.5	30.0	2.7	33.3	22.5	2.7
TransE-Jointly(Bi-GRU+AVG)	**34.2**	**40.9**	**5.3**	**40.6**	**29.5**	**6.1**
TransH-Jointly(Bi-GRU+AVG)	32.8	39.3	5.1	39.0	27.8	5.8

textual representations and Single(AVG) only utilizes structural representations. Single(CNN) is a simplified DKRL [23] that only utilizes entity descriptions to build representations for entities. From the result, we observe that:

1. Only our joint models outperform all baselines, which indicates that the combination of textual representation and structural representation is necessary. In the majority of cases, Single(Bi-GRU) achieves better performance than Single(CNN). It demonstrates that Bi-GRU is more suitable to encode entity descriptions in KGC.
2. On "d-e"/"e-d" set, "Head"/"Tail" represents predicting Out-of-KG entities. From Table 3 we can know that the MRR of predicting Out-of-KG entities is very low, it indicates that the representations of Out-of-KG entities are not particularly appropriate and the ability to discriminate between In-KG entities and Out-of-KG entities remains to be strengthened.

Table 4. MRR results of relation prediction on FB20K

Test set	e-e	d-e	e-d
Metric	Filter	Filter	Filter
DKRL(Partial-CNN)	-	66.9	61.6
DKRL(CNN)	92.7	67.7	63.0
TransE-Single(CNN, des-only)	91.2	65.9	61.9
TransE-Single(Bi-GRU, des-only)	92.1	67.8	63.9
TransE-Single(AVG, struct-only)	92.8	**80.8**	**79.3**
TransH-Single(AVG, struct-only)	92.7	80.7	79.2
TransE-Jointly(Bi-GRU+AVG)	**93.2**	78.1	75.6
TransH-Jointly(Bi-GRU+AVG)	91.7	76.0	73.4

4.3   Relation Prediction

The goal of relation prediction is to infer the possible relation when given two entities, i.e., we predict r given $(h, *, t)$.

Table 5. HITS@1 Results of relation prediction on FB20K

Test set	e-e	d-e	e-d
Metric	Filter	Filter	Filter
DKRL(Partial-CBOW)	-	49.0	42.2
DKRL(CBOW)	82.7	52.2	47.9
DKRL(Partial-CNN)	-	56.6	52.4
DKRL(CNN)	89.0	60.4	55.5
TransE-Single(CNN, des-only)	86.6	57.6	53.6
TransE-Single(Bi-GRU, des-only)	87.8	59.7	56.0
TransE-Single(AVG, struct-only)	88.8	**74.2**	**73.4**
TransH-Single(AVG, struct-only)	88.6	74.1	72.9
TransE-Jointly(Bi-GRU+AVG)	**89.4**	71.3	69.3
TransH-Jointly(Bi-GRU+AVG)	87.4	68.9	66.9

Results. Evaluation results on FB20K are shown in Tables 4 and 5. From the results, we observe that:

1. Compared with DKRL(CNN), as for HITS@1, TransE-Single(AVG) improves by 13.8%/17.9% on "d-e"/"e-d" set. The improvements of our models on relation prediction are larger than that on entity prediction. Since for the **golden triplets**, the characteristic of head/tail entity are highly related to corresponding relation. Thus, one possible reason of this situation is that with combination structural representations and textual representations, our models can aggregate entities with similar semantics or concept.
2. According to the performance of our models, we can know that the improvements on relation prediction are highly relying on structural representation. Interestingly, Single(AVG) achieves best results on relation prediction but almost worst results on entity prediction, it indicates that Single(AVG) may make the structural representations of similar entities too similar and lowers the discrimination between them.

4.4   The Impact of Parameter β

We first divide entity prediction into two types: In-KG prediction and Out-of-KG prediction, which depend on the ascription of missing entity. Figure 4 shows the results of TransE-Jointly(Bi-GRU+AVG) on the "total" set, which is the combination of "*d-e*" and "*e-d*". β represents the weight of descriptions in entity similarity algorithm. When $\beta = 1$, the proportion of descriptions in entity similarity algorithm is 100%, almost all predictions achieve worst performance. However, compared with DKRL(CNN), only utilizing descriptions to calculate entity similarity still obtains 2.4%/4.5% improvements on entity/relation prediction. It indicates that although the descriptions used in our models are kind of noisy, they still could capture potential correlations between In-KG entities and Out-of-KG entities. In addition, the best results of all predictions are achieved among different combinations for types and descriptions information, it shows that this two kind of information is a complementary relationship and the combination for them is necessary.

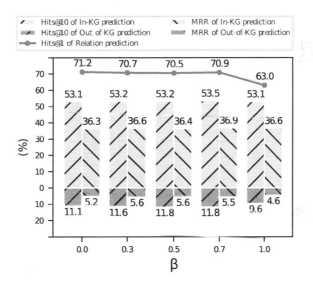

Fig. 4. The impacts of parameter β on TransE-Jointly(Bi-GRU+AVG)

5   Conclusion and Future Work

In this paper, we propose a joint representation for knowledge graph with structural information and textual information. We construct a new structural feature space and build combination structural representations for entities through their most similar base entities. Specifically, we implement Bi-GRU network with pooling to obtain textual representations. Experimental results on entity and relation prediction show that our models could enhance the correlations between In-KG entities and Out-of-KG entities, which benefit knowledge graph completion in zero-shot scenario.

Since our models are based on TransE and TransH, which are the most fundamental translation-based models, in the future, we will explore some more effective translation-based models, such as TransR and PTransE to extend our models.

382   J. Ding et al.

Acknowledgements. This work is supported by DCT-MoST Joint-project No. (025/2015/AMJ); University of Macau Funds Nos: CPG2018-00032-FST & SRG2018-00111-FST; Chinese National Research Fund (NSFC) Key Project No. 61532013; National China 973 Project No. 2015CB352401; Shanghai Scientific Innovation Act of STCSM No.15JC1402400 and 985 Project of Shanghai Jiao Tong University: WF220103001.

References

1. Bollacker, K.D., Evans, C., Paritosh, P., Sturge, T., Taylor, J.: Freebase: a collaboratively created graph database for structuring human knowledge. In: Proceedings of the ACM SIGMOD International Conference on Management of Data, SIGMOD 2008, Vancouver, BC, Canada, 10–12 June 2008, pp. 1247–1250 (2008)
2. Bordes, A., Glorot, X., Weston, J., Bengio, Y.: Joint learning of words and meaning representations for open-text semantic parsing. In: Proceedings of the Fifteenth International Conference on Artificial Intelligence and Statistics, AISTATS 2012, La Palma, Canary Islands, 21–23 April 2012, pp. 127–135 (2012)
3. Bordes, A., Usunier, N., García-Durán, A., Weston, J., Yakhnenko, O.: Translating embeddings for modeling multi-relational data. In: Advances in Neural Information Processing Systems 26: 27th Annual Conference on Neural Information Processing Systems 2013, Proceedings of a Meeting Held, Lake Tahoe, Nevada, USA, 5–8 December 2013, pp. 2787–2795 (2013)
4. Cao, Y., Huang, L., Ji, H., Chen, X., Li, J.: Bridge text and knowledge by learning multi-prototype entity mention embedding. In: Proceedings of the 55th Annual Meeting of the Association for Computational Linguistics, ACL 2017, Vancouver, Canada, 30 July–4 August, vol. 1: Long Papers, pp. 1623–1633 (2017)
5. Carlson, A., Betteridge, J., Kisiel, B., Settles, B., Hruschka Jr., E.R., Mitchell, T.M.: Toward an architecture for never-ending language learning. In: Proceedings of the Twenty-Fourth AAAI Conference on Artificial Intelligence, AAAI 2010, Atlanta, Georgia, USA, 11–15 July 2010 (2010)
6. Chung, J., Gulcehre, C., Cho, K., Bengio, Y.: Empirical evaluation of gated recurrent neural networks on sequence modeling. arXiv preprint arXiv:1412.3555 (2014)
7. Dai, Z., Li, L., Xu, W.: CFO: conditional focused neural question answering with large-scale knowledge bases. In: Proceedings of the 54th Annual Meeting of the Association for Computational Linguistics, ACL 2016, Berlin, Germany, 7–12 August 2016, vol. 1: Long Papers (2016)
8. Hamaguchi, T., Oiwa, H., Shimbo, M., Matsumoto, Y.: Knowledge transfer for out-of-knowledge-base entities: a graph neural network approach. In: Proceedings of the Twenty-Sixth International Joint Conference on Artificial Intelligence, IJCAI 2017, Melbourne, Australia, 19–25 August 2017, pp. 1802–1808 (2017)
9. Jaccard, P.: Distribution de la Flore Alpine: dans le Bassin des dranses et dans quelques régions voisines. Rouge (1901)
10. Kingma, D.P., Ba, J.: Adam: a method for stochastic optimization. arXiv preprint arXiv:1412.6980 (2014)
11. Lin, Y., Liu, Z., Luan, H., Sun, M., Rao, S., Liu, S.: Modeling relation paths for representation learning of knowledge bases. In: Proceedings of the 2015 Conference on Empirical Methods in Natural Language Processing, EMNLP 2015, Lisbon, Portugal, 17–21 September 2015, pp. 705–714 (2015)

12. Lin, Y., Liu, Z., Sun, M., Liu, Y., Zhu, X.: Learning entity and relation embeddings for knowledge graph completion. In: Proceedings of the Twenty-Ninth AAAI Conference on Artificial Intelligence, Austin, Texas, USA, 25–30 January 2015, pp. 2181–2187 (2015)
13. Luong, T., Pham, H., Manning, C.D.: Effective approaches to attention-based neural machine translation. In: Proceedings of the 2015 Conference on Empirical Methods in Natural Language Processing, EMNLP 2015, Lisbon, Portugal, 17–21 September 2015, pp. 1412–1421 (2015)
14. Mikolov, T., Chen, K., Corrado, G., Dean, J.: Efficient estimation of word representations in vector space. arXiv preprint arXiv:1301.3781 (2013)
15. Miller, G.A.: WordNet: a lexical database for English. Commun. ACM **38**(11), 39–41 (1995)
16. Neelakantan, A., Roth, B., McCallum, A.: Compositional vector space models for knowledge base inference. In: 2015 AAAI Spring Symposium Series (2015)
17. Palatucci, M., Pomerleau, D., Hinton, G.E., Mitchell, T.M.: Zero-shot learning with semantic output codes. In: Advances in Neural Information Processing Systems 22: 23rd Annual Conference on Neural Information Processing Systems 2009, Proceedings of a Meeting Held, Vancouver, British Columbia, Canada, 7–10 December 2009, pp. 1410–1418 (2009)
18. Socher, R., Chen, D., Manning, C.D., Ng, A.Y.: Reasoning with neural tensor networks for knowledge base completion. In: Advances in Neural Information Processing Systems 26: 27th Annual Conference on Neural Information Processing Systems 2013, Proceedings of a Meeting Held, Lake Tahoe, Nevada, USA, 5 8 December 2013, pp. 926–934 (2013)
19. Toutanova, K., Chen, D., Pantel, P., Poon, H., Choudhury, P., Gamon, M.: Representing text for joint embedding of text and knowledge bases. In: Proceedings of the 2015 Conference on Empirical Methods in Natural Language Processing, EMNLP 2015, Lisbon, Portugal, 17–21 September 2015, pp. 1499–1509 (2015)
20. Wang, Z., Zhang, J., Feng, J., Chen, Z.: Knowledge graph and text jointly embedding. In: Proceedings of the 2014 Conference on Empirical Methods in Natural Language Processing, EMNLP 2014, Doha, Qatar, 25–29 October 2014, a Meeting of SIGDAT, a Special Interest Group of the ACL, pp. 1591–1601 (2014)
21. Wang, Z., Zhang, J., Feng, J., Chen, Z.: Knowledge graph embedding by translating on hyperplanes. In: Proceedings of the Twenty-Eighth AAAI Conference on Artificial Intelligence, Québec City, Québec, Canada, 27–31 July 2014, pp. 1112–1119 (2014)
22. Xiao, H., Huang, M., Meng, L., Zhu, X.: SSP: semantic space projection for knowledge graph embedding with text descriptions. In: Proceedings of the Thirty-First AAAI Conference on Artificial Intelligence, San Francisco, California, USA, 4–9 February 2017, pp. 3104–3110 (2017)
23. Xie, R., Liu, Z., Jia, J., Luan, H., Sun, M.: Representation learning of knowledge graphs with entity descriptions. In: Proceedings of the Thirtieth AAAI Conference on Artificial Intelligence, Phoenix, Arizona, USA, 12–17 February 2016, pp. 2659–2665 (2016)
24. Xu, J., Qiu, X., Chen, K., Huang, X.: Knowledge graph representation with jointly structural and textual encoding. In: Proceedings of the Twenty-Sixth International Joint Conference on Artificial Intelligence, IJCAI 2017, Melbourne, Australia, 19–25 August 2017, pp. 1318–1324 (2017)

25. Zhang, D., Yuan, B., Wang, D., Liu, R.: Joint semantic relevance learning with text data and graph knowledge. In: ACL-IJCNLP 2015, p. 32 (2015)
26. Zhong, H., Zhang, J., Wang, Z., Wan, H., Chen, Z.: Aligning knowledge and text embeddings by entity descriptions. In: Proceedings of the 2015 Conference on Empirical Methods in Natural Language Processing, EMNLP 2015, Lisbon, Portugal, 17–21 September 2015, pp. 267–272 (2015)

Neural Typing Entities in Chinese-Pedia

Yongjian You[1], Shaohua Zhang[1], Jiong Lou[1], Xinsong Zhang[1],
and Weijia Jia[1,2(✉)]

[1] Shanghai Jiao Tong University, Shanghai 200240, China
{youyongjian,zhangsh950618,lj1994,xszhang0320}@sjtu.edu.cn,
jia-wj@cs.sjtu.edu.cn
[2] University of Macau, Macau SAR 999078, China

Abstract. Typing entities in structured sources such as Wikipedia has
been well studied to construct English knowledge bases automatically.
However, there still remain two tough challenges in typing entities in
Chinese-pedia. The first one is that structured information from Chinese-
pedia cannot assign entities fine-grained types due to its inaccuracy and
coarseness. The other challenge is the incompletion of Chinese-pedia,
which means we can only use limited attribute fields to type entities.
In this paper, we propose a novel Hierarchical Neural System (HNS) to
infer fine-grained types for entities in Chinese-pedia. The HNS contains
three main models which are hierarchical attention model, feature fusion
model and hierarchical classification model. The hierarchical attention
model extracts features from entity description based on a bi-LSTM
network with hierarchical attention mechanism to break the limitation
of inaccurate Chinese-pedia. To deal with the incompletion of Chinese-
pedia, the feature fusion model is presented to obtain type features from
multi-source such as descriptions, info-boxes, and categories. Through
this model, we fuse all the features from different sources together and
reduce the features to low-dimensional and dense vectors. Finally, the
hierarchical classification model is designed to infer fine-grained types
for entities in Chinese-pedia with features obtained from the other two
models. The experiments illustrate that HNS outperforms the start-of-
art work by 15.6% on f1-score.

1 Introduction

With the continuous evolution and development of information technology, the
Web has become a vast treasure house of human knowledge. Many researchers
have made great efforts to construct knowledge bases which translate the human-
readable content on the Web into well-constructed information. Nowadays, there
are many well-established English knowledge bases such as DBpedia and Yago,
but high-quality Chinese knowledge bases are still very rare. English knowledge
bases are always constructed by well typing entities of structured sources such
as Wikipedia. Therefore, we should focus on the task of entity typing to build a
high-quality Chinese knowledge base.

Type information is important to indicate the attributes of entities and it
has a clear hierarchical structure to present different abstract levels of attributes.

© Springer International Publishing AG, part of Springer Nature 2018
Y. Cai et al. (Eds.): APWeb-WAIM 2018, LNCS 10987, pp. 385–399, 2018.
https://doi.org/10.1007/978-3-319-96890-2_32

For example, the entity **Leonardo DiCaprio** has types **Person, Artist** and **Actor**. Entity typing is to find the most fine-grained types for entities. Fine-grained types are valuable to depict entities precisely and vital to the future construction works such as representation learning [1,10] and knowledge fusion [3,9]. Therefore typing entities in Chinese-pedia is a foundation task for constructing Chinese knowledge bases.

To type an entity in Chinese-pedia correctly, as much as possible information of the entity is indispensable. The BaiduBaike[1], largest Chinese-pedia, contains info-boxes, categories and descriptions, all of which can help to type its entities in different ways. In this paper, we focus on typing entities in the BaiduBaike with existing type paradigm such as DBpedia types. However, due to the inaccuracy and incompletion of the BaiduBaike, there still remain two serious challenges.

- **The challenge of fine-grained typing from inaccurate data.** The type paradigm has a hierarchical structure and fine-grained types tend to depict entities more precisely. Fine-grained types should be inferred with accurate and specific information, while Chinese-pedia can only provide inaccurate and coarse structured data, such as info-boxes and categories. Statistics indicate that more than 30% entities in BaiduBaike only have names in info-boxes. Under such circumstance, it is harder to assign fine-grained types such as **Actor** than coursed-grained types like **Person**.
- **The challenge of incomplete Chinese-pedia.** The Chinese-pedia is incomplete due to the fact that it is constructed by crowd-sourcing with low quality of supervision. For instance, more than 15% entities in BaiduBaike have no descriptions and 20% entities are lacking of info-boxes. The incompletion of Chinese-pedia makes it very difficult to infer the types for entities.

To infer fine-grained types, it is necessary to make use of more specific information of the entities such as descriptions, which can provide sufficient semantic information. As for the challenge of incomplete Chinese-pedia, different features can be fused together to complement each other and weaken the influence of missing data. In this paper, we propose a novel system named Hierarchical Neural System (HNS) to infer fine-grained types for entities in Chinese-pedia. This system consists of three models, which are hierarchical attention model, feature fusion model and hierarchical classification model. The hierarchical attention model employs a bi-directional Long Short-Term Memory (bi-LSTM) network with hierarchical attention mechanism to extract features from descriptions. The feature fusion model obtains type features from multi-source and fuses them together to be a low-dimensional and dense vector. With the vector, the hierarchical classification model infers the fine-grained types for Chinese-pedia entities. The experimental results show that our system increases the f1-score by 0.121 as against the state-of-art work.

In summary, the contributions in this paper are listed as follows.

- We utilize the semantic information extracted from descriptions to infer fine-grained types for Chinese-pedia entities.

[1] baike.baidu.com.

- We fuse the features from multi-source together to weaken the influence of incompleteness and sparseness in Chinese-pedia.
- Our system outperforms all the baselines and achieves the precision of 89.3% and recall of 89.1%, which are higher by 2.9% and 19.4% respectively than the state-of-art work.

The rest of this paper is organized as follows. Section 2 discusses the related works, and Sect. 3 introduces preliminaries. Section 4 presents our approach in details. In Sect. 5, we present the experimental results. At last, we conclude the paper in Sect. 6.

2   Related Work

Related work can be outlined with respect to the fine-grained mention typing and fine-grained entity typing.

Fine-Grained Mention Typing. Lee *et al.* was the first to address the task of fine-grained type classification for entity mentions [4]. They manually defined 147 fine-grained types and evaluated their conditional random field model on the Korean dataset. Yosef *et al.* manually organized 505 types in an ontology and developed a multi-label hierarchical classification system [18]. Corro *et al.* presented a fine-grained mention type classification system, which operated on the WordNet hierarchy with more than 16,000 types. On this basis, Shimaoka *et al.* introduced attention mechanism to allow their model to focus on relevant expressions [11]. However, these approaches mainly extract features on sentence-level and infer types for entity mentions, but we focus on the entity-level features and infer types for entities in Chinese-pedia.

Fine-Grained Entity Typing. DBpedia [5] built the mappings between the Wikipedia templates and types through crowd-sourcing, and the entities created with these templates should have the corresponding types. Yago [13] proposed a holistic method for knowledge base construction based on Wikipedias in different languages. They used natural language processing tools to determine whether a category generated by Wikipedia should belong to a type. Unfortunately, most of English entity typing methods cannot be applied to type Chinese-pedia entities. On one hand, due to language specific features, there is no effective way to apply the language-dependent methods; on the other hand, the performance of the language-independent methods [8,15] is unsatisfactory.

As for Chinese-pedia entity typing, Zhishi.me is the pioneer in building a Chinese knowledge base [7]. They used the inter-language links provided by Wikipedia to assign Yago types for Chinese-pedia entities. Wu *et al.* obtained type information from the description, info-box, and category of an entity [14]. They presented an attribute propagation algorithm to generate attributes for types and a graph-based random walk method to assign types for entities. CUTE is the prominent representative of Chinese-pedia entity typing [16]. They reused the ontology tree constructed by DBpedia directly and extracted features from info-box and category. Finally, a hierarchical classifier was used to infer the types. Whereas, all the above methods haven't apply neural method to precisely extract type features from semantic information and multi-source data.

3   Preliminaries

Before introducing our system, it is necessary to give relevant definitions and present the structured type features used in the system.

3.1   Definitions

In this section, we formally define three relevant concepts of our task and finally formalize the Chinese-pedia entity typing problem.

Ontology. The ontology is a collection of structured types provided by the knowledge base. It can be formally represented as $O = <T, R>$, where T is the set of types and R is the set of relation instances. A relation instance is a directed connection between two types represented as $[head, tail]$. The head type is more generalized than the tail type such as [**Person**, **Actor**].

The DBpedia types are organized as a tree[2], which means a type may have several subtypes but only directly connect to one supertype. For example, **Person** has many subtypes **Arist**, **Lawyer**, **Scientist**, etc, but it only connect to supertype **Agent**.

Entity. The entity is the abstraction of a real-world object. It can be formally represented as e_i and $E = \{e_i\}_{i=1}^m$, where E is the collection of all entities and m is the size of E.

In BaiduBaike, the information of an entity e_i can be represented by a five-tuple $(N(e_i), D(e_i), I(e_i), C(e_i), T(e_i))$, where $N(e_i)$ denotes the name of e_i; $D(e_i)$ is the description of e_i; $I(e_i)$ is the info-box of e_i, the $C(e_i)$ is the categories and $T(e_i)$ is the type. Figure 1 shows an example of BaiduBaike entity.

Description. For almost every BaiduBaike entity e, there is a description $D(e)$ which gives e a brief introduction. Because $D(e)$ contains directional and explicit information, the $D(e)$ is well recognized as the most valuable text information to infer the types.

We formally define the description of entity e as follows:

$$D(e) = (s_1, s_2, ..., s_L) \tag{1}$$

$$s_j = (w_j^1, w_j^2, ..., w_j^Q) \tag{2}$$

where L is the number of sentences in a description, s_j is the j-th sentence of $D(e)$, Q is the number of words in the j-th sentence and w_j^k is the k-th word.

---

[2] The latest version of DBpedia ontology is organized as a directed-acyclic graph.

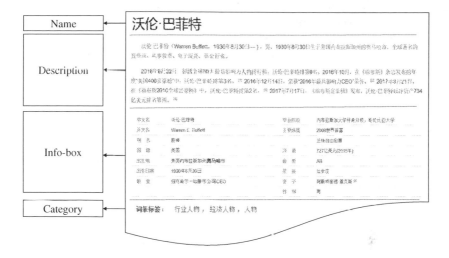

Fig. 1. An example of BaiduBaike entity

Chinese-Pedia Entity Typing. Given a collection of Chinese-pedia entities E and a type set T of ontology O, Chinese-pedia entity typing is the process of finding a corresponding type $T(e_i) \in T$ for each entity $e_i \in E$.

Since the structure of DBpedia types is a tree, our Chinese-pedia entity typing is a tree-based classification task.

3.2 Structured Features

In this section, we introduce the info-box and category features used to characterize BaiduBaike entities and present the process of structured feature extraction.

Info-Box Features. Typically, the info-box includes the most frequently-needed information of an entity which is organized as attribute-value pairs. It is obvious that these attributes are useful for entity typing. For example, an entity with type **Person** should have attributes *Name, Gender, Nationality*, etc. in the info-box, while an entity with type **Settlement** does not have these attributes. What's more, the values of attributes are sometimes helpful too. For example, the values of attribute *Occupation* are significant for fine-grained typing.

The info-box features of entity e are represented by a N-dimension vector as follows:

$$f_i(e) = (i_1, i_2, ..., i_N) \tag{3}$$

where N is the size of the info-box set. This info-box set consists of two parts which are attributes and attribute-value pairs. i_j denotes the j-th item, which can be an attribute or a pair. i_j is set to 1 if the j-th item appears, otherwise it is set to 0. The attributes and pairs in our system should fit the following definitions.

- **Attributes.** The attributes which are shared by at least 100 entities are considered as the first part of the info-box set.
- **Pairs.** The attribute-value pairs which are shared by at least 10 entities are considered as the second part of the info-box set.

Category Features. Although the categories only indicate the relevant topics, they are also vital to type entities. For the above example, **Warren Buffett** has categories *industry figure, economic figure*. Similar to the info-box, the category feature of entity e is represented by a M-dimension vector. The formal definition is as follows:

$$f_c(e) = (c_1, c_2, ..., c_M) \tag{4}$$

where M is the size of the category set. c_j is set to 1 if and only if the j-th category appears, otherwise it is set to 0.

4   Methodology

In this section, we present our Hierarchical Neural System (HNS) and introduce its three models in detail.

4.1   Overview

As shown in Fig. 2, our system consists of three models, which are hierarchical attention model, feature fusion model, and hierarchical classification model. To utilize the semantic information, the hierarchical attention model employs a bi-directional Long Short-Term Memory (bi-LSTM) network with hierarchical attention mechanism to obtain the type features from descriptions. Given a description for a sentence, the bi-LSTM network with hierarchical attention mechanism recurrently performs semantic compositions over the words to obtain the sentence vector, and the feature vector of description is recurrently computed by the sentence vectors. To tackle with the challenge of incomplete Chinese-pedia, the features fusion model exploits a compression module to compress the type features from multi-source into low-dimensional and dense vectors. Then the fusion module fuses these features together and obtains the final type features. Since DBpedia ontology is organized as tree-structure, the hierarchical classification model is designed with a group of Double-Layer Perceptron (DLP) networks to infer types from coarse-grained to fine-grained. With the help of multitask learning technique [2], hierarchical classification model shares the hidden states of previous two models and feeds the type features into each network as inputs. The output layer is a softmax classifier which provides the inference result. The whole system is jointly trained, and these three models are introduced in following subsections.

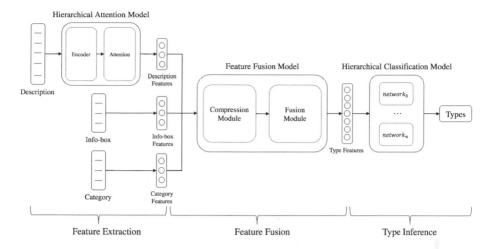

Fig. 2. The architecture of hierarchical neural system.

4.2 Hierarchical Attention Model

The architecture of hierarchical attention model is shown in Fig. 3. The hierarchical attention model exploits a bi-LSTM network with hierarchical attention mechanism proposed in [17] to capture the semantic information of description.

At the word level, each word of sentence s_i is embedded into a low dimensional semantic space with the word2vec method [6], which means each word w_i^j is mapped to its embedding x_i^j. With the help of bi-LSTM encoder, we obtain the summarized hidden state of word w_i^j, $h_i^j = [\overrightarrow{h_i^j}, \overleftarrow{h_i^j}]$, where $\overrightarrow{h_i^j} = \overrightarrow{LSTM}(x_i^j)$ and $\overleftarrow{h_i^j} = \overleftarrow{LSTM}(x_i^j)$. Considering that not all words contribute equally to the representation of meaning, it is necessary to introduce the attention mechanism. Specifically,

$$u_i^j = \tanh(W_w h_i^j + b_w) \tag{5}$$

$$\alpha_i^j = \frac{\exp(u_w u_i^j)}{\sum_k \exp(u_w u_i^k)} \tag{6}$$

$$s_i = \sum_j \alpha_i^j h_i^j \tag{7}$$

where u_i^j is the non-linear transformation of hidden state h_i^j, W_i and b_w are the parameters. α_i^j is the normalized importance weight for word w_i^j. The sentence vector for sentence s_i is obtained as a weighted sum of word vectors.

The similar characteristic also exists at sentence level and attention mechanism is used to reward the sentences which are clues to classification. h_i is used to represent the hidden state of sentence s_i which encoded by bi-LSTM encoder, and $h_i = [\overrightarrow{h_i}, \overleftarrow{h_i}]$, where $\overrightarrow{h_i} = \overrightarrow{LSTM}(s_i)$ and $\overleftarrow{h_i} = \overleftarrow{LSTM}(s_i)$. This yields

$$u_i = \tanh(W_s h_i + b_s) \tag{8}$$

$$\alpha_i = \frac{\exp(u_s u_i)}{\sum_k \exp(u_s u_k)} \tag{9}$$

$$v = \sum_i \alpha_i h_i \tag{10}$$

where u_i is the non-linear transformation of hidden state h_i, W_s and b_s are the parameters. α_i is the normalized importance weight for sentence s_i. The vector v is the description vector which contains all the information, and the vector is considered as the description features.

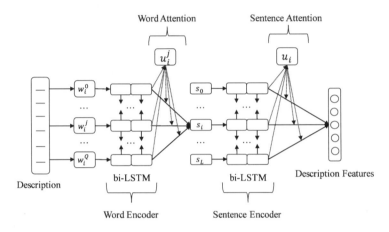

Fig. 3. The architecture of hierarchical attention model.

4.3   Feature Fusion Model

The feature fusion model in our system consists of two parts named compression module and fusion module. The compression module compresses the description features, info-box features, and category features into three low-dimensional and dense vectors. Then the fusion module fuses all these features together as the final type features.

Compression Module. To fuse multi-source features efficiently, it is necessary to compress all the features into low-dimensional and dense vectors first. The compression module employs a nonlinear transformation to compress features. The m-dimensional compressed features are obtained via

$$f^c = g(W^T f + b) \tag{11}$$

where $f \in \mathbb{R}^{p \times 1}$ is the original features, and $W \in \mathbb{R}^{q \times p}$ is the weight matrix, $b \in \mathbb{R}^{q \times 1}$ is the bias term, and g is a nonlinearity function.

Fusion Module. The concatenation vector of the compressed description features, compressed info-box features, and compressed category features are considered as the final type features. The formal definition is given as follows:

$$f - [f_d^c, f_i^c, f_c^c] \tag{12}$$

where f denotes the type feature vector, and f_d^c, f_i^c, f_c^c presents the compressed description features, compressed info-box features, and compressed category features respectively.

4.4 Hierarchical Classification Model

The hierarchical classification model employs a group of DLP networks to accomplish the Chinese-pedia entity typing task and the number of networks equals to the layers of ontology tree[3]. Each network infers the most appropriate type for the corresponding layer. During the process of entity typing, there may be no appropriate type for an entity in lower type layers. For example, the types of the entity **Las Vegas** can be **Settlement** and **City**, while it cannot be the subtypes **Capital** or **CapitalOfRegion**.

To assign types for entities in each layer, the original ontology tree is expanded by appending a type to their subtypes until the lowest layer. With the expanded ontology tree, the hierarchical classification model infers fine-grained types in each layer.

We take a simple ontology tree as an example, which contains types **Settlement, City, Capital**, and **CapitalOfRegion**. With the ontology tree, the architecture of corresponding hierarchical classification model is shown in Fig. 4.

The input layer of each network is fed by the type features, and the output layer is a softmax classifier, which computes the probability distribution over types. For entity e_i, the distribution is calculated by

$$y_i = softmax(Ux_i) \tag{13}$$

where U is the parameter matrix for classification and x_i is the type features of e_i.

With this hierarchical classification model, a group of types are obtained from coarse-grained to fine-grained for each entity. Finally, we consider the most fine-grained type as the final result of Chinese-pedia entity typing.

4.5 Model Training

Above three models of our system are jointly-trained by the back-propagation algorithm. The softmax classifiers of hierarchical classification model are employed to compute probability distribution over N types and the predicted distributions y_i are compared with the ground truth label t_i, where $y_i \in \mathbb{R}^{N \times 1}$,

---

[3] Layer-by-layer classification model structure can be also employed in DAG ontology, such as Wikipedia.

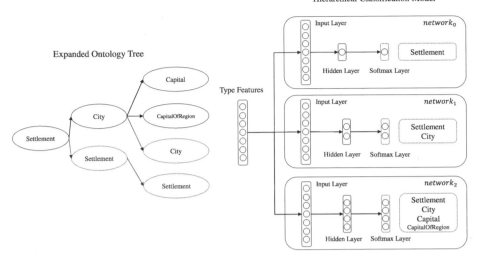

Fig. 4. This hierarchical classification model consists of three DLP networks, which correspond to the three layers of expanded ontology tree. The softmax layer of the first network computes the probability distribution over type **Settlement**. The second computes the probability distribution over types **Settlement, City**. And the third computes the probability distribution over types **Settlement, City, Capital, CapitalOfRegion**.

$t_i \in \mathbb{R}^{N \times 1}$. The value of t_i^j is set to 1 if the type j is correct and the others are set to 0. The cross-entropy objective function of one network is as follows:

$$\mathcal{L} = -\sum_i \sum_j t_i^j \log(y_i^j) \tag{14}$$

The final loss function of the hierarchical classification model is the sum of all the cross-entropy objective functions.

5   Experiments

In this section, we present the experimental results to verify the effect of our system. First, we construct a dataset and manually set up some details in the experiment. Then, our system is compared with baselines on the dataset. Finally, our system is applied to type large-scale Chinese-pedia entities.

5.1   Dataset

In order to verify the effect of our system, we need a dataset with BaiduBaike entities and their corresponding types.

DBpedia provides a file to link the same entities between different languages. This file is exploited to establish the **sameAs** relations between DBpedia entities and BaiduBaike entities. Specifically, we utilize the Chinese name of DBpedia entities to match BaiduBaike entities. With these **sameAs** relations, it is convenient to assign the types of DBpedia entities to their corresponding BaiduBaike entities.

Since two entities may have the same name in the knowledge bases, the name matching method may bring the polysemy issues. For example, when we discuss the entity **The Great Wall**, we mean the ancient military structure or the film starring Matt Damon. Therefore, the type of BaiduBaike entity **The Great Wall** can be either **Film** or **Book**.

In order to address the polysemy issue caused by name matching, the process of data cleaning is necessary. There are two common mistakes in the dataset. The first one is that the adapted films based on books are mistaken for books. On this account, more than 40% entities with type **Book** are actually a film. The second mistake is that a person is mistaken for another person with the same name and the person will be assigned the wrong type. The further analysis shows that the most wrong-labeled persons have incomplete information in the info-box. In this paper, two rules are set manually for data cleaning as follows[4].

- **Rule 1:** The entity with type **Book** must have *author* attribute in the info-box.
- **Rule 2:** The entity with type **Person** or the subtypes of **Person** shall have two attribute-value pairs at least in its info-box.

Statistical analysis shows that not all DBpedia types are assigned to entities. In the DBpedia knowledge base, more than 40% types have no instances such as **LunarCrater**. Since our dataset is a subset of DBpedia in essence, it suffers from the same problem. Here, we stipulate that the effective types in our system should have at least 150 instances. The instances of ineffective types will be considered as the instances of their effective supertypes. Under such stipulation, there are 112 effective types participate in classification. The first layer of expanded ontology tree has only one type named **Thing**, the second layer has 15 types, the third layer has 35 types, the fourth layer has 81 types, the fifth layer has 107 types and the last layer has 112 types.

After all the above processes, the final dataset contains 141553 BaiduBaike entities with DBpedia type. We take 80% data for training and 20% for test.

5.2   Setups

In this section, we present the metrics for evaluation and the comparison baselines.

Metrics. The metrics of hierarchical precision (hP), hierarchical recall (hR) and hierarchical f-measure (hF) are used to evaluate the performance of hierarchical classification models [12]. These metrics are defined as follows:

---

[4] The entities with type **Book** and **Person** accounted for 16% of the dataset.

$$hP = \frac{\sum_i |P(e_i) \cap T(e_i)|}{\sum_i |P(e_i)|} \tag{15}$$

$$hR = \frac{\sum_i |P(e_i) \cap T(e_i)|}{\sum_i |T(e_i)|} \tag{16}$$

$$hF = \frac{2 \times hP \times hR}{hP + hR} \tag{17}$$

where $P(e_i)$ is the set of the type inferred for entity e_i and all its supertypes, and $T(e_i)$ is the set of labeled type and all its supertypes.

Baselines. The HNS compares with two baselines and these baselines are introduced as follows.

- **DLP.** We replace the hierarchical classification model of our system with a DLP network, and the output layer of this DLP network is a softmax classifier which produces the final distribution directly.
- **CUTE.** CUTE is the start-of-art work in the field of Chinese-pedia entity typing [16], which only takes the info-box features and category features as the input.

5.3   Results Analysis

Several tests are carried out on the dataset, the overall performances of our system and the other baselines are shown in Table 1.

In the table, the symbol "basic" denotes the basic implementation of DLP and HNS, which only takes the info-box features and category features as the input. The "des" implementation contains the sematic information of description and the "des+att" append the hierarchical attention mechanism on this basis.

From the table, we can find that our system achieves the highest precision of 89.3% and recall of 89.1%. Compared with the baseline DLP, our system has clear improvements in the result of various circumstances. These improvements indicate the effect of our hierarchical classification model, which introduce the implicit hierarchical information in the process of classification. The CUTE has the worst performance among all the methods especially at low recall, since it does not introduce the semantic information and hierarchical information.

Effectiveness of Semantic Information. In addition to comparing the overall performance, we also evaluate the result on different layers. We compare the hP and hR on six layers of expended ontology tree with those of HNS.

As the results in Table 2 reveal, our system outperforms CUTE on almost every layer of expended ontology tree at both hP, and hR and the system with semantic information performs better than the basic HNS. As for the fine-grained typing, it is evident that our system with hierarchical attention model obtains excellent performance on lower layers. Since the lower layers contain fine-grained types, this results highlight the effectiveness of semantic information to infer fine-grained types.

Table 1. Comparison results on test dataset.

	Model	hP	hR	hF
CUTE	-	0.864	0.697	0.771
DLP	basic	0.837	0.851	0.843
	Des	0.852	0.865	0.858
	des+att	0.877	0.886	0.881
HNS	basic	0.839	0.860	0.849
	des	0.868	0.879	0.873
	des+att	**0.893**	**0.891**	**0.892**

Table 2. Comparison on different layers between CUTE and HNS

	hP			hR		
	CUTE	HNS		CUTE	HNS	
		basic	des+att		basic	des+att
Layer1	0.694	0.767	**0.812**	0.572	0.734	**0.822**
Layer2	0.809	0.818	**0.898**	0.734	0.823	**0.886**
Layer3	0.823	0.852	**0.851**	0.707	0.816	**0.837**
Layer4	**0.876**	0.828	0.868	0.743	0.880	**0.919**
Layer5	0.868	0.882	**0.899**	0.538	0.940	**0.959**
Layer6	0.829	0.657	**0.876**	0.288	0.842	**0.974**

Effectiveness of Feature Fusion. In our system, the feature fusion model is designed to fuse features from different sources and weaken the influence of incomplete information in Chinese-pedia. To evaluate the effect of feature fusion model, we select incomplete entities from the test dataset to form the incomplete dataset. This dataset contains 8534 incomplete entities such as without description or info-box. As shown in Table 3, the HNS performs well on the incomplete dataset. The results demonstrate the effectiveness of feature fusion.

Table 3. Comparison results on incomplete dataset.

	hP	hR	hF
CUTE	0.864	0.410	0.556
HNS	**0.877**	**0.632**	**0.734**

6   Conclusion

In this paper, we propose a novel hierarchical neural system to infer fine-grained types for Chinese-pedia entities. To infer fine-grained types, the semantic information of description is introduced to depict Chinese-pedia entities precisely. We present the hierarchical attention model, which exploits a bi-directional LSTM network with hierarchical attention mechanism to obtain the vector representations of descriptions. To deal with the challenge of incomplete information, the feature fusion model is designed to fuse the features from different sources into low-dimensional and dense vectors. Finally, the hierarchical classification model infers the fine-grained types for Chinese-pedia entities.

Our system is evaluated on 112 DBpedia types and 141 thousand BaiduBaike entities. The experimental results show that our system has respectively achieved the precision of 89.3% and the recall of 89.1%, which are higher than the start-of-art work by 2.9% and 19.4%.

Acknowledgements. This work is supported by DCT-MoST Joint-project No. (025/2015/AMJ); University of Macau Funds Nos: CPG2018-00032-FST & SRG2018-00111-FST; Chinese National Research Fund (NSFC) Key Project No. 61532013; National China 973 Project No. 2015CB352401; Shanghai Scientific Innovation Act of STCSM No. 15JC1402400 and 985 Project of Shanghai Jiao Tong University: WF220103001.

References

1. Bengio, Y., Courville, A., Vincent, P.: Representation learning: a review and new perspectives. IEEE Trans. Pattern Anal. Mach. Intell. **35**(8), 1798–1828 (2013)
2. Caruana, R.: Multitask learning. In: Thrun, S., Pratt, L. (eds.) Learning to Learn, pp. 95–133. Springer, Boston (1998). https://doi.org/10.1007/978-1-4615-5529-2_5
3. Dong, X., Gabrilovich, E., Heitz, G., Horn, W., Lao, N., Murphy, K., Strohmann, T., Sun, S., Zhang, W.: Knowledge vault: a web-scale approach to probabilistic knowledge fusion. In: Proceedings of the 20th ACM SIGKDD International Conference on Knowledge Discovery and Data Mining, pp. 601–610. ACM (2014)
4. Lee, C., et al.: Fine-grained named entity recognition using conditional random fields for question answering. In: Ng, H.T., Leong, M.-K., Kan, M.-Y., Ji, D. (eds.) AIRS 2006. LNCS, vol. 4182, pp. 581–587. Springer, Heidelberg (2006). https://doi.org/10.1007/11880592_49
5. Lehmann, J., Isele, R., Jakob, M., Jentzsch, A., Kontokostas, D., Mendes, P.N., Hellmann, S., Morsey, M., Van Kleef, P., Auer, S., et al.: DBpedia-a large-scale, multilingual knowledge base extracted from Wikipedia. Seman. Web **6**(2), 167–195 (2015)
6. Mikolov, T., Sutskever, I., Chen, K., Corrado, G.S., Dean, J.: Distributed representations of words and phrases and their compositionality. In: Advances in Neural Information Processing Systems, pp. 3111–3119 (2013)
7. Niu, X., Sun, X., Wang, H., Rong, S., Qi, G., Yu, Y.: Zhishi.me - weaving chinese linking open data. In: Aroyo, L., et al. (eds.) ISWC 2011. LNCS, vol. 7032, pp. 205–220. Springer, Heidelberg (2011). https://doi.org/10.1007/978-3-642-25093-4_14
8. Paulheim, H., Bizer, C.: Type inference on noisy RDF data. In: Alani, H., et al. (eds.) ISWC 2013. LNCS, vol. 8218, pp. 510–525. Springer, Heidelberg (2013). https://doi.org/10.1007/978-3-642-41335-3_32
9. Preece, A., Hui, K., Gray, A., Marti, P., Bench-Capon, T., Jones, D., Cui, Z.: The kraft architecture for knowledge fusion and transformation. Knowl.-Based Syst. **13**(2), 113–120 (2000)

10. Radford, A., Metz, L., Chintala, S.: Unsupervised representation learning with deep convolutional generative adversarial networks. arXiv preprint arXiv:1511.06434 (2015)
11. Shimaoka, S., Stenetorp, P., Inui, K., Riedel, S.: An attentive neural architecture for fine-grained entity type classification. arXiv preprint arXiv:1604.05525 (2016)
12. Silla Jr., C.N., Freitas, A.A.: A survey of hierarchical classification across different application domains. Data Min. Knowl. Disc. **22**(1–2), 31–72 (2011)
13. Suchanek, F.M., Kasneci, G., Weikum, G.: YAGO: a core of semantic knowledge. In: Proceedings of the 16th International Conference on World Wide Web, pp. 697–706. ACM (2007)
14. Wu, T., Ling, S., Qi, G., Wang, H.: Mining type information from Chinese online encyclopedias. In: Supnithi, T., Yamaguchi, T., Pan, J.Z., Wuwongse, V., Buranarach, M. (eds.) JIST 2014. LNCS, vol. 8943, pp. 213–229. Springer, Cham (2015). https://doi.org/10.1007/978-3-319-15615-6_16
15. Wu, W., Li, H., Wang, H., Zhu, K.Q.: Probase: a probabilistic taxonomy for text understanding. In: Proccedings of the 2012 ACM SIGMOD International Conference on Management of Data, PP. 481–492. ACM (2012)
16. Xu, B., Zhang, Y., Liang, J., Xiao, Y., Hwang, S., Wang, W.: Cross-lingual type inference. In: Navathe, S.B., Wu, W., Shekhar, S., Du, X., Wang, X.S., Xiong, H. (eds.) DASFAA 2016. LNCS, vol. 9642, pp. 447–462. Springer, Cham (2016). https://doi.org/10.1007/978-3-319-32025-0_28
17. Yang, Z., Yang, D., Dyer, C., He, X., Smola, A.J., Hovy, E.H.: Hierarchical attention networks for document classification. In: HLT-NAACL, pp. 1480–1489 (2016)
18. Yosef, M.A., Bauer, S., Hoffart, J., Spaniol, M., Weikum, G.: HYENA: hierarchical type classification for entity names (2012)

Knowledge Graph Embedding by Learning to Connect Entity with Relation

Zichao Huang, Bo Li$^{(\boxtimes)}$, and Jian Yin

Guangdong Key Laboratory of Big Data Analysis and Processing,
School of Data and Computer Science, Sun Yat-sen University, Guangzhou 510006,
People's Republic of China
huangzch7@mail2.sysu.edu.cn, {libo68,issjyin}@mail.sysu.edu.cn

Abstract. Knowledge graph embedding aims to learn low-dimensional embedding vector representations for entities and relations, which can be used in further machine learning tasks. However, previous knowledge graph embedding models perform poorly when dealing with unbalanced relations which occupy a large proportion in knowledge graphs. In addition, modeling connections between entities and relations accurately is still a big challenge. In this paper, we propose a novel knowledge graph embedding model called ConnectER. It can solve the above problems through a "Connection-Classification" architecture. Experiment results show consistent improvements compared with state-of-the-art baselines.

Keywords: Knowledge graph · Knowledge representation
Representation learning

1 Introduction

Recently, knowledge graphs such as Freebase [2], WordNet [14] and YAGO [18] are playing a crucial role in many machine learning applications such as question answering, Web search, relation extraction, etc. A typical knowledge graph is comprised of entities as nodes and relations as different types of edges, representing relationships between entities as triples ⟨*head entity, relation, tail entity*⟩ (abbreviated as ⟨*h, r, t*⟩). Although these knowledge graphs may contain a massive amount of relational facts (i.e. triples), they are still far from complete. Therefore, many knowledge graph completion models have been proposed to solve the problem by predicting missing relations between entities based on the existing triples.

In the past decade, many knowledge graph completion models based on logic and symbol have been proposed. However, these models become intractable as the volumes of existing knowledge graphs grow fast. Recently, embedding based models have been proposed and obtained state-of-the-art performances. Knowledge graph embedding models attempt to embed every element of a knowledge graph (i.e., every entity and every relation) into a low-dimensional embedding space while preserving certain important properties of the original multi-relational graph.

© Springer International Publishing AG, part of Springer Nature 2018
Y. Cai et al. (Eds.): APWeb-WAIM 2018, LNCS 10987, pp. 400–414, 2018.
https://doi.org/10.1007/978-3-319-96890-2_33

Generally, knowledge graph embedding models can be divided into three branches: translation models [5,6,8,9,12,20], neural network models [3,4,17] and other models [3,7,15,16,19]. Translation models consider the relation as translation from the head entity to the tail entity in the embedding space. Neural network models often incorporate a neural network to model the knowledge graph.

Among all above mentioned models, TransE [5] is a simple yet powerful baseline. For a golden triple $\langle h,\ r,\ t \rangle$, it considers relation r as translation from head entity h to tail entity t in the embedding space, modeling translation with vector addition, which leads to $\mathbf{h} + \mathbf{r} \approx \mathbf{t}$. TransE performs well when modeling 1-to-1 relations, yet appears to be inadequate for modeling unbalanced relations (i.e., 1-to-N, N-to-1 and N-to-N relations). Many translation models such as TransH [20] are proposed to alleviate these problems. We will discuss these translation models in detail in next section. Besides, neural network models such as NTN [17] have also attracted much attention. NTN relates head entity and tail entity through a tensor network, which can interact vectors across multiple dimensions. Neural network models such as NTN are very expressive, as they incorporate a neural network with large parameter size to capture internal structure of knowledge graphs.

Although these models have strong ability to model knowledge graphs, there are still flaws in them: (1) As illustrated in [9], a large proportion of relations in knowledge graphs are unbalanced (i.e., 1-to-N, N-to-1 and N-to-N relations). For instance, predicting head entities for N-to-1 relations is a typical unbalanced prediction task. Previous models such as TransH [20], TransR [12] and TransD [8] performs poorly on this task, only obtaining less than 50.0% on Hits@10[1] on benchmark datasets. TranSparse [9] alleviates this issue with sparse transfer matrix with relation-specific sparse degree, yielding slightly better result. One of the main reasons of their poor performances is that they only fit a single triple $\langle h,\ r,\ t \rangle$ at a time during training. However, unlike balanced relations (i.e., 1-to-1 relations), there are multiple valid tail entities t for a given h-r pair (or head entities h for a given t-r pair) for unbalanced relations. We should take into account all valid tail (or head) entities for a given head (or tail) entity and a given relation when it comes to unbalanced relations. More specifically, all previous models aim to fit $\langle h,\ r,\ t \rangle$, while what we ought to do is to fit $\langle h,\ r,\ \{t_1, t_2, \ldots, t_N\} \rangle$ for 1-to-N relations and $\langle \{h_1, h_2, \ldots, h_N\},\ r,\ t \rangle$ for N-to-1 relations. (2) Since a knowledge graph is comprised of triples (i.e., $\langle h,\ r,\ t \rangle$), modeling the connections between entities and relations is the main purpose of a knowledge graph embedding model. A simple model like TransE [5] models the connections between entities and relations with vector addition, while a complex model like NTN [17] models these connections with a complex neural network. Simple models like TransE can not achieve good performance owing to lacking of expressing ability, while complex models like NTN are not efficient when extending to large-scale knowledge graphs. There is a tradeoff between

---

[1] The proportion of correct entities ranked in the top 10.

model expressing ability which usually comes with more parameters, and model efficiency which requires less parameters.

To solve the above problems, we propose a novel knowledge graph embedding model, ConnectER. In ConnectER, we first define a connection layer for learning to connect a head (or tail) entity with a relation, and then define a classification layer for predicting all valid tail (or head) entities. We manage to solve the problem of unbalanced relations by considering all valid entities for 1-to-N, N-to-1 and N-to-N relations. We evaluate ConnectER on two benchmark datasets and the experiment results show that ConnectER outperforms state-of-the-art baselines.

Our contributions in this paper are:

1. We propose a novel knowledge graph embedding model called ConnectER, which incorporates a two-layer neural network to model entity-relation connections in knowledge graphs.
2. ConnectER alleviates the problem of unbalanced relations by taking into account all valid tail (or head) entities for a given head (or tail) entity and a given relation.
3. Experiment results show that ConnectER achieves state-of-the-art performance.

The rest of this paper is structured as follows. Section 2 summarizes the previous knowledge graph embedding models as well as discusses their relations and differences. Section 3 describes our proposed model ConnectER in detail. Section 4 presents evaluation tasks and compares ConnectER with several strong baselines. Section 5 concludes our work with a discussion of future research directions.

2   Related Work

We define our mathematical notations as follows: a triple is denoted by $\langle h, r, t \rangle$ and its embedding vectors are denoted by bold lower case letters $\mathbf{h}, \mathbf{r}, \mathbf{t}$; matrices are denoted by bold upper case letters such as $\mathbf{M}$; a score function is denoted by $f(h, r, t)$; L1 or L2 distance is denoted by $\|\cdot\|_{1/2}$.

2.1   Translation Models

Translation models share a common principle that $(\mathbf{h} + \mathbf{r})$ is supposed to be close to $\mathbf{t}$ if a triple $\langle h, r, t \rangle$ holds and far from $\mathbf{t}$ otherwise. These models differ in how they define the relation space where $\mathbf{h}$ connects with $\mathbf{t}$. Hence, their score functions measuring the loss of a triple are usually in the form of:

$$f(h, r, t) = \|\mathbf{h}_r + \mathbf{r} - \mathbf{t}_r\|_2^2 \qquad (1)$$

where $\mathbf{h}_r$ and $\mathbf{t}_r$ are respectively the head and tail entity embedding projection in the relation space. This function returns a low score for a golden triple and a high score otherwise.

TransE. TransE [5] is a simple yet strong baseline. It supposes that entities and relations are in the same embedding space. That is, $\mathbf{h}_r = \mathbf{h}$ and $\mathbf{t}_r = \mathbf{t}$. The score function of TransE is:

$$f(h, r, t) = \|\mathbf{h} + \mathbf{r} - \mathbf{t}\|_2^2 \tag{2}$$

Since TransE has flaws dealing with 1-to-N, N-to-1 and N-to-N relations, several models based on TransE are proposed to alleviate these problems.

TransH. TransH [20] considers relation r as translation from head entity h to tail entity t on a relation-specific hyperplane characterized by a normal vector $\mathbf{w}_r$. To obtain $\mathbf{h}_r$ and $\mathbf{t}_r$ in Eq. (1), it projects $\mathbf{h}$ and $\mathbf{t}$ to the relation-specific hyperplane, leading to $\mathbf{h}_r = \mathbf{h} - \mathbf{w}_r^\top \mathbf{h} \mathbf{w}_r$ and $\mathbf{t}_r = \mathbf{t} - \mathbf{w}_r^\top \mathbf{t} \mathbf{w}_r$. The score function of TransH is:

$$f(h, r, t) = \left\| (\mathbf{h} - \mathbf{w}_r^\top \mathbf{h} \mathbf{w}_r) + \mathbf{r} - (\mathbf{t} - \mathbf{w}_r^\top \mathbf{t} \mathbf{w}_r) \right\|_2^2 \tag{3}$$

TransR. TransR [12] defines different embedding spaces to model entity and relation respectively, considering relation r as translation from head entity h to tail entity t in the relation space. It maps $\mathbf{h}$ and $\mathbf{t}$ into the relation space using a relation-specific transfer matrix $\mathbf{M}_r$ to obtain $\mathbf{h}_r$ and $\mathbf{t}_r$ in Eq. (1), which leads to $\mathbf{h}_r = \mathbf{M}_r \mathbf{h}$ and $\mathbf{t}_r = \mathbf{M}_r \mathbf{t}$. The score function of TransR is:

$$f(h, r, t) = \|\mathbf{M}_r \mathbf{h} + \mathbf{r} - \mathbf{M}_r \mathbf{t}\|_2^2 \tag{4}$$

TransD. TransD [8] considers that a transfer matrix should not only be determined by relation r, but also by head entity h and tail entity t. Given a triple $\langle h, r, t \rangle$, besides the original embedding vectors $\mathbf{h}, \mathbf{r}, \mathbf{t}$, TransD defines three projection vectors $\mathbf{h}_p, \mathbf{r}_p, \mathbf{t}_p$ additionally. The transfer matrices of TransD are denoted as $\mathbf{M}_{h,r}$ and $\mathbf{M}_{t,r}$, defined as follows:

$$
\begin{aligned}
\mathbf{M}_{h,r} &= \mathbf{r}_p \mathbf{h}_p^\top + \mathbf{I} \\
\mathbf{M}_{t,r} &= \mathbf{r}_p \mathbf{t}_p^\top + \mathbf{I}
\end{aligned} \tag{5}
$$

where $\mathbf{I}$ represents the identity matrix. $\mathbf{h}_r$ and $\mathbf{t}_r$ can be derived by $\mathbf{h}_r = \mathbf{M}_{h,r} \mathbf{h}$ and $\mathbf{t}_r = \mathbf{M}_{t,r} \mathbf{t}$. The score function of TransD is:

$$f(h, r, t) = \|\mathbf{M}_{h,r} \mathbf{h} + \mathbf{r} - \mathbf{M}_{t,r} \mathbf{t}\|_2^2 \tag{6}$$

TranSparse(Share/Separate). TranSparse(share) [9] argues that modeling simple and complex relations with the same number of parameters may lead to overfitting on simple ones and underfitting on complex ones. Like TransR [12] and TransD [8], TranSparse(share) defines a transfer matrix $\mathbf{M}_r$ for each relation. Moreover, in order to distinguish all relations into simple and complex ones, it makes the transfer matrix $\mathbf{M}_r$ sparse and defines a relation-specific sparse degree

θ_r. The sparse transfer matrix of TranSparse(share) is denoted as $\mathbf{M}_r\left(\theta_r\right)$. $\mathbf{h}_r$ and $\mathbf{t}_r$ is obtained by $\mathbf{h}_r = \mathbf{M}_r\left(\theta_r\right)\mathbf{h}$ and $\mathbf{t}_r = \mathbf{M}_r\left(\theta_r\right)\mathbf{t}$. Its score function is:

$$f\left(h, r, t\right) = \left\|\mathbf{M}_r\left(\theta_r\right)\mathbf{h} + \mathbf{r} - \mathbf{M}_r\left(\theta_r\right)\mathbf{t}\right\|_2^2 \tag{7}$$

TranSparse(separate) [9] use two different sparse transfer matrices $\mathbf{M}_r^h\left(\theta_r^h\right)$ and $\mathbf{M}_r^t\left(\theta_r^t\right)$ for head entity h and tail entity t, respectively. The rest of the two models are identical. The score function of the separate version is:

$$f\left(h, r, t\right) = \left\|\mathbf{M}_r^h\left(\theta_r^h\right)\mathbf{h} + \mathbf{r} - \mathbf{M}_r^t\left(\theta_r^t\right)\mathbf{t}\right\|_2^2 \tag{8}$$

CirE. CirE [6] is based on the holographic projection with cross correlation between entities and relations. Instead of using common matrix like in TransR [12] or sparse matrix like in TranSparse [9] as transfer matrix, CirE uses a extended version of circulant matrix and achieves state-of-the-art performance on unbalanced relations.

2.2 Neural Network Models

Neural network models usually incorporate a neural network to model the knowledge graph and our model belongs to this branch.

Single Layer Model (SLM). SLM [17] models relation r with two matrices $\mathbf{M}_{h,r}$, $\mathbf{M}_{t,r}$ and two vectors $\mathbf{u}_r$, $\mathbf{b}_r$. It connects head entity h with tail entity t through a single layer neural network, which leads to:

$$f\left(h, r, t\right) = \mathbf{u}_r^\top \tanh\left(\mathbf{M}_{h,r}\mathbf{h} + \mathbf{M}_{t,r}\mathbf{t} + \mathbf{b}_r\right) \tag{9}$$

where $\tanh\left(\cdot\right)$ serves as the activation function. Although SLM incorporates a neural network, the two entity vectors do not interact with each other directly, which makes it hard to learn an effective representation of the relation between them.

Neural Tensor Network (NTN). NTN [17] is a generalized version of SLM. It adds a bilinear tensor layer that directly relates the head entity and tail entity embedding vectors across multiple dimensions. Its score function is:

$$f\left(h, r, t\right) = \mathbf{u}_r^\top \tanh\left(\mathbf{h}^\top \mathbf{W}_r^{[1:k]}\mathbf{t} + \mathbf{M}_{h,r}\mathbf{h} + \mathbf{M}_{t,r}\mathbf{t} + \mathbf{b}_r\right) \tag{10}$$

where $\mathbf{W}_r^{[1:k]}$ represents a relation-specific 3-way tensor composed of k matrices. Due to the relation-specific tensor mechanism, NTN brings in enormous amount of parameters, which makes it the most expressive model so far. However, NTN may suffer from overfitting on small-scale knowledge graphs owing to its parameter size. On the other hand, the high complexity of its tensor-vector multiplication makes NTN difficult to applied to large-scale knowledge graphs.

Semantic Matching Energy (SME). SME [3,4] defines a neural network to capture correlations between entities and relations with shared weight matrices. There are two forms of SME, the linear form:

$$f(h,r,t) = (\mathbf{M}_1\mathbf{h} \mid \mathbf{M}_2\mathbf{r} + \mathbf{b}_1)^\top (\mathbf{M}_3\mathbf{t} + \mathbf{M}_4\mathbf{r} + \mathbf{b}_2) \tag{11}$$

and the bilinear form:

$$f(h,r,t) = ((\mathbf{M}_1\mathbf{h}) \otimes (\mathbf{M}_2\mathbf{r}) + \mathbf{b}_1)^\top ((\mathbf{M}_3\mathbf{t}) \otimes (\mathbf{M}_4\mathbf{r}) + \mathbf{b}_2) \tag{12}$$

where $\otimes$ is the Hadamard product, $\mathbf{M}_{\{1,2,3,4\}}$ are the weight matrices and $\mathbf{b}_{\{1,2\}}$ are the bias vectors. Matrices of the bilinear form is replaced by 3-way tensors in [4]. Note that the neural network parameters of SME are shared by all relations.

2.3   Other Models

There are also many other knowledge graph embedding models and here we briefly introduce some typical ones, which will also serve as baselines in our experiments.

LFM [7] encodes every entity into a vector and models every relation with a relation-specific matrix $\mathbf{M}_r$. It models the interaction of two entities h, t and relation r in a triple $\langle h, r, t \rangle$ by a score function as $f(h,r,t) = \mathbf{h}^\top \mathbf{M}_r \mathbf{t}$. RESCAL [16] is a tensor factorization model. It consider a knowledge graph as a three-way tensor and factorize it into a core tensor and a factor matrix. Each entry in the factorization result would be considered as a measure to indicate whether a corresponding triple exists or not. HolE [15] uses dot product rather than tensor product like RESCAL. It employs circular correlation between head entity and tail entity to represent the entity pairs. Instead of using embeddings containing real numbers, ComplEx [19] discusses and demonstrates the capability of complex embeddings (i.e., vectors with entries in $\mathbb{C}$).

3   Methodology

To accurately model the connections between entities and relations as well as solving the problem of unbalanced relations for knowledge graph embedding, we propose a novel knowledge graph embedding model called ConnectER, which incorporates a two layer neural network with a relation-specific connection gate, treating entity prediction as a multi-label multi-class classification problem. The model architecture of ConnectER is presented in Fig. 1.

3.1   ConnectER

We treat the problem as a prediction task: predict the valid tail (or head) entities with a given head (or tail) entity and a given relation. As shown in Fig. 1, ConnectER consists of a connection layer and a classification layer. The insights

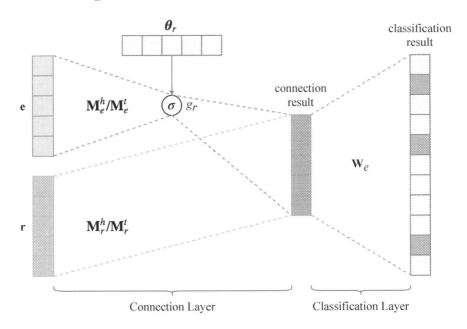

$\boldsymbol{\theta}_r$

classification result

e

$\mathbf{M}_e^h/\mathbf{M}_e^t$

σ g_r

connection result

$\mathbf{W}_e$

r

$\mathbf{M}_r^h/\mathbf{M}_r^t$

Connection Layer   Classification Layer

Fig. 1. The model architecture of ConnectER (Color figure online)

of ConnectER are: (1) We define a connection layer with a relation-specific connection gate, aiming to learn to connect two inputs (i.e., a head (or tail) entity and a relation); (2) The result of the connection layer is then fed to the following classification layer, which predicts the valid tail (or head) entities.

The math notations of ConnectER are as follows: k is the dimension of embedding vectors. $\mathbf{e}, \mathbf{r} \in \mathbb{R}^k$ are entity and relation embedding vectors, respectively. $\mathbf{M}_e^h, \mathbf{M}_e^t, \mathbf{M}_r^h, \mathbf{M}_r^t \in \mathbb{R}^{k \times k}$ are the connection weight matrices. $\mathbf{b}_c^h, \mathbf{b}_c^t \in \mathbb{R}^k$ are the connection bias vectors. $\boldsymbol{\theta}_r \in \mathbb{R}^k$ is the relation-specific connection gate vector and g_r denotes the relation-specific connection gate. $\mathbf{W}_e \in \mathbb{R}^{N_e \times k}$ is the set of all entity embedding vectors, which also serves as the classification weight matrix. N_e is the number of entities of the knowledge graph.

Connection Layer. TransE [5] models the entity-relation connection with simple unweighted vector addition, which may oversimplify the connection and limit its expressing ability. On the other hand, NTN [17] introduces a tensor layer which makes it a much more expressive model, yet the enormous parameter size makes NTN less efficient on large-scale knowledge graphs. In order to accurately model the entity-relation connection as well as limits the parameter size appropriately, we define a connection layer with a relation-specific connection gate.

We model connections at two different granularities. (1) Note that an entity e can be a head entity or a tail entity. At a coarse granularity, we model a connection by $\mathbf{M}_e^h \mathbf{e} + \mathbf{M}_r^h \mathbf{r} + \mathbf{b}_c^h$ if e is a head entity, and $\mathbf{M}_e^t \mathbf{e} + \mathbf{M}_r^t \mathbf{r} + \mathbf{b}_c^t$ if e is a tail entity. The connection weight matrices and bias vectors are shared by all entities and relations of the knowledge graph. We adopt a simplified version of structured sparse pattern in [9], which restricts the connection weight matrices to be diagonal. (2) At a finer granularity, we introduce a relation-specific connection gate g_r to capture instance level features, which allows an entity to have different representations when connecting with different relations. g_r takes the relation-specific connection gate vector $\boldsymbol{\theta}_r$ as input and apply it to the entity embedding $\mathbf{e}$, which leads to the final connection function:

$$f_{\text{connect}}\left(\mathbf{e}, \mathbf{r}, \boldsymbol{\theta}_r\right) = \mathbf{M}_e^h \mathbf{e} \odot g_r + \mathbf{M}_r^h \mathbf{r} + \mathbf{b}_c^h \qquad (13)$$

where $\odot$ denotes element-wise multiplication, $g_r = \sigma\left(\boldsymbol{\theta}_r\right)$ and $\sigma\left(\cdot\right)$ denotes the sigmoid function. Note that the above connection function is for head entities. We replace $\mathbf{M}_e^h, \mathbf{M}_r^h, \mathbf{b}_c^h$ with $\mathbf{M}_e^t, \mathbf{M}_r^t, \mathbf{b}_c^t$ for tail entities. After connecting the two inputs, we use $\tanh\left(\cdot\right)$ as the activation function of the connection layer.

Classification Layer. We aim to take into account all valid tail (or head) entities when dealing with 1-to-N, N-to-1 and N-to-N relations. Hence, we treat the connection result as the input of the classification layer and all entities as classification labels, which makes the prediction task a multi-label N_e-class classification problem. That is, for a given head (or tail) entity and a given relation, our target is to predict multiple potential valid tail (or head) entities. We first construct a label vector $\mathbf{y} \in \mathbb{R}^{N_e}$ where all valid entities have label 1 while others have label 0. For N_e-class classification, we construct the classification weight matrix $\mathbf{W}_e \in \mathbb{R}^{N_e \times k}$ with all entity embeddings. That is, $\mathbf{W}_e = [\mathbf{e}_1^\top; \mathbf{e}_2^\top; \ldots; \mathbf{e}_{N_e}^\top]$. We use softmax as the activation of the classification layer, which leads to the following classification function:

$$f_{\text{classify}}\left(\mathbf{e}, \mathbf{r}, \boldsymbol{\theta}_r\right) = \text{softmax}\left(\mathbf{W}_e \tanh\left(f_{\text{connect}}\left(\mathbf{e}, \mathbf{r}, \boldsymbol{\theta}_r\right)\right)\right) \qquad (14)$$

The result of the classification layer indicates the normalized probability distribution of the valid entities in the entity set, which means if entity e_i is valid, the value of th i^{th} entry of the classification result vector is relatively high, otherwise low. As illustrated in Fig. 1, the red cubes of the classification result represent the valid entities while the grey cubes represent the wrong entities.

3.2 Training Strategy

Training Objective. For maximizing the likelihood between the classification result and the label vector $\mathbf{y}$, we define the training objective function as:

$$
\begin{aligned}
\mathcal{L} &= -\sum_{k=1}^{N_e} \left(\mathbb{1}\{\mathbf{y}_k = 1\} \log f_{\text{classify}}\left(\mathbf{e}, \mathbf{r}, \boldsymbol{\theta}_r\right)_k\right) \\
&= -\sum_{k=1}^{N_e} \left(\mathbb{1}\{\mathbf{y}_k = 1\} \log \frac{\exp\left(\mathbf{W}_e \tanh\left(f_{\text{connect}}\left(\mathbf{e}, \mathbf{r}, \boldsymbol{\theta}_r\right)\right)\right)_k}{\sum_i^{N_e} \exp\left(\mathbf{W}_e \tanh\left(f_{\text{connect}}\left(\mathbf{e}, \mathbf{r}, \boldsymbol{\theta}_r\right)\right)\right)_i}\right)
\end{aligned} \qquad (15)
$$

where $1\{\cdot\}$ denotes the indicator function where $1\{\text{a true statement}\} = 1$, and $1\{\text{a false statement}\} = 0$. We optimize the objective function with mini-batch Adam [10] optimizer.

Negative Sampling. For a large-scale knowledge graph with millions of entities, the softmax operation of the classification layer could be quite slow because it involves an $\exp(\cdot)$ operation on every entity. To expedite the training of ConnectER, we adopt a simplified version of negative sampling strategy in [13]. Instead of considering all invalid entities when constructing $\mathbf{y}$, we sample a subset of invalid entities from a binomial distribution $\mathcal{B}(1, p_s)$, where p_s is the probability that an invalid entity is sampled. In the classification layer, we only consider the valid entities and the sampled invalid entities, which accelerates model training and preserves accuracy at the same time.

Skew Training. We introduce a skew training strategy for unbalanced relations. Normally, given a triple $\langle h, r, t \rangle$ during training, we would randomly choose to: (1) predict all valid tail entities using head entity h and relation r as inputs, or (2) predict all valid head entities using tail entity t and relation r as inputs. However, for unbalanced relations (i.e., 1-to-N, N-to-1 and N-to-N relations), it is much harder to learn the representation of entities on the "N" side than that on the "1" side. For example, gender is a typical N-to-1 relation as it connects many person names as its head entities and only connects male and female as its tail entities. For gender, all its head entities (i.e., person names) are informative for its tail entities, while only two tail entities (i.e., male and female) are informative for its head entities, which makes it much harder to learn the representation of its head entities than that of its tail entities. Skew training makes a model pay more attention to the "N" side. For skew training, given a triple $\langle h, r, t \rangle$ during training, instead of randomly choosing, with probability $\frac{N_t}{N_h + N_t}$ we choose to predict all valid tail entities, and with probability $\frac{N_h}{N_h + N_t}$ we choose to predict all valid head entities. N_h and N_t are respectively the number of head entities and tail entities of relation r. This setting shows consistent performance improvements on benchmark datasets in our experiments.

3.3   Complexity Analysis

In Table 1, we list the complexity of certain previous models and ConnectER, measured by model parameter size. N_e represents the number of entities and N_r represents the number of relations of a knowledge graph. k_e and k_r represent the dimension of the entity embedding space and the relation embedding space, respectively. Note that k_e and k_r are identical for some models (as indicated in Table 1). For TranSparse models, θ, θ_h, and θ_t ($0 \leq \theta, \theta_h, \theta_t \leq 1$) represent the sparse degree of their sparse transfer matrices. For neural network models, h represents the number of neural cells of a hidden layer and s represents the number of slice of a tensor. We divide Table 1 into four parts. The first three parts list parameter size of all three branches of knowledge graph embedding

Table 1. Complexity of knowledge graph embedding models

Model	#Parameters
TransE [5]	$\mathcal{O}(N_e k_e + N_r k_r)(k_e = k_r)$
TransH [20]	$\mathcal{O}(N_e k_e + 2N_r k_r)(k_e = k_r)$
TransR [12]	$\mathcal{O}(N_e k_e + N_r k_r + N_r k_e k_r)$
TransD [8]	$\mathcal{O}(2N_e k_e + 2N_r k_r)$
TranSparse(share) [9]	$\mathcal{O}(N_e k_e + N_r k_r + (1-\theta)N_r k_e k_r)$
TranSparse(separate) [9]	$\mathcal{O}(N_e k_e + N_r k_r + (2-(\theta_h+\theta_t))N_r k_e k_r)$
CirE [6]	$\mathcal{O}(N_e k_e + 2N_r k_r)$
SLM [17]	$\mathcal{O}(N_e k_e + 2N_r k_r h + 2N_r h)(k_e = k_r)$
NTN [17]	$\mathcal{O}(N_e k_e + N_r k_r^2 s + 2N_r k_r s + 2N_r s)(k_e = k_r)$
SME(linear) [3,4]	$\mathcal{O}(N_e k_e + N_r k_r + 4k_e h + 2h)(k_e = k_r)$
SME(bilinear) [3,4]	$\mathcal{O}(N_e k_e + N_r k_r + 4k_e h s + 2h)(k_e = k_r)$
LFM [7]	$\mathcal{O}(N_e k_e + N_r k_r^2)(k_e = k_r)$
RESCAL [16]	$\mathcal{O}(N_e k_e + N_r k_r^2)(k_e = k_r)$
HolE [15]	$\mathcal{O}(N_e k_e + N_r k_r)(k_e = k_r)$
ConnectER(this paper)	$\mathcal{O}(N_e k_e + 2N_r k_r + 6k_e)(k_e = k_r)$

models mentioned in **Related Work**. The last part shows the parameter size of ConnectER.

The parameter size of ConnectER is much less than all of the other neural network models (listed in the second part of Table 1). For other models, ConnectER only has more parameters than TransE, TransH, CirE and HolE, which indicates that ConnectER is more efficient than most of the knowledge graph embedding models. Compared with TransH [20], ConnectER has only $\mathcal{O}(6k)$ more parameters yet obtain tremendous improvements, which indicates the connection layer of ConnectER models connections between entities and relations more effectively.

4   Experiments

We evaluate ConnectER on two benchmark datasets, which are subsets of Word-Net [14] and Freebase [2]. Both datasets are divided into three parts for training, validation and test. The detailed statistics of datasets are given in Table 2. We compare ConnectER with strong baseline models listed in **Related Work**, showing that ConnectER achieves state-of-the-art performance.

4.1   Datasets

WordNet [14] is a knowledge graph on lexical semantic of words. An entity in WordNet is a synset, consisting of a set of (one or more) synonyms, corresponding to a distinct word sense. A relation in WordNet defines a distinct

Table 2. Statistics of datasets

Dataset	#Rel	#Ent	#Train	#Valid	#Test
WN18	18	40,943	141,442	5,000	5,000
FB15k	1,345	14,951	483,142	50,000	59,071

semantic or lexical relation between two entities. An example triple of WordNet is ⟨_house_3, _has_part, _porch_1⟩. In our experiment, we employ WN18, a subset of WordNet which contains 40,943 entities and 18 types of relations.

Freebase [2] is a large collaborative knowledge graph on general world facts. For instance, a triple ⟨william_whewell, profession, scientist⟩ from Freebase indicates that the profession of william_whewell is scientist. In our experiment, we employ FB15k, a subset of Freebase which contains 14,951 entities and 1,345 types of relations.

4.2   Experiment Settings

For evaluation on entity prediction, we use the same metrics as in TransE [5]: (1) the average rank of all correct entities (MeanRank) and (2) the proportion of correct entities ranked in the top 10 (Hits@10).

As explained in [5], these two metrics are indicative yet can be flawed when some of the negative samples are actually valid ones, which may underestimate model performance. To avoid such a misleading behavior, following the protocol in [5], we remove all negative samples which have appeared in training, validation or test sets before ranking, ensuring that all negative samples do not belong to the current knowledge graph. We report MeanRank and Hits@10 under both settings. The former setting is called "Raw" and the latter one is called "Filter". Note that a good knowledge graph embedding model should achieve low Mean-Rank and high Hits@10. Since we use the same datasets and evaluation metrics, we report the performance of baseline models from their original papers.

We implement ConnectER with TensorFlow r1.2 [1]. For training ConnectER, we select the initial learning rate λ of the Adam [10] optimizer among {0.001, 0.003, 0.01}, the embedding dimension k among {50, 100, 200}, the mini-batch size B among {100, 200, 400, 800}, the negative sampling probability p_s among {0.25, 0.5, 0.75}, the weight decay α among {10^{-4}, 10^{-5}, 10^{-6}} and regularization among {L_1, L_2}. The optimal configuration obtained by validation set on Hits@10 under the "Filter" setting are: $\lambda = 0.003, k = 200, B = 800, p_s = 0.5, \alpha = 10^{-5}$, and L_1 regularization on FB15k; $\lambda = 0.003, k = 100, B = 200, p_s = 0.5, \alpha = 10^{-5}$ and L_1 regularization on WN18.

4.3   Entity Prediction

Entity prediction aims to predict the missing head entity h or tail entity t of a given relational fact triple ⟨h, r, t⟩. For each test triple, we remove its head

Table 3. Experiment result of entity prediction

Datasets	WN18				FB15k			
Metric	MeanRank		Hits@10(%)		MeanRank		Hits@10(%)	
	Raw	Filter	Raw	Filter	Raw	Filter	Raw	Filter
LFM [7]	469	456	71.4	81.6	283	164	26.0	33.1
RESCAL [16]	1,180	1,163	37.2	52.8	828	683	28.4	44.1
SME(linear) [3]	545	533	65.1	74.1	274	154	30.7	40.8
SME(bilinear) [3]	526	509	57.4	61.3	284	158	31.3	41.3
TransE [5]	263	251	75.4	89.2	243	125	34.9	47.1
TransH(unif) [20]	318	303	75.4	86.7	211	84	42.5	58.5
TransH(bern) [20]	401	388	73.0	82.3	212	87	45.7	64.4
TransR(unif) [12]	232	219	78.3	91.7	226	78	43.8	65.5
TransR(bern) [12]	238	225	79.8	92.0	198	77	48.2	68.7
TransD(unif) [8]	242	229	79.2	92.5	211	67	49.4	74.2
TransD(bern) [8]	224	212	79.6	92.2	194	91	53.4	77.3
TranSparse(share)[a] [9]	233	221	80.5	93.9	191	86	53.5	78.3
TranSparse(separate)[b] [9]	223	211	80.1	93.4	187	63	53.7	79.9
HolE [15]	-	-	-	94.9	-	-	-	73.9
CirE(unif) [6]	228	220	81.3	94.2	203	68	52.4	80.3
CirE(bern) [6]	221	213	**82.1**	94.6	**163**	85	**54.1**	80.5
ComplEx [19]	-	-	-	94.7	-	-	-	84.0
ConnectER(this paper)	**213**	**196**	80.6	**95.6**	166	**34**	53.2	**86.6**

[a],[b]We only report the best performance among all the TranSparse variations.

entity or tail entity, then, for each entity in the knowledge graph, we fill it into the triple to replace the removed entity and calculate the probability that this reconstructed triple is true. Finally, we rank the entities in descending order according to their corresponding probabilities.

Overall Result. Experiment results on both WN18 and FB15k are shown in Table 3. As expected, lower MeanRank and higher Hits@10 can be achieved under the "Filter" setting consistently. As shown in Table 3, ConnectER outperforms all baselines on both datasets under the "Filter" setting, which is a more indicative evaluation setting of model performance as mentioned in [5].

More particularly, on FB15k, ConnectER improves MeanRank by 29 (compared with TranSparse(separate) [9]) and Hits@10 by 2.6% (compared with ComplEx [19]) under the "Filter" setting, compared with the state-of-the-art baseline. On WN18, ConnectER outperforms all baselines under the "Filer" setting and reaches 95.6% on Hits@10. Compared with TransH [20], the parameter size of ConnectER is slightly larger, yet its MeanRank and Hits@10 are improved by 53 and 22.2% on FB15k, respectively, which shows that ConnectER models knowledge graphs in a more accurate way. Consistent performance improvements on both benchmark datasets show that the "Connection-Classification" architecture of ConnectER is effective.

Table 4. Experiment results of FB15k by mapping properties of relations (%)

Tasks	Predicting Head(Hits@10)				Predicting Tail(Hits@10)			
Relation category	1-1	1-N	N-1	N-N	1-1	1-N	N-1	N-N
SME(linear) [3]	35.1	53.7	19.0	40.3	32.7	14.9	61.6	43.3
SME(bilinear) [3]	30.9	69.6	19.9	38.6	28.2	13.1	76.0	41.8
TransE [5]	43.7	65.7	18.2	47.2	43.7	19.7	66.7	50.0
TransH(unif) [20]	66.7	81.7	30.2	57.4	63.7	30.1	83.2	60.8
TransH(bern) [20]	66.8	87.6	28.7	64.5	65.5	39.8	83.3	67.2
TransR(unif) [12]	76.9	77.9	38.1	66.9	76.2	38.4	76.2	69.1
TransR(bern) [12]	78.8	89.2	34.1	69.2	79.2	37.4	90.4	72.1
TransD(unif) [8]	80.7	85.8	47.1	75.6	80.0	54.5	80.7	77.9
TransD(bern) [8]	86.1	95.5	39.8	78.5	85.4	50.6	94.4	81.2
TranSparse(share)[a] [9]	87.5	95.9	50.3	79.1	87.6	57.7	94.1	81.7
TranSparse(separate)[b] [9]	87.1	95.8	51.8	81.2	87.5	60.0	94.5	83.7
CirE(unif) [6]	84.8	85.5	54.6	82.0	85.5	62.3	93.8	84.0
CirE(bern) [6]	87.8	96.1	50.2	83.0	88.0	60.0	94.5	84.3
ConnectER(this paper)	**92.3**	**97.0**	**63.8**	**86.6**	**92.2**	**68.6**	**96.2**	**89.5**

[a,b]We only report the best performance among all the TranSparse variations.

Detailed Result. Table 4 shows Hits@10 under the "Filter" setting for different types[2] of relations in FB15k. We can see that ConnectER significantly outperforms all baselines on all different types of relations (i.e., 1-to-1, 1-to-N, N-to-1 and N-to-N).

The detailed results reveal more specifics of the model behavior: (1) It is easier for a model to predict the head entity of a 1-to-N relation or the tail entity of a N-to-1 relation, as most models reach their highest Hits@10 on these predictions. (2) Compared with other models, ConnectER is much better at predicting the "N" side. Specifically, compared with CirE [6], when predicting head entities for N-to-1 and N-to-N relations, ConnectER improves Hits@10 by 13.6% and 3.6%, respectively. When predicting tail entities for 1-to-N and N-to-N relations, the corresponding improvements are 8.6% and 5.2%. These detailed results show that ConnectER can alleviate the problem of unbalanced relations effectively. The reason is that ConnectER trains all valid head entities and tail entities jointly for N-to-1 and 1-to-N relations and it adopts skew training described before.

---

[2] Mapping properties of relations follows the same protocol in TransE [5].

5   Conclusion and Future Work

In this paper, we propose a novel knowledge graph embedding model called ConnectER. Incorporating a two-layer neural network, ConnectER embeds a knowledge graph into a low-dimensional embedding space by learning to connect entities with relations. In order to model entity-relation connections at finer granularity, ConnectER introduces a relation-specific connection gate, which allows entities to have different representations when connecting with different relations. Extensive experiments show that ConnectER achieves state-of-the-art performance.

Besides triples of knowledge graphs, ConnectER does not need any additional information like text description in DKRL [21], paths between multiple entities and relations like PTransE [11]. We will try to incorporate other useful information with ConnectER in the future.

Acknowledgements. This work is supported by the Research Foundation of Science and Technology Plan Project in Guangdong Province and Guangzhou City (2014B030301007, 2015A030401057, 2016B030307002, 2014SY000013, 2017B030308007) and CCF-Tencent Open Research Fund.

References

1. Abadi, M., Agarwal, A., Barham, P., Brevdo, E., Chen, Z., Citro, C., Corrado, G.S., Davis, A., Dean, J., Devin, M., Ghemawat, S., Goodfellow, I.J., Harp, A., Irving, G., Isard, M., Jia, Y., Józefowicz, R., Kaiser, L., Kudlur, M., Levenberg, J., Mané, D., Monga, R., Moore, S., Murray, D.G., Olah, C., Schuster, M., Shlens, J., Steiner, B., Sutskever, I., Talwar, K., Tucker, P.A., Vanhoucke, V., Vasudevan, V., Viégas, F.B., Vinyals, O., Warden, P., Wattenberg, M., Wicke, M., Yu, Y., Zheng, X.: TensorFlow: large-scale machine learning on heterogeneous distributed systems. arXiv preprint arXiv:1603.04467 (2016)
2. Bollacker, K.D., Evans, C., Paritosh, P., Sturge, T., Taylor, J.: Freebase: a collaboratively created graph database for structuring human knowledge. In: Proceedings of the 2008 ACM SIGMOD International Conference on Management of Data, SIGMOD 2008, Vancouver, Canada, pp. 1247–1250 (2008)
3. Bordes, A., Glorot, X., Weston, J., Bengio, Y.: Joint learning of words and meaning representations for open-text semantic parsing. In: Proceedings of the Fifteenth International Conference on Artificial Intelligence and Statistics, AISTATS 2012, La Palma, Canary Islands, pp. 127–135 (2012)
4. Bordes, A., Glorot, X., Weston, J., Bengio, Y.: A semantic matching energy function for learning with multi-relational data - application to word-sense disambiguation. Mach. Learn. **94**(2), 233–259 (2014)
5. Bordes, A., Usunier, N., García-Durán, A., Weston, J., Yakhnenko, O.: Translating embeddings for modeling multi-relational data. In: Proceedings of the 27th International Conference on Neural Information Processing Systems, NIPS 2013, Lake Tahoe, Nevada, vol. 2, pp. 2787–2795 (2013)
6. Du, Z., Hao, Z., Meng, X., Wang, Q.: CirE: circular embeddings of knowledge graphs. In: Candan, S., Chen, L., Pedersen, T.B., Chang, L., Hua, W. (eds.) DASFAA 2017. LNCS, vol. 10177, pp. 148–162. Springer, Cham (2017). https://doi.org/10.1007/978-3-319-55753-3_10

7. Jenatton, R., Roux, N.L., Bordes, A., Obozinski, G.: A latent factor model for highly multi-relational data. In: Proceedings of the 26th International Conference on Neural Information Processing Systems, NIPS 2012, Lake Tahoe, Nevada, pp. 3176–3184 (2012)
8. Ji, G., He, S., Xu, L., Liu, K., Zhao, J.: Knowledge graph embedding via dynamic mapping matrix. In: Proceedings of the 53rd Annual Meeting of the Association for Computational Linguistics and the 7th International Joint Conference on Natural Language Processing, Beijing, China, vol. 1: Long Papers, pp. 687–696 (2015)
9. Ji, G., Liu, K., He, S., Zhao, J.: Knowledge graph completion with adaptive sparse transfer matrix. In: Proceedings of the Thirtieth AAAI Conference on Artificial Intelligence, AAAI 2016, Phoenix, Arizona, pp. 985–991 (2016)
10. Kingma, D.P., Ba, J.: Adam: a method for stochastic optimization. arXiv preprint arXiv:1412.6980 (2014)
11. Lin, Y., Liu, Z., Luan, H., Sun, M., Rao, S., Liu, S.: Modeling relation paths for representation learning of knowledge bases. In: Proceedings of the 2015 Conference on Empirical Methods in Natural Language Processing, Lisbon, Portugal, pp. 705–714 (2015)
12. Lin, Y., Liu, Z., Sun, M., Liu, Y., Zhu, X.: Learning entity and relation embeddings for knowledge graph completion. In: Proceedings of the Twenty-Ninth AAAI Conference on Artificial Intelligence, AAAI 2015, Austin, Texas, pp. 2181–2187 (2015)
13. Mikolov, T., Sutskever, I., Chen, K., Corrado, G.S., Dean, J.: Distributed representations of words and phrases and their compositionality. In: Proceedings of the 27th International Conference on Neural Information Processing Systems, NIPS 2013, Lake Tahoe, Nevada, vol. 2, pp. 3111–3119 (2013)
14. Miller, G.A.: WordNet: a lexical database for English. Commun. ACM **38**(11), 39–41 (1995)
15. Nickel, M., Rosasco, L., Poggio, T.A.: Holographic embeddings of knowledge graphs. In: Proceedings of the Thirtieth AAAI Conference on Artificial Intelligence, AAAI 2016, Phoenix, Arizona, pp. 1955–1961 (2016)
16. Nickel, M., Tresp, V., Kriegel, H.: A three-way model for collective learning on multi-relational data. In: Proceedings of the 28th International Conference on Machine Learning, ICML 2011, Bellevue, Washington, USA, pp. 809–816 (2011)
17. Socher, R., Chen, D., Manning, C.D., Ng, A.Y.: Reasoning with neural tensor networks for knowledge base completion. In: Proceedings of the 27th International Conference on Neural Information Processing Systems, NIPS 2013, Lake Tahoe, Nevada, vol. 1, pp. 926–934 (2013)
18. Suchanek, F.M., Kasneci, G., Weikum, G.: YAGO: a core of semantic knowledge. In: Proceedings of the 16th International Conference on World Wide Web, WWW 2007, Banff, Alberta, Canada, pp. 697–706 (2007)
19. Trouillon, T., Welbl, J., Riedel, S., Gaussier, É., Bouchard, G.: Complex embeddings for simple link prediction. In: Proceedings of the 33rd International Conference on Machine Learning, ICML 2016, New York, NY, USA, vol. 48, pp. 2071–2080 (2016)
20. Wang, Z., Zhang, J., Feng, J., Chen, Z.: Knowledge graph embedding by translating on hyperplanes. In: Proceedings of the Twenty-Eighth AAAI Conference on Artificial Intelligence, AAAI 2014, Québec City, Québec, Canada, pp. 1112–1119 (2014)
21. Xie, R., Liu, Z., Jia, J., Luan, H., Sun, M.: Representation learning of knowledge graphs with entity descriptions. In: Proceedings of the Thirtieth AAAI Conference on Artificial Intelligence, AAAI 2016, Phoenix, Arizona, pp. 2659–2665 (2016)

StarMR: An Efficient Star-Decomposition Based Query Processor for SPARQL Basic Graph Patterns Using MapReduce

Qiang Xu[1], Xin Wang[1,2(✉)], Jianxin Li[3], Ying Gan[1], Lele Chai[1], and Junhu Wang[4]

[1] School of Computer Science and Technology, Tianjin University, Tianjin, China
{xuqiang3,wangx,yinggan,lelechai}@tju.edu.cn
[2] Tianjin Key Laboratory of Cognitive Computing and Application, Tianjin, China
[3] The Department of Computer Science and Software Engineering,
The University of Western Australia, Perth, Australia
jianxin.li@uwa.edu.au
[4] School of Information and Communication Technology, Griffith University,
Brisbane, Australia
j.wang@griffith.edu.au

Abstract. With the proliferation of knowledge graphs, large amounts of RDF graphs have been released, which raises the need for addressing the challenge of distributed SPARQL queries. In this paper, we propose an efficient distributed method, called StarMR, to answer the SPARQL basic graph pattern (BGP) queries on big RDF graphs using MapReduce. In our method, query graphs are decomposed into a set of *stars* that utilize the semantic and structural information embedded RDF graphs as heuristics. Two optimization techniques are proposed to further improve the efficiency of our algorithms. One filters out invalid input data, the other postpones the Cartesian product operations. The extensive experiments on both synthetic and real-world datasets show that our StarMR method outperforms the state-of-the-art method S2X by an order of magnitude.

Keywords: Star decomposition · SPARQL · BGP · MapReduce
RDF graphs

1 Introduction

The *Resource Description Framework* (RDF), a graph-based data model, is commonly used to represent and organize resources in *knowledge graphs* because of its flexibility. An RDF data is a collection of triples (s, p, o), each of which represents a statement of a predicate p between a subject s and an object o. An RDF triple can be naturally viewed as an edge with s and o as vertices. Thus, an RDF graph can be represented as a labeled directed graph, e.g., the example RDF graph G_1 excerpted from DBpedia dataset in Fig. 1. Due to the flexibility

© Springer International Publishing AG, part of Springer Nature 2018
Y. Cai et al. (Eds.): APWeb-WAIM 2018, LNCS 10987, pp. 415–430, 2018.
https://doi.org/10.1007/978-3-319-96890-2_34

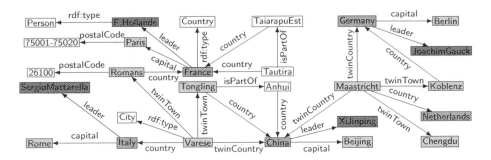

Fig. 1. An example RDF graph G_1 excerpted from DBpedia dataset

of RDF data, they are widely applied in various fields, such as science, bioinformatics, business intelligence, and social networks [6]. In real world, the size of RDF data often reaches hundreds of millions of triples.

SPARQL is the standard query language for RDF recommended by W3C, where *basic graph pattern* (BGP) is the most common query pattern used in SPARQL. Essentially, the core of each SPARQL query is a BGP query. For instance, the following BGP query Q_1 consists of six triple patterns over G_1. It is known that the semantics of SPARQL corresponds to graph homomorphism [10] which is NP-complete [2]. Therefore, how to efficiently answer SPARQL queries over big RDF graphs has been widely recognized as a challenging problem.

```
SELECT ?x ?y
WHERE { ?x country ?y . ?x twinTown ?z . ?x twinCountry ?w .
        ?y capital ?c . ?y leader ?p . ?z country ?w . }
```

Currently, there has been some research works on SPARQL queries over RDF data in a **distributed** environment. One category of methods is based on the relational schema [3,5,7,11,13], in which RDF data is modeled as a set of triples and stored in relational tables or a variant relational schema. All of these methods do not consider inherent graph-like structures of RDF data. When processing complex SPARQL queries, excessive join operations over relational tables are needed, which may incur expensive cost. In contrast, the other category of methods manages RDF data in native graph formats [9,12,16] and represents SPARQL queries as query graphs, which typically employs adjacency lists to store RDF data. Thus, the evaluation of SPARQL queries can be converted into a subgraph matching problem, in which how to reduce the enormous intermediate results is crucial.

In [14], query graphs are decomposed into stars (trees of depth 1). Lai et al. pointed out that the star-join algorithm in [14] suffers from scalability problems due to the generation of a large number of matches when evaluating a star with multiple edges [8]. The reason for this issue is that in unlabeled, undirected graphs they focused on it is very likely that the large combination of intermediate results are generated due to the lack of distinguishable information on vertices and edges. Thus, they proposed the so-called TwinTwigJoin MapReduce algorithm, where a TwinTwig is either a single edge or two incident edges of a vertex. Unlike unlabeled and undirected graphs in [8,14], RDF graphs have URIs as the unique vertex labels and directed edges. Thus, the problem concerned in [8] does not exist over RDF graphs. Therefore, it is reasonably safe to exploit the more holistic star-shaped structures other than just twin twigs as decomposition units of query graphs to minimize the amount of intermediate results.

To this end, we propose a new star-based query decomposition strategy, in which the star retains more holistic graph structures of query graphs than the TwinTwig method [8]. Thus, our approach can be completed in fewer MapReduce rounds. In our method, in order to evaluate SPARQL BGP queries more efficiently, query graphs are decomposed into a set of stars by using the semantic and structural information embedded in RDF graph as heuristics (i.e., h-values defined in this paper), to evaluate SPARQL BGP queries in MapReduce [1]. In addition, in order to reduce the intermediate results, the matching order of stars are determined by a greedy strategy.

Our main contributions include: (1) we propose an efficient and scalable distributed algorithm based on star decomposition, called StarMR, for answering SPARQL BGP queries on RDF graphs; (2) two optimization strategies of StarMR are devised, one of which employing the properties in RDF graphs to filter out invalid input data in MapReduce iterations, the other postponing part of Cartesian product operations to the final step of MapReduce to reduce a part of unpromising Cartesian product operations; and (3) extensive experiments on both synthetic and real-world RDF graphs have been conducted to verify the efficiency and scalability of our method. On average, the experimental results show that StarMR outperforms the state-of-the-art method by an order of magnitude.

The rest of this paper is organized as follows. Section 2 briefly reviews related work. In Sect. 3, we introduce preliminary definitions on RDF graphs and SPARQL BGP queries. In Sect. 4, we describe in detail how to decompose BGP queries, determine the matching order of stars, and match BGP queries using MapReduce. We then present two optimization strategies in Sect. 5. Section 6 shows experimental results, and we conclude in Sect. 7.

2   Related Work

The existing research work on ***distributed/parallel*** SPARQL queries over large-scale RDF graphs can be classified as follows:

Relational Schema Approach. SHARD [11] handles SPARQL queries over RDF data for triple-stores which needs to iterate over query clauses to bind variables to vertices in data graphs while satisfying all of the query constraints. Each round of MapReduce only adds one query clause with the join operation in [11]. Similarly, HadoopRDF [7], using MapReduce, divides RDF triples based on the predicates into multiple smaller files. A triple pattern in SPARQL queries cannot simultaneously take part in more than one join in a single Hadoop job.

These two methods do not use any structural information of query graphs, thus a large number of join operations may incur expensive costs. Furthermore, Virtuoso [3], supporting RDF in a native RDBMS, also model RDF data as a set of triples. TriAD [5], using a custom MPI protocol, employs six SPO permutation indexes, partitions RDF triples into those indexes, and use a locality-based summary graph to speed up queries. S2RDF [13] introduces the relational partitioning model ExtVP to store RDF data over the Spark [15] parallel framework, by which it can effectively minimize the query input size. Nevertheless, the cost of the semi-join preprocessing in [13] is prohibitively expensive.

Native Graph Approach. In [16], RDF data is modeled in its native graph form, a key-value store which saves node identifiers as the keys, and the adjacency lists of nodes as the values. Trinity. RDF [16] leverages graph exploration to reduce the volume of intermediate results, while the final results need to be enumerated at the single master node using a single thread. S2X [12] builds on GraphX [4], a distributed graph processing framework in top of Spark [15], to implement BGP matching of SPARQL. In S2X, a query graph is also decomposed into triple patterns which is similar to the methods in [7,11]. All of these triple patterns are matched first; then intermediate results are gradually discarded by iterative computation; finally, the remaining matching results are joined, which may lead to potentially large intermediate results. In addition, Peng et al. adopt a partial evaluation and assembly framework to perform SPARQL queries based on gStore [17], a graph-based SPARQL query engine using VS*-tree indexes [9]. In their method, each slave machine evaluates the query in the partial computation phase, and then in the assembly phase a large number of local partial matches are sent to the coordinator and joined together to obtain the final results, which may become a performance bottleneck when the amount of partial matches are large.

In this paper, we focus on the analytical processing scenario of RDF graphs using MapReduce which does not take advantage of any prebuilt indexes. Though building indexes can definitely accelerate lookups with high selectivity, it will not benefit analytical processing in which almost all data are accessed. So, it is unfair to compare our approach with those based on intensive indexes, such as S2RDF [13], the distributed gStore system [9], etc. In our method, (1) we store RDF triples using the adjacency list scheme; (2) a star decomposition strategy with heuristic information is proposed, which is able to keep more holistic structures of query graphs; (3) as to optimization startegies, we employ RDF properties to filter out unpromising input data and postpone Cartesian product operations.

3   Preliminaries

In this section, we introduce several basic background definitions about RDF graphs and SPARQL BGP queries which are used in our algorithms.

Definition 1 (RDF graph). *Let U and L be the disjoint infinite sets of URIs and literals, respectively. A tuple $(s, p, o) \in U \times U \times (U \cup L)$ is called an* RDF triple, *where s is the* subject, *p is the* predicate, *and o is the* object. *A finite set of RDF triples is called an* RDF graph.

Given an RDF graph G, let V, E, Σ denote the set of vertices, edges and edge labels, respectively. Formally, $V = \{s \mid (s, p, o) \in G\} \cup \{o \mid (s, p, o) \in G\}$, $E \subseteq V \times V$, and $\Sigma = \{p \mid (s, p, o) \in G\}$. The function $\mathsf{lab}{:}E \to \Sigma$ returns the labels of edges in G.

Definition 2 (Query graph). *Let Var be an infinite set of variables disjoint from U and L, and the name of every element in Var starts with the character? conventionally (e.g., $?x \in Var$). A SPARQL BGP query Q over an RDF graph G is defined as: $Q = \{(s_i, p_i, o_i) \mid (s_i, p_i, o_i) \in (V \cup Var) \times (\Sigma \cup Var) \times (V \cup Var)\}$, where $tp_i = (s_i, p_i, o_i)$ is a triple pattern. A SPARQL BGP query Q is also referred to as a query graph G_Q.*

Let $V(Q), E(Q)$ be the set of vertices and edges in G_Q, respectively. For each vertex $u \in V(Q)$, if $u \in Var$, then u can match any vertex $v \in V$; otherwise u only can match the vertex $v \in V$ whose label is the same as that of u.

Definition 3 (SPARQL BGP match). *The semantics of a SPARQL BGP query Q over an RDF graph G is defined as: (1) μ is a mapping from vertices in $V(Q)$ to vertices in V; (2) $(G, \mu) \vDash Q$ iff for any $(s_i, p_i, o_i) \in Q$ satisfies that i) s_i and o_i can match $\mu(s_i)$ and $\mu(o_i)$, respectively, ii) $(\mu(s_i), \mu(o_i)) \in E$, and iii) $p_i \in Var$ or $\mathsf{lab}((\mu(s_i), \mu(o_i)))$ is the same as p_i; (3) $\Omega(Q)$, the set of μ such that $(G, \mu) \vDash Q$, is the answer set of the SPARQL BGP query G_Q over G.*

Some definitions about *mappings* are needed. Two mappings μ_1 and μ_2 are called *compatible* denoted as $\mu_1 \sim \mu_2$, iff every element $v \in dom(\mu_1) \cap dom(\mu_2)$ satisfies $\mu_1(v) = \mu_2(v)$, where $dom(\mu_i)$ is the domain of μ_i. Furthermore, the set-union of two compatible mappings, i.e., $\mu_1 \cup \mu_2$, is also a mapping.

4   The StarMR Algorithm

In this section, we present how to decompose the query graph into a set of stars and determine the matching order of these stars. Then, we describe in detail how to match SPARQL BGP query using MapReduce in a left-deep-join framework.

4.1   Star Matching

A star is a tree of height one, denoted by $T = (r, L)$, where (1) r is the root of T; and (2) L is a set of 2-tuples (p_i, l_i), i.e., l_i is a leaf of T and (r, p_i, l_i) is an edge from r to l_i. Then, let $V(T), E(T)$ be the set of nodes and edges in T, respectively. In this paper, the minimum matching unit is a star in our method, and the RDF graph G is stored in adjacency lists. For each vertex $v \in V$, we use $N(v)$ to denote

---

Algorithm 1. STARMATCH($T, N(v)$)

Input : Star: $T = (r, L)$, where $L = \{(p_1, l_1), \ldots, (p_t, l_t)\}$, $N(v), v \in V$
Output: Matching results of T over $N(v)$: $\Omega_v(T) = \{\mu_1, \mu_2, \ldots, \mu_n\}$

1 $\Omega_v(T) \leftarrow \emptyset$;
2 **if** $T.r$ *matches vertex* v **then**
3   **foreach** $(p_i, l_i) \in T.L$ **do**   // get the candidate set $S(l_i)$ of leaf l_i
4   **if** $p_i \notin Var$ **then** $S(l_i) \leftarrow \{v' \mid (p_i, v') \in N(v) \wedge l_i \text{ matches } v'\}$;
5   **else** $S(l_i) \leftarrow \{v' \mid (\_, v') \in N(v) \wedge l_i \text{ matches } v'\}$; // _ is a wildcard
       // do Cartesian product operation $\{v\} \times S(l_1) \ldots S(l_t)$ to get $\Omega_v(T)$
6      $\Omega_v(T) \leftarrow \{\mu \cup \mu_1 \ldots \mu_t \mid \mu = \{(T.r, v)\} \wedge \mu_i = \{(l_i, v')\}, l_i \text{in } T.L \wedge v' \in S(l_i)\}$;
7 **return** $\Omega_v(T)$;

---

the neighbor information of vertex v, where $N(v) = \{(p_i, v_i') \mid (v, p_i, v_i') \in G\}$. We give the star matching algorithm as follows.

Algorithm 1 will be run in the following steps: (1) first matches the root $T.r$ with the vertex v (line 2); (2) then obtains the candidate matching set of every leaf (lines 3–5); (3) next does the Cartesian product operations on the candidate matching sets of vertices in the star T to get matching results (line 6). Finally, StarMatch($T, N(v)$) returns the matching results of the star T over $N(v)$ (line 7), and $\Omega(T)$ is the union of $\Omega_v(T)$, where $v \in V$.

4.2   Star Decomposition of Query Graphs

Definition 4 (Star decomposition). *The* star decomposition *of a SPARQL BGP query* $Q = \{tp_1, \ldots, tp_n\}$ *is denoted as* $D = \{T_1, \ldots, T_m\}$, *where (1)* T_i *is a star; (2)* $T_i.r \neq T_j.r, T_i, T_j \in D \wedge i \neq j$; *(3)* $E(T_i) \cap E(T_j) = \emptyset, T_i, T_j \in D \wedge i \neq j$; *and (4)* $\bigcup_{1 \leq i \leq m} E(T_i) = E(Q)$.

Example 1. Consider the example query Q_1 over the RDF graph G_1 in Sect. 1, the query graph G_{Q_1} of Q_1 is shown in Fig. 2. Moreover, D is the star decomposition of G_{Q_1} which contains three stars, T_1, T_2, and T_3. $\square$

After obtaining the query decomposition D of Q_1, there exist six matching orders. According to Algorithm 1, stars T_1, T_2, and T_3 over G_1 have 4, 6, and 8 matching results, respectively. Consider the matching order $T_1 T_3 T_2$, there exists 32 intermediate results by joining the matching results of star T_1 and T_3, because these two stars do not share any common vertex. However, another matching order $T_2 T_1 T_3$ only generates 2 intermediate results. In other words, the matching order of stars has a significant effect on the performance of queries.

We leverage the structure information and semantics in RDF graphs to decompose the query graph into stars and give a matching order to reduce the number of intermediate results using a greedy strategy. In particular, we define h-value as the heuristic information. The function fre: $\Sigma \rightarrow \mathbb{N}$ gets the frequency of a predicate p in an RDF graph G, where $\mathbb{N}$ is the set of natural

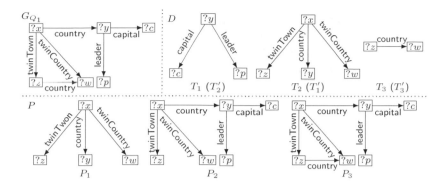

Fig. 2. The query graph and star decomposition of BGP query Q_1

Algorithm 2. STARDECOMPOSE(Q)

Input : A BGP Q: $\{tp_1, tp_2, ..., tp_n\}$
Output: A Queue of stars D: $\{T_1, ..., T_m\}$

1 $D \leftarrow \emptyset$;   // D: the queue of stars, $V(D)$: the set of vertices in D
2 $Q_c \leftarrow \{s \mid s \in Sub(Q) \wedge s \notin Var\}$;   // $Sub(Q)$: the set of subjects in Q
3 **if** $Q_c \neq \emptyset$ **then** $r \leftarrow \arg\max_{v \in Q_c} \mathsf{h}(v)$;
4 **else** $r \leftarrow \arg\max_{v \in Sub(Q)} \mathsf{h}(v)$;
5 genStar(r, Q, D);   // generate the star rooted at vertex r
6 **while** $Q \neq \emptyset$ **do**
7 $\quad M_v \leftarrow \{s \mid s \in Sub(Q) \wedge s \in V(D)\} \cup \{s \mid (s,p,o) \in Q \wedge o \in V(D)\}$;
8 $\quad r \leftarrow \arg\max_{v \in M_v} \mathsf{h}(v)$;
9 $\quad$ genStar(r, Q, D);
10 **return** D : $\{T_1, ..., T_m\}$;
11 **Function** genStar (r, Q, D) // generate a star
12 $\quad T.r \leftarrow r, T.L \leftarrow \{(p_i, l_i) \mid (r, p_i, l_i) \in Q\}$;
13 $\quad D$.enqueue(T);
14 $\quad Q \leftarrow Q \setminus \{(r, p_i, l_i) \mid (p_i, l_i) \in T.L\}$;

---

numbers and $\mathsf{fre}(p) = |\{(s_i, p, o_i) \mid (s_i, p, o_i) \in G\}|$. Then, for a query Q over G, let $P(u)$ be the set of properties (a.k.a., predicates) of vertex u in Q, i.e., $P(u) = \{p_i \mid (u, p_i, u_i') \in Q\}$. The h-value of each vertex $u \in V(Q)$ is defined as $\mathsf{h}(u) = \frac{|outDeg|}{\min_{p \in P(u)} \mathsf{fre}(p)}$, where $outDeg$ is the out degree of vertex u. The h-value is determined by two factors: (1) the more out degrees a vertex u has, the more variables may be bound when the star rooted at u is matched; and (2) the smaller $\mathsf{fre}(p)$, $p \in P(u)$ is, the higher selectivity of vertex u has. If all properties of vertex u are variables, $\mathsf{h}(u) = 0$. Our star decomposition algorithm guided by h-values is shown in Algorithm 2.

In Algorithm 2, a constant vertex in Q_c having the maximum h-value is selected as the root of the first star (line 3). If Q_c is an empty set, the algorithm picks up a vertex in $Sub(Q)$ whose h-value is the maximum (line 4). The star rooted at the selected vertex is generated (line 5) by calling the function genStar (lines 11–14). Then, we use M_v to denote the candidate set of root nodes which can guarantee that the star to be generated and the stars that have been generated share at least one common vertex (line 7). Similarly, after obtaining the vertex r w.r.t. the h-value, a new star is generated (lines 8–9). This process (lines 6–9) terminates until the set Q is empty.

For a SPARQL BGP query Q, Algorithm 2 can produce a star decomposition D and determine an order of these stars, $T_1 \ldots T_m$, such that $\bigcup_{1 \le i < j} V(T_i) \cap V(T_j) \ne \emptyset$, $1 \le j \le m$. Based on this matching order, we further introduce the concept of the partial query graph: the *partial query graph* $P_j, 1 \le j \le m$ is a subgraph of G_Q, where (1) $V(P_j) = \bigcup_{1 \le i \le j} V(T_i)$; and (2) $E(P_j) = \bigcup_{1 \le i \le j} E(T_i)$. Obviously, $P_1 = T_1$ and $P_m = G_Q$. Let $\Omega(T_i)$ and $\Omega(P_i)$ be the set of matching results for star T_i and partial query graph P_i, respectively. We have $\Omega(P_1) = \Omega(T_1)$, $\Omega(P_t) = \Omega(P_{t-1}) \bowtie \Omega(T_t)$, and $\Omega(P_m) = \Omega(Q)$.

Example 2. Consider the query Q_1 over G_1, where $\mathsf{h}(?y) = \frac{2}{4}$, $\mathsf{h}(?x) = \frac{3}{3}$, and $\mathsf{h}(?z) = \frac{1}{8}$. According to Algorithm 2, the first selected vertex is $?x$ and the corresponding star is T_2 (T_1') in Fig. 2. Then stars T_1 (T_2') and T_3 (T_3') are generated. Based on this order, P_1, P_2, and P_3 in Fig. 2 are the partial query graphs of Q_1. Obviously, P_3 is exactly the original query graph G_{Q_1}.   □

4.3   SPARQL BGP Matching Algorithm Using MapReduce

Next, we show how to use MapReduce to answer a SPARQL BGP query in a left-deep-join framework and demonstrate the pseudocode of our StarMR algorithm.

Algorithm 3 decomposes the query Q into a queue K of stars (line 1) and matches these stars in MapReduce iterations (lines 3–9). Each round of the MapReduce iteration joins one star with the partial results until all stars are matched. Map function consists of two parts: (1) when the input value is $N(v)$ (lines 12–18), the function matches the star T_t over every neighbor information $N(v)$ in RDF graph G *in parallel* (line 13), then let the matching results of intersection of vertex sets of star T_t and partial query graph P_{t-1}, i.e., μ_{key}, be keys (line 18); and (2) when the input value is a mapping in $\Omega(P_{t-1})$ (lines 19–20), similarly, let μ_{key} be keys (line 20). Every mapping μ in $\Omega_v(T_t)$ and $\Omega(P_{t-1})$ is transformed into a key-value pair (μ_{key}, μ). Note that when $t = 1$, the output of Map is $(\emptyset, \mu)$ (lines 14–15). Reduce function joins the matching results of T_t and P_{t-1} to generate the matching results of P_t with μ_{key} as the keys (lines 21–22).

Theorem 1. *Given an RDF graph G and a query graph G_Q, we assume that Algorithm 2 decomposes G_Q into a queue of stars $D = \{T_1, \ldots, T_m\}$. The time complexity of Algorithm 3 is bounded by $O(|V|^m |N_{max}|^{m|L_{max}|})$, where $|V|$ is the size of G, $|N_{max}|$ is the largest out degree in G, and $|L_{max}|$ is the largest out degree in G_Q.*

Proof (Sketch). In Algorithm 3, each round of the MapReduce iteration matches one star, therefore, it can evaluate the query Q in m rounds. The time complexity

---

Algorithm 3. StarMR

Input : RDF graph $G = (V, E, \Sigma)$, A SPARQL BGP Q: $\{tp_1, tp_2, ..., tp_n\}$
Output: The answer set: $\Omega(Q)$

1 $K \leftarrow$ STARDECOMPOSE(Q);   // decompose query graph
2 $\Omega(Q) \leftarrow \emptyset, t \leftarrow 1$;
3 **while** K *is not empty* **do**   // MapReduce iterations
4     $T_t \leftarrow K$.dequeue();
5   Map$(\emptyset, N(v))$;
6   **if** $t > 1$ **then**
7   Map$(\emptyset, \mu)$ s.t. $\mu \in \Omega(P_{t-1})$;
8   Reduce$(\mu_{key}, (\Omega_1, \Omega_2)$ s.t. $\Omega_1 \subseteq \Omega(T_t) \wedge \Omega_2 \subseteq \Omega(P_{t-1}))$;
9     $t \leftarrow t + 1$;

10 **return** $\Omega(P_m)$;
11 **Function** Map $(\emptyset, N(v)$ *or* $\mu)$
12   **if** *value is an adjacency list* $N(v)$ *in* G **then**   // match the star T_t
13        $\Omega_v(T_t) \leftarrow$ STARMATCH$(T_t, N(v))$;
14   **if** $t = 1$ **then**
15   **foreach** $\mu \in \Omega_v(T_1)$ **do return** $(\emptyset, \mu)$;
16   **else**
17   **foreach** $\mu \in \Omega_v(T_t)$ **do**   // get the mapping μ_{key} as key
18              $\mu_{key} \leftarrow \{(u_k, \mu(u_k)) \mid u_k \in V(P_{t-1}) \cap V(T_t)\}$; **return** (μ_{key}, μ);

19   **else**   // value is a mapping
20        $\mu_{key} \leftarrow \{(u_k, \mu(u_k)) \mid u_k \in V(P_{t-1}) \cap V(T_t)\}$; **return** (μ_{key}, μ);

21 **Function** Reduce $(\mu_{key}, (\Omega_1, \Omega_2))$
22   **foreach** $(\mu, \mu') \in \Omega_1 \times \Omega_2$ **do return** $(\emptyset, \mu \cup \mu')$;   // $\mu \cup \mu' \in \Omega(P_t)$

---

of Algorithm 3 consists of two parts: (1) the time complexity of star matching is $\sum_{1 \leq t \leq m} \sum_{v \in V}(|N(v)| + |\Omega(T_t)|)$; and (2) the time complexity of join operations is $\sum_{1 < t \leq m} |\Omega(P_{t-1})| \times |\Omega(T_t)|$. In the worst case, every leaf in T_t can match all neighbouring vertices of a vertex v, $v \in V$, i.e., $|\Omega_v(T_t)| = |N(v)|^{|T_t.L|}$. Thus, the time complexity of Algorithm 3 is bounded by $O(|V|^m |N_{max}|^{m|L_{max}|})$. □

5   Two Optimization Strategies

In this section, two optimization strategies are proposed to improve the efficiency of the StarMR algorithm.

5.1   RDF Property Filtering

We take advantage of the inherent semantics embedded in RDF graphs to filter out unpromising computations. The RDF Schema[1] can be used to define

---
[1] https://www.w3.org/TR/rdf-schema/.

classes of resources (i.e., vertices in RDF graphs). For example, the RDF triple $(s, \texttt{rdf:type}, C)$ declares that resource s is an instance of the class C, denoted by $s \in C$. We assume that for each subject s in an RDF graph G there exists at least a triple $(s, \texttt{rdf:type}, C) \in G$. We believe that this assumption is reasonable since every resource should belong to at least one type in the real world.

Example 3. As shown in Fig. 1, the triple $(\texttt{Varese}, \texttt{rdf:type}, \texttt{City})$ denotes that $\texttt{Varese}$ is an instance of the class $\texttt{City}$. Other such triples in Fig. 1 are omitted. Moreover, all instances of class $\texttt{City}$ are highlighted in green. Similarly, there exist other classes in G_1, e.g., $\texttt{Person}$ in blue and $\texttt{Country}$ in red.   □

Given an RDF graph $G = (V, E, \Sigma)$, let $P'(C) = \{p \mid (s, p, o) \in G \wedge s \in C\}$ be the set of RDF properties of class C. Note that the size of classes in an RDF graph is much less than vertices in the corresponding RDF graph. When matching a star T in the function $\mathsf{Map}(T, N(v))$, RDF properties of C can be used to filter out input data as follows: if $v \in C \wedge P(T.r) \not\subseteq P'(C)$, the procedure STARMATCH$(T, N(v))$ in Map function can be pruned.

Example 4. Consider matching star T_2 rooted at $?x$ in Fig. 2 over G_1 in Fig. 1, we have $P(?x) = \{\texttt{twinTown}, \texttt{country}, \texttt{twinCountry}\}$ and $P'(\texttt{Country}) = \{\texttt{capital}, \texttt{leader}\}$. Due to $\texttt{Italy} \in \texttt{Country} \wedge P(?x) \not\subseteq P'(\texttt{Country})$, the computation of STARMATCH$(T_2, N(\texttt{Italy}))$ can be pruned.   □

5.2 Postponing Cartesian Product Operations

In our optimization method, the initial star matching phase only needs to calculate the candidate matching sets of leaves of a star. Let f be a mapping from vertices in a star to the candidate matching sets of the corresponding vertices. We use $\Omega'(T)$ and $\Omega'(P)$ to denote the matching results of star T and partial query graph P, respectively.

Unlike Algorithm 1, for every vertex $u \in V(T)$, $\mathsf{StarMatch}_{opt}(T, N(v))$ adds $(u, S(u))$ to a mapping f (line 6). Similarly, the functions Map and Reduce are also changed. In the t-th MapReduce iteration, the Map function does the Cartesian product operation $f(u_1) \times \cdots \times f(u_k)$, where $u_k \in V(P_{t-1}) \cap V(T_t)$ (lines 10–18). For every $f \in \Omega'(P_m)$, we do the remaining Cartesian product operation $f(u'_1) \times \cdots \times f(u'_k)$ to get the final matching results $\Omega(Q)$, where $u'_k \in V(Q) \setminus \bigcup_{1 < t \le m}(P_{t-1} \cap V(T_t))$. Although the strategy does not change the complexity of our algorithm, they can improve query efficiency significantly.

Example 5. When answering Q_1 over G_1, STARMATCH$(T'_1, N(\texttt{Masstricht}))$ obtains the *candidate set* $S(l_i)$ of every leaf l_i in T'_1, e.g., $S(?y) = \{\texttt{Netherlands}\}$. We have $V(P_1) \cap V(T'_2) = \{?y\}$. Next, star T'_2 is matched, but the candidate set of $?y$ in all $N(v)$ in G_1 does not contain vertex $\texttt{Netherlands}$. Thus, we do not need to execute the Cartesian product operation $S(?z) \times S(?y) \times S(?w)$ to get $\Omega_{\texttt{Masstricht}}(T'_1)$, whose cost is prohibitively expensive.   □

Algorithm 4. StarMR$_{opt}$

---

1 **Function** StarMatch $_{opt}(T, N(v))$
2   **if** $T.r$ *matches vertex* v **then**
3   **foreach** $(p_i, l_i) \in T.L$ **do**
4   **if** $p_i \notin Var$ **then** $S(l_i) \leftarrow \{v' \mid (p_i, v') \in N(v) \land l_i \text{ matches } v'\}$;
5   **else** $S(l_i) \leftarrow \{v' \mid (\_, v') \in N(v) \land l_i \text{ matches } v'\}$;
6        $f \leftarrow \{(u, S(u)) \mid u \in V(T)\}$;
7   **return** f;

8 **Function** Map ($\emptyset$, $N(v)$ or f' s.t. $t > 1$)
9   **if** $t = 1$ **then** **return** $(\emptyset, \text{STARMATCH}_{opt}(T_1, N(v)))$;
10   **else**
11        $V_{key} : \{u_1, \ldots, u_k\} \leftarrow V(P_{t-1}) \cap V(T_t)$;
12   **if** *value is an adjacency list* $N(v)$ *in* G **then**
13           $f \leftarrow \text{STARMATCH}_{opt}(T_t, N(v))$;
          // do $f(u_1) \times \cdots \times f(u_k)$ to get Ω_{key}
14           $\Omega_{key} \leftarrow \{f_1 \cup \cdots \cup f_k \mid f_i = (u_i, \{v\})\} \land u_i \in V_{key} \land v \in f(u_i)\}$;
15   **foreach** $f_{key} \in \Omega_{key}$ **do** **return** $(f_{key}, f - V_{key})$;   // $f \in \Omega_f$
16   **else**
17           $\Omega_{key} \leftarrow \{f'_1 \cup \cdots \cup f'_k \mid f'_i = (u_i, \{v\})\} \land u_i \in V_{key} \land v \in f'(u_i)\}$;
18   **foreach** $f_{key} \in \Omega_{key}$ **do** **return** $(f_{key}, f' - V_{key})$;   // $f' \in \Omega_{f'}$

19 **Function** Reduce $(f_{key}, (\Omega_f, \Omega_{f'}))$
20   **foreach** $(f, f') \in \Omega_f \times \Omega_{f'}$ **do** **return** $(\emptyset, f_{key} \cup f \cup f')$;

---

6 Experiments

We have carried out extensive experiments on both synthetic and real-world RDF graphs to verify the efficiency and scalability of StarMR and compared it with the optimization method StarMR$_{opt}$ and the state-of-the-art S2X. Our algorithm is orthogonal to the graph partitioning and placement strategies in the cluster environment. In particular, for the implementation of StarMR, we use the default partitioner employed by the Hadoop Distributed File System (HDFS).

6.1 Settings

The prototype program, which is implemented in Scala using Spark, is deployed on an 8-site cluster connected by a gigabit Ethernet. Each site has a Intel(R) Core(TM) i7-7700 CPU with 4 cores of 3.60 GHz, 16 GB memory, and 500 GB disk. We used Hadoop 2.7.4 and Spark 2.2.0. All the experiments are carried out on Linux (64-bit CentOS) operating systems.

We used two RDF datasets, including synthetic dataset WatDiv[2] and real-world dataset DBpedia[3] in our experiments. (1) WatDiv is an RDF benchmark,

---

[2] http://dsg.uwaterloo.ca/watdiv/.
[3] http://wiki.dbpedia.org/downloads-2016-10.

Table 1. Datasets

Dataset	|Vertices|	|Edges|
WatDiv1M	158,118	1,109,678
WatDiv10M	1,052,571	10,916,457
WatDiv100M	10,250,947	108,997,714
DBpedia	6,060,648	23,509,250

Table 2. Queries

Query	L	S	F	C
WatDiv	$L_1, L_2, L_3,$ L_4, L_5	$S_1, S_2, S_3,$ $S_4, S_5, S_6,$ S_7	$F_1, F_2, F_3,$ F_4, F_5	$C_1, C_2,$ C_3
DBpedia	L_1, L_2	S_1, S_2	F_1, F_2	C_1, C_2

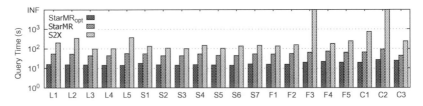

(a) The results on different queries over WatDiv100M

Query	$S1$	$S2$	$S3$	$S4$	$S5$	$S6$	$S7$	$C1$	$C2$	$C3$
StarMR$_{opt}$	**17.773**	**14.973**	**14.606**	**15.331**	**14.698**	**14.096**	**16.565**	**19.667**	**27.144**	**24.712**
StarMR	54.741	43.882	43.884	53.696	45.313	44.165	53.858	66.647	93.888	44.820
S2X	134.380	108.325	100.568	147.862	104.754	137.311	147.112	747.289	INF	244.216

Query	$L1$	$L2$	$L3$	$L4$	$L5$	$F1$	$F2$	$F3$	$F4$	$F5$
StarMR$_{opt}$	**15.987**	**15.403**	**14.998**	**14.497**	**14.240**	**16.092**	**16.081**	**19.954**	**22.118**	**19.987**
StarMR	54.636	54.379	44.281	44.476	56.826	52.733	55.855	64.719	75.001	64.818
S2X	206.564	347.468	100.222	101.839	375.845	137.796	157.659	INF	180.965	248.616

(b) The query times (in s) of StarMR$_{opt}$, StarMR, and S2X on WatDiv100M

Fig. 3. Efficiency on the 8-site cluster

which allows users to define their own datasets and generate test datasets with different sizes; and (2) DBpedia is a real-world dataset extracted from Wikipedia. As listed in Table 1, we summarize the statistics of these datasets. For RDF queries, we group them into four categories according to their shapes, including *linear* queries (L), *star* queries (S), *snowflake* queries (F), and *complex* queries (C), which are listed in Table 2. Regarding WatDiv, it gives 20 basic query templates. Due to the absence of query templates on DBpedia, we designed 8 queries, covering the 4 query categories mentioned above.

6.2 Efficiency

Experiments were conducted on WatDiv100M to verify the query efficiency of our method. As shown in Fig. 3(a), our optimization method StarMR$_{opt}$ has the best query efficiency on all 20 queries and StarMR is also much better than S2X. The query execution times of these 20 queries are shown in Fig. 3(b). For the query F_3 (resp. C_2), it can be observed that S2X cannot finish in the time limit (1×10^4 s), denoted by INF, while StarMR$_{opt}$ and StarMR can return the answers within 20 s (resp. 28 s) and 65 s (resp. 94 s), respectively. Furthermore, the average execution speed of the remaining 18 queries in StarMR$_{opt}$ is about 11 times faster than of that in S2X, i.e., our optimization method on average, outperforms S2X by an order of magnitude over WatDiv100M.

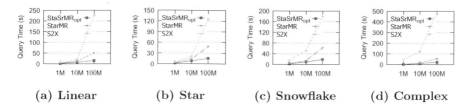

(a) Linear   (b) Star   (c) Snowflake   (d) Complex

Fig. 4. Scalability on WatDiv datasets

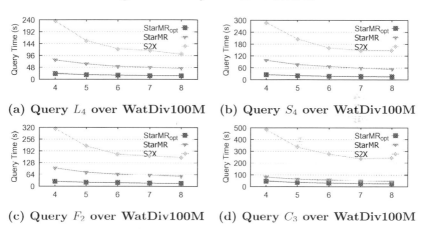

(a) Query L_4 over WatDiv100M   (b) Query S_4 over WatDiv100M

(c) Query F_2 over WatDiv100M   (d) Query C_3 over WatDiv100M

Fig. 5. Scalability on cluster sites

In addition, compared with StarMR, the time of StarMR$_{opt}$ is reduced from 44.86% to 74.94%, as listed in Fig. 3(b). Thus, StarMR$_{opt}$ is able to evaluate the query more efficiently.

The experimental result demonstrates that the effect of optimization strategies in StarMR$_{opt}$ is significant. We analyze that the reasons include: (1) S2X joined the intermediate matching results of all triple patterns in BGP, incuring expensive cost; (2) StarMR did Cartesian product operations in the star matching phase, which may lead to expensive cost; and (3) in StarMR$_{opt}$, a part of invalid input data were pruned by utilizing RDF properties embedded in RDF graphs, and a large number of Cartesian operations were postponed and reduced.

6.3   Scalability

In this section, we compared StarMR$_{opt}$ with StarMR and S2X. The comparison experiments were conducted on WatDiv datasets to verify the scalability.

Different Size of Datasets. We conducted experiments on WatDiv datasets over the 18 queries except F_3 and C_2. Moreover, the average times of each query category were calculated, as shown in Fig. 4. When changing the size of

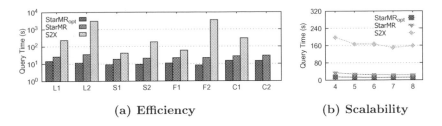

(a) Efficiency   (b) Scalability

Fig. 6. Query results on DBpedia

datasets from WatDiv1M to WatDiv100M, query times of all three methods
have increased and StarMR$_{opt}$ is always the best one. We can observe that the
growth rate of query times in S2X is much higher than that of StarMR$_{opt}$ and
StarMR. Compared with StarMR$_{opt}$, the growth rate of query times in StarMR
is higher than that of StarMR$_{opt}$. As shown in Fig. 4, the performance of S2X
dropped significantly with the size of datasets increasing. For complex queries
in S2X, the growth rate of query time is much higher than that of other query
categories.

Different Numbers of Cluster Sites. Extensive experiments were carried
out on the WatDiv100M dataset with the number of cluster sites varying from 4
to 8. During these experiments, we randomly selected one query from each of the
4 query categories, i.e., L_4, S_4, F_2, and C_3. As shown in Fig. 5, the experimental
results verify our intuition that query times of all these three methods decreased
as the number of cluster sites increased. This is because when the number of sites
increased, the degree of parallelism also increased. Although, with the number
of sites increasing, the speedup ratios of S2X and StarMR$_{opt}$ are comparative,
the performance of StarMR$_{opt}$ is stable for selective queries and the query times
of StarMR$_{opt}$ are far less than that of S2X.

6.4   Experiments on the Real-World Dataset

We also carried out experiments on a real-world dataset DBpedia to verify the
efficiency and scalability of our methods and S2X. These experimental results
are shown in Fig. 6.

As shown in Fig. 6(a), StarMR$_{opt}$ also demonstrates the best query efficiency
on all queries over DBpedia. When answering C_2, i.e., the query Q_1 mentioned in
Sect. 1, S2X reported errors. Thus, S2X cannot efficiently evaluate the complex
query involving a large number of intermediate results. For the remaining 7
queries, the execution speeds in StarMR$_{opt}$ are about 4 to 469 times faster than
that in S2X. Compared with StarMR, the time of StarMR$_{opt}$ is reduced from
44.19% to 67.48%, i.e., the optimization effect on DBpedia is prominent. So
in summary, the experimental results in Fig. 6(a) demonstrate that StarMR$_{opt}$
reduces invalid input data and postpones Cartesian product operations by a large

margin. In addition, the scalbility experiments were conducted on 4 queries, i.e., L_1, S_1, F_1, and C_1 over DBpedia. Similarly, query times of these three methods decreased with the number of cluster sites varying from 4 to 8, as shown in Fig. 6(b). Moreover, the speedup ratio of StarMR is about 1.1 times of S2X.

7   Conclusion

In this paper, we proposed the StarMR star-decomposition based query processor for efficiently answering SPARQL BGP queries on big RDF graph data using MapReduce. Moreover, we also developed two optimization strategies, including RDF property filtering and postponing Cartesian product operations, to improve the basic StarMR algorithm. Our extensive experimental results on both synthetic and real-world datasets have verified the efficiency and scalability of our method, which outperforms S2X by one order of magnitude.

Acknowledgments. This work is supported by the National Natural Science Foundation of China (61572353), the National High-tech R&D Program of China (863 Program) (2013AA013204), and the Natural Science Foundation of Tianjin (17JCY-BJC15400).

References

1. Dean, J., Ghemawat, S.: MapReduce: simplified data processing on large clusters. Commun. ACM **51**(1), 107–113 (2008)
2. Dyer, M., Greenhill, C.: The complexity of counting graph homomorphisms. Random Struct. Algorithms **17**(3–4), 260–289 (2000)
3. Erling, O., Mikhailov, I.: Virtuoso: RDF support in a native RDBMS. In: de Virgilio, R., Giunchiglia, F., Tanca, L. (eds.) Semantic Web Information Management, pp. 501–519. Springer, Heidelberg (2010). https://doi.org/10.1007/978-3-642-04329-1_21
4. Gonzalez, J.E., Xin, R.S., Dave, A., Crankshaw, D., Franklin, M.J., Stoica, I.: GraphX: graph processing in a distributed dataflow framework. In: OSDI, vol. 14, pp. 599–613 (2014)
5. Gurajada, S., Seufert, S., Miliaraki, I., Theobald, M.: TriAD: a distributed shared-nothing RDF engine based on asynchronous message passing. In: Proceedings of the 2014 ACM SIGMOD International Conference on Management of Data, pp. 289–300. ACM (2014)
6. Hammoud, M., Rabbou, D.A., Nouri, R., Beheshti, S.M.R., Sakr, S.: DREAM: distributed RDF engine with adaptive query planner and minimal communication. Proc. VLDB Endow. **8**(6), 654–665 (2015)
7. Husain, M., McGlothlin, J., Masud, M.M., Khan, L., Thuraisingham, B.M.: Heuristics-based query processing for large RDF graphs using cloud computing. IEEE Trans. Knowl. Data Eng. **23**(9), 1312–1327 (2011)
8. Lai, L., Qin, L., Lin, X., Chang, L.: Scalable subgraph enumeration in MapReduce. Proc. VLDB Endow. **8**(10), 974–985 (2015)
9. Peng, P., Zou, L., Özsu, M.T., Chen, L., Zhao, D.: Processing SPARQL queries over distributed RDF graphs. VLDB J. **25**(2), 243–268 (2016)

10. Pérez, J., Arenas, M., Gutierrez, C.: Semantics and complexity of SPARQL. In: Cruz, I., et al. (eds.) ISWC 2006. LNCS, vol. 4273, pp. 30–43. Springer, Heidelberg (2006). https://doi.org/10.1007/11926078_3

11. Rohloff, K., Schantz, R.E.: High-performance, massively scalable distributed systems using the MapReduce software framework: the SHARD triple-store. In: Programming Support Innovations for Emerging Distributed Applications, p. 4. ACM (2010)

12. Schätzle, A., Przyjaciel-Zablocki, M., Berberich, T., Lausen, G.: S2X: graph-parallel querying of RDF with GraphX. In: Wang, F., Luo, G., Weng, C., Khan, A., Mitra, P., Yu, C. (eds.) Big-O(Q)/DMAH -2015. LNCS, vol. 9579, pp. 155–168. Springer, Cham (2016). https://doi.org/10.1007/978-3-319-41576-5_12

13. Schätzle, A., Przyjaciel-Zablocki, M., Skilevic, S., Lausen, G.: S2RDF: RDF querying with SPARQL on spark. Proc. VLDB Endow. 9(10), 804–815 (2016)

14. Sun, Z., Wang, H., Wang, H., Shao, B., Li, J.: Efficient subgraph matching on billion node graphs. Proc. VLDB Endow. 5(9), 788–799 (2012)

15. Zaharia, M., Chowdhury, M., Franklin, M.J., Shenker, S., Stoica, I.: Spark: cluster computing with working sets. HotCloud 10(10–10), 95 (2010)

16. Zeng, K., Yang, J., Wang, H., Shao, B., Wang, Z.: A distributed graph engine for web scale RDF data. Proc. VLDB Endow. 6, 265–276 (2013). VLDB Endowment

17. Zou, L., Özsu, M.T., Chen, L., Shen, X., Huang, R., Zhao, D.: gStore: a graph-based SPARQL query engine. VLDB J. 23(4), 565–590 (2014)

DAVE: Extracting Domain Attributes and Values from Text Corpus

Yongxin Shen[1], Zhixu Li[1(✉)], Wenling Zhang[1], An Liu[1], and Xiaofang Zhou[1,2]

[1] School of Computer Science and Technology,
Soochow University, Suzhou 215006, China
`zhixuli@suda.edu.cn`
[2] The University of Queensland, Brisbane, QLD 4067, Australia

Abstract. Open Information Extraction (OpenIE) has been studied extensively, targeting at extracting structured information from free text. In this paper, we work on a novel OpenIE problem defined as Domain-specified Attribute-Value Extraction: Given a text corpus and a domain Knowledge Base (KB) with a number of domain attributes and corresponding attribute values, the task is to extend the KB by identifying more domain attributes and attribute values from the text corpus. Existing solutions adopted from the other OpenIE problems rely heavily on either using deep linguistic parsing or identifying effective lexical patterns. However, linguistic parsing does not always work well especially on short texts, while learning lexical patterns is too strict to reach a high extraction recall. In this paper, we propose an effective graph-based iterative extraction approach based on the cooccurrence between attribute terms and attribute value terms in the same sentences. Our experiments performed on two large real world data collections demonstrate that our method outperforms state-of-the art approaches in reaching 10% higher extraction precision and recall.

1 Introduction

Open Information Extraction (OpenIE) has been studied extensively, targeting at extracting structured information from free text [5,6,9]. In this paper, we study a novel OpenIE problem defined as *Domain-specified Attribute-Value Extraction*, or *DAVE* for short. Given an initial domain Knowledge Base (KB) with a number of domain attributes and corresponding attribute values such as the one given in Fig. 1(a) as well as a text corpus such as the one shown in Fig. 1(b), the task of DAVE is to extend the KB by identifying more domain attributes and values from the text corpus as depicted in Fig. 1(c).

Solutions to the DAVE problem could be adopted from the approaches to the other OpenIE problems. For instance, most iterative OpenIE systems, such as KnowItAll [5], Snowball [1] and TextRunner [13], learn syntactic and lexical patterns to identify more extraction results. Similar intuition can be adopted in doing DAVE. Besides, to resolve the problem of incoherent and uninformative relations caused by systems above, some verb phrase-based relation extraction

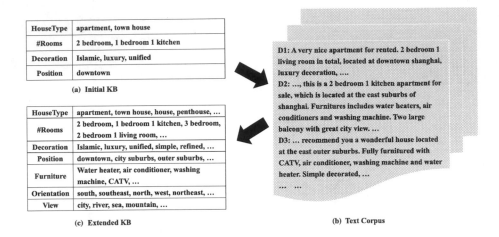

HouseType	apartment, town house
#Rooms	2 bedroom, 1 bedroom 1 kitchen
Decoration	Islamic, luxury, unified
Position	downtown

(a) Initial KB

HouseType	apartment, town house, house, penthouse, ...
#Rooms	2 bedroom, 1 bedroom 1 kitchen, 3 bedroom, 2 bedroom 1 living room, ...
Decoration	Islamic, luxury, unified, simple, refined, ...
Position	downtown, city suburbs, outer suburbs, ...
Furniture	Water heater, air conditioner, washing machine, CATV, ...
Orientation	south, southeast, north, west, northeast, ...
View	city, river, sea, mountain, ...

(c) Extended KB

D1: A very nice apartment for rented. 2 bedroom 1 living room in total, located at downtown shanghai, luxury decoration, ...

D2: ..., this is a 2 bedroom 1 kitchen apartment for sale, which is located at the east suburbs of shanghai. Furnitures includes water heaters, air conditioners and washing machine. Two large balcony with great city view. ...

D3: ... recommend you a wonderful house located at the east outer suburbs. Fully furnitured with CATV, air conditioner, washing machine and water heater. Simple decorated, ...

... ...

(b) Text Corpus

Fig. 1. An example scenario of DAVE

systems are also developed such as ReVerb [6]. The latter improves the former by performing deep analysis and then all relations mediated by verbs, nouns, adjectives and others are extracted by the system. However, while a small number of high-quality patterns can bring limited recall to the extraction results, the rest patterns usually bring as many noises as new instances.

In this paper, we propose an effective graph-based iterative extraction approach for DAVE based on the cooccurrence between attribute terms and attribute value terms in the same sentences. According to our observation, an attribute value term is usually mentioned with a corresponding attribute term in the same sentence in texts. For instance, "air conditioner" and "water heaters" are mentioned with "furnitures" in the same sentence as shown in Fig. 1(b). Through light-weight part-of-speech tagging to the whole corpus, a cooccurrence graph could be drawn, where each vertex corresponds to either an attribute term or an attribute value term, while an edge between two vertices denotes an cooccurrence relation between two terms. In this cooccurrence graph, the existing attributes and attribute values in the KB will help identify new ones that cooccur with the existing ones into the KB. Then in later iterations, we keep on updating the cooccurrence graph with the new involved ones, which will bring further new attributes and values into the KB. We perform this process iteratively until no more attributes and values could be detected from the texts.

A challenge in realizing our approach lies on how we find out the topic that each attribute/value belongs to from the graph, where two different types of vertices as well as plenty of noises are contained in the graph and sometimes an attribute value could belong to different topics of the domain. Although the existing works on community detection on social network [7,8] give us inspirations, these approaches could not be directly adopted given the stricter constraints and problem settings in our case.

To find out the topics for each attribute/value, a novel graph-based community detection algorithm is proposed, which finds out the core of each community first and then extend the community from its core to its boundary step by step. We could also allow overlaps between communities by allowing vertices on boundaries be involved into more than one communities. Since we are detecting a community from its higher-quality core to its lower-quality boundary, we could minimize the influence brought by noises as much as possible.

2   Problem Definition

Domain-Specified Attribute-Value Extraction (DAVE) Problem: Given an initial domain KB denoted by $K^0 = \{(A_1, V_1^0), (A_2, V_2^0), ..., (A_p, V_p^0))\}$, where A_i $(1 \leq i \leq p)$ is an attribute, and V_i^0 $(1 \leq i \leq p)$ is a set of initial attribute values of A_i. Let $\mathbf{C} = \{D_1, D_2, ..., D_q\}$ denote a text corpus, where each D_i $(1 \leq i \leq q)$ is a free text document. The DAVE task extends K^0 into a KB $K^f = \{(A_1, V_1), (A_2, V_2), ..., (A_p, V_p), (A_{p+1}, V_{p+1}), ..., (A_n, V_n)\}$ in both horizontal and vertical dimensions with $\mathbf{C}$ through: (1) identifying more attribute values for attribute A_i $(1 \leq i \leq p)$ from $\mathbf{C}$ to expand V_i^0 $(1 \leq i \leq p)$ into V_i; (2) identifying more attributes and their corresponding attribute values $\{(A_{p+1}, V_{p+1}), (A_{p+2}, V_{p+2}), ..., (A_n, V_n)\}$ from $\mathbf{C}$ to expand the KB in vertical dimension.

3   Methodology

In this section, we present our innovative solution to the DAVE problem.

(1) **Preprocessing Step.** As a preprocessing step, each document in the given corpus is divided into a number of *Text Segments* according to its punctuations such as ",", ".", "?", "-", and some predefined stop words etc. We then do part-of-speech tagging to every document to obtain pos tag of each phrase, and only keep those text segments having at least one candidate attribute and at least one candidate attribute values, where we assume that a candidate attribute is usually a noun phrase, while a candidate attribute value is usually a noun or adjective phrase after removing stop words.

(2) **CoGraph Construction.** Based on an observation that an attribute value term is usually mentioned in the same sentence with its corresponding attribute term, we would like to construct a *Cooccurrence Graph (or CoGraph for short)*, where each vertex on this graph corresponds to an attribute or attribute value, while each edge on this graph corresponds to an cooccurrence between a pair of attribute and (or) attribute value. To guarantee the quality of the CoGraph, an attribute or an attribute value will be put into the CoGraph if and only if (1) it is already in the KB; or (2) it is co-occurred with an existing attribute or an attribute value in the KB in the same text segment.

(3) **Topic Detection.** To perform attributes and values extraction from a CoGraph, we need to find out to which topic that each attribute/value belongs from the CoGraph. Inspired by the community detection task on social network [7, 8] targeting at detecting social communities from the social network, we would like to detect "communities" of attributes and attribute values from the CoGraph, where each community corresponds to a *topic* of the domain, where a topic refers to all the information of a specific aspect, which should have an attribute and its relevant attribute values such as "HouseType (Apartment/Flat/House/...)". From each of these communities (or topics), we then figure out which are attributes, and which are attribute values. Informally, we would like to call our task as *Topic-Oriented Community Detection.*

We propose a novel topic-oriented community detection approach based on CoGraph. Given a CoGraph with multiple communities, the nodes with the largest degree within a community are so called *Core Nodes* of the community, while those nodes with the smallest degree connecting to the other nodes of the community are so called *Boundary Nodes* of a community. While the core nodes of a community in CoGraph correspond to the attribute of a topic, the boundary nodes may correspond to its attribute values which might be polysemous phrases or commonly-used adjectives like "excellent", "nice", "fantastic", "good" and "bad". Given the intuition above, our approach would like to find out the core of each community first, and then extend each community from its core to its boundary step by step. Our approach would allow overlaps between communities by allowing attribute value vertices on boundaries be involved into more than one communities. Since we are detecting a community from its higher-quality core to its lower-quality boundary, we could minimize the influence brought by noises as much as possible. We call our algorithm as Core-Boundary Overlapable topic-oriented community detection approach based on a CoGraph, or *CBO* algorithm for short.

The workflow of CBO algorithm is described as follows:

(1) *Generating Initial Communities.* Given a CoGraph $G(V, E)$, we calculate the weighted degree denoted by $degree(v)$ for every vertex $v \in V$ and then take k nodes with the highest degree as core nodes, where k is a user-defined number case by case. Each of these core nodes will be taken as an initial community, which could be quickly extended by involving the other nodes that have connections with its core node into the community. Note that a node is allowed to be involved into multiple communities in this step.

(2) *Handling the Left Nodes.* To prevent from missing important but small communities (that could not be identified in the first step), CBO expects to identify more communities from the left nodes outside the initial communities. Given an user-defined threshold τ_D, from these left nodes outside the initial communities, CBO takes the node with the largest degree as a core node of a new community, as long as its degree is larger than τ_D. We then involve those nodes that have connections with this core node into the community. We keep on doing this until no more communities could be generated.

(3) *Merging Similar Communities.* Finally, we consider to merge similar communities according to the measured overlap between these communities. Particularly, we first measure the overlap between two communities, say C_i and C_j, as follows:

$$OL(C_i, C_j) = \frac{|C_i \cap C_j|}{|min(C_i, C_j)|} \qquad (1)$$

where $min(C_i, C_j)$ gets the smaller community between C_i and C_j.

We calculate the overlap between each pair of communities. We say two communities C_i and C_j have the qualification to be merged if and only if $OL(C_i, C_j) > \theta_O$, where the threshold θ_O should be decided carefully in order to ensure the overlaps between communities are within a certain range. However, a pair of communities having the qualification only would not be merged. Instead, among all pairs of communities, each time we only select the most promising pair to merge according to the change it brings to the *modularity* of the graph G. Particularly, the modularity of a graph G could reflect the goodness of a community partitioning result to G, which could be calculated according to [10,11] as follows:

$$EQ = \frac{1}{2m} \sum_i \sum_{u,v \in C_i, u \neq v} \frac{1}{N_C(u)N_C(v)} [A_{uv} - \frac{k(u)k(v)}{2m}] \qquad (2)$$

where each C_i denotes a community, u, v are two different vertices in C_i, $N_C(u)$ (or $N_C(v)$) is the number of communities to which a vertex u (or v) belongs, $k(u)$ (or $k(v)$) is the degree of a vertex u (or v), A_{uv} is the weight of edges between the u-th community and the v-th community, and $m = \frac{1}{2} \sum_{u,v} A_{uv}$ is the total weight of all the edges in the graph.

To find out the most promising pairs of communities to merge at a step, our CBO algorithm calculates the EQ score for every possible merging way, and only the merging way with the highest EQ score, which should be larger than the original EQ score of the graph would be adopted. The CBO community detection algorithm is finalized when no more communities could be merged.

(4) **Recognizing New Identified Attributes/Attribute Values.** Finally, a CNN-based recognition model could be built to prevent from introducing noises into the KB as well as recognizing the new identified attributes and attribute values for every topic after each iteration. Due to the limitation of space, we won't go to the details of this part.

4   Experiments

We report our experimental study results on two data collections. The **House** database contains 53.7 k house renting information in five large-medium cities of

China collected from three house renting information websites, $Ganji^1$, $Anjuke^2$, and 58 tongcheng[3]. Each house renting record contains both a structured info-box data and a free text description about the house. The **Car** database contains the information of about 19.7 k second-hand cars for sale crawled from Ganji website and "The home of used-car" website[4], which contains the information of second-hand cars including structured data and CText information.

Table 1. Comparing the extraction quality on the two datasets

(a) House

Attribute	Pattern	Parsing	LDA	CoGraphExt
Appliances	0.4035	0.1103	0.1731	**0.4525**
Orientation	**0.5121**	0.2171	0.2046	0.4401
#Rooms	**0.6017**	0.1900	0.1108	0.5601
Decoration	0.2109	0.1346	0.2566	**0.4240**
Payment	0.3578	0.2033	0.1347	**0.4342**
View	0.5607	0.1333	0.2769	**0.7755**
Surroundings	0.3016	0.1766	0.1574	**0.5242**
Housetype	0.3901	0.2135	0.1623	**0.4187**
Position	0.4198	0.2043	0.1832	**0.5148**
Average	0.4103	0.1723	0.1845	**0.5036**

(b) Car

Attribute	Pattern	Parsing	LDA	CoGraphExt
Band	**0.5473**	0.1667	0.1596	0.4761
Colour	0.4355	0.2343	0.2506	**0.5430**
Body	0.1467	0.1655	0.1751	**0.6023**
DriveType	0.3376	0.1135	0.1667	**0.4802**
Engine	0.4012	0.1470	0.1205	**0.4165**
Interior	0.3793	0.1911	0.2387	**0.5641**
Exterior	0.2713	0.1509	0.2476	**0.3708**
Average	0.3598	0.1670	0.1941	**0.4932**

We basically use three metrics to evaluate the effectiveness of the methods: **Extraction Precision:** the percentage of correctly extracted attributes and values among all extracted results, **Extraction Recall:** the percentage of correctly extracted attributes and values among all attributes and values that should be extracted from the text corpus, and $F1$ **Score:** a combination of precision and recall, which is calculated by $F1 = \frac{2*precision*recall}{precision+recall}$. Besides, we use **Time Cost** of an algorithm for evaluating the efficiency of a method.

We compare the effectiveness of our approach (CoGraphExt for short) with several adapted state-of-the-art approaches as follows:

- The *Pattern-based* method relies on identifying syntactic patterns to detect new attributes/values from the text corpus. The pattern generation and the iteration extraction process are adopted from the literature [1,5].
- The *Parsing-based* method relies on NLP techniques [9] to analyze the synthetic structure of a text with expect to find out new attributes/values.
- The *LDA-based* method relies on the LDA topic model [3] to mine hidden topics for each text segment, and then identify attribute/values from the segments.

(1) **Extraction Quality Comparison.** We compare the extraction quality of these methods on the two data sets. For a comprehensive comparision, we not only compare their F-measure on every attribute, but also the average F-measure score of all the attributes included in knowledge base in each data

---

[1] www.ganji.com.

[2] www.anjuke.com.

[3] www.58.com.

[4] www.carsales.com.au.

set. Note that the initial KB used in the experiments only covers no more than 30% of the whole knowledge base. As we could observed in Table 1(a) and (b), our method outperforms the three adapted state-of-art methods: Pattern-based method, Parsing-based method and LDA-based in extracting more attributes and values on the two data sets. Averagely, CoGraphExt improves the extraction quality of the state-of-art methods about 10% on the two data sets.

(2) **Efficiency Comparison**. We also compare the average time cost of three methods used in processing the same data set. As can be observed in Table 2, CoGraph uses much less time in dealing with inputs than LDA and Pattern-based method in which case CNN model training can be done offline.

Table 2. Comparing the time cost of different methods

Datasets	Pattern	Parsing	LDA	CoGraphExt
House	1430 ms	1790 ms	2356 ms	1371 ms
Car	375 ms	876 ms	750 ms	253 ms

5   Related Work

There are already a number of OpenIE systems and approaches developed to extract various kinds of structured information such as isA pairs or other instance pairs in specified types of relations from free texts. Some of these work and their ideas could be adopted in solving the DAVE problem.

A mainstream OpenIE systems and approaches including KnowItAll [5], Snowball [1], TextRunner [13], WOE [12] and Reverb [6] rely on syntactic and lexical patterns, that is, each iteration finds additional patterns that can be used for information extraction. They can be roughly organized in two categories: those systems based on self-supervision, in which labeled training data are generated by automatic heuristics, and those based on specific relations or heuristics. Similar idea can be adopted in doing DAVE. However, while a small number of high-quality patterns can bring limited recall to the extraction results, the rest patterns usually bring as many noises as new instances.

To help improve the quality of the extraction results, some approaches like StatSnowball [9] and ClausIE [4] introduce NLP techniques such as deep linguistic parsing to identify more instances from free texts. More specifically, StatSnowball iterates over the initial seed attributes in bootstrapping until new trusted templates or knowledge are no longer generated, which automatically generates and selects templates to form the extractor. As a more recent OIE system, ClausIE uses dependency parsing and a set of rules for domain-independent lexica to detect clauses without any requirement for training data. Some other approaches also use NLP techniques to help remove noises from the iterative pattern-based extraction results. For instance, KRAKEN [2], an N-ary OIE fact extraction system, subtly introduces syntactic features, which can be accurately detected by dependency paths. However, deep linguistic parsing brings about reduction of efficiency, which does not always work well especially on short texts.

6   Conclusions

In this paper, we define a novel OpenIE problem DAVE, and then propose a cooccurrence-based iterative extraction approach to solve the problem. Experiments conducted on two real world data collections demonstrate that our proposed approaches could outperform state-of-the-art ones by reaching 10% higher extraction precision and recall.

Acknowledgements. This research is partially supported by National Natural Science Foundation of China (Grant No. 61632016, 61572336, 61472263, 61232006), the Postdoctoral scientific research funding of Jiangsu Province (No. 1501090B), the National Postdoctoral Funding (No. 2015M581859, 2016T90493) and the Natural Science Research Project of Jiangsu Higher Education Institution (No. 17KJA520003).

References

1. Agichtein, E., Gravano, L., Pavel, J., Sokolova, V., Voskoboynik, A.: Snowball: a prototype system for extracting relations from large text collections. In: ACM SIGMOD International Conference on Management of Data, p. 612 (2001)
2. Akbik, A.: KRAKEN: N-ary facts in open information extraction. In: Joint Workshop on Automatic Knowledge Base Construction and Web-Scale Knowledge Extraction, pp. 52–56 (2012)
3. Blei, D.M., Ng, A.Y., Jordan, M.I.: Latent dirichlet allocation. J. Mach. Learn. Res. **3**, 993–1022 (2003)
4. Del Corro, L., Gemulla, R.: ClausIE: clause-based open information extraction. In: International Conference on World Wide Web, pp. 355–366 (2013)
5. Etzioni, O., Cafarella, M., Downey, D., Popescu, A.M., Shaked, T., Soderland, S., Weld, D.S., Yates, A.: Unsupervised named-entity extraction from the web: an experimental study. Artif. Intell. **165**(1), 91–134 (2005)
6. Fader, A., Soderland, S., Etzioni, O.: Identifying relations for open information extraction. In: Conference on Empirical Methods in Natural Language Processing, pp. 1535–1545 (2011)
7. Fortunato, S.: Community detection in graphs. Phys. Rep. **486**(3), 75–174 (2010)
8. Fortunato, S., Hric, D.: Community detection in networks: a user guide. Phys. Rep. **659**, 1–44 (2016)
9. Min, B., Shi, S., Grishman, R., Lin, C.Y.: Ensemble semantics for large-scale unsupervised relation extraction. In: Joint Conference on Empirical Methods in Natural Language Processing and Computational Natural Language Learning, pp. 1027–1037 (2012)
10. Newman, M.E.J., Girvan, M.: Finding and evaluating community structure in networks. Phys. Rev. E Stat. Nonlinear Soft. Matter Phys. **69**(2 Pt 2), 026113 (2004)
11. Shen, H., Cheng, X., Cai, K., Hu, M.B.: Detect overlapping and hierarchical community structure in networks. Phys. A Stat. Mech. Appl. **388**(8), 1706–1712 (2008)

12. Wu, F., Weld, D.S.: Open information extraction using Wikipedia. In: ACL 2010, Proceedings of the Meeting of the Association for Computational Linguistics, 11–16 July 2010, Uppsala, Sweden, pp. 118–127 (2010)
13. Yates, A., Cafarella, M., Banko, M., Etzioni, O., Broadhead, M., Soderland, S.: TextRunner: open information extraction on the web. In: Human Language Technologies: the Conference of the North American Chapter of the Association for Computational Linguistics: Demonstrations, pp. 25–26 (2007)

PRSPR: An Adaptive Framework for Massive RDF Stream Reasoning

Guozheng Rao[1,3], Bo Zhao[1,3], Xiaowang Zhang[1,3(✉)], Zhiyong Feng[2,3], and Guohui Xiao[4]

[1] School of Computer Science and Technology, Tianjin University,
Tianjin 300350, China
xiaowangzhang@tju.edu.cn
[2] School of Computer Software, Tianjin University, Tianjin 300350, China
[3] Tianjin Key Laboratory of Cognitive Computing and Application,
Tianjin 300350, China
[4] Faculty of Computer Science, Free University of Bozen-Bolzano,
39100 Bolzano, Italy

Abstract. In this paper, we propose a plugin-based framework for massive RDF stream reasoning to support complicated tasks on RDF stream in an adaptive and flexible way. Within this framework, the problem of RDF stream reasoning can be equivalently reduced to the combination problem of SPARQL querying and rule-based reasoning. Take advantage of the plug-in method, we can apply various off-the-shelf SPARQL query engines and rule-based reasoners in a simple way. Moreover, to efficiently support real-time reasoning on massive RDF stream, we develop a multi-threaded batch processing approach to manage resources and an adaptive reasoning plan for dynamically managing inference rules in the stream reasoning. Finally, our experiments evaluate on dataset built on the benchmark LUBM and DBpedia. The experimental results show that our framework is effective and efficient.

1 Introduction

Driven by the application of massive real-time RDF streaming data processing, obtaining more sound and complete answers for a given continuous query becomes an emerging research trend. This has many potential applications, e.g. monitoring and smart cities [11]. The RDF management system capable of supporting massive stream data, real-time processing and stream reasoning becomes more and more important and urgent. However, there are a few prototype implementations in stream reasoning (e.g. Strider [10,13]). Corresponding to the field of stream reasoning, there exist many state-of-the-art distributed static RDF reasoning engines based on RDFS/OWL rules such as RORS [9], Cichlid [4] and so on. They perform well on massive static RDF data inference. However, they are not directly applicable to RDF stream reasoning. How to apply these RDF reasoning engines to perform stream reasoning becomes an interesting problem.

© Springer International Publishing AG, part of Springer Nature 2018
Y. Cai et al. (Eds.): APWeb-WAIM 2018, LNCS 10987, pp. 440–448, 2018.
https://doi.org/10.1007/978-3-319-96890-2_36

A framework PRSP [7] is presented to process C-SPARQL queries on RDF streams by exploiting various SPARQL query engines in a unified way. Although PRSP can not be used for stream reasoning, it provides a new idea. In this paper, we propose an adaptive framework for RDF stream reasoning (PRSPR) in a plug-in method. This paper extends our previous proceedings [8]. PRSPR can adopt various inference engines for RDFS/OWL for massive RDF stream reasoning to get richer knowledge. Compared with the PRSP, we increase the ability to perform stream data inference and maintain the original high performance of inference engine and query engine. Compared with the static reasoning engines, we make it possible to reason the stream data.

For the purpose of guaranteeing the real-time demand of massive stream data reasoning through the use of static reasoning engine. We optimize the framework from the two perspectives of computing resource allocation and reasoning rules selection. Firstly, we adopt a multi-threaded batch processing approach to guarantee the full use of inference resources and query resources. When the inference process is over, we start a new thread to execute the SPARQL query. When the next window data arrives, we can immediately perform the inference process without waiting for the end of SPARQL query. Secondly, the number of rules has a huge impact on the reasoning time of rule-based reasoning engines. Therefore, we add a reason planning module in the reasoning stage. This module dynamically adjusts the number of inference rules used in the current window according to the reasoning situation of the previous windows, which makes our framework more flexible and efficient.

The main contributions of this paper are summarised as follows:

- We propose a framework for massive RDF stream reasoning, which could use state-of-the-art distributed reasoning engine to perform stream reasoning based on RDFS/OWL rules.
- We optimize our framework from the perspective of computational resource allocation and selection of inference rules.
- We implement the framework PRSPR by constructing six new modules, namely, *query preprocessing, data transformer, RDFS/OWL reasoner, reason planning, query execution, query planning.* Finally, we evaluate the overall performance of our stream reasoning framework in the LUBM [3] dataset and DBpedia[1] dataset. The experiments show that PRSPR enables the inference of massive streaming data in real time.

2   Background Knowledge

In this section, we introduce the RDF stream, C-SPARQL and RDFS/OWL Reasoning.

RDF Stream. RDF (Resource Description Framework) is the W3C-recommended data model for integrating and representing semantic information on the Web. An RDF stream S is defined as an ordered sequence of pairs

---

[1] http://wiki.dbpedia.org/develop/datasets.

Table 1. S_{Lubm}: an RDF stream of LUBM

Subject (sub)	Predicate (pre)	Object (obj)	Timestamp
UndergradStudent0	rdf:type	un:UndergraduateStudent	1483850233586
UndergraduateStudent0	un:name	UgStudent0	1483850233586
UndergradStudent0	un:memberOf	Depart0.University0.edu	1483850233586
UndergradStudent0	un:emailAddress	UgStudent0@D0.U0.edu	1483850233586
...	...	...	...

which are quadruples, and each pair is made of an RDF triple and a timestamp. An RDF stream S_{Lubm} generated from the LUBM Benchmark is shown in the Table 1.

C-SPARQL. As the firstly proposed and implemented RDF stream query language, C-SPARQL (Continuous SPARQL) [1] extends SPARQL by adding new operators, namely, *registration* and *windows*, to support processing RDF streams. Q_{LUBM} denote a C-SPARQL query *LubmQuery*.

REGISTER QUERY *LubmQuery* AS
SELECT ?X ?Y
FROM STREAM *LubmStream* [RANGE 105s STEP 105s]
WHERE { ?X rdf:type ?Y . }

RDFS/OWL Reasoning. RDF Schema (RDFS) [14] is an extension of RDF that allows the users to define the vocabulary used in RDF documents. OWL is more expressive than RDFS, and it is a merge of two unstandardized languages DAML and OIL [6]. By combining RDF data with RDFS/OWL rules for reasoning, we can get a lot of implicit knowledge. In this paper, we choose OWL-Horst [2] rule for OWL reasoning.

3   PRSPR: An Adaptive Framework for RDF Stream Reasoning

In this section, we introduce PRSPR, an adaptive framework for massive RDF stream reasoning. Finally, we use an instance to execute our processing flow.

The system framework is shown in Fig. 1, containing six main modules: query preprocessing, data transformer, RDFS/OWL reasoner, reason planning, query execution and query planning. Both as the input of PRSPR, C-SPARQL queries and RDF streams are transformed by query preprocessing module and data transformer module, respectively. Query preprocessing module processes C-SPARQL queries and outputs two types of queries, namely, window selectors and SPARQL queries. Data transformer module periodically transforms the RDF streams into RDF graphs and outputs the RDF graph to the RDFS/OWL reasoner module. RDFS/OWL reasoner module performs inference on the RDF graph and outputs the inferred RDF(S) graph to the query execution module. Query execution module performs query processing and outputs the final query results.

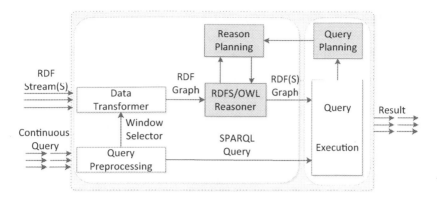

Fig. 1. The framework of PRSPR

Query Preprocessing. Query preprocessing module is responsible for the C-SPARQL query preprocessing, it receives the input continuous queries and generates two types of queries, namely, SPARQL query $S(Q)$ and window selector P, which can be addressed in the query execution module and data transformer module respectively. The window selector is a 4-tuple $P(Req, S, w, s)$.

Data Transformer. Data transformer module is responsible for the management of RDF streams. Via Esper or another Data Stream Management System (DSMS), the data transformer module transforms the RDF stream S declared in the C-SPARQL query to capture snapshots according to the window selector P obtained from query preprocessing module, then removes the timestamp of each quadruple in the window. So the data transformer module can periodically generate RDF graphs and output the RDF graphs to the RDFS/OWL reasoner module.

RDFS/OWL Reasoner. RDFS/OWL reasoner module calls the rules-based reasoning engine to reason RDF graphs based on the rule-based knowledge ontology. The reasoner module computes the deductive closure of an ontology by applying RDF/RDFS and OWL entailment rules in order to make implicit knowledge explicit.

Reason Planning. Reason planning module is responsible for dynamically adjusting the number of rules used in the RDFS/OWL reasoner module. It maintains main reasoning information of all previous windows such as the size and reasoning time of RDF graph. In addition, it can select the appropriate number of rules for the current window based on the inference information of the previous windows.

Query Execution. Query execution module which contains a series of SPARQL processing engine interfaces is responsible for static RDF graph data processing.

This module receives the RDF(S) graphs from the RDFS/OWL reasoner module and SPARQL query $S(Q)$ from the query preparation module, then it calls the state-of-the-art SPARQL processing engine(centralized and distributed) to execute the query and output the query results to the user.

Query Planning. Query planning module aims at dynamically adjusting the query time. Compared with reasoning time, the query time is relatively short. Considering very few cases that query time is greater than reasoning time, from the integrity point of view, we designed query planning module. Query planning module records the current window data query time, once the current window query time exceeds the window size of a given continuous query, it will feedback this information to reason planning. When this situation is likely to reappear, the query time is indirectly adjusted by the reasoning module reducing the number of rules. Although this situation has not happened in our experiments, the design of query planning module is necessary.

Considering the C-SPARQL query q_1 and RDF stream S_{LUBM} mentioned in Sect. 2. According to our framework, the query is processed as follows:

Firstly, query preprocessing module receives the C-SPARQL query and parses it into a 4-tuple P [Req, S, w, s] and a SPARQL query Q_{SPARQL}, where

- Req = REGISTER QUERY LubmQuery AS;
- S = LubmStream;
- w = $105s$;
- s = $105s$;
- Q_{SPARQL} is the SPARQL query shown as follow:

SELECT ?X ?Y
WHERE { ?X rdf:type ?Y. }

The P constitutes window selector parameters and inputs to the data transformer module. Via Esper, data transformer module begins to periodically convert stream data according to the window selector. Taking the first window as an example, the first window data is shown in Table 2.

The timestamp of each quadruple is removed into RDF graph data and fed into the reasoning module. By accessing the reason planning module, the RDFS/OWL inference module determines the OWL-Horst rules used in the inference stag and begins to perform inference. For the first window, we use full rules to reason. The RDF(S) graph data is shown in Table 3 after reasoning. The reasoning module sends the reasoned RDF(s) graph to the query execution module, and feeds back information such as the inference time and the size of the RDF graph to the reason planning module.

Finally, the query execution module calls the SPARQL engine such as RDF-3X to execute the query and feeds back information such as query time and size of RDF(S) graph to the query planning. The final results are shown in Table 4.

Table 2. The first window data

Subject (sub)	Predicate (pre)	Object (obj)	Timestamp
UndergradStudent0	rdf:type	un:UndergraduateStudent	1483850233586
...	...	...	...
UndergradStudent0	un:emailAddress	UgStudent0@D0.U0.edu	1483850233586

Table 3. The first window data after reasoning

Subject (sub)	Predicate (pre)	Object (obj)
UndergradStudent0	rdf:type	un:UndergraduateStudent
...	...	...
UndergradStudent0	un:emailAddress	UgStudent0@D0.U0.edu
UndergradStudent0	rdf:type	un:Student

Table 4. The final results

No	?X	?Y
1	UndergradStudent0	un:UndergraduateStudent
2	...	...

4 Experiments and Evaluations

4.1 Experimental Setup

Implementations and Running Environment. For the reasoning engine, we choose the SPARK-based reasoning engine RORS. For the query engine, we use RDF-3X [12] and TriAD [5] as the SPARQL query engines for subgraph matching. Centralized experiments are carried out on a machine running Ubuntu 14.04.5 LTS, which has 4 CPUs with 6 cores (E5-4607) and 64 GB memory. And a cluster of 4 machines (1 master and 3 workers) with the same performance as the former are used for distributed experiments. For the SPARK environment. We set up a cluster with 1 master and 3 worker nodes. The configuration of each node is the same as before. The version of Spark is 1.5.2.

Datasets and Continuous Queries. There are some well-established RSP benchmarks datasets such as Srbench [15] and so on, but they lack support for the task of stream reasoning. As a result, we are unable to use these datasets. For experimental evaluation, we choose the synthetic benchmark dataset LUBM and the real DBpedia dataset. In our framework, we choose five sizes of LUBM dataset (i.e., LUBM200, LUBM300, LUBM400, LUBM500 and LUBM1000) and three types of DBpedia dataset (i.e., mappingbased literals, mappingbased Objects and instance types) from DBpedia version 2016-10. The LUBM data sizes increase from 27.6 million to 138.3 million. The DBpedia's data size is

52 million. For the continuous queries, we choose the standard query $Q8$ as $q1$ provided by LUBM benchmark and the query $q2$ mentioned in the Sect. 2.

Experiment Metrics. We mainly measure the performance of our framework by comparing four metrics, i.e., data reasoning time (DRT), data load time (DLT), query response time (QRT) and query execution time (QET).

4.2   Performance Analysis

In order to verify the performance of our framework at different stream data throughput and queries of varying complexity, we select the query $q1$ and $q2$ and perform tumbling windows with a 470-s-window which slides every 470 s, and these two queries have different levels of complexity. We choose TriAD as the SPARQL engine. In addition, we select streaming data with different data input rates, from LUBM200 (i.e. minimum input rate) to LUBM1000 (i.e. maximum input rate). The experimental results are shown in Fig. 2. We compare the DRT, DLT, QRT and QET in adaptive inference and non-adaptive inference. It can be seen from Fig. 2(a)–(b), when the data input rate is increased to LUBM400, since the reasoning time exceed the window time, a large amount of data is lost, so the reasoning time and data load time of non-adaptive is greatly decreased. Adaptive reasoning avoids the massive data loss by using optimization strategies. From Fig. 2(b)–(d), we can see that our framework still maintains better query performance with different query complexity.

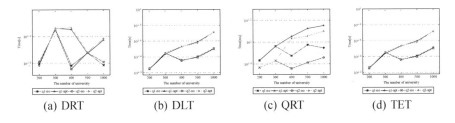

(a) DRT   (b) DLT   (c) QRT   (d) TET

Fig. 2. The processing time in different LUBM data size within PRSPR

To verify our framework still has good inference performance on different dataset, we choose DBpedia dataset at different data throughput rates from DB1 to DB4 and choose RDF-3x and TriAD as the SPARQL engin. In addition, we design the query q_2 and perform tumbling windows with a 560-s-window which slides every 560 s. The experimental results are shown in Fig. 3. After the data input rate reaches DB3, we can see that the adaptive reasoning strategy begins to show obvious advantages. Compared with the LUBM dataset, there is some decline in inference performance. It can be seen that the reasoning efficiency of our framework is sensitive to dataset. From the perspective of query processing, we can also see that the query performance of TriAD is better than that of RDF-3X.

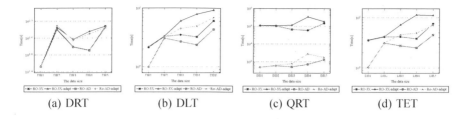

| (a) DRT | (b) DLT | (c) QRT | (d) TET |

Fig. 3. The processing time in different DBpedia data size within PRSPR

5   Conclusions and Future Work

In this paper, we present a framework PRSPR for RDF stream reasoning in an adaptive and simple way. Our proposal opens a novel way to support complicated reasoning tasks. Moreover, we believe that the findings about the relation between querying and reasoning are interesting to those researchers and engineers in desiging and optimizing reasoning tasks over RDF stream. As a future work, we further optimize PRSPR to support more complicated reasoning tasks such as complicated event reasoning.

Acknowledgments. We would like to thank Qiong Li for constructive comments. This work is supported by the National Natural Science Foundation of China (61373165, 61672377), the National Key R&D Program of China (2016YFB1000603, 2017YFC0908401), and the Key Technology R&D Program of Tianjin (16YFZCGX00210).

References

1. Barbieri, D.F., Braga, D., Ceri, S., Della Valle, E., Grossniklaus, M.: Querying RDF streams with C-SPARQL. SIGMOD REC. **39**(1), 20–26 (2010)
2. Liu, C., Qi, G., Wang, H., Yu, Y.: Large scale fuzzy pD^* reasoning using MapReduce. In: Aroyo, L., et al. (eds.) ISWC 2011. LNCS, vol. 7031, pp. 405–420. Springer, Heidelberg (2011). https://doi.org/10.1007/978-3-642-25073-6_26
3. Guo, Y., Pan, Z., Heflin, J.: LUBM: a benchmark for OWL knowledge base systems. J. Web Sem. **3**(2–3), 158–182 (2005)
4. Gu, R., Wang, S., Wang, F., Yuan, C., Huang, Y.: Cichlid: efficient large scale RDFS/OWL reasoning with Spark. In: Proceedings of IPDPS 2015, pp. 700–709 (2015)
5. Gurajada, S., Seufert, S., Miliaraki, I., Theobald, M.X.: TriAD: A distributed shared-nothing RDF engine based on asynchronous message passing. In: Proceedings of SIGMOD 2014, pp. 289–300 (2014)
6. Antoniou, G., van Harmelen, F.: A Semantic Web Primer. The MIT Press, Cambridge (2004)
7. Li, Q., Zhang, X., Feng, Z.: PRSP: a plugin-based framework for RDF stream processing. In: Proceedings of WWW 2017, pp. 815–816 (2017)
8. Li, Q., Zhang, X., Feng, Z., Xiao, G.: An adaptive framework for RDF stream reasoning. In: Proceedings of ISWC 2017 (2017)

9. Liu, Z., Feng, Z., Zhang, X., Wang, X., Rao, G.: RORS: enhanced rule-based OWL reasoning on spark. In: Li, F., Shim, K., Zheng, K., Liu, G. (eds.) APWeb 2016. LNCS, vol. 9932, pp. 444–448. Springer, Cham (2016). https://doi.org/10.1007/978-3-319-45817-5_43

10. Liu, C., Urbani, J., Qi, G.: Efficient RDF stream reasoning with graphics processingunits (GPUs). In: Proceedings of WWW 2014, pp. 343–344 (2014)

11. Margara, A., Cugola, G.: Processing flows of information: from data stream to complex event processing. In: Proceedings of DEBS 2011, pp. 359–360 (2011)

12. Neumann, T., Weikum, G.: The RDF-3X engine for scalable management of RDF data. VLDB J. **19**(1), 91–113 (2010)

13. Ren, X., Curé, O., Ke, L., Lhez, J., Belabbess, B., Randriamalala, T., Zheng, Y.: Strider: an adaptive, inference-enabled distributed RDF stream processing engine. PVLDB **10**(12), 1905–1908 (2017)

14. Urbani, J.: RDFS/OWL reasoning using the MapReduce framework. Master's thesis, Vrije Universiteit - Faculty of Sciences, Department of Computer Science (2009)

15. Zhang, Y., Duc, P.M., Corcho, O., Calbimonte, J.-P.: SRBench: a streaming RDF/SPARQL benchmark. In: Cudré-Mauroux, P., et al. (eds.) ISWC 2012. LNCS, vol. 7649, pp. 641–657. Springer, Heidelberg (2012). https://doi.org/10.1007/978-3-642-35176-1_40

Demo Papers

TSRS: Trip Service Recommended System Based on Summarized Co-location Patterns

Peizhong Yang, Tao Zhang, and Lizhen Wang[✉]

School of Information Science and Engineering,
Yunnan University, Kunming 650091, China
pzyang0924@163.com, taozhangcoder@163.com,
lzhwang@ynu.edu.cn

Abstract. Co-location patterns, whose instances are frequently located together, are particularly valuable for many applications. With co-location patterns, the location-based service recommendation can be made to give guidance to the user's trip. However, the number of co-location patterns is typically huge, thus it is restricted for practical applications. Based on summarized co-location patterns, we design a trip service recommended system, named TSRS. In TSRS, a large number of co-location patterns are compressed into a small quantity of summarized co-location patterns and their instances are stored into the retrieval tree for fast querying. Furthermore, TSRS provides the service point recommendation according to summarized co-location patterns, and route planning is given to help the user get to service points conveniently.

Keywords: Spatial data mining · Co-location pattern · Summarized pattern
Service recommendation

1 Introduction

As one of the spatial knowledge discovery technologies, the co-location pattern mining is intended to discover a subset of spatial features whose instances are frequently located together in geography. Spatial co-location patterns may yield important insights in various applications, such as public health, transportation, and various locations based services [1].

Co-location patterns reveal the association relationship among spatial features. Profited from the spatial dependencies, some location based service recommendations can be realized, such as commercial area recommendation in urban construction [3]. However, lots of co-location patterns are discovered in the approaches which use the participation index [2] to measure the prevalence, especially on the massive spatial data. A large number of co-location patterns make the user confused and it is difficult to find useful information from them. In order to provide less and more constructive co-location patterns for the user, some methods were proposed to compress co-location patterns [4–6], for example, the summarized co-location pattern.

We usually encounter trip planning issues, but it is hard when there is no available information to support. Based on the summarized co-location pattern [6], a kind of compressed pattern, we developed a trip service recommended system (TSRS) to provide trip guide to the user, such as service point recommendation, trip route planning. With the view of responding quickly to the user's querying, the retrieval tree is built for storing summarized co-location patterns and their instances.

2 System Overview

Figure 1 shows the description of TSRS. Firstly, TSRS discovers summarized co-location patterns from the user-specified spatial area under the user-specified parameters. Then, summarized co-location patterns and their instances are stored in the retrieval tree for the decision support. Lastly, TSRS provides suggestions for the service requirements of the user depending on the retrieval tree. TSRS contains five modules and each module is described as below.

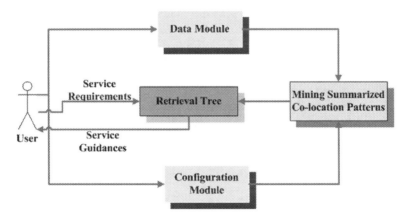

Fig. 1. Framework description

Data Module: The initial input of TSRS is the map information, and the user needs to specify the area of interest on the map. TSRS provides tools to help the user complete the designation of the spatial area.

Configuration Module: The configuration module allows the user to set parameters required for mining summarized co-location patterns, such as the distance threshold, the minimum prevalence threshold.

Mining Module: For the purpose of providing the user with a small number of available co-location patterns, TSRS discovers summarized co-location patterns which not only provide a satisfactory compression rate but also preserve reasonable prevalence information. A summarized co-location pattern c is the centralized representation of the co-location patterns covered by the pattern c.

Storage Module: The retrieval tree is built to storage summarized co-location patterns and their instances so that the user can query the required information quickly. Moreover, the covered relationship between co-location patterns can be obtained easily from the retrieval tree.

Service Module: TSRS offers some trip guide for the user depending on summarized co-location patterns. According to the user's activities plan (*a co-location pattern*), such as {*parking, shopping*}, TSRS recommends some service points (*co-location instances*) for the user. In addition, some other service points, which the user may have an interest in and frequently locate together with the planning activities but not in the plan, can be recommended also. Associated with the user's location information, TSRS plans the route, and provides driving or walking route to the user for arriving at service points conveniently.

3   Demonstration Scenarios

TSRS is encapsulated well with a friendly interface, and what the user faces is just a simple user interface. In this demonstration, the data from points of interests (POI data) in Kunming are used to show the demonstration of TSRS.

Fig. 2. Interface of TSRS

Figure 2 shows the user interface of TSRS. At first, the user specifies the interested spatial area on the map. Some tools are available to the user for choosing a different shape of the area, such as round, rectangular. Figure 2(a) displays a chosen circular area. In order to mine summarized co-location patterns, some parameters required to be set in Fig. 2(b), and the *Mine* button starts the mining task. The consequence is

Fig. 3. Service recommendation

presented in Fig. 2(c) and all the patterns are the summarized co-location pattern. Clicking on one, the retrieval tree which expresses the covered relationship among patterns will be exhibited in Fig. 2(d).

After summarized co-location patterns are discovered, TSRS can implement the service recommendation. Assuming such a scenario, the user goes out for dinner first, and then purchases cosmetics, and buys glasses at last. The user's activities can be abstracted into a co-location pattern {*restaurant, cosmetic store, optical store*}. For providing the user with recommended services, the user needs to locate his/her position (or starting position) on the map like Fig. 3(a). Then, the activities plan {*restaurant, cosmetic store, optical store*} is entered in the query box in Fig. 3(b) and clicking on the *Search* button to start the recommendation task. The recommended result that consists of some service points is displayed in Fig. 3(c) and multiple recommendations are delivered to the user. The user could choose one of them and decide to drive or walk. Route planning is shown on the map and the navigation information is illustrated in Fig. 3(d) to help the user go to the service points.

4 Conclusion

The co-location pattern reveals the spatial association relationship, and it can provide assistance to our life. The compressed co-location pattern is more valuable for practical applications. In this demonstration, we design a trip service recommended system based on summarized co-location patterns to give some guidance for the trip service requirement of the user. The demonstration scenarios indicate the feasibility of our system in trip service recommendation.

Acknowledgement. This work is supported by the National Natural Science Foundation of China (61472346, 61662086, 61762090), the Natural Science Foundation of Yunnan Province (2015FB114, 2016FA026), the Project of Innovative Research Team of Yunnan Province, and the Project of Yunnan University Graduate Student Scientific Research (YDY17110).

References

1. Shekhar, S., Huang, Y.: Discovering spatial co-location patterns: a summary of results. In: Jensen, C.S., Schneider, M., Seeger, B., Tsotras, V.J. (eds.) SSTD 2001. LNCS, vol. 2121, pp. 236–256. Springer, Heidelberg (2001). https://doi.org/10.1007/3-540-47724-1_13
2. Huang, Y., Shekhar, S., Xiong, H.: Discovering colocation patterns from spatial data sets: a general approach. IEEE Trans. Knowl. Data Eng. **16**(12), 1472–1485 (2004)
3. Wang, X., Chen, H., Xiao, Q.: MVUC: an interactive system for mining and visualizing urban co-locations. In: WAIM, pp. 524–526 (2016)
4. Wang, L., Bao, X., Zhou, L.: Redundancy reduction for prevalent co-location patterns. IEEE Trans. Knowl. Data Eng. **30**(1), 142–155 (2018)
5. Yoo, J.S., Bow, M.: Mining top-k closed co-location patterns. In: IEEE International Conference on Spatial Data Mining and Geographical Knowledge Services (ICSDM), pp. 100–105 (2011)

6. Liu, B., Chen, L., Liu, C., Zhang, C., Qiu, W.: RCP mining: towards the summarization of spatial co-location patterns. In: Claramunt, C., Schneider, M., Wong, R.C.-W., Xiong, L., Loh, W.-K., Shahabi, C., Li, K.-J. (eds.) SSTD 2015. LNCS, vol. 9239, pp. 451–469. Springer, Cham (2015). https://doi.org/10.1007/978-3-319-22363-6_24

DFCPM: A Dominant Feature Co-location Pattern Miner

Yuan Fang, Lizhen Wang$^{(\boxtimes)}$, Teng Hu, and Xiaoxuan Wang

School of Information Science and Engineering, Yunnan University,
Kunming 650091, China
{fangyuan,lzhwang}@ynu.edu.cn,
{hutengann,wangxiaoxuan1037}@163.com

Abstract. Co-location pattern mining is an important task in spatial data mining. However, the availability of the discovered co-location patterns is limited due to lack of specific target. Unlike existing works, we consider the dominant relation as a specific target in co-location pattern mining process. This demonstration presents DFCPM (Dominant Feature Co-location Pattern Miner), a system for users who not only take an interest in the prevalence of a feature set, but also concern which features play the dominant role in a pattern. Given a set of POIs (Point of Interest) data, we evaluate and identify the co-location patterns which are prevalent and contain dominant features. Also, DFCPM extracts the dominant features from each DFCP (Dominant Feature Co-location Pattern) to provide more information and help the decision making.

Keywords: Spatial co-location pattern · Dominant feature · POI data

1 Introduction

Co-location pattern mining [1] discovers the subsets of spatial features whose instances are located together frequently in geography. For example, {Hospital, Pharmacy, Florist} is a co-location pattern means that their instances always appear together in the same places. Spatial co-location pattern mining has been a problem of great practical importance due to its broad applications at environmental protection, public transportation, location-based service and urban public-service, etc. Finding available and interesting patterns for users with specific needs is a tough task due to large collections of results which make people hardly understand and identify the targeted ones. Thus, many researchers did a lot of works to improve the availability of result co-location patterns such as co-location pattern concise representations [2], redundancy reduction [3], co-location pattern mining based on domain knowledge [4], etc. In some applications (e.g. Urban Planning, Commercial Site Selection), users are not only interested in the prevalence of a feature set, but also concern which features play the dominant role in a pattern. Dominant-Feature Co-location Pattern (DFCP) is a subset of spatial features that their instances are located together frequently and exists a dominant relationship. For example, for prevalent co-location pattern {Hospital, Pharmacy, Florist}, there are many instances of "Florist" or "Pharmacy" close to "Hospital" individually, but there is no additional neighborhood relationship between "Florist" and

Y. Cai et al. (Eds.): APWeb-WAIM 2018, LNCS 10987, pp. 456–460, 2018.
https://doi.org/10.1007/978-3-319-96890-2_38

"Pharmacy" without "Hospital". However, the prevalence metric is failed to reveal such dominant relationship between "Hospital" and other features in the pattern and to extract the dominant feature like "Hospital". Thus, [5] gave a framework to mine the DFCP and extract the dominant features. On the one hand, a DFCP contains more information to support specific decision making. Thus, it guarantees that the recommendation has practical significance. On the other hand, identifying the DFCP can help reduce the number of prevalent patterns. Thus, it improves the availability of patterns. Finding co-location pattern with a dominant relationship is significant and practical in some applications such as urban planning and commercial site selection.

In this paper, we develop a DFCPM (Dominant-Feature Co-location Pattern Miner) system for users, which take dominant relationship among features into account. Given a set of spatial data (e.g., urban POI data), we aim to find DFCPs with dominant features. We firstly discover prevalent co-location patterns, then identify DFCPs from prevalent patterns, further, extract corresponding dominant features of each DFCP. At last, the system will provide a visual analysis.

2 System Overview

As Fig. 1 illustrates, our system contains three major parts: (1) Data acquisition and pre-processing, (2) DFCP mining process and (3) dominant feature extracting process. DFCPM takes a set of spatial point with location (e.g., urban POI data) as initial input.

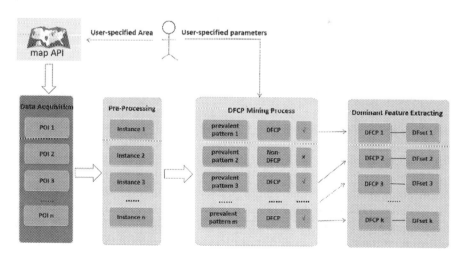

Fig. 1. The process of finding dominant features

Urban facility data mostly exist in point form. In this demonstration, we collect POI (Point of Interest) data which in user-specified area from the public map application API (e.g., Google API or Amap API) as initial data. In data pre-processing, we firstly extract the type information and location information of each POI, then we translate the initial POI data into general input format of co-location mining as {Feature Type,

Instance ID, Location} to instantiate each POI as an input of DFCP mining. In DFCP mining process, the user-specified parameters include distance threshold, prevalence threshold and disparity threshold. We firstly take all instances as input. Next, we build neighborhood relationship between instances by the distance threshold. Then, we mine the prevalent co-location patterns by prevalence metrics (i.e. Participation Index) and determine whether there exists a dominant relationship among features by a new measure, namely disparity, to identify DFCP. In dominant feature extracting process, if a pattern is a DFCP, we calculate the disparity between a feature and all the rest of features in a pattern then extract the dominant feature as a set for each DFCP. The DFCPs with dominant feature set is an output as mining results.

3   Demonstration Scenarios

DFCPM is well encapsulated with a friendly interface. In this demonstration, we use POI data from public Amap API in Beijing to show the demonstration of DFCPM.

Figure 2 shows a map interface which applies Amap API. Figure 2(a) is a map GUI allows users to choose several points on the map to delimit an area, Fig. 2(b) shows the location of each point which was selected in map GUI by user. Figure 2(c) shows a POI selection button. The function of the POI selection button is to delimit the area based on the selected points. Figure 2(d) shows a clear button, it allows the user to clear all selected points on the map. Figure 2(e) shows the number of POIs in the selected area. All POIs in this area are stored in a text file. The text file can be the input of DFCP mining process. Figure 3(a) displays the interface of DFCPM in the DFCP mining process. The Input data are the points of interest (POI) in Beijing which consists of 26,546 spatial instances and 16 spatial feature types. The spatial distance threshold is 50 by default (meaning 50 m in the real world) and we set the prevalence threshold as $min\_prev = 0.3$ and the disparity threshold as $min\_fd = 0.2$. Given the input file which obtained from user-specified area on the map and the path of output text file, the DFCP mining process can be performed once pressing the *running* button. Figure 3(b) shows the mining result interface after the data acquisition and pre-processing, Non-DFCP only provided with Participation Index (i.e., prevalence metrics), DFCP is provided with Participation Index, Disparity Index (i.e., disparity metrics) and corresponding dominant features. Figure 3(c) shows a pie chart of the mining results. We can notice that after the DFCP mining process, the number of prevalent co-location patterns is 63 and the number of DFCP is 18. The DFCP mining results based on POI data shows that the DFCPs can offer targeted and abundant information. For example, {Chinese Food*, Parking*, Clothing Shop} is a DFCP, "Chinese Food" and "Parking" dominate "Clothing Shop". This DFCP infers that "It is a good idea to open a clothing store nearby parking lots and Chinese restaurants". Therefore, the DFCP can better explain the correlation of co-location patterns and further apply in some significant applications.

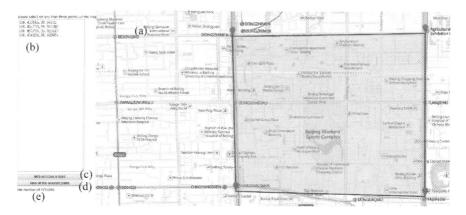

Fig. 2. A map interface of DFCPM

Fig. 3. Interface of DFCPM in the processing and results

4   Conclusions

In this demonstration, we designed a system to discover DFCPs to reveal the dominant relation between features of a pattern and reduce the number of result prevalent co-location patterns. The demonstration scenarios showed the effectiveness of our system. The DFCPM based on POI data presents the significance in practice and can be further applied in some applications such as urban planning, commercial location site recommendation.

Acknowledgements. This work is supported by the National Natural Science Foundation of China (61472346, 61662086, 61762090), the Natural Science Foundation of Yunnan Province (2015FB114, 2016FA026), and the Project of Innovative Research Team of Yunnan Province and the Project of Yunnan University Graduate Student Scientific Research (YDY17110).

References

1. Huang, Y., Shekhar, S., Xiong, H.: Discovering colocation patterns from spatial data sets: a general approach. IEEE Trans. Knowl. Data Eng. (TKDE 2004) **16**(12), 1472–1485 (2004)
2. Wang, L., Zhou, L., Lu, J., Yip, J.: An order-clique based approach for mining maximal co-locations. Inf. Sci. **179**(19), 3370–3382 (2009)
3. Wang, L., Bao, X., Zhou, L.: Redundancy reduction for prevalent co-location patterns. IEEE Trans. Knowl. Data Eng. **30**(1), 142–155 (2018)
4. Flouvat, F., Soc, J., Desmier, E.: Domain-driven co-location mining. GeoInformatica **19**(1), 147–183 (2015)
5. Fang, Y., Wang, L., Wang, X., Zhou, L.: Mining co-location patterns with dominant features. In: Bouguettaya, A., et al. (eds.) WISE 2017. LNCS, vol. 10569, pp. 183–198. Springer, Cham (2017). https://doi.org/10.1007/978-3-319-68783-4_13

CUTE: Querying Knowledge Graphs by Tabular Examples

Zichen Wang[1], Tian Li[1], Yingxia Shao[1(✉)], and Bin Cui[1,2]

[1] School of EECS, Key Lab of HCST (MOE), Peking University, Beijing, China
{wang.zichen,tian.li,shao.yingxia}@pku.edu.cn
[2] ECE, Shenzhen Graduate School, Peking University, Shenzhen, China
bin.cui@pku.edu.cn

Abstract. Knowledge graphs and the query language SPARQL have opened up the possibility of retrieving information, acquiring knowledge and building applications over large linked data. However, due to the unfamiliarity with both SPARQL and the datasets, users always struggle to write well-expressed queries. To increase the usability of knowledge graphs, we develop a query-by example system CUTE, which supports complex query intent. CUTE takes tabular examples as input, and returns high-quality results via continuous user interaction.

1 Introduction

Knowledge graphs are one of the fundamental data sources for AI applications. They are represented as RDF and queried by SPARQL. However, it is not easy to write a well-expressed query. The user needs to not only know the details of RDF datasets, but also master the complex SPARQL syntax. This has become one of the chief obstacles to fully realizing the potential of knowledge graphs.

So far many efforts have been devoted to improving the usability of knowledge graphs. One paradigm is to automatically construct SPARQL queries based on user-provided examples [3]. But previous work mainly focuses on simple inputs, such as a single entity or a pair of entities [1,2]. With the advent of domain-specific AI applications, including those in finance, public security and education, simply querying by one or two entities may not be satisfactory anymore.

In this demo, we develop an interactive system CUTE to easily query RDF in complex scenarios by *tabular* examples. A tabular example consists of a set of tuples. A single tuple is one of the evaluation results of an implicit SPARQL query. The whole tabular is a partial view of the complete results. CUTE first maps the inputs to entities in the knowledge graph, then automatically constructs a SPARQL query by analyzing the relations among the entities and finally returns all related results with the query. In addition, users can interactively label the returned results as negative examples, and CUTE will improve the quality of results iteratively. The formal definition of the problem is given as below.

T. Li—Equal contribution with the first author.

© Springer International Publishing AG, part of Springer Nature 2018
Y. Cai et al. (Eds.): APWeb-WAIM 2018, LNCS 10987, pp. 461–465, 2018.
https://doi.org/10.1007/978-3-319-96890-2_39

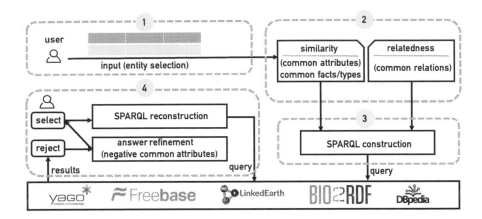

Fig. 1. The high-level execution flow of CUTE

Problem Definition. **Input:** A tabular example $E_{m \times n}$ has m rows and n columns. The ith row $[e_{i1}, \ldots, e_{in}]$ of $E_{m \times n}$ is an instance of the results of an implicit SPARQL query over a knowledge graph G. Further, we assume that entities in a column have the same types, and different rows have common implicit relations. **Output:** A SPARQL query Q and the results R after the evaluation of Q, s.t. $E_{m \times n} \subseteq R$.

2 System Overview

Figure 1 shows the high-level execution flow of CUTE. It is built on public SPARQL endpoints[1,2] with four main components.

(1) *Entity Recommendation.* Considering that CUTE receives ad-hoc input entities with non-standard representations, it generates a list of top-k candidates for each entity with a string-based similarity measure [5]. The user can pick up the desired ones from the candidates when they appear in the recommendations. Therefore, CUTE identifies the accurate entities in the knowledge graph.

(2) *Common Attributes and Relations Discovering.* This is a critical component in CUTE, as it discovers common attributes and common relations among input entities.

Discovering common attributes is to detect ***similarities*** between entities in the same column. This process consists of two parts: (a) inferring *common types* of entities based on the ontology of the dataset, and (b) discovering *common facts* of entities considering the following two cases:

---

[1] https://linkeddata1.calcul.u-psud.fr/sparql.
[2] https://dbpedia.org/sparql.

- Given a column j ($j = 1, 2, ..., n$), from triples $\{\langle e_{ij}, ?p, ?o\rangle\}$, it detects the same '$?p, ?o$' shared by $\{e_{ij}\}, i = 1, 2, ..., m$.
- Given a column j ($j = 1, 2, ..., n$), from triples $\{\langle ?s, ?p, e_{ij}\rangle\}$, it detects the same '$?s, ?p$' shared by $\{e_{ij}\}, i = 1, 2, ..., m$.

With all discovered common attributes of entities in the same column, CUTE ranks those attributes [4] and lets users select from the top-k candidates.

Discovering common relations is to analyze **relatedness** among entities in the same row. First, for each row, CUTE constructs a subgraph of G containing all the entities in that row. Specifically, for every two entities, it finds a shortest path between them. If there are multiple shortest paths, it picks the one with the maximal *Predicate Frequency Inverse Triple Frequency* [4]. After that it merges all these paths to form a pattern for this row. Next with m patterns, CUTE computes the maximal common substructure of them, which can be regarded as the relatedness between input entities.

(3) *SPARQL Construction.* Given triples representing the **similarities** and **relatedness** of entities, CUTE constructs a SPARQL query by directly combining them together and replacing entities in the examples with variables.

(4) *Answer Refinement and SPARQL Re-construction.* CUTE also provides an interface for the user to label those results against her intention, which are used as 'negative examples' to refine the previous answers. With the negative examples, CUTE reconstructs the SPARQL query by using the *FILTER NOT EXISTS* and *FILTER* expressions to represent the common attributes that only belong to the negative examples. Then it executes the new query after adding the new constraints to the previous SPARQL and generates refined results. This process will not finish until the user is satisfied.

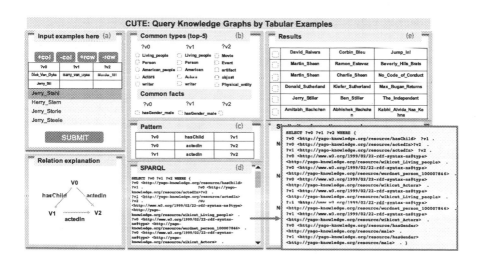

Fig. 2. The main interface of CUTE

3   Demonstration Scenarios

In this demo, the audiences can experience the following three scenarios.

SPARQL Generation. In this scenario, users can check the SPARQL queries generated by CUTE. Assume a user is interested in *"Which actors of two generations have acted in the same piece of art work?"*. In Fig. 2, (a) she provides CUTE with two tuples and selects the exact name from the recommendations; (b–c) CUTE discovers *hasChild* and *actedIn* relations among entities, along with the common attributes; (d–e) CUTE constructs the query and returns related results.

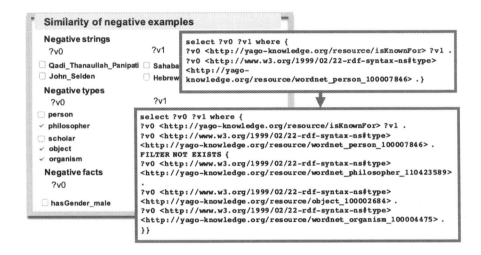

Fig. 3. The panel for answer refinement and SPARQL re-construction.

Answer Refinement. Users can improve the quality of results by interactively refining the answers. Assume a user wants to know about scientists and their inventions. First, she feeds an example *'Alvin_Hansen, IS_LM_model'* into CUTE and obtains all related results. Then she can reject undesirable answers (e.g., philosophers and their writings) by labelling them as negative records, and CUTE automatically displays all common attributes of those negative examples (Fig. 3). Next she selects *philosopher*, *object* and *organism* as negative types to generate a new query with *'FILTER NOT EXIST'* patterns. Finally, the answers are refined by the new query.

Various Tabular Examples. In this scenario, we show that CUTE can handle almost all kinds of tabular examples effectively. **Input 1: a single entity.** Users can query CUTE with an example containing a single entity. For instance, a user retrieves "European capital cities" by providing an example *'Moscow'*. **Input 2: multiple entities.** Users can query CUTE with an example containing multiple entities. E.g., a user queries the "US politicians and their graduate schools" with an example *'Bush, Yale_University'*. **Input 3: a full tabular example.** Users can query CUTE with multiple tuples containing multiple entities. After inputting two examples *'Alvin_Hansen, IS_LM_model'* and *'Adam_Smith, Free_market'*, CUTE returns the results about the "scientists and their inventions".

4   Conclusion

We have demonstrated CUTE, an example-based querying system for knowledge graphs. CUTE takes tabular examples as input, and refines the results iteratively with users' feedback. Through this demo, we show that CUTE can handle complex queries on knowledge graphs flexibly with a little human effort in the loop.

Acknowledgements. This research is funded by China Postdoctoral Science Foundation (No. 2017M610020), National Natural Science Foundation of China (No. 61702015), Shenzhen Gov Research Project (No. CYJ20151014093505032).

References

1. Diaz, G., Arenas, M., Benedikt, M.: SPARQLByE: querying RDF data by example. PVLDB **9**(13), 1533–1536 (2016)
2. Fionda, V., Pirrò, G.: Explaining and querying knowledge graphs by relatedness. PVLDB **10**(12), 1913–1916 (2017)
3. Jayaram, N., Khan, A., Li, C., Yan, X., Elmasri, R.: Querying knowledge graphs by example entity tuples. TKDE **27**(10), 2797–2811 (2015)
4. Pirró, G.: Reword: semantic relatedness in the web of data. In: AAAI, pp. 129–135 (2012)
5. Winkler, W.E.: String comparator metrics and enhanced decision rules in the Fellegi-Sunter model of record linkage (1990)

ALTAS: An Intelligent Text Analysis System Based on Knowledge Graphs

Xiaoli Wang, Chuchu Gao, Jiangjiang Cao, Kunhui Lin$^{(\boxtimes)}$,
Wenyuan Du, and Zixiang Yang

Software School of Xiamen University,
Xiamen 361005, People's Republic of China
{xlwang,khlin}@xmu.edu.cn, 596626434@qq.com,
892368206@qq.com, 849358686@qq.com, 541892963@qq.com

Abstract. This paper presents an intelligent text analysis system, called ALTAS, to support various text analysis tasks such as statistics analysis, sentiment analysis, text classification, and text clustering. The system contains four main components: knowledge graphs, text processing, text analysis and intelligent report. First, the system has built a semantic-rich knowledge base using several knowledge graph resources. A novel text processing and analysis framework based on knowledge graphs is developed and implemented. Given a text dataset, the text processing phase will do data cleaning, word segmentation and feature extraction for it. With the extracted features, the text analysis phase allows users to select a text mining task. We have implemented the proposed novel algorithm and several typical algorithms for each task. If users select multiple algorithms for the task, the intelligent report phase will automatically generate comparison results for users. Especially, the intelligent report phase also provides users a paper summary generating function on text mining problems.

Keywords: Text analysis · Knowledge graphs · Text mining
Intelligent report

1 Introduction

In the information age, the number of text data doubled rapidly. How to analyze such massive text data has taken much attention from the research community. Existing researchers have focused on proposing a more effective algorithm for a text mining task (e.g., [1, 2, 3]). To our knowledge, none of typical text mining algorithms could outperform all the others. In this paper, we focus on developing a novel text processing and analysis framework based on knowledge graphs. We aim to improve existing text mining algorithms using the semantic-rich knowledge graphs, and provide researchers a powerful tool for doing comparison experiments and automatically generating experiment report or paper summary.

The system has implemented four main components. First, a semantic-rich knowledge base is built using various resources (e.g., [4]). Based on the knowledge graphs, we then develop a novel framework to do text processing and analysis. In the data processing component, the system will do data cleaning, word segmentation and

Y. Cai et al. (Eds.): APWeb-WAIM 2018, LNCS 10987, pp. 466–470, 2018.
https://doi.org/10.1007/978-3-319-96890-2_40

feature extraction for the text data. Then in the text analysis component, users can further select text mining algorithms to do the comparison experiments. The system will finally automatically generate experimental results or paper summary for users. Our main contribution can be summarized as follows.

- We develop novel text processing and analysis algorithms based on knowledge graphs to enhance the performance of typical text mining algorithms.
- We develop a powerful intelligent report tool for researchers on text mining to conveniently generate an experiment report or paper summary by several easy clicks on the system.

2   System Architecture and Demonstration

Figure 1 shows our system framework with four main components. It is implemented as a J2EE project and deployed on the Apache Tomcat. We omit the details of the proposed algorithms based on knowledge graphs, as we have submitted several papers on the detailed techniques, and the technical reports will be published online later.

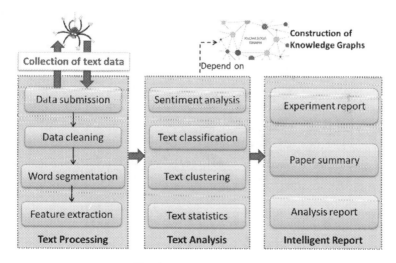

Fig. 1. System architecture

2.1   Construction of Knowledge Graphs

To build a semantic-rich knowledge base, we use multiple high-quality resources such as Wikipedia, Baidupeida, Zhihu and Q&A sites. For Wikipedia, we directly use the knowledge graphs. For other resources, we extract the related entities and the relationships between entities using the methods proposed in our previous work [5].

2.2 Text Processing

We have crawled several text datasets of high-quality contents from web sites like Xmfish [6], as shown in Fig. 2(a). Users can select these datasets to do the text analysis tasks. The system also allows users to submit their own text datasets in terms of a type-in text paragraph, a text file or a zipped package in Fig. 2(b).

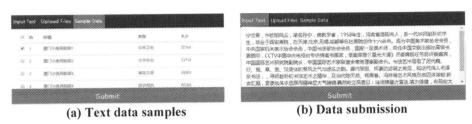

| (a) Text data samples | (b) Data submission |

Fig. 2. Data submission

The selected dataset from users is submitted as input of the text processing phase, and the system will continue to do data cleaning, word segmentation and feature extraction. Here, the system allow users to select various word segmentation tools, such as SmartAnalyzer, IKAnalyzer, Ansj, HanLP, and the proposed algorithm based on knowledge graphs. For each step, the system will show the results using a visualization tool, and users also can select to download the results as a file or a zip package as shown in Fig. 3(a) and (b).

| (b) Word segmentation | (b) Feature extraction |

Fig. 3. Text processing

2.3 Text Analysis

The text processing results are directly used as the input for the text analysis phase. The system contains four types of text analysis tasks: statistics analysis, sentiment analysis, text classification and text clustering. For statistics analysis, various results are visualized and discussed in the analysis report. For sentiment analysis and text classification, we implement the proposed algorithm based on knowledge graphs and typical algorithms such as KNN, SVM, linear regression, Naive Bayes and deep learning for comparison experiments. For text clustering, we implement typical k-means algorithm and an improved k-means algorithm based on knowledge graphs. The snapshots of four tasks are shown in Fig. 4(a)–(d).

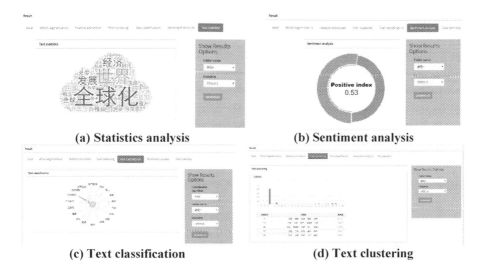

(a) Statistics analysis   **(b) Sentiment analysis**

(c) Text classification   **(d) Text clustering**

Fig. 4. Text analysis

2.4   Intelligent Report

The system allow users to view and download the automatically generated experiment report and paper summary, as shown in Fig. 5(a) and (b).

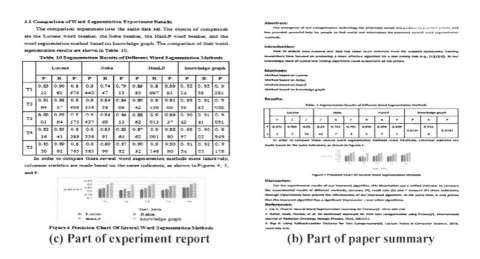

(c) Part of experiment report   **(b) Part of paper summary**

Fig. 5. Intelligent report

Acknowledgment. This work is supported by the National Natural Science Foundation of China under Grant No. 61702432, and the International Cooperation Projects of Fujian Province in China under Grant No. 2018I0016.

References

1. Yu, L., Zheng, J., Shen, W.C., et al.: BC-PDM: data mining, social network analysis and text mining system based on cloud computing. In: ACM SIGKDD, pp. 1496–1499 (2012)
2. Bansal, N., Koudas, N.: BlogScope: a system for online analysis of high volume text streams. In: VLDB, pp. 1410–1413 (2007)
3. Alpaydin, E.: Introduction to Machine Learning. MIT Press, Cambridge (2010)
4. Quora. https://www.quora.com/sitemap/topics
5. Lin, K., Wu, M., Wang, X., Yang, P.: MEDLedge: a Q&A based system for constructing medical knowledge base. In: ICCSE, pp. 485–489 (2016)
6. Xmfish. http://www.xmfish.com/

SPARQLVis: An Interactive Visualization Tool for Knowledge Graphs

Chaozhou Yang[1,2], Xin Wang[1,2(✉)], Qiang Xu[1,2], and Weixi Li[1,2]

[1] School of Computer Science and Technology, Tianjin University, Tianjin, China
{yangchaozhou,wangx,xuqiang3,bbFUC}@tju.edu.cn
[2] Tianjin Key Laboratory of Cognitive Computing and Application, Tianjin, China

Abstract. With the rapid development of knowledge graphs, an increasingly large volume of RDF graph data has been published on the Web. However, those data cannot be well consumed because most users have no background knowledge about SPARQL, a structural query language for RDF database. To increase the availability of knowledge graphs, we developed an interactive visualization tool for multiple SPARQL endpoints. We as well introduced advanced features like pattern matching and property path query to enhance its functionality.

Keywords: Knowledge graph · Interactive tool · Visualization

1 Introduction

Resource Description Framework (RDF), the de facto standard of constructing knowledge graph, has developed and accumulated a large amount of knowledge graph like DBpedia [1], YAGO [2] and WikiData [3]. The introduction of SPARQL comes up with a method to query knowledge graph, yet it as well causes another problem: users have to study the SRARQL grammar to query what they want, which reduces the availability of knowledge graphs for most users. Though there are some visualization tools, they only apply to the data of specific domains that they are not compatible with every knowledge graphs. In some cases, users would like to query certain entities and relations that meet given conditions, which need to use SPARQL features like pattern matching query and property path query [4]. However, queries based on them are not supported by current tools, which prevents users from acquiring specific data. Pattern matching query is composed of several conditional clauses. For example, if we want to know all the physicists born in German Empire, we are supposed to write two SPARQL `WHERE` clauses and full URIs or abbreviation with prefixes. Both of them are difficult for ordinary users to remember so that people tend to use keywords instead of RDF triples. Property path query is another query pattern to retrieve data with complex relationship, but it is based on regular expressions which is not an intuitive method for interaction. Moreover, there are a huge number of relations and entities in several RDF database such as

DBpedia; however, it is not necessary to show all of them in the query result so that a method to filter trivial data should be implemented.

To bridge the gap between end users and knowledge graphs, we developed an interactive visualization tools, called SPARQLVis, which enables people without background knowledge about SPARQL to query visualized information in multiple languages from several SPARQL endpoints.

2   Demonstration

This tool supports multilingual versions of DBpedia, WikiData, and arbitrary SPARQL endpoints. We demonstrate the tool by the following use cases based on the DBpedia dataset in English by changing the endpoint and language setting as shown in Fig. 1.

Fig. 1. Endpoint and language setting

Use Case 1: Searching One Entity and Exploring It. This tools can be used as Wikipedia that users is able to search an entity to get its relations with others and a brief introduction about it. Type some of the letters of the entity's name in the search box, choose one item in combo box, and click search button. For example, we search Socrates and get all related entities as shown in Fig. 2a. All the nodes are linked by directed edges, which denote the predicates. While hovering over one node, Aristotle, the abstract and thumbnail (if available) of him will be shown in the right box. In Fig. 2b, we can get the related entities and relations of Aristotle by clicking the node.

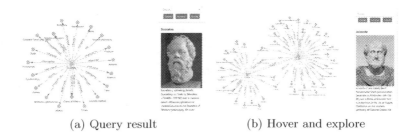

(a) Query result   (b) Hover and explore

Fig. 2. Searching one entity and exploring it

Use Case 2: Query with Pattern Matching. This function of our tool provides a method to help users complete conditional clauses to query all qualified entities and relations. For example, if we want to know the physicists born in German Empire, then the conditional clauses should be whose field is physics and whose birth place is German Empire. Thus, click the Advance Button to switch to Advanced Search Mode. Input *physics* in the Entity text box as the object of the first clause, check Object radio. Then choose *field* as the relation and add the condition clause. Repeat these operations to add the second clause as shown in Fig. 3a. Click Search button to get the query result as shown in Fig. 3b. Users can also change the view of results from graph to table if they do not want to see a complex graph with too many nodes.

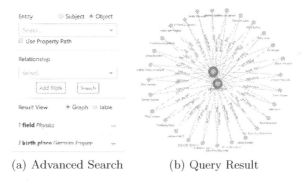

(a) Advanced Search   (b) Query Result

Fig. 3. Query with pattern matching

Use Case 3: Property Path Query. In some cases, a conditional clause cannot be expressed by a triple with a simple predicate. Instead, property path query is able to describe complex relations based on regular expressions. First, switch to Advanced Search Mode and check the Use Property Path option. Then input the name of the entity related to what we would like to get. To complete a syntax tree, we are supposed to choose an operator and add it. Then we have a tree with one leaf which indicates an unitary operator, or two leaves which indicates a binary operator. In regular expression syntax tree, each node is a literal i.e., predicate, or an operator so that we can change a leaf to an operator or a predicate by clicking Add button or Set button, respectively. When the syntax tree is complete, click Add button to add this triple with property path into conditional clause list and click Search button to get the query result. With property path query, users is able to concatenate and combine multiple predicates with several operators. For example, when we want to know who was the father or predecessor of Elizabeth II, we need to add an *at least one* node and an *or* node, which denotes + and | operator in regular expressions, respectively, and set two predicates as shown in Fig. 4a. Then we get the result as shown in Fig. 4b.

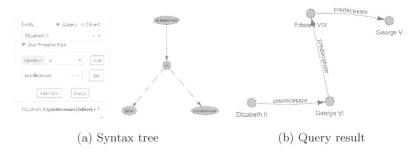

(a) Syntax tree   (b) Query result

Fig. 4. Property path query

3   Conclusion

We developed an interactive visualization tool to help users retrieve information from knowledge graphs and explore the whole knowledge graph even if they do not know about SPARQL. It also supports pattern matching query and property path query to help users acquire specific entities and relations. Another feature of this tool is compatibility with different SPARQL endpoints and multiple language.

Acknowledgments. This work is supported by the National Natural Science Foundation of China (61572353), the Natural Science Foundation of Tianjin (17JCY-BJC15400), and the National Training Programs of Innovation and Entrepreneurship for Undergraduates (201710056091).

References

1. Lehmann, J., Isele, R., Jakob, M., Jentzsch, A., Kontokostas, D., Mendes, P.N., Hellmann, S., Morsey, M., van Kleef, P., Auer, S., Bizer, C.: DBpedia - a large-scale, multilingual knowledge base extracted from Wikipedia. Semant. Web J. **6**(2), 167–195 (2015)
2. Mahdisoltani, F., Biega, J., Suchanek, F.: YAGO3: a knowledge base from multilingual Wikipedias. In: 7th Biennial Conference on Innovative Data Systems Research CIDR 2015, Asilomar, CA (2014)
3. Vrandečić, D., Krötzsch, M.: Wikidata: a free collaborative knowledgebase. Commun. ACM **57**(10), 78–85 (2014)
4. Wang, X., Wang, J.: ProvRPQ: an interactive tool for provenance-aware regular path queries on RDF graphs. In: Cheema, M.A., Zhang, W., Chang, L. (eds.) ADC 2016. LNCS, vol. 9877, pp. 480–484. Springer, Cham (2016). https://doi.org/10.1007/978-3-319-46922-5_44

PBR: A Personalized Book Resource Recommendation System

Yajie Zhu, Feng Xiong, Qing Xie$^{(\boxtimes)}$, Lin Li, and Yongjian Liu

School of Computer Sceince and Technology, Wuhan University of Technology,
Wuhan 430070, China
{zyjwhcn,xfeng,felixxq,cathylilin}@whut.edu.cn, liuyj626@163.com

Abstract. Recommendation system is widely applied for online resource retrieval, especially in digital publishing industry. A good recommendation system can help the users to efficiently find the desirable reading materials from the massive online resources. However, the conventional recommendation systems are always facing the cold-start problem, and it is difficult to provide the personalized service in an efficient way, since the users' preference may change sometimes. To address the problems above, this work introduces a personalized book resource recommendation system, which well utilizes the tag information of book resources to interact with the users. The user feedback will deliver their real-time preference, and the system can provide more precise recommendation candidates to improve the service quality. In this demo, we will introduce the overall framework and some important modules of the recommendation system, with relevant technical details. We will show the system functions by providing the visual results of the actual book resource recommendation.

Keywords: Recommendation system · Interactive recommendation
Tag association analysis · Cold start

1 Introduction

Digital publishing is a fast growing industry nowadays. Providing the various digital book resources available, and the large number of online customers, it is important to meet the customers' reading need from the massive book resources, and deliver the personalized knowledge service. Therefore, it is necessary to design a good recommendation system for users to retrieve their desired reading materials, so as to improve the users' reading experience and guarantee the system magnetism to the users. Generally speaking, the recommendation system needs to provide precise and personalized service. However, it is difficult to provide knowledge services to users when they are new to the system, because

This work is partially supported by Natural Science Foundation of China (Grant No. 61602353), National Social Science Foundation of China (Grant No. 15BGL048), and Natural Science Foundation of Hubei Province (Grant No. 2017CFB505).

the system has little knowledge about the characteristics of the new users and cannot recognize their reading preference, i.e., the cold start problem [2]. In addition, the accuracy of the traditional recommendation algorithm will be reduced if the users changed their interest greatly, so the system needs to keep high adaptiveness and adjusts the recommendation dynamically.

Considering the problems above, we propose to design an interactive book recommendation system named PBR, which can deal with cold start issue and meet the user's real-time reading demand. In this work, the most significant feature of the recommendation system is the tag-based user interaction, through which we are able to explore the user's preference by his feedback in keyword selection, and adjust the recommendation candidates dynamically.

There are some challenges in this recommendation system to achieve the expectant performance. The first issue is how to recommend book resources to the new users. We propose to primely analyze the new user by his basic information, and the recommendation candidates can be initialized following a decision tree based approach. When the new user accesses the system, his basic information will match a path in the decision tree and the initial recommendation candidates can be generated. The second issue is how to guarantee the efficiency of the system. We propose to optimize the tag interaction by entropy analysis. The tag set of book resources is analyzed by exploring the probability of tag occurrence. When the system interacts with users by tag information, the recommendation candidates will be dynamically reduced, and the recommendation list can be updated accordingly.

2   System Description

The PBR system is designed to extend an existing digital publishing platform RAYS[1], by which all the book, book tag and user information can be collected. Figure 1 shows the general framework of PBR system, which is divided into three parts. The first part is the analysis module, which mainly analyzes the tag distribution, the user and resource characteristics. The second part is the interaction module [3], which interacts with the users by providing various tag keywords, and refines the recommendation candidates according to users' tag selection. The third part is the recommendation module that applies collaborative filtering [1] algorithm to generate the final recommendation list for users to browse. Then we will introduce the relevant technical details of each module.

Analysis Module. The analysis module includes functions of user analysis, book resource analysis and tag distribution analysis. The user analysis runs to obtain the users' characteristics based on the book tag vector weighted by the users' historical behaviors on the book resources. The role of the resource analysis is to generate the feature vectors of the resources based on the assigned tag information, and the similarity between different resources can be estimated. The tag distribution analysis aims to analyze the tag occurrence probability

---

[1] http://dcrays.cn/.

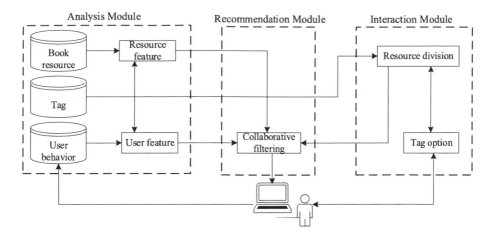

Fig. 1. System overview.

of different tags based on the resource-tag matrix. Given the categorized tags of the RAYS platform, the entropy values of different tag categories can be estimated, i.e., the information to be deliver by each tag category, which will be further utilized for quick identification of the desirable resource in the Interaction Module.

Recommendation Module. The PBR system adopts the tag-based collaborative filtering algorithm. For existing users, we are able to get the user and resource characteristics from Analysis Module. Based on the user feature, the system can find out the similar users and provide the resource recommendation following collaborative filtering. If the user is unsatisfied with the recommendation, the system will interact with the user to refine the recommendation results, which will be further explained in Interaction Module. For any new user, the system will deal with the cold start problem by proposing an initial recommendation list. Primely, the system analyzes the existing user information, and constructs a decision tree using the users' age, gender, reading location and reading time, and the output will be the user's theme tag. When a new user enters the system, once he provides some basic information, the matched path in the decision tree will suggest a set of book resources of certain subject (theme tag).

Interaction Module. The Interaction Module is the most important part, which aims to reduce the resource candidate set according to the user interaction, so as to improve the recommendation efficiency. During the recommendation procedure, The PBR system will first provide an initial recommendation list suggested by Recommendation Module. If the users are not satisfied with the recommended resources, the system will provide tag options to interact with the users. The tag options are determined according to the entropy results of Analysis Module. When the user selects certain tag words, the range of recommendation candidates can be reduced, and the system will generate a reordered recommendation list from the updated candidates. Such interaction will continue until the user stops the interaction. Meanwhile, the process of interaction can also improve the user's tag feature model by real-time updating.

Through the analysis of user, tag and resource, the system is able to recommend book resources to different users, and achieve the personalized knowledge service. At the same time, it considers the cold start problem and user interest transfer. The tag information is effectively used to interact with the users, so as to achieve precise and personalized knowledge service.

3   Demonstration

In the demonstration part, we will show the functions of the proposed book resource recommendation system PBR. We will display the user information in the analysis module, where the user's characteristics can be generated according to the historical behaviors in the system. In addition, the resource recommendation module is mainly displayed. From a user entering the system, PBR will determine whether it is a new user, and initialize the recommendation list. The user interaction process will be exemplified, together with the recommendation list updating after interaction.

(a) user feature   (b) recommendation   (c) user interaction   (d) list update

Fig. 2. Demonstration of the system workflow.

First we analyze the users' characteristics, as shown in Fig. 2(a). The users' operation behavior in the system will be collected to build the users' tag features. The system will recommend the book resources for the users through a tag-based collaborative filtering algorithm. Figure 2(b) shows the initial book resources recommended to a user after entering the system. If the user is not satisfied with the recommended resources, the tag interaction is carried out by the interaction module shown in Fig. 2(c). The system then reduces the resource candidates according to the user's tag choice, and reorders the resource recommendation list. The Fig. 2(d) shows the recommended book resources after user interaction.

References

1. Sarwar, B.M., Karypis, G., Konstan, J.A., Riedl, J.: Item-based collaborative filtering recommendation algorithms. In: The 10th International Conference on World Wide Web, pp. 285–295 (2001)
2. Shi, L., Zhao, W.X., Shen, Y.D.: Local representative-based matrix factorization for cold-start recommendation. ACM Trans. Inf. Syst. **36**, 22:1–22:28 (2017)
3. Xie, Q., Xiong, F., Han, T., Liu, Y., Li, L., Bao, Z.: Interactive resource recommendation algorithm based on tag information. World Wide Web - Internet Web Inf. Syst. 1–19 (2018). https://doi.org/10.1007/s11280-018-0532-y

Author Index

Printed in the United States
By Bookmasters